GUIDE

TO THE
HISTORY OF SCIENCE

9th Edition, 2002

Edited by
Roger Turner
History of Science Society

Use the *Guide* online at
http://www.hssonline.org

A PUBLICATION OF THE
HISTORY OF SCIENCE SOCIETY

The *Guide to the History of Science* was prepared by the History of Science Society Executive Office at the University of Washington. Melissa Cheng designed the print edition.

A paperbound copy is mailed free of charge to individual members of the History of Science Society. The *Guide* is also available to nonmembers and institutions from the University of Chicago Press, Journals Division, P.O. Box 37005, Chicago, IL 60637, USA.

For information on the History of Science Society and its publications, including the journals *Isis* and *Osiris*, send e-mail to info@hssonline.org, or write to the HSS Executive Office, Box 351330, University of Washington, Seattle, WA 98195, USA. Information about the Society is available online at http://www.hssonline.org.

Members: Send subscription address changes to Subscription Fulfillment, University of Chicago Press, P.O. Box 37005, Chicago, Illinois 60637, USA. Changes of address can also be submitted online at http://www.journals.uchicago.edu/HSS-PSA/.

Library edition, ISBN 0-226-81708-3
Paperbound edition, ISBN 0-226-81709-1

INTRODUCTION

GUIDE TO THE PROFESSION

PREFACE

This ninth edition of the *Guide* represents a sea change from the last edition, published in 1992. From creation to production to maintenance, the 2002 *Guide* differs from its predecessor. Thanks to the dogged efforts of the *Guide*'s editor, Roger Turner, the Executive Office was able to collect, edit, and produce the new *Guide* almost entirely through the use of online databases. While construction of the databases introduced innumerable delays and many unexpected problems, a structure is now in place that will facilitate annual updates to the *Guide* and make it possible to produce a triennial print edition, as was originally intended. More importantly, the Society will now be able to offer an electronic version that will be available to anyone with Web access, making our profession's vitality apparent to the world.

Although we recognize that the printed *Guide* is an important tool in the field, and we hope that it will enhance communication throughout the discipline, the nature of the *Guide*'s information is better suited for electronic searches. The electronic version, available on the HSS Web site (hssonline.org), will not only simplify searches (no more reliance on the whims of indexers), its information will remain more accurate as we make yearly updates to the electronic *Guide*.

In addition to its electronic innovations, the 2002 *Guide* features our efforts at codifying the field. Long-time members of the HSS will remember that various categories of research interests have been introduced and dropped over the years, with little effort devoted to their systematization. In order to measure how the field is growing, and to enhance electronic searches, standard-ized fields have been introduced. (Further information on these efforts can be found under the section "Research Interests and Program Strengths.") While there is considerable risk in trying to force such categories—one writer recently complained that such efforts "suck some of the lifeblood out of the living for the sake of convenience and fast retrieval"—a standardized approach can provide valuable information. As one supporter of the HSS recently commented, if the History of Science Society does not have a firm grasp of what is being done in the field, then who does?

Because reference works like the *Guide* become obsolete as soon as they are published, be they print or electronic, we recognize that this is an ongoing project and encourage readers to alert us to any errors or omissions by writing to us at Guide@hssonline.org. Since the directory information on HSS members is kept by the University of Chicago Press, it is especially important that members review and update their membership information, which can be found at http://www.journals.uchicago.edu/HSS-PSA/.

Finally, it was a distinct privilege to have been appointed as the first Executive Director of the HSS. No one could ask for kinder colleagues or more inspired individuals. I am continually amazed, and grateful, for the thousands of hours that our members devote to the furtherance of the Society and its mission. It is my hope that this publication will encourage others to join our Society and help us as we further interest and research in the history of science.

JAY MALONE
Executive Director

ACKNOWLEDGEMENTS

It has been said that the *Guide to the History of Science* carries a curse. The pitfalls inherent in the process can be fully appreciated by those who have devoted their time to such a project. The fact that you now have the *Guide* in your hands, or are viewing it online, reflects the enormous effort that literally hundreds of individuals have devoted to this important publication. The staff in the HSS Executive Office have contributed beyond measure: Roger Turner, Chris Pearson, and Carson Burrington are to be thanked. Melissa Cheng, Mason Malone, Constance Malpas, and Alyssa Machle, provided crucial support in the final hours; Christina Kilbourn and the technical support staff at Digital Forest gave important tips and insight on web-enabled databases. The University of Chicago Press, especially Julie Noblitt, Jane Seiberling, and Tess Mullin, was most helpful.

The members of the HSS Executive Committee, Ronald L. Numbers, John W. Servos, Marc Rothenberg, Margaret Rossiter, Michael Sokal, and Margaret Osler were unbending in their support, even in the darkest times, and the significant contributions they made to the entire project were much appreciated. Likewise, members of the HSS Committee on Publications offered their support and contributions: Karen Reeds, Susan Lindee, Spencer Weart, Ted Porter, and Bruce Hunt gave advice and encouragement throughout the *Guide's* preparation.

Because much of the *Guide's* production had never been tried before, the scores of volunteers who were willing to test the beta version of the online databases were crucial in launching a successful effort. We owe them our deepest thanks.

The international scope of the *Guide* required that we recruit membes from around the world in assembling the information. Our sincere thanks to those who offered to serve as international editors: Daniel Alexandrov, Maher Abd. El-Kader Ali, Rachel Ankeny, Dimitry Bayuk, S. Pausek Bazdar, Jenny Beckman, Jim Bennett, Alexandru Bologa, Peter Bowler, Maurelia Coutinho, Naomi Darr, Liu Dun, Hamed Ead, Bernardino Fantini, Tore Frangsmyr, Robert Marc Friedman, Jean Gayon, Hae-Gyung Geong, Tom Glick, Loren Graham, Christiane Groeben, Bob Hall, Ikmeleddin Ihsanoglu, Jan Janko, Hsu Kuang-Tai, Deepak Kumar, Sushil Kumar, Bernard Lightman, Janet McCalman, Alessandro Minelli, Hazime Mizoguchi, Pap Ndiaye, Ronald Numbers, Margaret Osler, Giuliano Pancaldi, Howard Phillips, John Pickstone, Mercedes Planta, Nicolaas Rupke, Jan Sadlak, Juan Jose Saldana, Ida Stamhuis, Klaas Van Berkel, Albert Van Helden, Fung Kam Wing, and Kim Yung Sik.

Finally, the *Guide* would not have been possible without the help and patience of its many contributors. Our sincere thanks to each of you.

HSS Executive Office

PRODUCING THE
GUIDE TO THE HISTORY OF SCIENCE

Producing the 2002 *Guide* involved communication among more than a thousand people around the world. Technologies of all kinds, social as well as electronic, played important roles, as did serendipity and personal relationships. Over 900 organizations, serial publications, and institutions were contacted, and 578 made timely replies with sufficient information to be included in print. Although the printed *Guide* will quickly become obsolete, the 2002 edition possesses significantly greater utility because of the Internet-based system used to produce it. Integrating database files, e-mail communication, and data submission via the World Wide Web, this system enables the *Guide* to comfortably inhabit both the online and print worlds. The system will make possible comprehensive annual updates, thus improving the *Guide's* reliability and insuring a more regular appearance of the printed version. Most valuably, the *Guide* now includes an online edition searchable by subject area and by keyword, as well as by geography and institution name or publication title. Visit http://www.hssonline.org to use the online *Guide*.

The production of the *Guide* started with a list of organizations, institutions and serial publications of potential interest to historians of science, broadly defined. This list began with the entries in the 1992 *Guide*, supplemented by extensive Web searches, various print and electronic reference directories, and fortuitous finds. Many groups heard about the *Guide* and asked to be included. The list was then divided into geographic regions and sent to approximately fifty

"international editors" for review. These scholars, particularly knowledgeable about the history of science in their regions, reviewed the contact information, suggested additional entries, and informed HSS of new and defunct programs.

While the international editors reviewed potential entries, the Executive Office devised a strategy for managing and collecting data. Budgetary and schedule restrictions demanded that e-mail and the Internet serve as the primary methods of communication. Internet-based data entry improved accuracy and reduced transcription errors by immediately displaying the information received. Web-enabled databases and e-mail communication thus emerged as the principal strategy for producing an Internet-era *Guide*.

This strategy's technical demands initially proved overwhelming. Integrating relational databases with the Web stymied the computer consultant who had assisted with the 2001 redesign of the HSS Web site. Particularly difficult was cleanly capturing extended characters, such as umlauts and accents—essential components of an international directory.

Short sessions with more experienced (and much more expensive) consultants suggested that Web data collection was technically feasible using a system of Filemaker Pro databases and Lasso CGI/server software. Lasso, a "Web data engine," enables databases, Web servers, and Internet users to communicate by utilizing special tags embedded within html documents. However, the Society could not afford the fees that experienced Web developers charge to construct a "solution," as consultants call these systems. Foundations were unresponsive to grant requests. Consequently, a staff member began experimenting and learning Lasso in December 2001.

After a homegrown Lasso/Filemaker solution successfully handled the paper submissions for the 2002 HSS Annual Meeting, the Executive Office decided to implement the *Guide* using Lasso/Filemaker. This system used four software programs to take in data over the Internet and create a printed, book-length output. An html editor, DreamWeaver UltraDev 1.0, was used to construct Web-based forms and enter Lasso tags. Data entered into these forms was processed by Lasso 3.6 and transferred to five Filemaker Pro 5.0 relational databases. Extensive scripting and calculations within Filemaker organized and arranged the data, while InDesign 2 was used to create a print-ready layout. Lasso/Filemaker also powers the searchable, online edition of the *Guide*.

After preliminary tests of the system, data collection ran between April 16th and June 21st, 2002. During that time, each history of science entity was sent three e-mail solicitations with instructions on how to submit data. In the absence of an e-mail address, instructions were sent by postal mail. Members of the HSS Committee on Publications also personally encouraged

many institutions to respond during the final weeks of data collection. Among the challenges encountered were server outages, browser incompatibility, and the inability of some people to use the forms. Responses to these difficulties included the postal and electronic mailing of questionnaires, telephone support, and occasionally, a little handholding while encouraging users to try the Web again. The membership directory was produced using data maintained by the University of Chicago Press, helpfully and cheerfully supplied by Rich Connelly. Melissa Cheng, who adds new features to Adobe's InDesign at her day job, used her software to design and index the print edition.

Looking down the road, the Lasso/Filemaker system should ease the maintenance of a current, accurate *Guide to the History of Science*. The added search capabilities and broader audience offered by the Internet make the 2002 *Guide* a far more valuable tool for understanding and promoting the history of science. Balanced against these benefits is the expensive challenge of maintaining expertise in the technologies which make possible an Internet-era *Guide*. These increased costs could be partially offset by partnering with similar societies, and perhaps expanding the *Guide* to include all aspects of the social studies of science.

ROGER TURNER
HSS Information Manager

RESEARCH INTERESTS
AND PROGRAM STRENGTHS

This edition introduces a new system for coding and organizing the subjects studied by historians of science. Previously, the Society had no standard nomenclature for describing research topics. This made it difficult to track scholarly interests and measure trends within the field. In order to capture macroscopic trends, a system for broadly organizing research topics has been adopted. The system uses four large categories to describe the breadth of the profession, and employs combinations of those choices to achieve a moderate degree of specificity in particular cases. The research interests, subject focus and program strengths of people and institutions are classified using four areas of specialization: Chronology, Geography, Scientific Genus (for example, "biological sciences"), and Topic. Though far more manageable than the 100-plus categories used in the membership directory, this broad system does not attempt to capture the fine detail that some scholars might prefer in describing their personal studies.

The list aims to avoid culturally and methodologically based categories. As this taxonomy will be used over the long term to measure changes in the field, it emphasizes relatively enduring entities.

Thus, it prefers continents to nations or regions, groups of centuries to named eras, and wide fields of scientific pursuit to narrower disciplinary names. A great deal of consideration and thought went into the choices for each category. For example, the list of eleven Topics was created by culling the research interests used in the 1992 *Guide* and comparing these to topics used in the 1999 *Isis Current Bibliography*, and a 1971 issue and a 2001 issue of *Isis*.

Most valuably, a standardized organizational schema will significantly improve the *Guide's* utility as a reference source. While programming challenges have hindered the collection of research interests for all individuals listed in the *Guide*, future editions will fully capitalize on the advantages offered by standardization This system for organizing research interests enables searches across multiple databases in order to find people—and associations, graduate programs, museums, archives, research centers, journals and newsletters—addressing various subjects in the history of science. To make the most of this capability, examine the online edition of the *Guide*, available at http://www.hssonline.org.

CHRONOLOGY
20th and 21st Century
19th Century
18th Century
17th Century
14th-16th Century
5th-13th Century
400 B.C.E.-400 C.E.
Pre-400 B.C.E.

GEOGRAPHY
Africa
Asia
Australia and Oceania
Europe
North America
Polar Regions
South America
Transcontinental

SCIENTIFIC GENUS
Astronomical Sciences
Biological Sciences
Cognitive Sciences
  (including Psychology)
Earth Sciences
Mathematics
Medical Sciences
Natural and Human History
Physical Sciences
Social Sciences
Technology

TOPICS
Biography
Education
Exploration, Navigation and Expeditions
Gender and Science
Humanistic Relations of Science
Institutions
Instruments and Techniques
Philosophy and Philosophy of Science
Science and Religion
Science Policy
Social Relations of Science

HISTORY OF SCIENCE SOCIETY
INTEREST GROUPS

Early Science
Contact: Liba Taub, LCT1001@hermes.cam.ac.uk

Earth and Environment Forum
http://www.cieq.uqtr.ca:591/EEF.htm
Contact: Stephane Castonguay, Stephane_Castonguay@uqtr.uquebec.ca

Forum for the History of Human Science
http://www.fhhs.org
Contact: Paul Croce, pcroce@stetson.edu

Forum for the History of Science in America
Contact: Julie Newell, jnewell@spsu.edu

History of Astronomy
Contact: Marc Rothenberg, Josephhenr@aol.com

THE HISTORY OF SCIENCE SOCIETY

ORGANIZATION AND PURPOSE

The History of Science Society is the world's largest society dedicated to understanding science, technology, medicine, and their interactions with society in their historical context. Founded over seventy-five years ago, it is the oldest such society. The Society is an international organization with nearly thirty-five percent of its individual members residing outside of the United States. In North America, the Society fosters cooperation with government agencies and private foundations concerned with science and the role of science in society. It does this through its affiliation with the American Council of Learned Societies, the American Association for the Advancement of Science, the International Union for the History and Philosophy of Science, and other organizations. Through its publications and other activities, the Society provides scholars, decision makers and the public with historical perspectives on science policy and on the potentials, achievements, and the limitations of basic and applied science.

Anyone may join the History of Science Society. It has members in university, college, and high-school history departments and science departments; in museums of science and technology; in government agencies; in archives, libraries, and foundations; in the medical, scientific, and engineering professions; and among interested amateurs. The Society is both a learned society and a professional association serving over 3,500 individual members and institutions around the world.

Publications enable the Society to carry out its primary role of advancing research and teaching in the history of science. The Society was incorporated in 1924 to secure the future of *Isis*, the international review that George Sarton (1884-1956) founded in Belgium in 1912. The four quarterly issues of *Isis* each year contain editorials, scholarly articles, essay reviews, book reviews, research notes, documents, discussions, and news of the profession. The fifth number of each volume, the *Isis Current Bibliography*, lists over 3,500 publications in all aspects of the history of science. *The Guide to the History of Science* is issued periodically to all members of HSS. Since 1972, the Society has also circulated a quarterly *Newsletter*, which provides not only news of the Society, but also information on professional meetings; announcements of fellowships, prizes, and awards; a list of books received by *Isis*; and notices of employment opportunities. In 1995, with the help of graduate students at the University of Minnesota, the HSS launched its Web site, which has grown considerably over the past seven years. The Web site now serves as the main link among the Society, its members, and the world. In 1985, the Society revived its research journal *Osiris*. Founded by George Sarton in 1936 as a companion to *Isis*, *Osiris* is now devoted to the thematic volumes on topics of wide interest to the history of science community. The Society also publishes or sponsors other research and teaching tools in the field such as the

Isis Cumulative Bibliography and the *Dictionary of Scientific Biography*.

The regular, formal set of communications is complemented by the Society's meetings. In recent years the annual meeting has involved eight concurrent sets of sessions, spread over two and a half days, on topics from ancient times to the present, from the pharmacopoeia of Galen to the politics of space science. The elected Council of the Society holds regular sessions in conjunction with each annual meeting.

THE EXECUTIVE OFFICE

The Executive Office emerged as part of Professor Gerald Holton's drive to establish an endowment for the History of Science Society. Professor Holton of Harvard University, with the active assistance of then-HSS President William Coleman, was well aware that foundations and philanthropic agencies do not look favorably upon organizations that lack permanent governing and institutionalized structures. His plan was realized when Professor Michael Sokal of Worcester Polytechnic Institute was elected to the new position of Executive Secretary in 1988. From that time through 1992, Professor Sokal threw himself selflessly and tirelessly into the task of organizing and centralizing the operations of the Society.

His successor, Professor Keith R. Benson at the University of Washington was elected in 1993 for a five-year term. Operations for the Executive Office moved to Seattle with the beginning of Professor Benson's tenure. Under his direction many exciting initiatives, such as the Sponsor-A-Scholar Program, were begun. In 1998, the Society hired its first Executive Director, Robert Jay Malone. The new director assumed many of the responsibilities that had been so ably shouldered by Keith Benson. The position of Executive Secretary was subsequently changed to "Secretary" and Professor Benson continued in this new role until 31 December 2000. He was succeeded by Professor Margaret J. Osler of the University of Calgary. The office in Seattle is staffed by a full-time information manager, Roger Turner, and two graduate student assistants. All assist Dr. Malone with the daily tasks of running a dynamic organization.

In addition to handling the ongoing activities of HSS committees, the office routinely fields myriad requests for information, ranging from queries about joining HSS to information about science in the Islamic world. The Executive Office coordinates numerous activities throughout the year, including the "Sponsor-A-Scholar" project; travel grants and reservations; handling the registration for the annual Society meeting and the preparation of the program; prizes; and serving as the home for Society publications, including a variety of guides and reference materials.

HSS is not a small society. In membership, it ranks in the top 25% of the constituent societies in the American Council of Learned Societies. The half million dollar annual operating budget helps the Society foster worldwide interest in the history of science and its cultural influences. The Executive Office strives to manage the organization with efficiency and courtesy.

For additional information about HSS or the Executive Office, please call 206-543-9366 during regular business hours, Pacific Coast time or write to us via email at info@hssonline.org.

Our mailing address is:
HSS Executive Office
University of Washington
Box 351330
Seattle, WA 98195-1330

FORMER PRESIDENTS OF THE SOCIETY

Lawrence J. Henderson (1924-1925)
James Harvey Breasted (1926)
David Eugene Smith (1927)
Edgar Fahs Smith (1928)
Lynn Thorndike (1929)
Henry Crew (1930)
William H. Welch (1931)
Berthold Laufer (1932)
James Playfair McMurrich (1933)
Harvey Cushing (1934)
Charles A. Browne (1935-1936)
Chauncey D. Leake (1937-1938)
Henry Ernest Sigerist (1939)
Richard H. Shryock (1940-1942)
Louis C. Karpinski (1943-1944)
Isaiah Bowman (1944)
Vilhjalmur Stefansson (1945-1946)
John F. Fulton (1947-1950)
Harcourt Brown (1951-1952)
Dorothy Stimson (1953-1956)
Henry Guerlac (1957-60)

I. Bernard Cohen (1961-1962)
Marshall Claggett (1963-1964)
Charles C. Gillispie (1965-1966)
C. D. O'Malley (1967-1968)
Thomas S. Kuhn (1969-1970)
Lynn White, Jr. (1971-1972)
Erwin N. Hiebert (1973-1974)
John C. Greene (1975-1976)
Richard S. Westfall (1977-1978)
Robert P. Multhauf (1979-1980)
Frederic L. Holmes (1981-1982)
Gerald Holton (1983-1984)
Edward Grant (1985-1986)
William Coleman (1987)
Mary Jo Nye (1988-1989)
Stephen G. Brush (1990-1991)
Sally Gregory Kohlstedt (1992-1993)
David C. Lindberg (1994-1995)
Frederick Gregory (1996-1997)
Albert Van Helden (1998-1999)
Ronald Numbers (2000-2001)

PRIZES AWARDED BY THE SOCIETY

The Society awards seven different prizes to encourage high-quality scholarship in the history of science and its cultural influences. Many of the prizes carry cash awards funded by generous donors and the Society's endowment. The Society also sponsors two public lectures given annually by outstanding scholars. More information about prizes, as well as nomination forms, can be found on the Society's Web site (http://www.hssonline.org).

The Sarton Medal, the History of Science Society's most prestigious award honors George Sarton, the founder of *Isis* and one of the founders of the modern phase in the history of science. It has been awarded annually since 1955 to an outstanding historian of science selected from the international scholarly community. The medal honors a scholar for lifetime scholarly achievement. Designed by Bern Dibner, the medal is donated each year by the Dibner Fund.

Winners of the Sarton Medal
1955 George Sarton
1956 Charles and Dorothea
   Waley Singer
1957 Lynn Thorndike
1958 John F. Fulton
1959 Richard Shryock
1960 Owsei Temkin
1961 Alexandre Koyré
1962 E. J. Dijksterhuis
1963 Vassili Zoubov
1964 Not awarded
1965 J. R. Partington
1966 Anneliese Maier
1967 Not awarded
1968 Joseph Needham
1969 Kurt Vogel
1970 Walter Pagel
1971 Willy Hartner
1972 Kiyosi Yabuuti
1973 Henry Guerlac
1974 I. Bernard Cohen
1975 René Taton
1976 Bern Dibner
1977 Derek T. Whiteside
1978 A.P. Youschkevitch
1979 Maria Luisa Righini-Bonelli
1980 Marshall Clagett
1981 A. Rupert Hall
   Marie Boas Hall
1982 Thomas S. Kuhn
1983 Georges Canguilhem
1984 Charles Coulston Gillispie
1985 Paolo Rossi
   Richard S. Westfall
1986 Ernst Mayr
1987 G.E.R. Lloyd
1988 Stillman Drake
1989 Gerald Holton
1990 A. Hunter Dupree
1991 Mirko D. Grmek
1992 Edward Grant
1993 John L. Heilbron
1994 Allen G. Debus
1995 Charles Rosenberg
1996 Loren Graham

1997 Betty Jo Teeter Dobbs
1998 Thomas L. Hankins
1999 David C. Lindberg
2000 Frederick L. Holmes
2001 Daniel J. Kevles

The *Ida and Henry Schuman Prize* was established in 1955 by Ida and Henry Schuman of New York City. It recognizes an outstanding graduate student essay in the history of science and its cultural influences. Today, the Prize carries a $500 cash award and up to $500 in travel reimbursement to enable the winner to attend the annual meeting.

Winners of the Schuman Prize
1956 Chandler Fulton (Brown
   University), "Vinegar Flies,
   T. H. Morgan, and Columbia
   University: Some Fundamental
   Studies in Genetics"
1958 Robert Wohl (Princeton
   University), "Buffon and his
   Project for a New Science"
1960 H. L. Burstyn (Harvard
   University), "Galileo's Attempt to
   Prove That the Earth Moves"
1961 Frederic L. Holmes (Harvard
   University), "Elementary Analysis
   and the Origins of Physiological
   Chemistry"
1962 Robert H. Silliman (Princeton
   University), "William Thomson:
   Smoke Rings and Nineteenth-
   Century Atomism"
1963 Roy MacLeod (St. Catherine's
   College, Cambridge), "Richard
   Owen and Evolutionism"
1964 Jerry B. Cough (Cornell
   University), "Turgot, Lavoisier,
   and the Role of Heat in the
   Chemical Revolution"
1965 Timothy O. Lipman (College
   of Physicians and Surgeons,
   Columbia University), "Vitalism

and Reductionism in Liebig's Physiological Thought"

1966 Paul Forman (University of California, Berkeley), "The Doublet Riddle and Atomic Physics circa 1924"

1967 Gerald Geison (Yale University), "The Physical Basis of Life: The Concept of Protoplasm 1835-1870"

1968 Ronald S. Calinger (University of Chicago), "The Newtonian-Wolffian Controversy in St. Petersburg, 1725-1756"

1969 Park Teter (Princeton University), "Bacon's Use of the History of Science for Scientific Revolution"

1970 Daniel Siegel (Yale University), "Balfour Stewart and Gustav Kirchhoff: Two Independent Approaches to 'Kirchhoff's Radiation Law'"

1971 Philip Kitcher (Princeton University), "Fluxions, Limits, and Infinite Littlenesse"

1972 John E. Lesch (Princeton University), "George John Romanes and Physiological Selection: A Post-Darwinian Debate and its Consequences"

1973 Robert M. Friedman (Johns Hopkins University), "The Methodology of Joseph Fourier and the Problematic of Analysis"

1974 Philip F. Rehbock (Johns Hopkins University), "Huxley, Haeckel, and the Oceanographers: The Case of Bathybius haeckelii"

1975 Lorraine J. Daston (Columbia University), "British Responses to Psycho-physiology"

1976 Richard F. Hirsh (University of Wisconsin), "The Riddle of the Gaseous Nebulae: What Are They Made of?"

1977 Thomas Jobe, M.D. (University of Chicago), "The Role of the Devil in Restoration Science: The Webster-Ward Witchcraft Debate"

1978 Robert Scott Bernstein (Princeton University), "Pasteur's Cosmic Asymmetric Force: The Public Image and the Private Mind"

1979 Geoffrey V. Sutton (Princeton University), "Electric Medicine and Mesmerism: The Spirit of Systems in the Enlightenment"

1980 Bruce J. Hunt (Johns Hopkins University), "Theory Invades Practice: The British Response to Hertz"

1981 Larry Owens (Princeton University), "Pure and Sound Government: Laboratories, Lecture Halls, and Playing Fields in Nineteenth-Century American Science"

1982 Richard Gillespie (University of Pennsylvania), "Aerostation and Adventurism: Ballooning in France and Britain, 1783-1786"

1983 Alexander Jones (Brown University), "The Development and Transmission of 248-Day Schemes for Lunar Motion in Astronomy"

1984 Pauline Carpenter Dear (Princeton University), "Richard Owen and the Invention of the Dinosaur"

1985 Lynn Nyhart (University of Pennsylvania), "The Intellectual Geography of German Morphology, 1870-1900"

1986 William Newman (Harvard University), "The Defense of Technology: Alchemical Debate in the Late Middle Ages"

1987 Marcos Cueto (Columbia University), "Excellence, Institutional Continuity, and Scientific Styles in the Periphery: Andean Biology in Peru"

1988 M. Susan Lindee (Cornell

University), "Sexual Politics of a Textbook: The American Career of Jane Marcet's Conversations on Chemistry, 1806-1853"

1989 Richard J. Sorrenson (Princeton University), "Making a Living out of Science: John Dolland and the Achromatic Lens"

1990 Michael Aaron Dennis (Johns Hopkins University), "Reconstructing Technical Practice: The Johns Hopkins University Applied Physics Laboratory and the Massachusetts Institute of Technology Instrumentation Laboratory after World War II"

1991 Alex Soojung-Kim Pang (University of Pennsylvania), "The Social Event of the Season: Solar Eclipse Expeditions and 19th-century Scientific Culture"

1992 Sungook Hong (University of Toronto), "Making a New Role for Scientist Engineer: John Ambrose Fleming (1849-1945) and the 'Ferranti Effect'"

1993 Paul Lucier (Princeton University), "Commercial Interest and Scientific Disinterestedness: Geological Consultants in Antebellum America"

1994 James Strick (Princeton University), "Swimming against the Tide: Adrianus Pijper and the debate over Bacterial Flagella, 1946-1956"

1995 Helen Rozwadowski (University of Pennsylvania), "Small World: Forging a Scientific Maritime Culture"

1996 James Spiller, "Re-Imagining Antarctica and the United States Antarctica Research Program: Enduring Representations of a Redemptive Science"

1997 Not awarded.

1998 Michael D. Gordin (Harvard University), "The Importation of Being Earnest"

1999 James Endersby (Cambridge University), "Putting Plants in their Place"

2000 Not awarded.

2001 Joshua Buhs (University of Pennsylvania), "The Fire Ant Wars: Nature and Science in the Pesticide Controversies of the Late Twentieth Century."

The **Pfizer Award** was established in 1958 through the generosity of Pfizer, Inc., a diversified research-based company. The award consists of a medal and $2,500. It is now given each year to the author of an outstanding book in the history of science published in English during the previous three years (initially to the best book published in the previous year).

Winners of the Pfizer Award

1959 Marie Boas Hall, *Robert Boyle and Seventeenth-Century Chemistry* (New York: Cambridge University Press, 1958).

1960 Marshall Clagett, *The Science of Mechanics in the Middle Ages* (Madison: University of Wisconsin Press, 1959).

1961 Cyril Stanley Smith, *A History of Metallography: The Development of Ideas on the Structure of Metal before 1890* (Chicago: University of Chicago Press, 1960).

1962 Henry Guerlac, *Lavoisier, The Crucial Year: The Background and Origin of His First Experiments on Combustion in 1772* (Ithaca, N.Y.: Cornell University Press, 1961)

1963 Lynn White, Jr., *Medieval*

Technology and Social Change (New York: Oxford University Press, 1962).

1964 Robert E. Schofield, *The Lunar Society of Birmingham: A Social History of Provincial Science and Industry in Eighteenth-Century England* (London: Oxford University Press, 1963).

1965 Charles D. O'Malley, *Andreas Vesalius of Brussels, 1514-1564* (Berkeley: University of California Press, 1964).

1966 L. Pearce Williams, *Michael Faraday: A Biography* (New York: Basic Books, 1965).

1967 Howard B. Adelmann, *Marcello Malpighi and the Evolution of Embryology* (Ithaca, N.Y.: Cornell University Press, 1966).

1968 Edward Rosen, *Kepler's Somnium* (Madison: University of Wisconsin Press, 1967).

1969 Margaret T. May, *Galen on the Usefulness of the Parts of the Body* (Ithaca. N.Y.: Cornell University Press, 1968).

1970 Michael Ghiselin, *The Triumph of the Darwinian Method* (Berkeley: University of California Press, 1969).

1971 David Joravsky, *The Lysenko Affair* (Cambridge, Mass.: Harvard University Press, 1970).

1972 Richard S. Westfall, *Force in Newton's Physics: The Science of Dynamics in the Seventeenth Century* (New York: American Elsevier, 1971).

1973 Joseph Fruton, *Molecules and Life: Historical Essays on the Interplay of Chemistry and Biology* (New York: John Wiley, 1972).

1974 Susan Schlee, *The Edge of an Unfamiliar World: A History of Oceanography* (New York: Dutton, 1973).

1975 Frederic L. Holmes, *Claude Bernard and Animal Chemistry: The Emergence of a Scientist* (Cambridge: Harvard University Press, 1974).

1976 Otto Neugebauer, *A History of Ancient Mathematical Astronomy* (3 vols.) (New York: Springer-Verlag, 1975).

1977 Stephen G. Brush, *The Kind of Motion We Call Heat* (Amsterdam/New York: North-Holland, 1976).

1978 (Two winners) Allen G. Debus, *The Chemical Philosophy: Paracelsian Science and Medicine in the Sixteenth and Seventeenth Centuries* (New York: Science History Publications, 1977). Merritt Roe Smith, *Harpers Ferry Armory and the New Technology: The Challenge of Change* (Ithaca, N.Y./London: Cornell University Press, 1977).

1979 Susan F. Cannon, *Science in Culture: The Early Victorian Period* (New York: Science History Publications, 1978).

1980 Frank J. Sulloway, *Freud, Biologist of the Mind: Beyond the Psychoanalytic Legend* (New York: Basic Books, 1979).

1981 Charles Coulston Gillispie, *Science and Polity in France at the End of the Old Regime* (Princeton, N.J.: Princeton University Press, 1980).

1982 Thomas Goldstein, *Dawn of Modern Science: From the Arabs to Leonardo da Vinci* (New York: Hougbton Mifllin, 1980).

1983 Richard S. Westfall, *Never at Rest: A Biography of Isaac Newton* (Cambridge: Cambridge

University Press, 1980).

1984 Kenneth R. Manning, *Black Apollo of Science: The Life of Ernest Everett Just* (Oxford: Oxford University Press, 1983).

1985 Noel Swerdlow and Otto Neugebauer, *Mathematical Astronomy in Copernicus's De Revolutionibus* (New York: Springer-Verlag, 1984).

1986 I. Bernard Cohen, *Rcvolution in Science* (Cambridge, Mass.: Belknap Press of Harvard University Press, 1985).

1987 Christa Jungnickel and Russell McCormmach, *Intellectual Mastery of Nature: Theoretical Physics from Ohm to Einstein; Volume I: The Torch of Mathematics, 1800-1870; Volume II: The Now Mighty Theoretical Physics, 1870-1925* (Chicago: University of Chicago Press, 1986).

1988 Robert J. Richards, *Darwin and the Emergence of Evolutionnry Theories of Mind and Behavior* (Chicago: University of Chicago Press, 1987).

1989 Lorraine J. Daston, *Classical Probability in the Enlightenment* (Princeton, NJ.: Princeton University Press, 1988).

1990 Crosbie Smith and M. Norton Wise, *Energy and Empire: A Biographical Study of Lord Kelvin* (Cambridge: Cambridge University Press, 1989).

1991 (Two winners) Adrian Desmond, *The Politics of Evolution: Morphology, Medicine, and Reform in Radical London* (Chicago: University of Chicago Press, 1989). John W. Servos, *Physical Chemistry from Ostwald to Pauling: The Making of a Science*

in America (Princeton, NJ: Princeton University Press, 1990).

1992 James R. Bartholomew, *The Formation of Science in Japan: Building a Research Tradition* (New Haven: Yale University Press, 1989).

1993 David Cassidy, *Uncertainty: The Life and Science of Werner Heisenberg* (New York: Freeman, 1992).

1994 Joan Cadden, *The Meanings of Sex Difference in the Middle Ages* (Cambridge: Cambridge University Press, 1993).

1995 Pamela H. Smith, *The Business of Alchemy: Science and Culture in the Holy Roman Empire* (Princeton, NJ: Princeton University Press, 1994).

1996 Paula Findlen, *Possessing Nature: Museums, Collecting, and Scientific Culture in Early Modern Italy* (Berkeley: University of California Press, 1995).

1997 Margaret W. Rossiter, *Women Scientists in America: Before Affirmative Action, 1940-1972* (Baltimore: Johns Hopkins University Press, 1995).

1998 Peter Galison, *Image and Logic: A Material Culture of Microphysics* (Chicago: University of Chicago Press, 1997).

1999 Lorraine Daston and Katharine Park, *Wonders and the Order of Nature, 1150-1750* (Zone Books, 1998).

2000 Crosbie Smith, *The Science of Energy: A Cultural History of Energy Physics* (University of Chicago Press, 1998),

2001 John Heilbron, *The Sun in the Church: Cathedrals as*

Solar Observatories (Harvard University Press, 1999).

The **Derek Price Award** was originally established as the Zeitlin-Ver Brugge Prize in 1978, to encourage the publication in *Isis* of original research of the highest standard. It is now supported by an endowment established through the generosity of an anonymous donor. Consisting of $1000 and a certificate, this prize is given annually to the author of an outstanding article in *Isis* in the three years prior to the year of the award.

Derek Price Award Winners:

1979 Robert Nye, "Heredity or Milieu: The Foundations of European Criminological Theory," *Isis*, 1976, 67: 335-355.

1980 Thomas L. Hankins, "Triplets and Triads: Sir William Rowan Hamilton on the Metaphysics of Mathematics," *Isis*, 1977, 68: 175-193.

1981 Linda E. Voigts, "Anglo-Saxon Plant Remedies and the Angle-Saxons," *Isis*, 1979, 70: 250-268.

1982 Timothy Lenoir, "Rant, Blumenbach, and Vital Materialism in German Biology," *Isis*, 1980, 71: 77-108.

1983 Alexander Vucinich, "Soviet Physicists and Philosophers in the 1930s: Dynamics of a Conflict," *Isis*, 1980, 71: 236-250.

1984 James Secord, "Nature's Fancy: Charles Darwin and the Breeding of Pigeons," *Isis*, 1981, 72: 163-186.

1985 Keith Hutchison, "What Happened to Occult Qualities in the Scientific Revolution?" *Isis*, 1982, 73: 233-253.

1986 Mara Beller, "Matrix Theory Before Schrodinger: Philosophy, Problems, Consequences," *Isis*, 1983, 74: 469-491.

1987 Richard S. Westfall, "Scientific Patronage: Galileo and the Telescope," *Isis*, 1985, 76: 11-30.

1988 Owen Hannaway, "Laboratory Design and the Aim of Science: Andreas Libavius versus Tycho Brahe," *Isis*, 1986, 77: 585-610.

1989 David C. Lindberg, "Science as Handmaiden: Roger Bacon and the Patristic Tradition," *Isis*, 1987, 78: 511-536.

1990 Steven Shapin, "The House of Experiment in Seventeenth-Century England," *Isis*, 1988, 79: 373-404.

1991 Mario Biagioli, "Galileo the Emblem Maker," *Isis*, 1990, 81: 230-258.

1992 Sharon Kingsland, "The Battling Botanist: Daniel Trembly MacDougal, Mutation Theory, and the Rise of Experimental Evolutionary Biology in America 1900-1912," *Isis*, 1991, 82: 479-209.

1993 John Harley Warner, "Ideals of Science and Their Discontents in late Nineteenth-century American Medicine," *Isis*, 1991, 82: 454-478.

1994 Mary Terrall, "Representing the Earth's Shape: The Polemics Surrounding Maupertuis's Expedition to Lapland," *Isis*, 1992, 83: 218-237.

1995 Paula Findlen, "Science as a Career in Enlightenment Italy: The Strategies of Laura Bassi," *Isis*, 1993, 84: 441-469.

1996 John Carson, "Army Alpha, Army Brass, and the Search for Army Intelligence," *Isis*, 1993, 84: 278-309.

1997 William J. Ashworth "Memory, Efficiency, and Symbolic Analysis: Charles Babbage, John Herschel, and the Industrial Mind," *Isis*, 1996, 87: 629-653.

1998 Deborah E. Harkness, "Managing an Experimental Household: The Dees of Mortlake and the Practice of Natural Philosophy," *Isis*, 1997, 88: 247-262.

1999 Lynn K. Nyhart, "Civic and Economic Zoology in Nineteenth-Century Germany: The "Living Communities" of Karl Möbius," *Isis*, 1998, 89: 605-630.

2000 Emily Thompson, "Dead Rooms and Live Wires: Harvard, Hollywood, and the Deconstruction of Architectural Acoustics, 1900-1930," *Isis*, 1997, 88, No.4: 597-626.

2001 Mary Henninger-Voss, "Working Machines and Noble Mechanics: Guidobaldo del Monte and the Translation of Knowledge." *Isis*, 2000, 91, No. 2: 232-259.

The **Watson Davis and Helen Miles Davis Prize**, named in honor of the longtime director of Science Service and his wife, was established in 1985 through a long-term pledge from Miles and Audrey Davis. The prize consists of $1000 and a certificate. It honors books in the history of science that are directed to wide public audiences or to undergraduate teaching. Books published during the three years prior to the award date are eligible for nomination.

Winners of the Watson Davis and Helen Miles Davis Prize

1986 Daniel J. Boorstin, *The Discoverers: A History of Man's Search to Know His World and Himself* (New York: Random House, 1983).

1987 Thomas L. Hankins, *Science in the Enlightenment* (Cambridge: Cambridge University Press, 1985).

1988 John L. Heilbron, *The Dilemmas of an Upright Man: Max Planck as Spokesman for German Science* (Berkeley: University of California Press, 1986).

1989 Joan Jacobs Brumberg, *Fasting Girls: The Emergence of Anorexia Nervosa as a Modern Disease* (Cambridge, Mass.: Harvard University Press, 1988).

1990 Robert W. Smith, *The Space Telescope: A Study of NASA Science, Technology, and Policy* (Cambridge: Cambridge University Press, 1989).

1991 Nancy G. Siraisi, *Medieval and Early Modern Medicine: An Introduction to Knowledge and Practice* (Chicago: University of Chicago Press, 1990).

1992 John Hedley Brooke, *Science and Religion: Some Historical Perspectives* (Cambridge: Cambridge University Press, 1991).

1993 James Moore and Adrian Desmond, *Darwin: The Life of a Tormented Evolutionist* (London: Michael Joseph, 1991).

1994 David C. Lindberg, *The Beginnings of Western Science: The European Scientific Tradition in Philosophical, Religious, and Institutional Context, 600 B.C. to A.D. 1450* (Chicago: University of Chicago Press, 1992).

1995 Victor J. Katz, *A History of Mathematics, An Introduction* (New York: Harper Collins, 1993).

1996 Betty Jo Teeter Dobbs and

Margaret C. Jacob, *Newton and the Culture of Newtonianism* (Humanities Press, 1995).

1997 Richard Rhodes, *Dark Sun: The Making of the Hydrogen Bomb* (Simon & Schuster, 1995).

1998 Ruth Lewin Sime, *Lise Meitner: A Life in Physics* (Berkeley: University of California Press, 1996).

1999 Daniel J. Kevles, *The Baltimore Case: A Trial of Politics, Science and Character* (W.W. Norton & Company, 1998).

2000 Gregg Mitman, *Reel Nature: America's Romance with Wildlife on Film* (Harvard University Press, 1999)

2001 Nancy Tomes, *The Gospel of Germs: Men, Women, and the Microbe in American Life* (Harvard University Press, 2000)

The **History of Women in Science Prize**, first awarded in 1987, honors an outstanding contribution to the history of women in science published in the preceding four years. It carries a $1,000 cash award. In odd numbered years, it honors a book; in even numbered years, an article.

Winners of the History of Women in Science Prize

1987 Regina Markell Morantz-Sanchez, *Sympathy and Science: Women Physicians in American Medicine* (Oxford: Oxford University Press, 1985).

1988 Pnina Abir-Am, "Synergy or Clash: Disciplinary and Marital Strategies in the Career of Mathematical Biologist Dorothy Wrinch," in *Uneasy Careers and Intimate Lives*, edited by Pnina Abir-Am and Dorinda Outram

(New Brunswick, N.J.: Rutgers University Press, 1987).

1989 Joan Mark, *A Stranger in Her Native Land: Alice Fletcher and the American Indians* (Lincoln: University of Nebraska Press, 1988).

1990 Ann Hibner Koblitz, "Science, Women, and the Russian Intelligentsia: The Generation of the 1860s," *Isis*, 1988, 79: 208-226.

1991 Martha H. Verbrugge, *Able-Bodied Womanhood: Personal Health and Social Change in Nineteenth-Century Boston* (New York: Oxford University Press, 1988).

1992 Judith Coffin, "Social Science Meets Sweated Labor: Reinterpreting Women's Work in Late Nineteenth-century France"

1993 Barbara Duden, *The Woman Beneath the Skin: A Doctor's Patients in Eighteenth-Century Germany* (Cambridge, Mass.: Harvard University Press, 1991).

1994 Londa Schiebinger, "Why Mammals Are Called Mammals: Gender Politics in Eighteenth-Century Natural History," *American Historical Review*, 1993, 98: 382-411.

1995 Elizabeth Lunbeck, *The Psychiatric Persuasion: Knowledge, Gender, and Power in Modern America* (Princeton, NJ: Princeton University Press, 1994).

1996 Ida Stamhuis, "A Female Contribution to Early Genetics: Tine Tammes and Mendel's Laws for Continuous Characters," *Journal of the History of Biology*, 1995, 28: 495-531.

1997 Margaret W. Rossiter, *Women Scientists in America: Before*

Affirmative Action, 1940-1972 (Baltimore: Johns Hopkins University Press, 1995).

1998 Mary Terrall "Émilie du Châtelet and the Gendering of Science," *History of Science*, 1995, 33: 283-310.

1999 Linda J. Lear, *Rachel Carson: Witness for Nature* (Henry Holt and Company, 1997).

2000 Naomi Oreskes, "Objectivity or Heroism? On the Invisibility of Women in Science," *Osiris*, 1996, 11: 87-113.

2001 Charlotte Furth, *A Flourishing Yin: Chinese Medical History, 960-1665* (University of California Press, 2000).

The *Joseph H. Hazen Education Prize* is awarded in recognition of outstanding contributions to the teaching of History of Science. Supported by a gift from the Hazen Polsky Foundation, it is the Society's newest prize. Education activities recognized by the award are to be construed in the broadest sense and should include but not be limited to the following: classroom teaching (K-12, undergraduate, graduate, or extended education), mentoring of young scholars, museum work, journalism, organization and administration of educational programs, influential writing, educational research, innovation in the methodology of instruction, preparation of pedagogical materials, or public outreach through non-print media.

Winners of the Joseph H. Hazen Education Prize

1998 Gerald Holton (Harvard University) and James Rutherford (American Association for the Advancement of Science)

1999 Liba Taub (Whipple Museum)

2000 Jane Maienschein (Arizona State University)

2001 Stephen Brush (University of Maryland)

The *George Sarton Memorial Lecture*, begun in 1960, is given annually at the American Association for the Advancement of Science meeting. The following is a list of the distinguished lecturers, chosen by the Executive Committee.

Sarton Memorial Lecturers

1960 Rene Dubos
1961 Joseph Kaplan
1962 Emilio Segre'
1963 Gerald Holton
1964 Lloyd Stevenson
1965 Stillman Drake
1966 George Wald
1967 Cyril Stanley Smith
1968 Oswei Temkin
1969 Martin Klein
1970 Evelyn Hutchinson
1971 Ernst Mayr
1972 Thomas Kuhn
1973-1975 no lectures
1976 Joseph Fruton
1977 Jane Oppenheimer
1978 I. Bernard Cohen
1979 George White
1980 Charles C. Gillispie
1981 Richard S. Westfall
1982 Henry Guerlac
1983 Derek de Solla Price
1984 Arnold Thackray
1985 Daniel J. Kevles
1986 Thomas Parke Hughes
1987 Frederic L. Holmes
1988 Stephen Jay Gould
1989 John L. Heilbron
1990 Margaret W. Rossiter
1991 Kenneth R. Manning
1992 Spencer Weart
1993 Gerald Geison

1994 Roy Porter
1995 Ronald Numbers
1996 Jane Maienschein
1997 Mott Greene
1998 Garland Allen
1999 Mary Jo Nye
2000 Edward Larson
2001 David Hollinger
2002 Loren Graham

The *History of Science Society Distinguished Lecture* series began in 1981 at the annual meeting in Los Angeles, California. In planning for that meeting, and in response to the proliferation of parallel sessions, program co-chairs David C. Lindberg and Ronald L. Numbers decided to create a plenary forum featuring a historian of science at the height of his or her career. Over the past 20 years this "Society Lecture" has evolved into a highlight of the annual meeting, drawing by far the largest attendance of any session. Through the generosity of Joseph H. Hazen, the renamed HSS Distinguished Lecture has been endowed, allowing the Society to cover the lecturer's expenses and honorarium.

HSS Distinguished Lecturers
1981 Charles C. Gillispie
1982 Charles E. Rosenberg
1983 Richard S. Westfall
1984 I. Bernard Cohen
1985 Frederic L. Holmes
1986 John L. Heilbron
1987 David C. Lindberg
1988 Sally Gregory Kohlstedt
1989 Jacques Roger
1990 Owen Hannaway
1991 Loren Graham
1992 Daniel J. Kevles
1993 Betty Jo Teeter Dobbs
1994 David Hollinger
1995 A. I. Sabra
1996 Allen G. Debus
1997 Thomas L. Hankins
1998 Martin Rudwick
1999 Charles C. Gillispie
2000 Mary Jo Nye
2001 John Hedley Brooke

Photo Credit: Don Dewsbury

Lorraine Daston receiving the Pfizer medal at the 1989 History of Science Society meeting in Gainesville, Florida. From left to right: Thomas Althuis (Pfizer Company), Lorraine Daston, Mary Jo Nye (HSS President), and Nicholas Steneck (Pfizer Prize Committee).

Membership Directory

The following pages list the members of the History of Science Society as of 15 June, 2002. The directory also includes long standing members who had not yet renewed their memberships by June of 2002. The data used for the membership directory is maintained by the University of Chicago Press (UCP), which coordinates membership and mailing address information for the Society. Members can update their information using UCP's online utility, at http://www.journals.uchicago.edu/HSS-PSA/.

Directory entries have some combination of the following information:

> Last Name, First Name, and Middle Name or Initial
> Mailing address (Office or business addresses, when available,
>   were preferred over home addresses)
> Telephone number
> Facsimile number
> E-mail address
> Highest degree and year earned
> Broad research interests (as translated from more precise listing
>   collected by UCP)

For the most current contact information and most precise listing of research interests, please see the online membership directory, at http://www.journals.uchicago.edu/HSS-PSA/.

Aaserud, Finn
Niels Bohr Archive
Blegdamsvej 17
Copenhagen DK-2100
Denmark
Phone: 45.3532.522
Fax: 45.3532.5428
aaserud@nbi.dk
Doctorate, 1984
20c, Physical Sciences;
20c, Europe, Social Relations of
Science

Aberman, Matthew
1416 San Andres St Apt A
Santa Barbara, CA 93101-6818
Phone: 1.805.966.3270
mca@umail.ussb.edu

Abiko, Seiya
Seirei Christopher College
3453 Mikatahara-Town
Hamamatsu-City 433-8558
Japan
Phone: 81.53.439.1400
Fax: 81.53.439.1406
abiko@ceres.dti.ne.jp

Abir-Am, Pnina G.
249 Orchard St
Belmont, MA 02178
Phone: 1.617.901.3109
Fax: 1.617.484.2709
pninaga@aol.com
Doctorate
20c, Europe, Biological Sciences;
20c, North America,
Gender and Science;
20c, Europe, Social Sciences;
20c, Europe, Science Policy

Abraham, Tara
Dibner Institute
MIT, E56-100
38 Memorial Dr.
Cambridge, MA 02139
Phone: 1.617.258.0507
Fax: 1.617.253.9858
tabraham@dibinst.mit.edu
Doctorate, 2000
20c, North America, Biological
Sciences

Abrams, Susan E.
University of Chicago Press
Books Division
1427 E 60th St

Chicago, IL 60637-2954
Phone: 1.773.702.7641
Fax: 1.773.702.9756
sabrams@press.uchicago.edu

Acampora, Renato
Oberer Huesliweg 23
Niederweningen CH-8166
Switzerland

Acherman, Simon Rei
A.A. 26228 Unicentro
Cali
Colombia

Acker, Caroline J.
Carnegie Mellon University
Department of Hisory
Baker Hall 240
Pittsburgh, PA 15213

Ackerberg-Hastings, Amy K.
5908 Halsey Rd
Rockville, MD 20851-2201
Phone: 1.301.515.8230
aackerbe@erols.com
Doctorate, 1999
19c, Europe, Mathematics;
19c, North America, Education;
19c, Europe, Education;
Social Sciences

Ackerman, Brian P.
University of Delaware
Department of Psychology
Newark, DE 19716
Phone: 1.302.831.2385
Fax: 1.302.831.3645
bpa@udel.edu
Doctorate, 1977
20c, Cognitive Sciences;
20c, Medical Sciences,
Instruments and Techniques;
20c, Medical Sciences

Ackerman, Michael
University of Virginia
History Department
Randall Hall
Charlottesville, VA 22903
Phone: 1.804.979.3700
Fax: 1.804.296.2860
ma2s@virginia.edu
Master's Level, 1999
20c, North America, Medical

Sciences; 20c, North America,
Biological Sciences; 20c, North
America, Science Policy

Ackert, Lloyd T.
3607 Chestnut Ave
Baltimore, MD 21211-2530
Phone: 1.410.554.9030
lloydack@bellatlantic.net

Adams, Mark B.
University of Pennsylvania
History & Sociology of Science
303 Logan Hall
249 S 36th St
Philadelphia, PA 19104-6304
Phone: 1.215.898.8406
Fax: 1.215.573.2231
madams@sas.upenn.edu
Doctorate, 1973
Biological Sciences; Europe,
Social Relations of Science;
20c, Europe

Agnoli, Paolo
Via Caterina Troiani 276
Rome 00144
Italy
Phone: 39.065.291034
p.agnoli@tin.it

Aieta, Joseph, III
165 Maple St
Framingham, MA 01702-5757
Phone: 1.617.243.2116
jaieta@lasell.edu

Ainley, Marianne G.
Univ of N British Columbia
Women's/Gender Studies/Hist
3333 University Way
Prince George, BC V2N4Z9
Canada
Phone: 1.250.960.6681
Fax: 1.250.960.5545
ainley@unbc.ca
Doctorate
20c, North America, Gender and
Science; 20c, North America,
Biography; 20c, North America,
Biological Sciences; 19c, North
America, Natural and Human
History

Akera, Atsushi
737 Taborton Rd
Sand Lake, NY 12153-2214

Phone: 1.518.276.2314
Fax: 1.518.276.2659
akeraa@rpi.edu

Akeroyd, F. Michael
Bradford University
Dept 1HS
Bradford
Great Britain
m.akeroyd@bilk.ac.uk

Alberti, Samuel J. M. M.
University of Manchester
CHSTM, Maths Tower
University of Manchester
Oxford Road
Manchester M13 9PL
Great Britain
Phone: 44.161.275.5861
Fax: 44.161.275.5699
sam.alberti@man.ac.uk
Master's Level, 2001
19c, Europe, Natural and
Human History; 19c, Europe,
Gender and Science;
19c, Europe, Institutions

Albree, Joe
Auburn Univ at Montgomery
Department of Mathematics
7300 University Blvd
PO Box 244023
Montgomery, AL 36124-4023
Phone: 1.334.244.3261
Fax: 1.334.244.3826
joe2@strudel.aum.edu
Master's Level, 1969
North America, Mathematics

Albury, W. R.
UNE
Faculty of Arts
Armidale NSW 2351
Australia

Alder, Ken
Northwestern University
Dept of History
1881 Sheridan Rd
Evanston, IL 60208-0818
Phone: 1.847.491.7260
Fax: 1.847.424.9384
k-alder@northwestern.edu
Doctorate, 1991
18c, Europe, Technology;
Social Relations of Science

Aldrich, Michele L.
Cornell University
726 University Avenue
Ithaca, NY 14850
Phone: 1.860.676.5577
Fax: 1.413.585.3553
73061.2420@compuserve.com
Doctorate, 1974
19c, North America, Earth
Sciences; 20c, North America,
Earth Sciences; 19c, Asia,
Earth Sciences

Alekseev, A.
Institute Economiki SO RAN
pr. Lavrentyeva 17
Novosibirsk 630090
Russia
Phone: 7.3832.344057
alekseev@ieie.nsc.ru
Doctorate, 1992
20c, Europe, Social Sciences;
20c, North America, Humanistic
Relations of Science;
20c, Europe, Philosophy
and Philosophy of Science;
20c, Europe, Humanistic
Relations of Science

Alexander, Amir R.
6941 Pomelo Drive
West Hills, CA 91307
Phone: 1.626.793.5100
Fax: 1.626.793.5228
amir@humnet.ucla.edu
Doctorate, 1996
17c, Europe, Mathematics;
17c, Humanistic Relations of
Science; 17c, Exploration,
Navigation and Expeditions

Alexander, Jennifer K.
University of Minnesota
Program in Hist of Sci & Tech
125 Mechanical Engineering
Minneapolis, MN 55455
Phone: 1.612.626.7309
Fax: 1.612.624.1398
jalexand@me.umn.edu
Doctorate
Europe, Technology;
Physical Sciences; Technology,
Instruments and Techniques

Alexandrov, Daniel A.
European Univ at St Petersburg
3, Gagarinskaia St

St Petersburg 199186
Russia
Phone: 7.812.2790695
Fax: 7.812.2755139
d_alexandrov@eu.spb.ru

Alfonso-Goldfarb, Ana M.
Pont Cath Univ of S. Paulo
Rua Marques de Paranagua 111
bl 1 sala 2
São Paulo, SP 03103 050
Brazil
Phone: 55.11.256.1622
Fax: 55.11.256.5039
belasartes@belasartes.com.br
Doctorate, 1986
Physical Sciences;
Medical Sciences

Alford, Robert R.
CUNY Graduate Center
PhD Program in Sociology
365 Fifth Ave
New York, NY 10016
Phone: 1.212.817.8779
ralford@gc.cuny.edu
Doctorate
20c, Europe, Institutions;
20c, Europe, Social Sciences;
19c, North America, Philosophy
and Philosophy of Science;
20c, North America, Natural and
Human History

Allchin, Douglas
2005 Carroll Ave
Saint Paul, MN 55104-5136
Phone: 1.651.603.8805
allchin@pclink.com
Doctorate, 1991
20c, Biological Sciences;
Education; Philosophy and
Philosophy of Science

Allen, Garland E.
Washington University
Biology Dept
Box 1137
1 Brookings Dr
Saint Louis, MO 63130-4899
Phone: 1.314.935.6808
Fax: 1.314.935.4432
allen@biology.wustl.edu
Doctorate, 1966
20c, North America, Biological
Sciences; 19c, Europe,
Biological Sciences

Allentuck, Marcia
5 W 86th St Apt 12B
New York, NY 10024-3665
Phone: 1.212.799.0132

Allington, R. W.
ISCO Inc
PO Box 82531
Lincoln, NE 68501-2531
Phone: 1.402.464.0231
Fax: 1.402.465.3905

Almgren, Beverly S.
Univ of the Sciences
600 S. 43rd Street
Philadelphia, PA 19118
balmgren@conknet.com
Doctorate, 1968
*19c, Europe, Humanistic
Relations of Science;
20c, Physical Sciences*

Almodovar, Carmen
Heredia 373 Apto 6
Entre San Mariano y Vista
Alegre, Vibora
Habana
Cuba

Alpher, Ralph A.
Dudley Observatory
107 Nott Ter Ste 201
Schenectady, NY 12308-3170

Alter, Stephen G.
Gordon College
Dept of History
255 Grapevine Rd
Wenham, MA 01984-1899
Phone: 1.978.927.2306
Fax: 1.978.524.3747
salter@faith.gordon.edu
Doctorate
*19c, Europe, Biological Sciences;
19c, North America, Biological
Sciences; 19c, North America,
Humanistic Relations of Science;
19c, Europe, Natural and
Human History*

Althuis, Thomas H.
Pfizer Inc
Central Research Division
Eastern Point Road
Groton, CT 06340

Altshuler, Jose B.
Calle 13 No 51, Esq N (piso 7)
10400 Ciudad de
La Habana
Cuba
Phone: 537.63.4823
jea@infomed.sld.cu
Doctorate, 1974
*20c, North America, Technology;
20c, Europe, Physical Sciences;
19c, Europe, Physical Sciences;
19c, North America, Technology,
Instruments and Techniques*

Alvarado, Carlos S.
PO Box 41
New York, NY 10021-0030
Phone: 1.212.396.0096
Fax: 1.212.628.1559
alvarado@parapsychology.org

Alvarez Zamayoa, Ma Luisa
Martinez de Navarrete #505
Frac las fuentes
59690 Zamora
Michoacan
Mexico

Ambedian, Lynn
University of Lethbridge
Faculty of Arts & Science
4401 University Drive
Lethbridge, AB T1K3M4
Canada
Phone: 1.403.329.5106
Fax: 1.403.380.1880
l.ambedian@uleth.ca

Amodeo, Ron
5033 Bryant Ave S
Minneapolis, MN 55419-1211
Phone: 1.612.822.8294
Doctorate, 1999
*20c, North America, Biological
Sciences; 19c, North America,
Education; 20c, Instruments and
Techniques; Biological Sciences*

Amparo Gomez, Luz
Colomos #2292/Providencia
44620 Guadalajara
Jalisco
Mexico

Amsterdamska, Olga
University of Amsterdam
Science Dynamics

Fac of Social & Behavioral Sci
Achterburgwal 237 k TO6
Amsterdam 1012 DL
The Netherlands
Phone: 31.20.5256597
Fax: 31.20.5252086
amsterdamska@pscw.uva.nl
Doctorate, 1984
*20c, Medical Sciences;
20c, Biological Sciences;
Social Relations of Science;
Social Sciences*

Analytis, P.
Hadjicosta 8B
Athens 11521
Greece
Phone: 30.1.644.6151

Anastasio, Salvatore
State Univ of NY - New Paltz
Department of Mathematics
New Paltz, NY 12561
Phone: 1.845.257.3530
Fax: 1.914.257.3571
anastasi@mcs.newpaltz.edu
Doctorate, 1964
*Mathematics; 18c, North
America, Mathematics;
Philosophy and Philosophy of
Science*

Anawati, Georges C.
Inst Dominicain d'Et Orient
1 Masna Al Tarabich Street
BP 18 Abbassiah
Cairo 11381
Egypt
Phone: 202.482.5509
Fax: 202.2820682
ideo@link.com.eg
Doctorate
*5c-13c, Africa, Philosophy
and Philosophy of
Science; 14c-16c, Asia,
Astronomical Sciences;
Asia, Physical Sciences;
Astronomical Sciences*

Andersen, Hanne
University of Copenhagen
Dept. Med. Philo. & Clin. Theo
Parnum, Room 22-3-14
Blegdamsvej 3
Copenhagen N DK 2200
Denmark
Phone: 45.35327931

Fax: 45.35327938
h.andersen@medphil.ku.dk
Doctorate, 1998
*Philosophy and Philosophy
of Science; Medical Sciences;
20c, Physical Sciences*

Andersen, Roy S.
Clark University
Department of Physics
Worcester, MA 01610
Phone: 1.508.793.7169
Fax: 1.508.793.8861
randersen@clarku.edu
Doctorate, 1951
*Astronomical Sciences;
Physical Sciences*

Anderson, John D., Jr.
1800 Billman Ln
Silver Spring, MD 20902-1422
Phone: 1.202.633.9890
Fax: 1.202.786.2447
john.anderson@nasm.di.edu

Anderson, Katharine M.
York University
4700 Keele St
Toronto, ON M3J1P3
Canada
Phone: 1.416.736.2100
kateya@yorku.ca
Doctorate
*19c, Europe, Earth Sciences;
Instruments and Techniques*

Anderson, Richard
28503 Rothrock Dr
Rncho Pls Vrd, CA 90275-3043
Phone: 1.310.377.0264
71234.2502@compuserve.com

Anderson, Ronald
Boston College
Dept of Philosophy
Chestnut Hill, MA 02467
Phone: 1.617.552.3866
Fax: 1.617.552.8219
ronald.anderson@bc.edu
Doctorate, 1991
*19c, Europe, Physical Sciences;
20c, Physical Sciences;
Philosophy and Philosophy of
Science*

Anderson, Warwick H.
UCSF
Dept of Anthro Hist & Soc Med
3333 California St Ste 485
San Francisco, CA 94143-0850
Phone: 1.415.476.1245
wanders@itsa.ucsf.edu
Doctorate
*20c, Asia, Medical Sciences;
19c, Australia and Oceania,
Medical Sciences; 20c, North
America, Medical Sciences*

Anderson-Meyer, Phyllis
St Xavier University
3700 W 103rd Street
Chicago, IL 60526
Phone: 1.773.298.3513
andermeyer@sxu.edu
Doctorate, 1983
Education; Physical Sciences

Andrei, Mary A.
University of Minnesota
3204 Chicago Ave
Minneapolis, MN 55407-2105
andr0309@tc.umn.edu

Andrews, Bridie J.
Harvard University
Hist of Science Dept
Science Center 235
Cambridge, MA 02138
Phone: 1.617.495.3550
bandrews@fas.harvard.edu
Doctorate, 1996
*Asia, Medical Sciences;
Exploration, Navigation and
Expeditions; Natural and
Human History*

Andrews, James T.
Iowa State University
Dept of History
603 Ross Hall
Ames, IA 50011-1202
Phone: 1.515.294.3828
Fax: 1.515.294.6390
andrewsj@iastate.edu
Doctorate, 1994
*20c, Europe, Social Relations of
Science*

Ankeny, Rachel A.
University of Sydney
Unit for History & Philosophy
of Science

Carslaw F07
Sydney NSW 2006
Australia
Phone: 61.2.9523.6196
Fax: 61.2.9351.2124
r.ankeny@scifac.usyd.edu.au
Doctorate, 1997
*Philosophy and Philosophy
of Science; 20c, Biological
Sciences; 20c, Medical Sciences*

Anker, Peder
University of Oslo
Box 1116 Blindern
Oslo 0317
Norway
Phone: 47.2285.8885
Fax: 47.2285.892
anker@post.harvard.edu
Doctorate, 1999
20c, Europe, Biological Sciences

Appel, John W.
Apartado Aereo #476
Popayan
Colombia
Phone: 57.28.232.792
jwappel@alum.mit.edu
Master's Level, 1977
South America; Physical Sciences

Appel, Toby A.
Yale University
Cushing/Whitney Med Library
333 Cedar St
PO Box 208014
New Haven, CT 06520-8014
Phone: 1.203.785.4354
Fax: 1.203.785.5636
toby.appel@yale.edu
Doctorate, 1976
*20c, North America, Biological
Sciences; 19c, North America,
Medical Sciences; 18c, North
America, Medical Sciences;
19c, North America, Natural and
Human History*

Apple, Rima D.
University of Wisconsin
School of Human Ecology
1300 Linden Drive
Madison, WI 53706
Phone: 1.608.263.9354
Fax: 1.608.262.5335
rdapple@consci.wisc.edu
Doctorate, 1981

20c, Gender and Science;
20c, Medical Sciences;
19c, Medical Sciences;
20c, North America

Applebaum, Wilbur
1124 Oak Ave
Evanston, IL 60202-1219
Phone: 1.847.492.1167
Fax: 1.847.492.1167
applebaum@iit.edu
Doctorate, 1968
17c, Europe, Astronomical
Sciences; 17c, Europe,
Social Relations of Science;
17c, Social Sciences

Apt, Adam J.
Colonial Management
Associates
One Financial Center
13th Floor
Boston, MA 02111
Phone: 1.617.657.7502
Fax: 1.617.350.6960
ajapt@world.std.com
Doctorate, 1982
17c, Europe, Astronomical
Sciences; 19c, North America,
Astronomical Sciences

Arabatzis, Theodore
University of Athens
Department of Philosophy and
History of Science
Panepistimioupolis (Ano Ilisia)
Athens 15771
Greece
Phone: 30.1.727.5524
Fax: 30.1.727.5530
tarabatz@cc.uoa.gr
Doctorate, 1995
19c, Physical Sciences;
20c, Physical Sciences;
Philosophy and Philosophy of
Science; Social Sciences

Arens, Katherine
University of Texas
Germanic Studies
EP Schoch 3.102
Austin, TX 78712-1190
Phone: 1.512.471.4123
Fax: 1.512.471.4025
k.arens@mail.utexas.edu
Doctorate, 1981
Europe, Philosophy and

Philosophy of Science;
Social Sciences; 19c, Cognitive
Sciences; Institutions

Arias, Pedro L.
Escuela Superior de Ingenieros
Alameda Urquijo S/N
Bilbao 48013
Spain
Phone: 34946014069
Fax: 34946014179
iaparerp@bi.ehu.es
Doctorate, 1984
20c, Technology, Instruments
and Techniques; Physical
Sciences; Biological Sciences;
Philosophy and Philosophy of
Science

Ariew, Roger
Virginia Tech
Dept of Philosophy
Major Williams Hall
Blacksburg, VA 24061-0126
Phone: 1.540.231.8490
Fax: 1.540.231.6367
ariew@vt.edu
Doctorate, 1976
17c, Europe, Astronomical
Sciences; 5c-13c, Europe,
Astronomical Sciences;
Pre-400 B.C.E., Europe,
Philosophy and Philosophy
of Science; 19c, Europe,
Physical Sciences

Arikha, Noga
32 Belsize Park
London NW3 4DX
Great Britain
Phone: 20.74311592
narikha@aol.com

Armijo, Maruxa
Instituto Investigaciones
Filosoficas
Coyoacan
Mexico City DF 04510
Mexico
Fax: 563.92.25
maruxa@servidor.unam.mx
Master's Level
5c-13c; Mathematics;
14c-16c, Education;
Social Sciences

Arns, Robert G.
University of Vermont
Physics Department
Cook Building
Burlington, VT 05405-0125
Phone: 1.802.656.0049
Fax: 1.802.658.0514
robert.arns@uvm.edu
Doctorate, 1960
20c, Physical Sciences;
20c, Technology; 20c, Education

Aronson, Jay D.
University of Minnesota
Program in Hist of Sci & Tech
435 Walter Library
Minneapolis, MN 55455
Phone: 1.617.666.4569
aron0031@tc.umn.edu

Arpaia, Paul M.
2592 45th St
Astoria, NY 11103-1126
Phone: 1.212.802.2852
paul_arpaia@baruch.cuny.edu

Arthur, Richard T. W.
University of Toronto
IHPST, 73 Queens Park Cres E
Room 316, Victoria College
Toronto, ON M5S1K7
Canada
Phone: 1.416.978.5020
Fax: 1.416.978.3003
arthur@middlebury.edu
Doctorate, 1981
17c, Astronomical Sciences,
Philosophy and Philosophy of
Science; Physical Sciences;
Philosophy and Philosophy of
Science; Astronomical Sciences

Arvin, Shelley
4943 Coventry Pkwy
Fort Wayne, IN 46804-7115
infopusher@earthlink.net

Ash, Caroline
Science International
Bateman House
82-88 Hills Road
Cambridge CB21LQ
Great Britain
Phone: 44.1223.510362
Fax: 44.1223.326501
cash@science.int.co.uk

Ash, Eric H.
Dibner Institute
Dibner Building
MIT E56-100
38 Memorial Drive
Cambridge, MA 02139
Phone: 1.617.253.8721
Fax: 1.617.253.9858
ericash@post.harvard.edu
Doctorate, 2000
14c-16c, Europe, Exploration,
Navigation and Expeditions;
14c-16c, Europe, Technology;
14c-16c, Europe, Earth Sciences;
14c-16c, Europe, Mathematics

Ash, Mitchell G.
University of Vienna
Inst of History
Dr. Karl-Lueger-Ring 1
Vienna 1010
Austria
Phone: 431427740837
Fax: 43142779408
mitchell.ash@univie.ac.at
Doctorate, 1982
20c, Europe, Social Relations
of Science; 19c, Europe,
Social Relations of Science;
20c, Cognitive Sciences

Ashurst, Frederick G.
1 Findon Hill - Sacriston
Durham DH7 6LR
Great Britain
Phone: 191.3710042
Bachelor's Level, 1971
Physical Sciences; Medical
Sciences; Biological Sciences

Ashworth, William J.
University of Liverpool
School of History
9 Abercromby Square
Liverpool L69 3BX
Great Britain
Phone: 0151.794.2374
Fax: 0151.794.2396
w.j.ashworth@liverpool.ac.uk

Asner, Glen
821 Maryland Ave NE Apt 305
Washington, DC 20002-5374
Phone: 1.202.544.0508
asner@andrew.cmu.edu

Assad, Arjang A.
University of Maryland
R H Smith School of Business
Van Munching Hall
College Park, MD 20742-0001
Phone: 1.301.405.5424

Assmus, Alexi
Polytechnic University
6 Metrotech Ctr
Brooklyn, NY 11201
Phone: 1.718.260.3787
aasmus@poly.edu

Astore, William J.
1623 N Cascade Ave
Colorado Spring, CO 80907
Phone: 1.719.333.8613
bill.astore@usafa.af.mil

Atkinson, James W.
1455 W Columbia Rd
Mason, MI 48854-9694
Phone: 1.517.353.2269
atkinso9@msu.edu

Atkinson-Grosjean, Janet
Green College Univ of BC
Interdicipilinary Studies
Vancouver, BC V6T1Z1
Canada
Phone: 1.604.486.0562
janetat@interchange.ubc.ca

Attis, David A.
419 Pershing Dr
Silver Spring, MD 20910-4254
Phone: 1.202.462.5174
daattis@alumni.princeton.edu
Doctorate, 1999
Europe, Institutions;
19c, Mathematics; Education;
19c, Technology, Instruments
and Techniques

Atzema, Eisso J.
University of Maine at Orono
Dept of Mathematics
328 Neville Hall
Orono, ME 04469-5752
Phone: 1.207.581.3928
Fax: 1.207.581.3902
atzema@math.umaine.edu
Doctorate, 1993
19c, Mathematics;
19c, North America, Education

Augeard, Alberic
sloesveldstraat, 37
Hoeilaart B-1560
Belgium
Phone: 32.2.722.08.34
Fax: 32.2.721.51.85
alberic_augeard@yahoo.com

Auger, Jean
CIRST-UQAM
CP 8888
Succ Centre-ville
Montreal, PQ H3C3P8
Canada
Phone: 1.514.987.3000
Fax: 1.514.987.7726
auger.jean-francois@uqam.ca
20c, North America, Technology;
20c, North America, Technology,
Instruments and Techniques;
20c, North America, Physical
Sciences

Azuma, Toru
2-2 Midorigaoka Kita Hannan
Osaka 599 0200
Japan

Baader, Gerhard
Freie Universitat Berlin
History of Medicine
Klingsorstr 119
Berlin 12203
Germany
gerhard.baader@
  medizin.fu-berlin.de

Baatz, Simon
Nlm\Hmd #38
History of Medicine Division
National Institutes of Health
8600 Rockville Pike
Bethesda, MD 20894-0001
Phone: 1.301.594.7895
Fax: 1.301.402.0872
simon_baats@nlm.nih.gov

Babkov, Vasilii
Institute of the History of
Science & Technology
Staropanski, 1/5
Moscow 103012
Russia
Phone: 7.095.9281307
Fax: 7.095.9259911
babkoff@ihst.ru
Doctorate, 1990

20c, Europe, Biological Sciences;
20c, Europe, Medical Sciences;
20c, Europe, Social Relations
of Science; 20c, Europe,
Institutions

Bach, Jose Alfredo
506 Deerwood Dr
San Marcos, TX 78666-1220
Phone: 1.512.392.4228
jabach@centurytel.net

Bacos, Remy
5242 Byron
Montreal, PQ H3W2E9
Canada
Phone: 1.514.484.1940
Fax: 1.514.848.1940
rodelia@videotron.ca

Badash, Lawrence
University of California
Department of History
Santa Barbara, CA 93106
Phone: 1.805.893.2665
badash@history.ucsd.edu
Doctorate, 1964
20c, Physical Sciences; North
America; 20c, Science Policy

Bair, Lorne
PO Box 828
Winchester, VA 22604-0828
Phone: 1.540.868.0222
satmind@mnsinc.com

Baird, Davis
University of South Carolina
Philosophy Dept
Columbia, SC 29208-0001
Phone: 1.803.777.4166
Fax: 1.803.777.9178
db@sc.edu
Doctorate, 1981
20c, North America, Physical
Sciences; 20c, North America,
Technology; 20c, North America,
Instruments and Techniques

Baker, Anthony
PO Box 123
Broadway NSW 2007
Australia
Phone: 61.2.9514.1764
Fax: 61.2.9514.1460
tony.baker@uts.edu.au

Baker, David B.
460 Delaware Ave
Akron, OH 44303
Phone: 1.330.835.1341
Fax: 1.330.972.2093
bakerd@uakron.edu

Baldasso, Renzo
3625 Falls Rd
Baltimore, MD 21211-1815
Phone: 1.410.366.1377
rb843@columbia.edu

Baldi, David
9224 Shady Tree Ct
Fair Oaks, CA 95628-4169
Phone: 1.916.987.6505
djbaldi@attbi.com

Baldwin, Martha R.
Harvard University
Science Center 235
One Oxford St
Cambridge, MA 02138
Fax: 1.617.495.3344
baldwin@fas.harvard.edu
Doctorate, 1987
17c, Europe, Medical Sciences;
14c-16c, Natural and Human
History; 18c, Physical Sciences;
Medical Sciences

Bales, Ellen
100 W 92nd St Apt 27H
New York, NY 10025-7546
Phone: 1.212.712.9127
ellenbales@earthlink.net

Balis, Andrea
301 W 110th St Apt 2N
New York, NY 10026-4059
Phone: 1.212.749.3861
andreabalis@hotmail.com

Bange, M. C.
Combardeux
69870 - Saint - Just
D'Avray F69870
France
Phone: 33.474715494
cbange@aol.com

Banks, Ronald E.
6325 Conner Rd
East Amherst, NY 14051-1522
Phone: 1.716.741.8505
Fax: 1.716.741.8014

Doctorate, 1999
20c, Biological Sciences;
Philosophy and Philosophy of
Science

Barber, Leslie A.
MIT
STS Program / E51-296B
77 Massachusetts Avenue
Cambridge, MA 02139
Doctorate
20c, North America, Biological
Sciences; 20c, North America,
Institutions; 20c, North America,
Social Relations of Science

Barbera, Keith
Johns Hopkins University
Dept of History of Science
and Medical Technology
3400 N Charles St
Baltimore, MD 21218-2680

Barberis, Daniela
Max Planck Institute for the
History of Science
Wilhelmstrasse 44
Berlin 10117
Germany
Phone: 4930.226.67178
Fax: 49630.226.67290
barberis@mpiwg-berlin.mpg.de
Master's Level, 1992
19c, Europe, Social Sciences;
19c, Europe, Philosophy
and Philosophy of Science;
19c, Europe, Medical Sciences;
19c, Europe, Cognitive Sciences

Barker, Peter
University of Oklahoma
Dept of History of Science
601 Elm, Room 622
Norman, OK 73019
Phone: 1.405.325.2242
Fax: 1.405.325.2363
barkerp@ou.edu
Doctorate, 1975
14c-16c, Astronomical Sciences;
20c, Physical Sciences;
Cognitive Sciences

Barker, Sinead
Universtiy of Bristrol
Sch of Biological Sciences
Woodland Road
Bristol BS8 1UG

Great Britain
Phone: 44.117.928.7478
Fax: 44.117.925.7473
sinead.barker@bristol.ac.uk

Barkley, Grant M.
Kent State University
4314 Mahoning Ave NW
Warren, OH 44483-1998
Phone: 1.330.675.8908
Fax: 1.330.847.6172
barkleyg33@aol.com
Doctorate, 1973
Biological Sciences

Barmore, Frank E.
University of Wisconsin
Physics Department
Cowley Hall
La Crosse, WI 54601-3788
Phone: 1.608.785.8432
Fax: 1.608.785.8403
barmore.fran@uwlax.edu
Doctorate, 1973
5c-13c, Astronomical Sciences;
20c, North America, Earth
Sciences; Education; Technology

Barnes, David S.
Harvard University
Dept of the History of Science
Science Center 235
Cambridge, MA 02138
Phone: 1.617.496.5184
Fax: 1.617.495.3344
dbarnes@fas.harvard.edu
Doctorate, 1992
19c, Europe, Medical Sciences;
Medical Sciences

Barnes, Michael H.
University of Dayton
Religious Studies Dept
Dayton, OH 45469-1530
Phone: 1.937.229.3490
barnes@udayton.edu
Doctorate, 1976
Natural and Human History;
Philosophy and Philosophy of
Science; Social Sciences

Barnett, Margaret
Univ of Southern Mississippi
Dept of History
Box 5047
Hattiesburg, MS 39406-5047
Phone: 1.601.266.4801

Fax: 1.601.266.4334
lbarnett@ocean.otr.usm.edu
Doctorate, 1982
Medical Sciences

Barow, Markus
Joh-Wolfgang-Goethe Univ
Inst für Geschichte de Nat
Robert Mayer Str. 1
Frankfurt D-60054
Germany
barow@iap.uni-frankfurt.de
Master's Level
Europe, Astronomical Sciences;
Europe, Instruments and
Techniques

Barr, Bernadine C.
6944 Flowing Springs Rd
Shenandoah Jctn, WV 25442
Phone: 1.304.876.2257
Fax: 1.304.876.2257
bbarr@citlink.net

Barrera, Antonio
Colgate University
History Department
Hamilton, NY 13346
Phone: 1.315.228.7549
abarrera@colgate.edu

Barrow, Mark V., Jr.
Virginia Tech
Department of History
Blacksburg, VA 24061-0117
Phone: 1.540.231.4099
Fax: 1.540.231.8724
barrow@vt.edu
Doctorate, 1992
Biological Sciences;
North America, Natural and
Human History; North America,
Biological Sciences

Barth, Kai H.
University of Minnesota
Prog in Hist of Science
435 Walter Library
Minneapolis, MN 55405
Phone: 1.202.544.8453
Fax: 1.202.687.5528
barth002@earthlink.net
Doctorate, 2000
20c, North America, Social
Relations of Science;

20c, North America, Science
Policy; 20c, Physical Sciences;
20c, Earth Sciences

Bartholomew, James R.
Ohio State University
History Dept
230 W 17th Ave
Columbus, OH 43210-1367
Phone: 1.614.292.8301
Fax: 1.614.292.2282
bartholomew.5@osu.edu
Doctorate, 1972
19c, Asia, Medical Sciences;
20c, Europe, Medical Sciences;
Social Relations of Science

Barton, Cathy
769 Rolling View Dr
Annapolis, MD 21401-4654

Barton, Ruth
University of Auckland
History Department
Private Bag 92019
Auckland
New Zealand
Phone: 64.9.373.7599
Fax: 64.9.3737.438
r.barton@auckland.ac.nz
Doctorate
19c, Europe, Social Relations
of Science; 19c, Australia and
Oceania

Bartz, Robert
2546 15th Ave
San Francisco, CA 94127-1312
Phone: 1.4153.502.0878
bartz@prodigy.net

Basalla, George
Univ of Delaware
Dept of History
Newark, DE 19716

Basargina, Ekaterina
Inst of History & Science Tech
St Petersburg Department
Universitetskaia NAB 5
St Petersburg 1
Russia

Bashaw, Charles E.
1053 Rifle Range Rd Apt 11G
Mount Pleasant, SC 29464-4638
Phone: 1.843.971.6029

bashawc@cofc.edu
Master's Level
17c, Europe, Philosophy
and Philosophy of Science;
14c-16c, Europe, Philosophy
and Philosophy of Science;
5c-13c, Europe, Philosophy
and Philosophy of Science;
400 B.C.E.-400 C.E., Europe,
Philosophy and Philosophy of
Science

Bassett, David A.
Boston Scientific
One Boston Scientific Pl
Mail Stop A4
Natick, MA 02331
Phone: 1.508.647.2420
bassettd@bsci.com
Master's Level, 1998
Pre-400 B.C.E., Physical
Sciences; 19c, Physical
Sciences; 14c-16c, Instruments
and Techniques;
18c, Instruments and Techniques

Bassett, Ross
North Carolina State Univ
PO Box 8108
Marrelson Hall
Raleigh, NC 27695-8108
Phone: 1.919.513.2230
Fax: 1.919.513.3886
ross_bassett@ncsu.edu
Doctorate, 1998
20c, North America, Technology;
20c, North America, Technology,
Instruments and Techniques

Bassi, Joseph P.
Joint Mil Intel College
DIA/MCE L
Washington, DC 20340
bas123@erols.com
Master's Level
20c, North America; 20c, North
America, Science Policy;
20c, Technology; North America,
Astronomical Sciences

Battimelli, Giovanni
Universita la Sapienza
Dipartimento di Fisica
Piazzale Aldo Moro 2
Roma 00185
Italy
Phone: 39.649913490

Fax: 39.64463158
battimelli@axrma.uniroma1.it
19c, Europe, Physical Sciences;
20c, Europe, Physical Sciences

Bauman, Harold
University of Utah
Dept of History
380 S 1400 E Rm 211
Salt Lake City, UT 84112-0311
Phone: 1.801.581.6672

Baxter, Alice L.
Loomis Chaffee School
44 Batchelder Rd
Windsor, CT 06095-3028
Phone: 1.860.687.6227
alicebaxter@loomis.org
Doctorate

Bayne, Martha
University of Chicago Press
Advertising Coordinator
1427 E 60th St
Chicago, IL 60637-2954

Bazerman, Charles
Univ of California
English Dept
Santa Barbara, CA 93106
Phone: 1.805.893.7543
Fax: 1.805.683.2337
bazerman@education.ucsb.edu
Doctorate, 1971
Humanistic Relations of Science;
Social Relations of Science;
Biological Sciences

Beatty, John
University of Minnesota
Department of Ecology,
Evolution and Behavior
100 Ecology Building
St Paul, MN 55108
Phone: 1.612.624.6749
Fax: 1.612.624.6777
beatty@tc.umn.edu
Doctorate, 1979
20c, Biological Sciences;
19c, Biological Sciences

Beaujouan, Guy
102 Ave du Grl Leclerc
Paris 75014
France

Beaulieu, Liliane
CRM Univ of Montreal
C.P. 6128, Succ A
Montreal, PQ H3C3J7
Canada
Phone: 1.514.343.7501
Fax: 1.514.343.2254
beaulieu@crm.umontreal.ca

Beaver, Donald deB.
Williams College
History of Science
Bronfman Science Center
18 Hoxsey St
Williamstown, MA 01267-2813
Phone: 1.413.597.2239
Fax: 1.413.597.4116
dbeaver@williams.edu
Doctorate
Social Relations of Science;
North America; 19c, Natural
and Human History; Technology

Bechtel, William
Univ of California - San Diego
Dept of Philosophy - 0119
9500 Gilman Dr
La Jolla, CA 92093-5004
bill@artsci.wustl.edu
Doctorate, 1977
Medical Sciences;
Instruments and Techniques;
Cognitive Sciences; Philosophy
and Philosophy of Science

Becker, Barbara J.
Univ of California Irvine
History Dept
Irvine, CA 92697
bjbecker@uci.edu
Doctorate, 1993
19c, Astronomical Sciences;
Astronomical Sciences;
Technology

Beetschen, Jean C.
Université Paul-Sabatier
118 Route de Narbonne
Toulouse Cedex 31062
France
Fax: 33.561556507
beetsche@cict.fr
Doctorate, 1960
19c, Europe, Biological Sciences;
19c, Biological Sciences

Beidleman, Richard G.
766 Bayview Ave
Pacific Grove, CA 93950
Phone: 1.831.375.1922
Doctorate, 1954
19c, North America, Exploration,
Navigation and Expeditions;
20c, North America, Biography;
19c, Australia and Oceania,
Biological Sciences;
19c, North America, Natural and
Human History

Bekasova, Alexandra V.
Russian Academy of Sciences
Inst History of Science & Tech
5 Universitetskaia Nab
St Petersburg 199034
Russia
Phone: 7.812.328.5924
Fax: 7.812.328.4667
bek@ab2352.spb.edu
Master's Level, 1990
Social Relations of Science;
Education; Medical Sciences

Bellon, Rich
University of Minnesota
Prog in the Hist of Sci & Tech
148 Tate Laboratory of Physics
116 Church St, SE
Minneapolis, MN 55455
Phone: 1.612.925.9062
Fax: 1.612.624.4578
bellon@physics.umn.edu
Master's Level, 2000
19c, Europe, Natural and
Human History;
19c, Europe, Biological Sciences

Beltran, Antonio
Universidad de Barcelona
Baldiri 1 Reixach
Barcelona 08028
Spain
Phone: 93.333.3466
Fax: 93.449.8510
abeltran@trivium.gh.ub.es
Doctorate, 1981

Ben-Zaken, Avner
9866 Vidor Dr
Los Angeles, CA 90035-1029
Phone: 1.310.825.4601
Fax: 1.310.206.9630
avner@ucla.edu

Benaroyo, Lazare
Chemin des Vignes 29
1009 Pully CH
Pully 1009
Switzerland
Phone: 41217290422
Fax: 41213147055
lazare.benaroyo@inst.hospvd.ch
Doctorate
20c, Europe, Philosophy
and Philosophy of Science;
19c, North America, Philosophy
and Philosophy of Science;
18c, Philosophy and Philosophy
of Science; Medical Sciences

Benbow, Peter K.
University of Oxford
St Edmund Hall
Oxford OX1 4AR
Great Britain
Phone: 1.805.965.7227
Fax: 1.805.965.6930
peterkandolyae@cs.com
Doctorate, 1999
5c-13c, Europe, Mathematics;
Astronomical Sciences;
Physical Sciences;
Medical Sciences

Benfey, Otto T.
909 Woodbrook Dr
Greensboro, NC 27410-3247
Phone: 1.336.854.2136
benfeyot@nr.infi.net
Doctorate, 1947
19c, Physical Sciences;
20c, Humanistic Relations of
Science; 20c, Philosophy and
Philosophy of Science

Benjamin, Ludy T., Jr.
Texas A & M University
Department of Psychology
College Station, TX 77843-4235
Phone: 1.979.845.2540
Fax: 1.979.845.4727
ltb@psyc.tamu.edu
Doctorate, 1971
19c, Cognitive Sciences;
20c, North America, Cognitive
Sciences; Social Sciences

Bennett, Joan W.
Tulane University
Dept of Cell & Molecular Bio
New Orleans, LA 70118

Phone: 1.504.862.8101
Fax: 1.504.865.6785
jbennett@tulane.edu

Benson, Etienne
4510 13th St NW
Washington, DC 20011-4404
Phone: 1.650.468.5782
ebenson@stanford.edu

Benson, John M.
Harvard Sch of Public Health
Dept of Health Policy & Mgt
677 Huntington Ave
Boston, MA 02115
Phone: 1.617.432.1094
Fax: 1.617.432.0092
jmbenson@hsph.harvard.edu
Master's Level
Earth Sciences; 5c-13c;
Education

Benson, Keith R.
University of Washington
Box 354330
Seattle, WA 98195
Phone: 1.206.543.6358
Fax: 1.206.543.7400
krbenson@u.washington.edu
Doctorate, 1979
19c, North America;
18c, Europe, Natural and
Human History; 20c, North
America, Earth Sciences;
18c, Europe, Biological Sciences

Beranek, Leo L.
975 Memorial Dr Apt 804
Cambridge, MA 02138-5755
Phone: 1.617.576.3141
Fax: 1.617.492.7816
Doctorate, 1940
Physical Sciences

Beretta, Francesco
16, ch. de Bethleem
Case Postale
Fribourg CH-1701
Switzerland
Phone: 41.26.321.55.77
Fax: 41.26.322.39.35
fr.beretta@bluewin.ch

Bergen, Edward F.
66 Orchard Park Blvd
Toronto, ON M4L3E2
Canada

Phone: 1.416.691.6786
j.e.bergen@sympatico.ca
Master's Level
Biological Sciences; Philosophy
and Philosophy of Science

Berger, Michael L.
Arcadia University
450 S Easton Rd
Glenside, PA 19038-3215
Phone: 1.215.572.2924
Fax: 1.215.572.2126
berger@arcadia.edu
Doctorate, 1972
20c, North America, Technology;
20c, North America, Education

Berger, Stephen
Springfield College
500 N Commercial St
Manchester, NH 03101-1131
Phone: 1.603.666.5700
Fax: 1.603.666.5705
sberger@spfldcol.edu

Bergeron, Nicolas
2747 Dupont Ave S Apt 302
Minneapolis, MN 55408-1220
Phone: 1.612.874.7208
berg1204@tc.umn.edu

Berggren, John L.
Simon Fraser University
Dept of Math & Statistics
8888 University Drive
Burnaby, BC V5A1S6
Canada
Phone: 1.604.291.3378
Fax: 1.604.291.4947
berggren@sfu.ca
Doctorate, 1966
400 B.C.E.-400 C.E., Asia,
Mathematics;5c-13c, Asia,
Mathematics; Asia

Bergstrom, Jon
Willey Hill Road Box 378
Norwich, VT 05055
Phone: 1.802.649.8866
Fax: 1.802.649.3804
j.bergstrom@valley.net

Berman, Bruce J.
Queen's University
Dept of Political Studies
Kingston, ON K7L3N6
Canada

Phone: 1.613.533.6242
Fax: 1.613.533.6848
bermanb@politics.queensu.ca

Berndt, W. O.
Univ of Nebraska Med Ctr
Vice Chancellor
987810 Nebraska Med Ctr
Omaha, NE 68198-7810
Phone: 1.402.559.5130
Fax: 1.402.559.7845

Bernick, Niles
287 Crispin Ln
Falling Waters, WV 25419-3855
Phone: 1.304.274.9310
bernickn@erols.com

Bernstein, Jeremy
c/o New Yorker
25 W 43rd St
New York, NY 10036-7406

Berol, David
420 Massachusette Ave, Apt #4
Arlington, MA 02474
Phone: 1.781.316.1525

Berrios - Ortiz, Angel
Univ of Puerto Rico - Mayaguez
Biology Department
Mayaguez, PR 00680
Phone: 1.787.265.3837
Doctorate, 1975
5c-13c, North America,
Biological Sciences;
18c, North America, Natural
and Human History; North
America, Biological Sciences;
South America, Social Sciences

Berry, Charles L.
9403 Ocala St
Silver Spring, MD 20901-3047

Bertol Domingues, Heloisa M.
Museu de Astronomia Ciencias
Rua General Bruce 586
Rio de Janeiro 20921 030
Brazil
Phone: 5521.580701
Fax: 5521.5804531
heloisa@mast.br
19c, South America, Biological
Sciences; 19c, South America,
Natural and Human History;
20c, South America, Natural

and Human History; 19c, South
America, Exploration,
Navigation and Expeditions

Bertoloni Meli, Domenico
Indiana University
Department of History and
Philosophy of Sciencs
Goodbody Hall 130
Bloomington, IN 47405
Phone: 1.812.855.8746
Fax: 1.812.855.3622
dbmeli@indiana.edu
Doctorate
17c, Europe, Physical Sciences;
17c, Europe, Medical Sciences;
17c, Europe, Mathematics;
18c, Europe, Medical Sciences

Beukers, Harm
V. D. Valk Boumanweg 40
Leiderdorp 2352
The Netherlands
Phone: 31.715891434
Fax: 31.715276567
h.beukers@lumc.nl

Bevan, William
Duke University
Talent Identification Program
1121 W Main St Ste 100
Durham, NC 27701
Phone: 1.919.683.1400
Doctorate, 1948
Cognitive Sciences;
Science Policy

Bevilacqua, Fabio
Universita di Pavia
Dipartimento di Fisica
Via Bassi 6
Pavia
Italy
Phone: 390382507495
bevilacqua@fisicavolta.unipv.it

Beyerchen, Alan D.
Ohio State University
History Department
230 W 17th Ave
Columbus, OH 43210-1367
Phone: 1.614.292.5478
Fax: 1.614.292.2282
beyerchen.1@osu.edu
Doctorate, 1973
20c, Europe, Social Relations
of Science; 20c, North America,

Technology; 20c, Europe,
Technology; 20c, Instruments
and Techniques

Beyler, Richard H.
Portland State University
Dept of History
PO Box 751
Portland, OR 97207-0751
Phone: 1.503.725.3996
Fax: 1.503.725.3953
beylerr@pdx.edu
Doctorate
Physical Sciences; Social
Relations of Science; Philosophy
and Philosophy of Science;
20c, Humanistic Relations of
Science

Biener, Zri
University of Pittsburgh
1017 Cathedral of Learning
Pittsburgh, PA 15260-6299
zvbl@pitt.edu

Biggs, James
1235 E Whittier Blvd
La Habra, CA 90631-4073
Phone: 1.562.690.6516
jrbiggs@aol.com

Bill, Alan
13 Wilmington Close
Newcastle Upon Tyne NE3 2SF
Great Britain
Phone: 1912424467
alanbill@globalnet.co.uk

Binnie, Anna E.
Macquarrie University
Physics Dept
NSW 2113
Australia
abinnie@laurel.ocs.mq.edu.au

Binns, James W.
10 Hunt Court
York Y01 7DE
Great Britain
Doctorate
17c, Education;
5c-13c, Humanistic Relations of
Science

Birn, Anne-Emanuelle
Milano Graduate School
New School for Social Research

72 5th Ave
New York, NY 10011-8802
Phone: 1.212.229.5339
Fax: 1.212.229.5404
aebirn@newschool.edu
Doctorate, 1993
20c, North America, Medical
Sciences; 20c, South America,
Science Policy

Bix, Amy S.
Iowa State University
History Department
633 Ross Hall
Ames, IA 50011-0001
Phone: 1.515.294.0122
Fax: 1.515.294.6390
abix@iastate.edu
Doctorate, 1994
Technology; Technology,
Instruments and Techniques;
Gender and Science;
Science Policy

Blackmore, John T.
4932 Sentinel Dr Apt 201
Bethesda, MD 20816-3546
Phone: 1.301.320.0256
Doctorate
20c, Europe, Philosophy
and Philosophy of Science;
20c, Europe, Biography;
20c, Europe, Physical Sciences

Blackwell, Richard J.
St Louis University
Philosophy Department
221 N Grand Blvd
Saint Louis, MO 63103-2097

Blaine, Bradford B.
586 W 11th St
Claremont, CA 91711-3720
Phone: 1.909.624.9912
Fax: 1.909.626.1023
Doctorate
5c-13c, Europe, Instruments and
Techniques; 400 B.C.E.-400 C.E.,
Technology; Pre-400 B.C.E.

Blair, Ann
Harvard University
Dept of History
Robinson Hall 200
Cambridge, MA 02138
Phone: 1.617.495.0752
Fax: 1.617.496.3425

amblair@fas.harvard.edu
Doctorate, 1990
Europe, Humanistic Relations
of Science; 17c, Europe,
Humanistic Relations of Science;
14c-16c, Europe, Natural and
Human History

Blair Bolles, Edmund
414 Amsterdam Ave Apt 4N
New York, NY 10024-6251
Phone: 1.212.595.0463
blair@ebbolles.com

Blaisdell, Muriel L.
Miami University
Sch Interdisciplinary Studies
Peabody Hall
Oxford, OH 45056
Phone: 1.513.529.5674
Fax: 1.513.529.5849
blaisdml@muohio.edu
Doctorate
19c, Biological Sciences;
Exploration, Navigation and
Expeditions; Biological Sciences;
19c, Natural and Human History

Blanchard, Leslie F.
947 NW 62nd St
Seattle, WA 98107-2844

Blancher, David
Queensborough Comm College
Dept of History
222-05 56th Ave
Bayside, NY 11364-1497
Phone: 1.718.631.6291
Fax: 1.718.631.6372
dblan24@aol.com
Doctorate
19c, North America, Medical
Sciences; 20c, North America,
Medical Sciences; 18c, North
America, Medical Sciences;
20c, North America, Social
Relations of Science

Blank, Bernard
596 E 8th St
Brooklyn, NY 11218-5906
Phone: 1.718.436.9529

Blay, Michel
CNRS
3 Rue Michel-Ange
Paris 16 75794

France
Phone: 331.42975068
Fax: 33.42974646
Doctorate, 1989
17c, Europe, Mathematics;
18c, Physical Sciences

Blazquez, Francisco
Calle Donana, 26
VVA De La Serena
Badajoz 06700
Spain
Phone: 34.924847087
Fax: 34.924845818
fblazq2@mimosa.pntic.mec.es
Bachelor's Level, 1987
20c, Europe, Biological Sciences;
20c, Biological Sciences

Bleichmar, Daniela
Princeton University
History Department
211 Dickinson Hall
Princeton, NJ 08540
dbleichm@princeton.edu

Blinn, Walter C.
Michigan State University
Integrative Studies/Gen Sci
Room 100, N Kedzie Lab
East Lansing, MI 48824-1031
Phone: 1.517.353.8158
Fax: 1.517.432.2175
blinnw@pilot.msu.edu
Doctorate, 1961
Education

Bloom, David A.
University of Michigan
Prof of Urology
1500 E Medical Center Dr
Ann Arbor, MI 48109-0330
Phone: 1.734.936.7025
Fax: 1.734.936.9127
dabloom@umich.edu

Blukis, Uldis
271 W End Ave
Brooklyn, NY 11235-4903
Phone: 1.718.934.6771
blukisu@aol.com

Blunt, John
19116 Palo Alto Ave
Hollis, NY 11423-1212

Phone: 1.718.368.5746
Fax: 1.718.368.4876
jblunt@kbcc.cuny.edu

Bock, Walter J.
Columbia University
Dept of Biological Sciences
1200 Amsterdam Ave
Mailbox 5521
New York, NY 10027-7003
Phone: 1.212.854.4487
Fax: 1.212.865.8246
wb4@columbia.edu
Doctorate, 1959
Philosophy and Philosophy of
Science; Biological Sciences

Bocking, Stephen A.
Trent University
Environmental Res Studies
Peterborough, ON K9J7B8
Canada
Phone: 1.705.748.1520
Fax: 1.705.748.1569
sbocking@trentu.ca
Doctorate, 1992
20c, Biological Sciences;
20c, Social Relations of Science

Bodson, Liliane
University of Liege
Place du Vingt Aout 32
B-4000 Liege
Belgium
Phone: 3243665579
Fax: 3243665723
liliane.bodson@ulg.ac.be
Doctorate, 1971
400 B.C.E.-400 C.E., Biological
Sciences

Boerlage, Daan
Nepveustraat 28-I
Amsterdam 1058 XP
The Netherlands
Phone: 31.612.3694
daanboer@xs4all.nl

Boersema, David
Pacific University
Dept of Philosophy
2043 College Way
Forest Grove, OR 97116-1797
Phone: 1.503.359.2150
Fax: 1.503.359.2242
boersema@pacificu.edu
Doctorate, 1985

20c, Philosophy and Philosophy
of Science; Philosophy
and Philosophy of Science;
Earth Sciences; Education

Bohan, William K.
104 Washington Rd
Scranton, PA 18509-2446

Bollens, Guy
Dorpsstraat 59
Kortenaken B-3470
Belgium
Phone: 3216805815
guy.bollens@tienen.vera.be

Bolt, Marvin P.
Adler Planetarium
History Department
1300 S Lake Shore Drive
Chicago, IL 60657
Phone: 1.312.322.0594
Fax: 1.312.341.9935
bolt@adlernet.org
Doctorate, 1998
Astronomical Sciences;
19c, Europe, Astronomical
Sciences; Instruments and
Techniques; 19c, Europe,
Social Relations of Science

Bolton Valencius, Conevery
Washington Univ in St.Louis
Dept of History
Campus Box 1062
One Brookings Drive
St. Louis, MO 63130
Phone: 1.314.935.7518
Fax: 1.314.935.4399
cvalenci@artsci.wustl.edu
Doctorate, 1998
19c, North America, Medical
Sciences; Biological Sciences;
19c, North America, Biological
Sciences; Exploration,
Navigation and Expeditions

Bonkalo, Ervin
11800 Lake Fraser Dr SE
Calgary, AB T2J7G8
Canada
Phone: 1.416.789.2423

Bonker - Vallon, Angelika
Westfalische Wilnems-Univ
Seminar für Philosophie
Domplatz 23

Munster D-48143
Germany
Fax: 08092.708491
a.boenker-vallon@t-online.de
Doctorate, 1995
17c, Europe, Mathematics;
17c, Europe, Physical Sciences;
18c, Europe, Mathematics;
Philosophy and Philosophy of
Science

Bonner, Thomas N.
10970 E San Salvador Dr
Scottsdale, AZ 85259-5726
Phone: 1.602.860.2606
Fax: 1.602.860.2286
sylbon@att.net
Doctorate, 1952
19c, North America, Medical
Sciences; 20c, Europe, Medical
Sciences

Bono, James J.
SUNY at Buffalo
Dept of History
Park Hall
Buffalo, NY 14260-4130
Phone: 1.716.645.2282
Fax: 1.716.645.5954
hischaos@acsu.buffalo.edu
Doctorate, 1981
14c-16c, Europe,
Medical Sciences; 17c, Europe,
Medical Sciences; 17c, Europe,
Humanistic Relations of Science;
Gender and Science

Bonoli, Fabrizio
Università degli Studi
di Bologna
via Ranzani, 1
Bologna I- 40127
Italy

Borck, Cornelius
Grimmstrasse 9
Berlin 10967
Germany
Phone: 49.30.8300.9222
Fax: 49.30.8300.9237
cornelius.borck@
  medizin.fu.berlin.de

Bordogna, Francesca M.
Max Planck Institute for the
History of Science
Wilhelmstrasse 44

Berlin 10117
Germany
Phone: 1.219.631.7540
bordogna.1@nd.edu
Doctorate, 1998
19c, North America, Philosophy
and Philosophy of Science;
19c, Cognitive Sciences;
19c, Mathematics

Borello, Mark E.
Michigan State University
Lyman Briggs School
E-30 Holmes Hall
East Lansing, MI 48825
Phone: 1.517.353.4504
borrell4@msu.edu
Master's Level
20c, North America, Biological
Sciences; 20c, North America,
Social Relations of Science

Borisov, Vasilii
Institute of the History of
Science & Technology
Russian Academy of Science
Staropanskii 1/5
Moscow 103012
Russia
Phone: 7.95.921.8061
Fax: 7.95.925.9911
borisov@history.ihst.ru
Doctorate, 1980
Technology; 20c, Europe,
Technology, Instruments and
Techniques; 20c, Europe,
Social Relations of Science

Bork, Kennard B.
Denison University
Dept of Geology & Geography
Granville, OH 43023
Phone: 1.740.587.6486
Fax: 1.740.587.6417
bork@cc.denison.edu
Doctorate, 1967
18c, Europe, Earth Sciences;
19c, Europe, Earth Sciences;
18c, Europe, Biological Sciences

Bortecen, Kerem
850 Orange St Apt 2
New Haven, CT 06511-2510
Phone: 1.203.777.2296
kerem.bortecen@yale.edu

Borut, Michael
67-66 Groton St
Forest Hills, NY 11375
Phone: 1.718.268.1301
Doctorate

Boschiero, Luciano
University of New South Wales
School of Science and Tech
Studies
Sydney 2052
Australia
luciano_boschiero@hotmail.com

Boulos, Pierre
University of Windsor
401 Sunset
Windsor, ON N9B3P4
Canada
Phone: 1.519.258.3851
boulos@uwindsor.ca

Boult, Lisa B.
University of Minnesota
Dept of History of Medicine
Mayo Bx 506 420 Delaware
St SE
Minneapolis, MN 55455
Phone: 1.612.818.5975
Fax: 1.612.625.7938
boult003@tc.umn.edu
Doctorate
Medical Sciences;
Natural and Human History

Bourdy, Franck
30 Boulevard Heurteloup
Tours 37000
France
Phone: 02.47.37.71.41
Fax: 02.47.39.06.95
f.bourdy@wanadoo.fr

Bourque, Monique
129 Saint Laurence Rd
Upper Darby, PA 19082-1313
mbourque@sas.upenn.edu

Bowden, Mary E.
Chemical Heritage Foundation
Senior Research Historian
315 Chestnut St
Philadelphia, PA 19106-2793
Phone: 1.215.925.2222
Fax: 1.215.925.1954
mebowden@chemheritage.org

Bowden, Thomas
3691 Notten Road
Grass Lake, MI 49240
Phone: 1.764.975.2800
Fax: 1.764.976.2787
tom@techdirections.com

Bowen, Alan C.
Inst for Research in Classical
Philosophy and Science
3 Nelson Ridge Rd
Princeton, NJ 08540-7423
Phone: 1.609.466.2098
Fax: 1.609.466.2098
acbowen@princeton.edu
Doctorate, 1977
400 B.C.E.-400 C.E.,
Astronomical Sciences;
400 B.C.E.-400 C.E.,
Mathematics;
400 B.C.E.-400 C.E., Philosophy
and Philosophy of Science

Bowen, William G.
The Andrew W Mellon
Foundation
President
140 E 62nd St
New York, NY 10021-8187

Bowler, Peter J.
Queen's Univ of Belfast
Anthropological Studies
Belfast BT7 1NN
Northern Ireland
Phone: 28.902.73882
Fax: 28.902.73700
pbowler@qub.ac.uk
Doctorate, 1971
19c, Europe, Biological Sciences,
Biography; Biological Sciences;
Earth Sciences

Bowser, Kenneth B.
1130 Jefferson St NE
Minneapolis, MN 55413-1401
Phone: 1.612.331.1475
k.b.bowser@worldnet.att.net

Boyd, Carole B.
125 W 1st St
Westover, WV 26501-3835
Phone: 1.304.296.3082
cbboyd@sbccom.com

Boyle, Eric
922 West Campus Lane
Goleta, CA 93117
Phone: 1.805.695.3984
eboyle@umail.ucsb.edu

Brace, C. Loring
Museum of Anthropology
University of Museums Bldg
Ann Arbor, MI 48109
Phone: 1.734.936.2951
clbrace@umich.edu

Brackenridge, J. Bruce
Lawrence University
Dept of Physics
Appleton, WI 54912
Phone: 1.920.832.6720
j.bruce.brackenridge@
   lawrence.edu

Bradlow, H. Leon
8625 Palo Alto St
Hollis, NY 11423-1203
Phone: 1.201.316.8104
Fax: 1.201.457.1882
leon@bradlow.net

Brame, Michael
9234 Stansberry Ave
Saint Louis, MO 63134-3634
Phone: 1.314.427.2615
stlscience@hotmail.com

Brammall, Kathryn M.
Truman State University
Dept of History
100 E Normal St
Kirksville, MO 63501-4211
Phone: 1.660.785.4665
Fax: 1.660.785.4480
brammall@truman.edu
Doctorate
14c-16c, Europe, Social
Relations of Science;
17c, Europe, Social Relations of
Science

Brandt, Allan M.
Harvard University
History of Science
Cambridge, MA 02138
Phone: 1.617.495.3532
Fax: 1.617.495.3344
brandt@fas.harvard.edu

Doctorate, 1983
20c, North America, Medical
Sciences

Brannon, James H.
193 Ely Pl
Palo Alto, CA 94306-4553
Phone: 1.408.256.4565
jbrannon@us.ibm.com

Braselmann, Sylvia
4828 19th St
San Francisco, CA 94114-2232

Brashear, Ronald S.
Smithsonian Institution
Dibner Library NMAH 1041
Washington, DC 20560-0672
Phone: 1.202.357.1568
Fax: 1.202.633.9102
brashearr@sil.si.edu
Master's Level, 1984
Astronomical Sciences;
Education; Mathematics;
Physical Sciences

Brauckmann, Sabine
Dept of Biological Sciences
Hanover, NH 03755-3576
Phone: 1.603.646.9149
Fax: 1.603.646.1347
sabine@dartmouth.edu

Brautigam, Jeffrey C.
Hanover College
Center for Free Inquiry
PO Box 108
Hanover, IN 47243-0108
Phone: 1.812.866.6848
Fax: 1.812.866.2164
brautgm@hanover.edu

Brayton, Jennifer
492 B George St
Fredericton, NB E3B1J9
Canada
Phone: 1.506.455.2025
i035g@unb.ca

Breitenberger, Ernst
45 Briarwood Dr
Athens, OH 45701-1302
Phone: 1.740.592.3699
Fax: 1.740.593.0433
brtbg@helios.phy.ohiou.edu

Doctorate, 1950
19c, Physical Sciences;
19c, Mathematics

Bremholm, Tony
835 Owl St
Norman, OK 73071
5og@ou.edu

Brenner, Anastasio A.
Univ de Toulouse le Mirail
5 Allees Antonio Machado
Toulouse 31058
France
Phone: 56.150.4737
Fax: 56.150.4422
brenner@univ-tlse2.fr
Doctorate, 1987
Philosophy and Philosophy of
Science; 19c, Physical Sciences;
20c, Physical Sciences;
Astronomical Sciences

Bresinsky, Tanya
7 Washburn Street
Watertown, MA 02742
Phone: 1.617.923.6242
tkbres@aol.com

Bresnahan, Jody C.
Harvard University
Women's Studies
Barker Ctr, 12 Quincy Street
Cambridge, MA 02138
Phone: 1.617.496.5519
jbresnah@fas.harvard.edu
Doctorate, 1999
20c, North America, Medical
Sciences; 20c, North America,
Gender and Science

Brewer, William F.
University of Illinois
Dept of Psychology
603 E Daniel St
Champaign, IL 61820-6232
Phone: 1.217.333.1548
Fax: 1.217.244.5876
w-brewer@uiuc.edu
Doctorate
Cognitive Sciences; Philosophy
and Philosophy of Science;
Education

Brey, Gerhard
57 Stone Street
Kent

Tunbridge Wells TN1 2QU
Great Britain
hss@brey.org.uk

Briggs, J. W.
38 Crofton Close
Purbrook
Waterlooville P07 5QA
Great Britain
Phone: 44.12.5239.4760
Fax: 44.12.5239.2622
jwbriggs@dera.gov.uk

Bright, Alison E.
Butler University
Noblesville, IN 46060-6827
abright@butler.edu
Master's Level, 1996
19c, Europe, Social Relations
of Science; 19c, Europe,
Biological Sciences;
19c, Europe; Philosophy and
Philosophy of Science

Brill, Robert H.
Corning Museum of Glass
Research Scientist
1 Museum Way
Corning, NY 14830-2253

Brinkman, Paul
University of Minnesota
Hist of Sci & Technology
Tate Laboratory of Physics
116 Church St SE
Minneapolis, MN 55455-0149
Phone: 1.612.626.8722
brin0142@tc.umn.edu
20c, North America, Natural and
Human History; Earth Sciences

Brinson Burke, Kimberlea
3100 Saint Paul Street
Apt. 614
Baltimore, MD 21218
Phone: 1.410.889.4574
kimberleab@yahoo.com

Brisigotti, Daniele
Via Marco Aurelio 31
00184 Roma
Italy

Broadley, J. A.
Alfaz del P1
Apartado 167

Alicante 03580
Spain
Fax: 34.96.588.1041

Broce, Gerald L.
University of Colorado
Department of Anthropology
PO Box 7150
Colorado Spgs, CO 80933-7150

Brock, David
Chemical Heritage Foundation
Associate Historian
316 Chestnut St
Philadelphia, PA 19106-2708
Phone: 1.215.925.2222
Fax: 1.215.925.2178
davidb@chemheritage.org

Brock, Emily K.
Princeton University
Dept of History
Princeton, NJ 08544
ebrock@princeton.edu

Brock, William H.
56 Fitzgerald Avenue
Seaford BN25 1AZ
Great Britain
Phone: 1323.891602
william.brock@btinternet.com
Doctorate
19c, Europe, Physical Sciences;
20c, Europe, Education;
Europe, Institutions;
Social Relations of Science

Brockley, Janice A.
Jackson State University
Room 340 School of Liberal Art
History Dept.
PO Box 17700
Jackson, MS 39217
Phone: 1.601.979.2504
janice.brockley@jsums.edu

Broecke, Steven V.
Johns Hopkins University
Dept of History of Science,
Medicine & Technology
3400 N Charles St
Baltimore, MD 21218-2680

Broemer, Rainer
Cultural History Group
Old Brewry
Aberdeen

Scotland AB243UB
Great Britain
Phone: 44.1224.272561
Fax: 44.1224.273262
rainer.broemer@gmx.de

Brokaw, James J.
Indiana University Sch of Med
8600 University Blvd
Evansville, IN 47712-3534
Phone: 1.812.465.1290
Fax: 1.812.465.1184
jbrokaw@iupui.edu
Doctorate, 1983
Medical Sciences

Broman, Thomas H.
University of Wisconsin
Dept of History of Science
7143 Social Science Bldg
1180 Observatory Dr
Madison, WI 53706-1393
Phone: 1.608.263.1562
Fax: 1.608.262.3984
thbroman@facstaff.wisc.edu
Doctorate
18c, Europe, Medical Sciences;
18c, Europe, Social Relations
of Science; 18c, Europe,
Philosophy and Philosophy of
Science

Bromberg, Joan Lisa
1933 Greenberry Rd
Baltimore, MD 21209-4555
Phone: 1.916.739.0544
Fax: 1.916.739.0544
joanlisa@jhunix.hcf.jhu.edu
Doctorate, 1966
19c, North America; 20c;
19c, Physical Sciences;
20c, Technology

Broncano, Fernando
Universidad Carlos III Madrid
Departamento de Humanidades
Madrid 129
Getafe
Madrid 28903
Spain
Fax: 34923294725
ibroncan@hum.uc3m.es
Philosophy and Philosophy of
Science; Biological Sciences;
Cognitive Sciences;
Humanistic Relations of Science

Brooke, John H.
Harris Manchester College
Oxford OX1 3TD
Great Britain
Phone: 1865.271019
Fax: 1865.271012
john.brooke@theology.
  oxford.ac.uk
Doctorate
19c, Europe, Physical Sciences;
19c, Europe, Biological Sciences;
Social Sciences;
Humanistic Relations of Science

Brooks, John I., III
Fayetteville State University
Dept of Government History
1200 Murchison Rd.
Fayetteville, NC 28301-4298
Phone: 1.910.672.1945
Fax: 1.910.670.1090
jibrooks@uncfsu.edu
Doctorate, 1990
19c, Europe, Social Sciences

Brooks, Nathan M.
New Mexico State University
Dept of History
MSC 3H, PO Box 30001
Las Cruces, NM 88003-8001
Phone: 1.505.646.1824
Fax: 1.505.646.8148
nbrooks@nmsu.edu
Doctorate, 1989
19c, Europe, Physical Sciences;
20c, Europe, Physical Sciences;
19c, Europe, Institutions;
20c, Europe, Social Relations of
Science

Brotons, Victor N.
University of Valencia
Av Vicente Blasco Ibanez 17
Valencia 46010
Spain
Phone: 34.96.3864164
Fax: 34.96.3613975
victor.navarro@uv.es
Doctorate
14c-16c, Europe, Astronomical
Sciences; 17c, Europe,
Physical Sciences; 17c, Europe,
Instruments and Techniques

Brown, Alexander
Massachusetts Institute of
Technology

E51-070
77 Massachusetts Ave
Cambridge, MA 02139-4307
afbrown@mit.edu

Brown, C.
873 Bernard Way
San Bernardino, CA 92404-2413

Brown, Gregory
University of Houston
Dept. of Philosophy
Houston, TX 77204-3004
gbrown@uh.edu

Brown, Harold I.
Northern Illinois University
Philosophy Department
Dekalb, IL 60115
Phone: 1.815.753.6406
Fax: 1.815.753.6302
hibrown@niu.edu
Doctorate, 1970
Philosophy and Philosophy
of Science; 17c, Philosophy
and Philosophy of Science;
18c, Philosophy and Philosophy
of Science

Brown, Laurie M.
Northwestern University
Dept of Physics & Astronomy
Evanston, IL 60208
Phone: 1.847.491.3236
Fax: 1.847.491.9982
lbrown@nwu.edu
Doctorate, 1951
20c, Europe, Physical Sciences;
20c, North America, Physical
Sciences; 20c, Asia, Physical
Sciences; 20c, Physical Sciences

Brown, Patricia S.
Siena College
Dept of Biology
515 Loudonville Rd
Loudonville, NY 12211-1462
Phone: 1.518.783.2458
Fax: 1.518.783.2986
brown@siena.edu
Doctorate, 1968
Biological Sciences;
Gender and Science

Browne, Janet
Wellcome Trust
Centre for History of Medicine

Euston House
24 Eversholt St
London NW1 1AD
Great Britain
Phone: 2.07.679.8144
ucgabro@ucl.ac.uk
Doctorate, 1978
Natural and Human History;
19c, Europe, Biological Sciences,
Biography; Medical Sciences;
Exploration, Navigation and
Expeditions

Brownstein, Dan
Humanities Consortium
UCLA #146102
310 Royce Hall
Los Angeles, CA 90095-1461
Phone: 1.323.653.5414
brownst@humnet.ucla.edu

Brozek, Josef M.
2353 Youngman Ave
Saint Paul, MN 55116-3063
Phone: 1.612.698.5325
Doctorate, 1937
19c, South America,
Cognitive Sciences;
20c, Europe, Social Sciences;
14c-16c, Europe, Education;
North America, Natural and
Human History

Brugger, Robert J.
2715 N Charles St
Baltimore, MD 21218-4319

Brush, Stephen G.
University of Maryland
History Department
College Park, MD 20742-7315
Phone: 1.301.405.4846
Fax: 1.301.314.9363
brush@ipst.umd.edu
Doctorate, 1958
20c, Physical Sciences;
20c, Biological Sciences;
20c, Social Sciences;
20c, Cognitive Sciences

Buchanan, Rex C.
Kansas Geological Survey
1930 Constant Ave.
Lawrence, KS 66047-9454
Phone: 1.785.864.2106
Fax: 1.785.864.5317
rex@kgs.ukans.edu

Master's Level, 1982
19c, North America,
Earth Sciences;
20c, Biological Sciences

Buck, A. T.
122 Skircoat Rd
West Yorkshire
Halifax HX1 2RE
Great Britain
Phone: 1422.25588
Fax: 1422.255881
andrew.buck1@tinyworld.co.uk

Bud, R. F.
The Science Museum
Exhibition Road
London SW7 2D
Great Britain
Phone: 44.207.942.4200
Fax: 44.207.942.4202
r.bud@nmsi.ac.uk

Buerki, Robert A.
Ohio State University
500 West 12th Ave
College of Pharmacy
Columbus, OH 43210-1291
Phone: 1.614.292.4722
Fax: 1.614.292.1335
buerki.1@osu.edu
Doctorate, 1972
19c, North America, Medical
Sciences; 20c, North America,
Philosophy and Philosophy of
Science; 19c, North America,
Education

Bugos, Glenn E.
188 King St
Redwood City, CA 94062-1940
Phone: 1.650.599.5033
isis@prologuegroup.com

Buhs, Joshua Blu
567 Countryside Dr
Vacaville, CA 95687-7307
Phone: 1.707.451.1102
Fax: 1.707.451.1102
jbbuhs@gtcinternet.com
Doctorate, 2001
20c, North America, Biological
Sciences; 19c, North America,
Natural and Human History

Buickerood, James G.
160 Hastings Way
Saint Charles, MO 63301-5506
Phone: 1.314.516.6791
JAMEINCL18@yahoo.com

Bullough, Vern L.
University of Southern Calif
Center for Health Professions
1540 Aalcazar St
Los Angeles, CA 90033
Fax: 1.805.449.8776
vbullough@csun.edu
Doctorate, 1954
Gender and Science

Bunner, Patricia A.
15 Devine Rd
Fairview, WV 26570
Phone: 1.304.798.3542
pbunner@wvu.edu
Doctorate, 2003
Pre-400 B.C.E., Astronomical
Sciences; 17c, Europe,
Astronomical Sciences;
North America, Biological
Sciences; Biological Sciences

Buonora, Paul
California State University
Department of Chemistry and
Biochemistry
Long Beach, CA 90840
Phone: 1.562.985.4946
pbuonora@csulb.edu

Burba, Juliet M.
University of Minnesota
Tate Laboratory of Physics
116 Church St SE
Minneapolis, MN 55455-0149
burb0006@umn.edu
Natural and Human History;
North America

Burchfield, Joe D.
157 Terrace Dr
Dekalb, IL 60115
Phone: 1.815.758.4970
Fax: 1.815.753.6302
burchfield@niu.edu
Doctorate, 1969
19c, Europe, Earth Sciences;
19c, Europe, Physical Sciences

Burgess, Helen J.
306-811 Helmcken St
Apt #1904
Vancouver, BC V6Z1B1
Canada
Phone: 1.604.899.0206
helen@burgess.net

Burian, Richard M.
Virginia Tech
Department of Philosophy
229 Major Williams
Blacksburg, VA 24061
Phone: 1.540.231.6760
Fax: 1.540.231.6367
rmburian@vt.edu
Doctorate, 1971
*Biological Sciences; Philosophy
and Philosophy of Science*

Burkhardt, Frederick
PO Box 1067
Bennington, VT 05201-1067
Phone: 1.802.442.9573
fhb@sover.net

Burkhardt, Richard W., Jr.
University of Illinois
309 Gregory Hall
810 S Wright Street
Urbana, IL 61801
Phone: 1.217.333.2450
burkhard@uiuc.edu
Doctorate
*Biological Sciences;
19c, Europe, Natural and
Human History; Institutions*

Burlingame, Leslie J.
187 S President Ave
Lancaster, PA 17603-4849
Phone: 1.717.299.1415
Fax: 1.717.291.4186
l_burlingame@acad.fandm.edu

Burmeister Blank, Rene
769 Greenwich Street
New York, NY 10014
Phone: 1.917.661.0240
mrb@walrus.com
Doctorate, 1999
*19c, Europe, Medical Sciences;
19c, Europe, Gender and
Science*

Burnett, D. Graham
Princeton University
G-27 Dickson Hall
Department of History
Princeton, NJ 08544
Phone: 1.609.258.7309
dburnett@princeton.edu
Doctorate, 1997
*Earth Sciences; Exploration,
Navigation and Expeditions;
19c, Europe, Biological Sciences,
Biography; Physical Sciences*

Burnett, John Nicholas
727 Virginia Rd
PO Box 238
Davidson, NC 28036-0238
Phone: 1.704.892.8998
Doctorate, 1965
19c, Europe, Physical Sciences

Burnham, John C.
Ohio State University
Department of History-106
230 W 17th Ave
Columbus, OH 43210-1367
Phone: 1.614.292.5465
Fax: 1.614.292.2282
burnham.2@osu.edu
Doctorate
*20c, Medical Sciences,
Instruments and Techniques;
20c, North America, Social
Sciences; 20c, Medical Sciences*

Burns, Chester R.
Inst For Med Humanities
UTMB at Galveston
Galveston, TX 77555
Phone: 1.409.772.9389
Fax: 1.409.772.5640
cburns@utmb.edu
Doctorate, 1969
*19c, North America, Medical
Sciences*

Burns, J. Conor
University of Toronto-Victoria
Inst for History & Philosophy
of Science & Technology Rm 316
91 Charles St W
Toronto, ON M5S1K7
Canada
Phone: 1.416.978.5047
Fax: 1.416.978.3003
jburns@chass.utoronto.ca

Burnside, Phillips B.
Ohio Wesleyan University
61 South Sandusky St
Delaware, OH 43015-1627
Phone: 1.740.368.3771
Fax: 1.740.368.3999
pbburnsi@cc.owu.edu
Doctorate, 1958
*Philosophy and Philosophy of
Science; Physical Sciences;
Social Relations of Science*

Burrington, Carson
102 E Gorham St
Madison, WI 53703-2130
csburrington@students.wisc.edu

Burstyn, Harold
216 Bradford Pkwy
Syracuse, NY 13224-1767
Phone: 1.315.445.0620
burstynh@iname.com

Burton, Dan
Auburn University
Department of History
Thach Hall Rm 309
Auburn, AL 36849-5207
Phone: 1.256.765.4539
Fax: 1.256.765.4536
deburton@una.edu
Doctorate, 1998
*5c-13c, Physical Sciences;
5c-13c, Astronomical Sciences*

Busard, H. L. L.
Herungerstraat 123
Venlo 5911 AK
The Netherlands
Phone: 77.35163

Bustamante, Juan A.
Facultad de Ciencas
Apartado Correos, 40
11510 Puerto Real
Cadiz
Spain

Butler, Loren J.
163 Lloyd Rd
Matawan, NJ 07747-1823
Phone: 1.732.566.66.4234
Fax: 1.732.566.4997
lbfeffer@alumni.princeton.edu

Butrica, Andrew J.
NASA
Code ZH, History Office
Washington, DC 20546
Phone: 1.301.486.1563
Fax: 1.301.486.4590
a.butricia@ieee.org
Doctorate, 1986
Technology, Instruments and
Techniques; Technology;
20c, North America,
Astronomical Sciences;
Physical Sciences

Bylebyl, Jerome
Johns Hopkins Univ
1900 E Monument St
Baltimore, MD 21218
Phone: 1.410.955.3037
jjbyleb@jhmi.edu

Bynum, William F.
Wellcome Trust Centre for the
History of Medicine at UCL
24 Eversloft St
London NW1 1AD
Great Britain
Phone: 0207.611.8550
Fax: 0207.611.08277
w.bynum@ucl.ac.uk
Doctorate
19c, Europe, Medical Sciences;
19c, Europe, Biological Sciences;
20c, Europe, Medical Sciences

Cadden, Joan
University of California
History Department
Davis, CA 95616
Phone: 1.530.752.2224
Fax: 1.530.752.5301
jcadden@ucdavis.edu
Doctorate
5c-13c, Europe;
Medical Sciences;
Social Relations of Science;
Humanistic Relations of Science

Cahan, David
University of Nebraska-Lincoln
Dept of History
610 Oldfather Hall
Lincoln, NE 68588-0327
Phone: 1.402.472.3238
Fax: 1.402.472.8839
dcahan@unlserve.unl.edu
Doctorate, 1980

Cahn, Robert W.
6 Storeys Way
Cambridge CB3 0DT
Great Britain
Phone: 44.1223.360143
Fax: 44.1223.334567
rwc12@cam.ac.uk

Cain, Joe
University College London
Science and Technology Studies
Gower Street
London WC1E 6BT
Great Britain
Phone: 44.207.679.3041
Fax: 44.207.916.2425
j.cain@ucl.ac.uk
Doctorate, 1995
20c, North America, Biological
Sciences

Calascibetta, Franco
Universita La Sapienza
Dipartimento di Chimica
Piazzale Aldo Moro 5
Roma 00185
Italy
Phone: 39.0649913339
Fax: 39.06490324
franco.calascibetta@uniromel.it
19c, Physical Sciences;
20c, Physical Sciences

Calinger, Ronald S.
Catholic University of America
620 Michigan Avenue
Department of History
Washington, DC 20064
Phone: 1.202.319.5484
Fax: 1.202.319.5569
calinger@cua.edu
Doctorate, 1971
18c, Europe, Biological Sciences;
17c, Europe, Biological Sciences;
17c, Europe, Social Sciences

Calkins, Laura M.
11420 US Highway 1 # 10
N. Palm Beach, FL 33408-3226
lmcalkins@aol.com
Doctorate
Science Policy; Technology;
Gender and Science

Cambrosio, Alberto
McGill University
Dept of Social Studies of Med

3647 Peel
Montreal, PQ H3A1X1
Canada
Phone: 1.514.398.4981
Fax: 1.514.398.1498
alberto.cambrosio@mcgill.ca
Doctorate
20c, Medical Sciences;
20c, Instruments and
Techniques; Social Sciences;
Science Policy

Camerini, Jane R.
University of Wisconsin
7143 Social Science Building
1180 Observatory Dr
Madison, WI 53706
jrcameri@facstaff.wisc.edu
Doctorate, 1987
20c, Exploration, Navigation
and Expeditions; 19c, Natural
and Human History;
20c, Biological Sciences;
Medical Sciences, Instruments
and Techniques

Cameron, Gary L.
1008 W Church St
Marshalltown, IA 50158-2519
Phone: 1.641.753.6192
marshallhistory@adiis.net

Camerota, Michele
Dipartimento Di Filosofia
Via Is Mirrionis 1
Loc. Sa Duchessa
Cagliari 09123
Italy
Phone: 39706757299
Fax: 396757302
camerota@unica.it
Doctorate, 1988
17c, Europe, Physical Sciences;
17c, Europe, Philosophy
and Philosophy of Science;
17c, Europe, Humanistic
Relations of Science;
20c, Europe, Social Sciences

Campbell, John T.
6942 28th St N
Arlington, VA 22213-1706

Campbell, Rod
PO Box 1349
Alief, TX 77411-1349

Campos, Luis
Harvard University
Dept of the History of Science
Science Center, 2nd Floor
1 Oxford St
Cambridge, MA 02138-2901
Fax: 1.617.495.8645
lcampos@fas.harvard.edu

Canales, Jimena
Harvard University
Science Center 235
Cambridge, MA 02138
jcanales@fas.harvard.edu
Master's Level, 2002
19c, Astronomical Sciences;
19c, Physical Sciences;
19c, Philosophy and Philosophy
of Science

Canaparo, Claudio
University of Exeter
Spanish Department
Queens Building
Exeter EX44 QH
Great Britain
Phone: 44.1392264245
Fax: 44.139226439
c.canaparo@exeter.ac.uk

Caneva, Kenneth L.
University of North Carolina
Department of History
219 McIver Bldg
PO Box 26170
Greensboro, NC 27402-6170
Phone: 1.336.334.5203
Fax: 1.336.334.5910
klcaneva@euler.uncg.edu
Doctorate
19c, Europe, Physical Sciences;
19c, Physical Sciences;
Social Sciences

Cannariato, Christy
1021 Scandia Ave Apt 56
Ventura, CA 93004-2484
Phone: 1.805.647.7502

Canose, Jeffrey
138 Berry Mountain Rd
Cramerton, NC 28032-1637
Phone: 1.704.824.3726
Fax: 1.704.834.2500
jcanose@carolina.rr.com

Canseco-Gomez, Juan I. C.
Consulate General of Mexico
in Milan
Via Cappuccini, 04
Milano 20122
Italy
Phone: 39.0276020541
juancanseco@mac.com

Canters, G. W.
Leiden Institute of Chemistry
Gorlaeus Laboratories
Leiden Univ Einsteinweg 55
PO Box 9502
Leiden 2300 RA
The Netherlands

Cantor, David
National Cancer Institute
Division of Cancer Prevention
Executive Plaza N, Suite 2025
6130 Executive Boulevard
Bethesda, MD 20892-7309
Phone: 1.301.594.1012
Fax: 1.301.480.4109
cantord@mail.nih.gov

Cantor, Geoffrey N.
University of Leeds
School of Philosophy
Leeds LS2 9JT
Great Britain
Phone: 44.1132.333.269
Fax: 44.1132.333.265
g.n.cantor@leeds.ac.uk
Doctorate, 1968
Europe; 19c, Europe,
Physical Sciences;
18c, Europe, Physical Sciences

Capshew, James H.
Indiana University
Dept of History & Phil Sci
Goodbody Hall 130
Bloomington, IN 47405
Phone: 1.812.855.3655
Fax: 1.812.855.3631
jcapshew@indiana.edu
Doctorate, 1986
Cognitive Sciences;
20c, North America; Social
Sciences; Biological Sciences

Cardoso, Joso Luos
ISEG
Department of Economics
Rua do Quelhas, 6

Lisbon 1200
Portugal
Phone: 351.21.3958356
Fax: 351.21.3967309
jcardoso@iseg.utl.pt

Carlsson, Carl L.
218 Cornwall Dr
Crete, IL 60417-1008
Phone: 1.708.672.5225

Carolino, Luis M.
Apartado Correios, 2030
Colares 2706-909
Portugal
Phone: 351.266706581
Fax: 351.266744677
carolino@uevora.pt

Carozzi, Albert V.
University of Illinois
Department of Geology NHB
245
1301 W Green St
Urbana, IL 61801-2999
Phone: 1.217.333.3008
Fax: 1.217.244.4996
acarozzi@uiuc.edu
Doctorate, 1948
18c, Europe, Earth Sciences

Carpenter, Kenneth
Univ. of California - Berkeley
Dept. of Nutritional Science
Berkeley, CA 94720-3104
Phone: 1.510.642.1038
Fax: 1.510.642.0535
kearp@uclink.berkeley
Doctorate
18c, Medical Sciences;
19c, Medical Sciences;
20c, Medical Sciences

Carrier, Richard
507 W 113th St Apt 41
New York, NY 10025-8047
Phone: 1.212.316.2105
rcc20@columbia.edu

Carroll, Patrick E.
2527 G St
Sacramento, CA 95816-3609
Phone: 1.530.752.5388
pcarroll@ucdavis.edu
Doctorate, 1999
17c, Europe, Social Relations
of Science; 18c, Europe,

Technology, Instruments and Techniques; 19c, Europe, Instruments and Techniques; 20c, North America, Medical Sciences

Carroll, William E.
Cornell College
Dept of History
600 1st St W
Mount Vernon, IA 52314-1098

Carson, Cathryn L.
University of California
Dept of History
3229 Dwinelle Hall
Berkeley, CA 94720-2550
Phone: 1.510.642.2118
Fax: 1.510.643.5323
clcarson@socrates.berkeley.edu
Doctorate, 1995
20c, Physical Sciences; Physical Sciences; Europe; North America

Carson, John
University of Michigan
Department of History
1029 Tisch Hall
Ann Arbor, MI 48109-1003
Phone: 1.734.647.7378
Fax: 1.734.647.4881
jscarson@umich.edu
Doctorate, 1994
North America; Social Sciences; North America, Cognitive Sciences; Social Relations of Science

Carter, Benjamin P.
1612 3rd St
Manhattan Beach, CA
90266-6304
Phone: 1.310.536.9190
Fax: 1.310.536.9067
bpc@gte.net

Casado de Otaola, Santos
Residencia de Estudiantes
Pinar 23
Madrid 28006
Spain
Phone: 34.91.396.7676
Fax: 34.1.5643890
santos.casado@uam.es
Doctorate, 1994
Biological Sciences;

20c, Biological Sciences; 18c, Europe, Natural and Human History; 20c, Europe, Natural and Human History

Casalini, Antonella
Casella Postale 12
Fiesole F 50014
Italy

Casalini, Venere
Via Benedetto da Maiano 3
Fiesole F 50014
Italy

Casalini, Vinicio
Casella Postale 12
Fiesole F 50014
Italy

Casanovas, Juan
Specola Vaticana
Vatican City 00120
Italy
Phone: 39.0669888242
Fax: 39.0669884671
jcc@specola.va
Doctorate
Astronomical Sciences; Philosophy and Philosophy of Science; Earth Sciences

Casas, Martha
University of Texas
Permian Basin
4901 E University Blvd
Odessa, TX 79762-8122
Phone: 1.915.552.2138
casas_m@utpb.edu

Cassedy, James H.
National Library of Medicine
8600 Rockville Pike
Bethesda, MD 20894
Phone: 1.301.594.0992
Fax: 1.301.402.0872
james_cassedy@nlm.nih.gov
Doctorate, 1959
North America, Physical Sciences; 20c, North America, Instruments and Techniques; North America, Social Sciences

Cassidy, David C.
Hofstra University
Nat Science Program

NSB106
Hempstead, NY 11549
Phone: 1.516.463.5537
chmdcc@hofstra.edu
Doctorate, 1976
20c, Europe, Physical Sciences; 18c, Europe, Physical Sciences

Cassidy, James G.
St Anslem College
100 Saint Anselms Dr
Manchester, NH 03102-1310
Phone: 1.603.641.7047
Fax: 1.603.641.7116
jcassidy@anselm.edu
Doctorate, 1991
19c, North America, Exploration, Navigation and Expeditions; North America, Medical Sciences; Biological Sciences

Casteel, Eric
11130 Rose Ave Apt 408
Los Angeles, CA 90034-6062
Phone: 1.310.737.0147
ecasteel@ucla.edu

Castelao-Lawless, Teresa
Grand Valley State University
Philosophy Dept LSH 215
1 Campus Dr
Allendale, MI 49401-9403
Phone: 1.616.895.3419
castelat@gvsu.edu
Doctorate
20c, Europe, Philosophy and Philosophy of Science; 20c, Europe, Physical Sciences; 17c, Europe, Social Sciences; Europe, Philosophy and Philosophy of Science

Castle, Clifton W.
Jefferson College
1000 Viking Drive
Hillsboro, MO 63050
Phone: 1.636.797.3000
Fax: 1.636.586.3942
ccastle@jeffco.ed
Master's Level, 1976
Physical Sciences

Castonguay, Stephane
Univ de Quebec a Trois-Riviere
Centre d'etudes quebecoises
C.P. 500
Trois-Rivieres, PQ G9A5H7

Canada
Phone: 1.819.376.5011
Fax: 1.819.376.5179
stephane_castonguay@
uqtr.uquebec.ca
Doctorate, 1999
20c, North America, Biological
Sciences; 20c, North America,
Science Policy; 19c, Europe,
Biological Sciences

Caswell, Lyman R.
6535 37th Ave NE
Seattle, WA 98115-7431
Phone: 1.206.527.2205
Fax: 1.206.527.0449
72712.757@compuserve.com

Catala-Gorgues, Jesus
Pinzon 11, 10
Valencia 46003
Spain
Phone: 654.884414
Fax: 963.3864091
jesus.i.catala@uv.es

Catania, Basilio
Via Torino 86
Fiano TO 10070
Italy
Phone: 39.011.925.4920
Fax: 39.011.925.4920
mark@esanet.it

Catt, Patrick A.
University of Chicago
Department of Mathematics
5734 University Ave
Chicago, IL 60637-1546
pacatt@midway.uchicago.edu
Doctorate, 1999
20c, North America; 20c, North
America, Social Relations of
Science; 20c, North America,
Institutions; 20c, North America,
Social Sciences

Catton, Philip E.
University of Canterbury
Department of Philosophy
Private Bag 4800
Christchurch 1
New Zealand
Phone: 64.3.364.2077
Fax: 64.3.364.2889
p.catton@phil.canterbury.ac.nz
Doctorate, 1991

Physical Sciences; Philosophy
and Philosophy of Science;
Astronomical Sciences,
Philosophy and Philosophy of
Science

Cavicchi, Elizabeth
Dibner Institute
MIT E56-100
38 Mem Dr
Cambridge, MA 02139
cavicchi@eecs.tufts.edu

Ceccatti, John S.
222 Church St Apt 5A
Philadelphia, PA 19106-4523
Phone: 1.215.629.8342
j-ceccatti@uchicago.edu
Master's Level, 2000
19c, Europe, Biological Sciences;
19c, Philosophy and Philosophy
of Science

Cech, Paul J.
Bethlehem - Center High Sch.
179 Crawford Road
Fredricktown, PA 15333
Phone: 1.724.267.4944
cechpj@helicon.net
Master's Level, 1987
19c, Europe, Biological Sciences,
Biography; Biological Sciences;
20c, North America, Cognitive
Sciences; Earth Sciences

Chabas Bergon, Jose
Universitat Pompeu Fabra
Rambla S. Monica
Barcelona
Spain
jose.chabas@trad.upf.es
Doctorate
5c-13c, Astronomical Sciences;
14c-16c, Astronomical Sciences;
Astronomical Sciences

Chakravorty, Ranes C.
Veterans Affairs Medical Center
Salem, VA 24153
Phone: 1.540.380.2362
Fax: 1.540.224.1961
rchakrav@vt.edu
Doctorate, 1949
5c-13c, Asia, Medical Sciences;
18c, Europe, Medical Sciences;

19c, Europe, Medical Sciences;
19c, North America, Medical
Sciences

Chalbaud Cardona, Pedro R.
Universidad de Los Andes
Facultad de Humanidades y
Educ. Escuela de Historia
Merida 5101A
Venezuela
Phone: 074.401732
Fax: 074.40851
chalbaud@ciens.ula.ve
Bachelor's Level, 1990
20c, South America,
Astronomical Sciences;
19c, Earth Sciences;
20c, Earth Sciences; 20c, Social
Relations of Science

Challey, James F.
Vassar College
Dept of Physics
124 Raymond Ave
Box 309
Poughkeepsie, NY 12604-0309
Phone: 1.845.437.7352
Fax: 1.845.437.5995
challey@vassar.edu
Master's Level, 1969
19c, Europe, Physical Sciences;
20c, North America, Technology;
20c, North America, Social
Relations of Science

Chalmers, Alan F.
Flinders University
Philosophy Dept
GPO Box 2100
Adelaide SA 5001
Australia
Phone: 8.8201.5092
Fax: 8.8359.2520
chalmers_alan@hotmail.com
Doctorate, 1971
19c, Europe, Physical Sciences;
17c, Europe, Physical Sciences;
20c, Europe, Philosophy
and Philosophy of Science;
18c, Europe, Physical Sciences

Champlin, Peggy
2169 Linda Flora Dr
Los Angeles, CA 90077-1408
Phone: 1.310.476.9795
champc@aol.com

Doctorate
19c, North America, Earth Sciences; 19c, North America

Chan, K. O.
39 Ma Tau Wai Rd
Rm 1015, Tower A
Hunghom Kowloon
Hong Kong

Chang, Hasok
University College London
Dept of Sci & Tech Studies
Gower Street
London WCIE 6BT
Great Britain
Phone: 44.20.76791324
Fax: 44.20.79162425
h.chang@ucl.ac.uk
Doctorate, 1993
Philosophy and Philosophy of Science; Physical Sciences; 20c, Physical Sciences

Chang, HsiaoNing
277 W 150th St Apt 10
New York, NY 10039-2344
Phone: 1.917.513.0610
hc682@columbia.edu

Chang, Ku-Ming
5639 S Maryland Ave BF
Chicago, IL 60637-1452
Phone: 1.773.643.7696
kchang@midway.uchicago.edu

Channell, David F.
University of Texas at Dallas
School of Arts & Humanities
Ms Jo 3.1
PO Box 830688
Richardson, TX 75083-0688
Phone: 1.972.883.2007
Fax: 1.972.883.2989
channell@utdallas.edu
Doctorate, 1975
19c, Europe, Technology; 19c, North America, Technology; 20c, Europe, Humanistic Relations of Science; 19c, Europe, Physical Sciences

Chaplin, Joyce E.
Harvard University
Dept of History
Robinson Hall
Cambridge, MA 02138

Phone: 1.617.495.2556
Fax: 1.617.496.3425
chaplin@fas.harvard.edu
Doctorate, 1986
17c, Exploration, Navigation and Expeditions; 18c; 14c-16c

Chapple, Ian L.
Austral. Soc. of Hist. of Med.
PO Box 8034
Mt Pleasant
Mackay QLD 4740
Australia
Fax: 7.49576382
lloydmc@ml30.aone.net.au
Bachelor's Level, 1950
20c, Australia and Oceania, Medical Sciences

Charette, Francois
Inst for History of Science
IW Goethe University FB-13
Robert-Mayer-Str 1
Frankfurt D-60054
Germany
charette@em.uni-frankfurt.de

Chastain, Ben B.
538 Hampton Dr
Birmingham, AL 35209-4340
Phone: 1.205.871.3859
bbchasta@samford.edu
Doctorate, 1967
18c, Europe, Physical Sciences; 19c, Europe, Physical Sciences; 19c, North America, Physical Sciences

Chazaro, Laura
Odontolgia 57-401
Colonia Copilco-Universidad
Mexico DF 04360
Mexico
Phone: 525.6598935
chazaro@colmich.edu.mx
Master's Level
19c, North America, Medical Sciences; 20c, North America, Science Policy

Chen, Cheng-Yih
Univ of California - San Diego
Dept of Physics 0319
La Jolla, CA 92093-0319
Phone: 1.858.534.2893
Fax: 1.858.534.0173
jchen@physics.ucsd.edu

Doctorate
Astronomical Sciences; Physical Sciences; Mathematics; Philosophy and Philosophy of Science

Chen, Xiang
California Lutheran University
Dept of Philosophy
60 W Olsen Rd
Thousand Oaks, CA 91360-2787
Phone: 1.805.371.9231
Fax: 1.805.493.3013
chenxi@clunet.edu

Chen-Morris, Raz
PO Box 31
Batzir St
Lion 99835
Israel
Fax: 972.2.9997432
razdov@post.tau.ac.il

Cheng, Sandra
143 N 7th St Apt 3L
Brooklyn, NY 11211-2965
Phone: 1.718.782.3194
scheng@udel.edu

Cheung, Tobias
Halker Zeile 29
Berlin 12305
Germany
Phone: 49307427311
tobias@paris7.jussieu.fr

Child, Mary
Cambridge University Press
History of Science, Editor
40 W 20th St
New York, NY 10011-4227
Phone: 1.212.924.3900
Fax: 1.212.691.3239
lbateman@cup.org

Chimisso, Cristina
Open University
Dept of Philosophy
Walton Hall
Milton Keynes MK7 6AA
Great Britain
Phone: 44.1908.659137
c.chimisso@open.ac.uk

Chinard, Francis P.
40 Warren Place
Montclair, NJ 07042-2534

Phone: 1.973.746.7847
chinard@umdnj.edu
Doctorate, 1941
17c, Europe, Medical Sciences;
18c, Europe, Physical Sciences;
19c, North America, Medical
Sciences; 20c, Medical Sciences

Chinenova, Vera
Moscow State University
Mechanis-Mathematic
Vorob'evy Gory, MGU
Moscow 119899
Russia
Phone: 959393860
Fax: 959392090
Doctorate
18c, Europe, Mathematics;
18c, Europe, Physical Sciences;
19c, Europe, Physical Sciences;
17c, Europe, Education

Chiu, Hsien-Po
131 Gien-Hsing St
Fu-Hsing Village
Bade City
Taoyuan County 334
Taiwan
Phone: 886.3.3682095
chiuhp@ms29.hinet.net

Christian, Kris B.
28 Fourth Ave
Pelham, NY 10803-1408
Phone: 1.914.738.8413
krisngnat@aol.com
Master's Level, 2001
5c-13c, Europe, Astronomical
Sciences

Christianson, Gale E.
Indiana State University
Stalker Hall
Department of History
Terre Haute, IN 47809
Phone: 1.812.237.2721
Fax: 1.812.237.7713
higalee@ruby.indstate.edu
Doctorate
Astronomical Sciences;
Earth Sciences

Christie, Bianca
1718 Welch St # E
Houston, TX 77006-1733
Phone: 1.713.807.1922
bianca@rice.edu

Chrostowski, Paul
7708 Takoma Avenue
Takoma Park, MD 20912
Phone: 1.301.585.8062
Fax: 1.301.585.2117
cpfassoc@aol.com

Chung, Yuehtsen J.
11800 Berans Rd
Lutherville Timonium, MD
21093-1501
Phone: 1.773.955.8191
ychung@fas.harvard.edu
Doctorate, 1999
17c, Asia, Biological Sciences;
18c, Biological Sciences;
19c, Medical Sciences;
20c, Social Sciences

Churchill, Frederick B.
Indiana University
Hist. & Philosophy of Science
Goodbody Hall 130
Bloomington, IN 47408
Phone: 1.812.855.3622
Fax: 1.812.855.3631
churchil@indiana.edu
Doctorate, 1967
19c, Europe, Biological Sciences;
20c, Europe, Medical Sciences;
North America, Natural and
Human History; 19c, Europe,
Biological Sciences, Biography

Ciancio, Luca
Via Coslop, 7
Rovereto 38068
Italy
Phone: 464432725
lciancio@seldati.it
Doctorate, 1993
18c, Europe, Earth Sciences;
18c, Europe, Natural and
Human History;
19c, Europe, Earth Sciences

Cirillo, Vincent J.
Rutgers University
New Brunswick, NJ 08903
vjeirillo@worldnet.att.net
Doctorate, 1999
19c, North America, Medical
Sciences

Cittadino, Eugene
31-28 36th St
Astoria, NY 11106-1002

Phone: 1.718.274.5488
ec15@nyu.edu
Doctorate, 1981
Biological Sciences;
Social Relations of Science

Clagett, Marshall
Institute for Advanced Study
School of Historical Studies
Princeton, NJ 08540
Phone: 1.609.734.8311

Clark, Charles W.
St Andrews Presbyterian Coll
History Program
1700 Dogwood Mile St
Laurinburg, NC 28352-5521
Phone: 1.910.277.5334
Fax: 1.910.277.5020
sclark@hotmail.com
Doctorate, 1979
5c-13c, Europe, Medical
Sciences; 5c-13c, Europe;
5c-13c, Europe, Astronomical
Sciences

Clark, Constance
1447 Sumac Ave
Boulder, CO 80304-0807
Fax: 1.303.444.7476
constance.clark@colorado.edu

Clark, Mark H.
Oregon Institute of Technology
Humanities & Social Sciences
3201 Campus Dr
Klamath Falls, OR 97601-8801
Phone: 1.541.885.1880
Fax: 1.541.885.1520
clarkm@oit.edu
Doctorate, 1992
20c, North America, Technology;
20c, Europe, Technology

Clarke, Adele E.
University of California
Box 0612, Soc. & Beh. Sci.
3333 California St., Ste. 455
San Francisco, CA 94143-0612
Phone: 1.415.476.0694
Fax: 1.415.476.6552
aclarke@itsa.ucsf.edu
Doctorate, 1985
20c, North America, Medical
Sciences; 20c, North America,
Gender and Science; 20c, North

America, Social Sciences;
20c, North America, Social
Relations of Science

Clarke, Desmond M.
University College
Philosophy Dept
Cork
Ireland
Phone: 353.21.902568
Fax: 353.21.276079
dclarke@ucc.ie
Doctorate, 1974
17c, Europe, Philosophy and
Philosophy of Science

Claro-Gomes, Jose M.
29 Rue Leopold Bellan
Paris 75002
France
Phone: 33.619931440
zeclaro22@hotmail.com

Clay, Landon T.
188 Old Street Rd
Peterborough, NH 03458-1644

Cliborn, James
21401 Lighthill Dr
Topanga, CA 90290-4444
Phone: 1.818.340.3688
jhcliborn@aol.com

Clulee, Nicholas H.
Frostburg State University
History Department
101 Braddock Road
Frostburg, MD 21532
Phone: 1.301.687.4428
Fax: 1.301.687.3099
nclulee@frostburg.edu
Doctorate, 1973
14c-16c, Europe, Philosophy
and Philosophy of Science;
14c-16c, Europe, Physical
Sciences

Coen, Deborah
Harvard University
Department Of History Of Sci
Science Center 235
Cambridge, MA 02138
coen@fas.harvard.edu

Coggon, Jennifer D.
620 Jarvis St #2408
Toronto, ON M4Y2R8

Canada
Phone: 1.416.323.0134
jennifer.coggon@utoronto.ca
19c, Europe, Biological Sciences;
19c, Europe, Medical Sciences,
Instruments and Techniques

Cohen, Benjamin
163 Lebanon Rd
Pembroke, VA 24136-3021
Phone: 1.540.626.5820
bcohen@vt.edu

Cohen, H. Floris
University of Twente
WMW Department of History
PO Box 217
Enschede 7500AE
The Netherlands
Phone: 31.53.4893300
Fax: 31.53.4892979
h.f.cohen@wmw.utwente.nl
Doctorate, 1974
17c; 17c, Social Sciences;
Europe; Humanistic Relations
of Science

Cohen, I. Bernard
Harvard University
Widener Library
Cambridge, MA 02138
Phone: 1.617.484.1221
Fax: 1.617.484.1211
ibcohen@fas.harvard.edu
Doctorate
18c, Physical Sciences;
17c, Mathematics;
Social Sciences

Cohen, Robert S.
Boston University
Philosophy Department
745 Commonwealth Ave
Boston, MA 02215-1401
Phone: 1.617.353.6395
Fax: 1.617.353.6805
atauber@acs.bu.edu
Doctorate, 1948
20c, North America, Philosophy
and Philosophy of Science;
19c, Europe, Physical Sciences;
19c, Europe, Humanistic
Relations of Science

Cohen, Seymour S.
10 Carrot Hill Rd
Woods Hole, MA 02543-1206

Phone: 1.508.548.7435
polyamin@cape.com
Doctorate, 1941
18c, Europe, Physical Sciences;
19c, North America, Physical
Sciences; 19c, North America,
Education

Cohen-Cole, Jamie
Princeton University
Program in History of Science
129 Dickinson Hall
Princeton, NJ 08544-0001
Phone: 1.617.258.5529
Fax: 1.617.495.3344
jamiecc@princeton.edu
Master's Level
20c, North America, Cognitive
Sciences; 20c, Social Sciences;
Mathematics

Cole, William A.
4251 Providence Point Dr SE
Issaquah, WA 98029-7217
Phone: 1.425.391.2668
Master's Level, 1950
18c, Physical Sciences;
Education

Coley, Noel G.
24 Kayemoor Road Sutton
Surrey SM2 5HT
Great Britain
Phone: 44.2.8642.7437
n.g.coley@surrey28.
  freeserve.co.uk
Doctorate
18c, Europe, Physical Sciences;
19c, Europe, Physical Sciences;
18c, Europe, Medical Sciences;
19c, Europe, Medical Sciences

Collins, Richard L.
52 Oak Cir
Princeton, MA 01541-1515

Collyns, C. N.
PO Box 8395
Emeryville, CA 94662-0395
collyns@gbn.com

Colnot, Thomas
Erbacher Strasse 99E
Brensbach
Stierbach D-64395

Germany
Phone: 49.9314.4326
thomas.colnot@merck.de

Colp, Ralph, Jr.
301 E 79th St Apt 12A
New York, NY 10021-0938
Phone: 1.212.737.1554
19c, Europe, Biological Sciences;
19c, Europe, Medical Sciences,
Instruments and Techniques

Coltham, Deborah
Pickering & Chatto
36 St George Street
London W 1R 9FA
Great Britain
Phone: 44.2.7491.2656
Fax: 44.2.7491.9161
d.colthan@pickering-chatto.com

Comastri, Alberto
V Salvaro 18
Pioppe Di Salva 40040
Italy
comastri@sga-storiageo.it

Comfort, Nathaniel
George Washington University
Dept of History
Center for Hist Recent Science
801 22nd St NW
Washington, DC 20037-2515
Phone: 1.202.994.3957
Fax: 1.202.994.6231
comfort@gwu.edu
Doctorate, 1997
20c, North America, Biological
Sciences

Conlin, Michael F.
Eastern Washington University
Dept of History
204 L Pattersoon Hall
526 5th St
Cheney, WA 99004-2431
Phone: 1.509.359.7851
mconlin@ewu.edu
Doctorate, 1999
19c, Europe, Physical Sciences;
18c, Europe, Physical Sciences;
19c, North America;
19c, North America, Exploration,
Navigation and Expeditions

Connaroe, Joel
Guggenheim Memorial
Foundation
President
90 Park Ave
New York, NY 10016-1478

Conner, Clifford D.
101 W 85th St Apt 5-3
New York, NY 10024-4458
Phone: 1.212.877.0853
drcdconner@aol.com

Connolly, David E.
Ohio State University
314 Cunz Hall
Dept. of Germanic Lang & Lit
Columbus, OH 43210
Phone: 1.614.447.3600
dconnolly@cas.org
Master's Level, 1987
5c-13c, Europe, Physical
Sciences; 14c-16c, Europe,
Physical Sciences;
17c, Humanistic Relations of
Science

Constant, Edward W.
Carnegie Mellon University
Dept of History
5000 Forbes Ave
Pittsburgh, PA 15213-3890
Phone: 1.412.268.8852
Fax: 1.412.268.1019
ec0a@andrew.cmw.edu
Doctorate, 1977

Conway, Erik M.
National Air & Space Museum
600 Independence Ave
Rm 3520D MRC 311
Washington, DC 20560
Phone: 1.757.593.3573
e.m.conway@larc.nasa.gov
Doctorate, 1998
20c, North America, Technology;
Technology, Instruments and
Techniques; North America

Conway, Gordon
Rockefeller Foundation
420 5th Ave
New York, NY 10018-2711

Cook, George M.
900 Yonge Street Apt. 601
Toronto, ON M4W3P5

Canada
Phone: 1.416.921.2342
george.cook@utoronto.ca
Master's Level, 1999
19c, Biological Sciences;
20c, North America, Earth
Sciences; Biological Sciences;
Social Sciences

Cook, Harold J.
University College London
Center for History of Medicine
Euston House
24 Eversholt St
London NW1 1AD
Great Britain
Fax: 44.2.0611.8277
h.cook@ucl.ac.uk
Doctorate, 1981
17c, Europe, Medical Sciences;
17c, Europe, Natural and
Human History; 17c, Europe,
Biological Sciences

Cook, Jason
14511 Markhurst Dr
Cypress, TX 77429-5364
Phone: 1.281.251.9875
threecooks@houston.rr.com

Cook, Margaret G.
983 Berkley Dr NW
Calgary, AB T3K1E3
Canada
Phone: 1.403.274.8716
cookmarg@shaw.ca

Cook, Ruth
6 Millbrook Road Figtree
Wollongong 2525
Australia

Cooke, Kathy J.
Quinnipiac College
P O - 077
275 Mt Carmel Ave
Hamden, CT 06518
Phone: 1.203.582.3475
Fax: 1.203.582.3471
kathy.cooke@quinnipiac.edu

Coon, Deborah J.
9684 Limar Way
San Diego, CA 92129
Phone: 1.858.538.2018
dcoon@post.harvard.edu
Doctorate, 1988

Cognitive Sciences;
North America, Social Relations
of Science; Instruments and
Techniques

Cooper, Alix
SUNY - Stony Brook
History Dept
S B S S-339
Stony Brook, NY 11794-4348
Phone: 1.631.632.7492
Fax: 1.631.632.7367
acooper@notes.cc.sunysb.edu
Doctorate, 1998
17c, Europe, Natural and
Human History; Earth Sciences;
Biological Sciences;
18c, Medical Sciences

Cooper, Brian
257 Concord Rd
Lincoln, MA 01773-5119
Phone: 1.781.259.0691
Fax: 1.781.259.0691
bcooper1@oswego.edu

Cooper, David Y.
Univ of Pennsylvania
Harrison Dpt Surgical Research
304 Medical Educ Bldg
Philadelphia, PA 19104
Phone: 1.215.662.2075
Fax: 1.215.614.1930

Cooper, Jill E.
Rutgers University
Institute for Health
30 College Avenue
New Brunswick, NJ 08901
Phone: 1.908.281.2132
jicooper@rci.rutgers.edu
Doctorate, 1998
20c, Biological Sciences

Cooper, Raymond D.
Key Books
PO Box 58097
St Petersburg, FL 33715
cooperrd@aol.com
Doctorate, 1967
20c, Physical Sciences;
20c, Astronomical Sciences;
20c, Biological Sciences;
20c, Technology

Coopersmith, Jonathan
Texas A & M University
Dept of History
College Station, TX 77843-0001
Phone: 1.979.845.7148
Fax: 1.979.862.4314
j-coopersmith@tamu.edu
Doctorate, 1989
North America, Technology;
North America, Earth Sciences;
Europe, Physical Sciences;
Earth Sciences

Copenhaver, Brian P.
UCLA
College of Letters & Science
1312 Murphy Hall, Box 951438
Los Angeles, CA 90095-9000
Phone: 1.310.825.4286
Fax: 1.310.825.9368
brianc@college.ucla.edu
Doctorate
14c-16c, Europe, Philosophy
and Philosophy of Science;
400 B.C.E.-400 C.E.;
Pre-400 B.C.E., Natural
and Human History;
17c, Medical Sciences

Corbett, Randall
146 Bridlewood Dr
Brandon, MS 39047-8480
Phone: 1.601.992.4569
Fax: 1.601.469.5119
rlcorbett@worldnet.att.net

Cormack, Lesley B.
University of Alberta
Dept of Hist & Classics
2-41 Tory Bldg
Edmonton, AB T6G2H4
Canada
Phone: 1.780.492.4269
Fax: 1.780.492.9125
lesley.cormack@ualberta.ca
Doctorate, 1988
14c-16c, Europe, Earth Sciences;
14c-16c, Europe, Education;
17c, Europe, Earth Sciences;
14c-16c, Exploration,
Navigation and Expeditions

Cornelius, Craig
13333 Landfair Rd
San Diego, CA 92130-1837

Phone: 1.858.259.1625
Fax: 1.858.259.1728
craigc@alumni.princeton.edu

Cornell, Thomas D.
107 Emerald Cir
Rochester, NY 14623-4425
Phone: 1.716.475.6029

Corson, David W.
Cornell University
213 Olin Library
Ithaca, NY 14853
Phone: 1.607.255.5068
Fax: 1.607.255.2493
dwc4@cornell.edu
Doctorate, 1974
Education; 18c, Europe,
Physical Sciences;
Natural and Human History

Cosandey, David A.
Credit Suisse Group
PO Box 100
Zurich 8070
Switzerland
Phone: 4113344981
Fax: 4113333616
dcosan@gmx.net
Doctorate
Asia; Europe

Costa, Albert B.
5 Bayard Rd Apt 402
Pittsburgh, PA 15213-1903
Doctorate, 1960
20c, North America, Physical
Sciences; 19c, Europe,
Physical Sciences

Costa, Shelley
106 Hosea Ave # 3
Cincinnati, OH 45220-1820
Phone: 1.513.281.4748
sac22@cornell.edu

Costanzo, Felice
Via Massimo D'azeglio 14A
Treviso 31100
Italy

Cottrell, Matthew J.
Applied Signal Technology, Inc
133 National Business Parkway
Ste 100
Columbia, MD 20701
Phone: 1.301.497.7120

Fax: 1.301.497.7135
matthew_cottrell@appsig.com
Doctorate, 1981
Astronomical Sciences;
Physical Sciences; Technology,
Instruments and Techniques

Coulston, Christopher
347 Beacon St
Somerville, MA 02143-3538
Phone: 1.617.497.3967
coulston@fas.harvard.edu

Counce, S. J.
3101 Camelot Ct
Durham, NC 27705-5405
Phone: 1.919.684.2018
s.counce@cellbio.duke.edu

Countway, Francis A.
Library of Medicine
10 Shattuck St
Boston, MA 02115-6011

Cowan, Ruth Schwartz
SUNY Stony Brook
History Dept
Stony Brook, NY 11542-3351
Phone: 1.631.632.4378
Fax: 1.631.632.4525
rcowan@notes.cc.sunysb.edu
Doctorate, 1969
19c, North America, Technology;
20c, Europe, Biological Sciences;
Biological Sciences; Gender and
Science

Cox, David J.
309 Crown Ln
Bellingham, WA 98226-5929
Phone: 1.360.756.7908
moody@gte.net

Crane, Gregory R.
Tufts University
Dept of Classics
Eaton Hall
Medford, MA 02155
Phone: 1.617.627.2435
Fax: 1.617.627.3032

Cravens, Hamilton
Iowa State University
Dept. of History
603 Ross Hall
Ames, IA 50011-1202
Phone: 1.515.294.1156

Fax: 1.515.294.6390
hcravens@iastate.edu
Doctorate, 1969
20c, North America; 20c, North
America, Social Sciences;
20c, North America, Medical
Sciences; Pre-400 B.C.E., North
America, Cognitive Sciences

Crawford, Cassandra
University of California
1443 Lyon St
San Francisco, CA 94115-2952
ccrawf1@itsa.ucsf.edu

Crawford, Matthew
University of California
Dept of History, 0104
9500 Gilman Dr
La Jolla, CA 92093-5004
Phone: 1.858.534.1996
mjcrawfo@ucsd.edu

Creager, Angela N. H.
Princeton University
Program in History of Science
129 Dickinson Hall
Princeton, NJ 08544-1017
Phone: 1.609.258.1680
Fax: 1.609.258.5326
creager@princeton.edu
Doctorate, 1991
20c, Biological Sciences;
Social Sciences;
Gender and Science

Creath, Richard J.
Arizona State University
Philosophy Department
Tempe, AZ 85287-2004
Phone: 1.480.965.6270
Fax: 1.480.965.0902
creath@asu.edu
Doctorate, 1975

Cremo, Michael A.
Bhaktivedanta Institute
9701 Venice Blvd Apt 5
Los Angeles, CA 90034-5170
Phone: 1.310.837.5283
Fax: 1.310.837.1056
mcremo@compuserve.com
Biological Sciences;
Natural and Human History;
Astronomical Sciences

Croce, Paul J.
Stetson University
Box 8274
421 N Woodland Blvd
Deland, FL 32720-3761
Phone: 1.904.822.7530
Fax: 1.904.822.7268
pcroce@stetson.edu
Doctorate
19c, North America; 19c, North
America, Biological Sciences;
19c, North America, Humanistic
Relations of Science; 19c, North
America, Cognitive Sciences

Crook, Jason
ACS-IIS
850 3rd Ave Fl 18
New York, NY 10022-6222
jcrook@acs.edu.lb

Crosby, Daniel
10020 Palms Blvd Apt 8
Los Angeles, CA 90034-3824
Phone: 1.310.709.9585
dcrosby@ucla.edu

Crossgrove, William
Brown University
PO Box 1857
Providence, RI 02912-1979
Phone: 1.401.863.7416
Fax: 1.401.863.1854
william_crossgrove@brown.edu
Doctorate
5c-13c, Europe, Medical
Sciences; Europe, Humanistic
Relations of Science; Technology

Crowe, Michael J.
University of Notre Dame
Program of Liberal Studies
Notre Dame, IN 46556
Phone: 1.574.631.6212
crowe.l@nd.edu
Doctorate, 1965
19c, Astronomical Sciences;
19c, Physical Sciences;
19c, Mathematics; Philosophy
and Philosophy of Science

Crowther-Heyck, Hunter A.
Johns Hopkins University
Dept Hist of Science, 216 Ames
3400 N Charles St
Baltimore, MD 21218
crowther-heyck@prodigy.net

20c, North America; 20c, North America, Cognitive Sciences; 20c, North America, Science Policy; 20c, North America, Social Sciences

Crumpton, Amy
American Association for the
Advancement of Science
Room 815
1200 New York Ave NW
Washington, DC 20005-3928
Phone: 1.202.326.6791
Fax: 1.202.289.4950
acrumpto@aaas.org
Doctorate, 1998
20c, North America, Social Relations of Science; 20c, North America, Biological Sciences; 20c, North America, Science Policy

Cunningham, A.
University of Cambridge
Dept of History & Philosophy
Free School Lane
Cambridge
Great Britain

Cunningham, Michael
82 White St
Manchester, CT 06040-3128
Phone: 1.860.643.4024
cunninghammichae@
  netscape.net

Cunsolo, Ronald C.
30 Bright St
Westbury, NY 11590-1004
Phone: 1.516.333.2356

Curtis, Anthony R.
Union Institute
440 E McMillan St
Cincinnati, OH 45206
acurtis@tui.edu

Cushing, James T.
University of Notre Dame
Department of Physics
225 Nieuwland Science Hall
Notre Dame, IN 46556
Phone: 1.219.631.6132
Fax: 1.219.631.5952
cushing.1@nd.edu
Doctorate, 1963
20c, Physical Sciences

Cushman, Gregory T.
3801 Manchaca Rd Apt 28
Austin, TX 78704-6750
Phone: 1.512.445.7313
Fax: 1.512.445.6475
gcushman@mail.utexas.edu

Cutcliffe, Stephen H.
Lehigh University
STS Program
9 W Packer Ave
Bethlehem, PA 18015-3082
Phone: 1.610.758.3350
Fax: 1.610.758.6554
shc0@lehigh.edu
Doctorate, 1976
North America, Technology; North America, Biological Sciences; North America, Social Relations of Science

D'Ambrosio, Ubiratan
Rua Peixoto Gomide, 1772
Apartamento 83
01409-002 Sao Paulo
Brazil
Phone: 55.11.3088.0266
Fax: 55.11.3082.5437
ubi@usp.br
Doctorate, 1963
*Social Sciences;
14c-16c, Humanistic Relations of Science; Mathematics; Philosophy and Philosophy of Science*

D'Antonio, Patricia O.
231 Avon Rd
Narberth, PA 19072-2307

Dadic, Zarko
Inst of Hist & Phil of Sci
Ante Kovacica 5
Zagreb 10000
Croatia
Phone: 385.1.4552619
Fax: 385.1.4856211
Doctorate
Europe, Astronomical Sciences; Europe, Mathematics; Europe, Physical Sciences

Dagenais, Fred
1401 Grizzly Peak Blvd
Berkeley, CA 94708-2201
freddage@uclink4.berkeley.edu

Dahl, Per F.
9 Commodore Dr # A211
Emeryville, CA 94608-1652
Phone: 1.510.601.7276
pfdahl@aol.com
Doctorate, 1960
*20c, Physical Sciences;
19c, Physical Sciences;
20c, Astronomical Sciences;
19c, Exploration, Navigation and Expeditions*

Daniell, Robert E., Jr.
Computational Physics Inc.
240 Bear Hill Rd Ste 202a
Waltham, MA 02451-1026
Phone: 1.781.762.2447
Fax: 1.781.762.2508
rob@fluffy.com
Doctorate, 1976
*19c, Europe, Physical Sciences;
20c, North America, Physical Sciences; 20c, North America; 20c, North America, Astronomical Sciences*

Darden, Lindley
University of Maryland
Department of Philosophy
1125A Skinner Building
College Park, MD 20742
Phone: 1.301.405.5699
Fax: 1.301.405.5690
darden@wam.umd.edu
Doctorate, 1974
*20c, Biological Sciences;
19c, Biological Sciences*

Dasgupta, Subrata
114 Claremont Cir
Lafayette, LA 70508-7300
Phone: 1.337.482.6607
Fax: 1.337.482.5991
dasgupta@cacs.louisiana.edu

Daston, Lorraine J.
MPI for History of Science
Wilhelmstrasse 44
Berlin 10117
Germany
Phone: 49.30.22667131
Fax: 49.30.22667293
ldaston@mpiwg-berlin.mpg.de
Doctorate, 1979
*18c, Europe, Social Sciences;
14c-16c, Europe, Natural and*

Human History; 17c, Europe,
Mathematics; Philosophy and
Philosophy of Science

Daub, Edward E.
4258 Manitou Way
Madison, WI 53711-3746
edaub@facstaff.wisc.edc

Dauben, Joseph W.
The Graduate Center, CUNY
Ph.D. Program in History
365 Fifth Avenue at 34th St
New York, NY 10016-4309
Phone: 1.212.817.8430
Fax: 1.212.817.1523
jdauben@worldnet.att.net
Doctorate, 1972

Daum, Andreas W.
Harvard University
Center for European Studies
27 Kirkland Street
Cambridge, MA 02138
Phone: 1.617.495.4303
Fax: 1.617.495.8509
adaum@fas.harvard.edu
Doctorate, 1996
19c, Natural and Human
History; 19c, Education;
Social Relations of Science;
North America

Davenport, Anne A.
96 Bedford St
Concord, MA 01742-1817
adavenport@cfa.harvard.edu
Doctorate, 1998
14c-16c, Europe, Philosophy
and Philosophy of Science;
14c-16c, Europe, Humanistic
Relations of Science;
5c-13c, Europe, Social
Relations of Science;
5c-13c, Europe, Mathematics

Davenport, Derek A.
Purdue University
Dept of Chemistry
1393 Brown Building
West Lafayette, IN 47907
Phone: 1.765.494.5465
Fax: 1.765.494.0239

Davidson, J. Keay
San Francisco Chronicle
Science Writer

901 Mission St
San Francisco, CA 94103-2972
Phone: 1.415.777.7793
jdavid2355@aol.com

Davidson, Jane
2990 Cahal Ct
Reno, NV 89523-2267
Phone: 1.775.747.2252
jdhexen@aol.com

Davis, Audrey B.
1214 Bolton St
Baltimore, MD 21217-4111
Phone: 1.410.728.4810
mabdavis@earthlink.net
Doctorate, 1969
20c, North America, Humanistic
Relations of Science; 19c, North
America, Medical Sciences;
19c, North America, Instruments
and Techniques; 20c, North
America, Gender and Science

Davis, Edward B.
Messiah College
One College Ave
Box 3030
Grantham, PA 17027-9799
Phone: 1.717.766.2511
Fax: 1.717.691.6002
tdavis@messiah.edu
Doctorate, 1984
17c, Europe; 20c, North America

Davis, Frederick
Rochester Inst of Technology
Science, Technology & Society
College of Liberal Arts
92 Lomb Memorial Dr
Rochester, NY 14623-5604
Phone: 1.585.475.4615
frdgsh@rit.edu

Davis, Henry F., Jr.
Duquesne University
Pittsburgh, PA 15282
hdavis@hypernet.com
Doctorate, 1998
Philosophy and Philosophy
of Science; Technology;
Social Relations of Science

Davis, Matthew
2427 Sandhurst Ave SW
Calgary, AB T3C2M5

Canada
Phone: 1.403.242.7187
matt0san@yahoo.com

Davis, Todd
321 W Yanonali St
Santa Barbara, CA 93101-3831
Phone: 1.805.884.1826
todd.davis@home.com

Dawson, John W., Jr.
Penn State University - York
1031 Edgewater Ave
York, PA 17403-3398
Phone: 1.717.771.4131
Fax: 1.717.771.8404
jwd7@psu.edu
Doctorate, 1972
20c, Philosophy and Philosophy
of Science; 19c, Mathematics;
20c, Mathematics

Dawson, Virginia P.
History Enterprises
11000 Cedar Ave
Cleveland, OH 44106-3052
Phone: 1.216.421.9622
vpd@historyenterprises.com
Doctorate, 1983
20c, North America, Technology;
18c, Europe, Biological Sciences

Day, Deborah A.
421 Santa Helena
Solana Beach, CA 92075
Phone: 1.858.793.1290
Fax: 1.858.534.5269
dday@ucsd.edu
Master's Level, 1977
20c, North America, Physical
Sciences; 19c, North America,
Biological Sciences;
20c, North America, Exploration,
Navigation and Expeditions;
19c, North America, Natural and
Human History

Day, Tammy
2806 E Knights Griffin Rd
Plant City, FL 33565-2346
Phone: 1.813.982.1181
kaywyne@tampabay.rr.com

De Andrade, Antonio A.
Univ Aveiro-Geociencias
Campus Univ Santiago
Aveiro 3810-193

Portugal
Phone: 351.234.370747
Fax: 51.234.370605
asandrade@geo.ua.pt

De Asua, Miguel
Casilla de Correo 5051
Correo Central
Buenos Aires 1000
Argentina
mdeasua@mail.retina.ar

De Boer, Jan
Parkboswyk 318
Doorn 3941AC
The Netherlands
Phone: 31.343.519318
106354.2756@compuserve.com
Doctorate
19c, Physical Sciences;
20c, Philosophy and Philosophy
of Science; 19c, Mathematics

de Ceglia, Francesco P.
viale Gramsci, 84
Molfetta (BA)
Italy
Phone: 39803977939
f.deceglia@filosofia.uniba.it

De Chadarevian, Soraya
University of Cambridge
Department of History &
Philosophy of Science
Free School Lane
Cambridge CB2 3RH
Great Britain
Phone: 44.1223.331105
Fax: 44.1223.334554
sd10016@hermes.cam.ac.uk
Doctorate, 1988
20c, Europe, Biological Sciences;
19c, Europe, Biological Sciences;
Instruments and Techniques;
Social Sciences

De Felice, Louis J.
Vanderbilt Univ Sch of Med
Department of Pharmacology
Medical Center
1161 21st Ave S # SO410
Nashville, TN 37212-2708
Phone: 1.615.343.6278
Fax: 1.615.343.1679
lou.defelice@mcmail.
  vanderbilt.edu
Doctorate

20c, Europe, Physical Sciences;
20c, Physical Sciences;
Medical Sciences; Philosophy
and Philosophy of Science

De Greiff, Alexis
Observatorio Astronomico Nac
Apartado Aero 2584
Santafe de Bogo DC
Colombia
Phone: 57.1.316.5222
Fax: 57.1.316.5383
alegreif@ciencias.unal.edu.co

De Groot, Jean C.
Catholic University of America
School of Philosophy
Washington, DC 20064-0001
Phone: 1.202.319.5636
degroot.cua.edu

De Knecht Vaneklen,
Annemarie
Rijksweg 55
Malden 6581EE
The Netherlands
Phone: 31243880436
Fax: 31243880437
deknecht@mailbox.kun.ne
Doctorate
19c, Europe, Medical Sciences;
20c, Europe, Medical Sciences;
19c, Asia, Earth Sciences;
20c, Europe, Gender and
Science

de Luca, Emilie P.
Peace College
15 E. Peace Street
Raleigh, NC 27604
Doctorate, 1974
5c-13c, Europe, Medical Sciences

de Mora, Maria Sol
Universidad Pais Vasco
Filosofia, Avda. Tolosa 70
20009
San Sebastian
Spain
Phone: 34.943.018262
Fax: 34.943.311056
marcharles@inicia.es
Doctorate, 1978
17c, Europe, Mathematics;
14c-16c, Europe, Astronomical
Sciences; 18c, Europe,
Philosophy and Philosophy

of Science; 17c, Europe,
Astronomical Sciences,
Philosophy and Philosophy of
Science

de Pater, Cornelis
Instituut Voor Geschiedenis
En Grondslagen Van De
Wiskunde En Natuurweten
Princetonplein 3584CC
The Netherlands
Phone: 31.30.253828
Fax: 31.30.2536313
c.depater@phys.uu.nl
Doctorate, 1979
18c, Europe, Physical Sciences

De Young, Gregg R.
American University in Cairo
Science Department
PO Box 2511
Cairo
Egypt
Fax: 1.011.202.3227565
gdeyoung@aucegypt.edu
Doctorate
Pre-400 B.C.E., Mathematics;
Pre-400 B.C.E., Africa,
Technology

Dean, Dennis R.
834 Washington St Apt 3W
Evanston, IL 60202-2254
Phone: 1.847.869.2232
Doctorate, 1968
19c, Europe, Earth Sciences;
19c, Europe, Humanistic
Relations of Science

Dean, Sheila A.
Cambridge University Library
Manuscripts Room
Darwin Correspondence Project
Cambridge CB3 9DR
Great Britain
Phone: 1223.333008
Fax: 1223.333008
sad21@cam.ac.uk
Doctorate
19c, Biological Sciences;
20c, Biological Sciences;
Natural and Human History

Dear, Peter R.
Cornell University
Dept. of Science & Tech Studies
632 Clark Hall

Ithaca, NY 14853-2501
Phone: 1.607.255.6049
Fax: 1.607.255.6044
prd3@cornell.edu
Doctorate, 1984
17c, Europe, Physical Sciences;
17c, Europe, Philosophy and
Philosophy of Science

Dearolph, Edward A.
Woodward Academy
1662 Rugby Ave.
College Park, GA 30337-1010
dearolph@mindspring.com
Master's Level, 1968
Philosophy and Philosophy of
Science; Physical Sciences;
Education

Deary, William P.
4205 Vicki Ct
Alexandria, VA 22312-1209
Phone: 1.703.354.8593
dearybill@aol.com

Debus, Allen G.
University of Chicago
1126 E 59th St
Chicago, IL 60637
Phone: 1.773.702.8391
Fax: 1.773.834.1299
adebus@midway.uchicago.edu
Doctorate, 1961
17c, Europe, Physical Sciences;
14c-16c, Europe,
Physical Sciences;
18c, Europe, Social Sciences;
Europe, Medical Sciences

Decarvalho, Roy J.
University of North Texas
Department of History
PO Box 310650
Denton, TX 76203-0650
Phone: 1.940.565.4209
Fax: 1.940.369.8838
roy@unt.edu
Doctorate
20c, Europe, Cognitive Sciences;
19c, Europe, Social Sciences;
18c, North America, Medical
Sciences, Instruments and
Techniques; 17c, Exploration,
Navigation and Expeditions

Dehue, Trudy C.
University of Groningen
Theory & History of Psychology
Grote Kruisstraat 2/1
Groningen 9712 TS
The Netherlands
Phone: 31.50.3636354
Fax: 31.50.3636304
g.c.g.dehue@ppsw.rug.nl
Doctorate
Social Sciences;
20c, Cognitive Sciences;
20c, Social Sciences

Deichmann, Ute
University of Köln
Institut für Genetik
Weyertal 121
Köln D 50931
Germany
Phone: 49.221.4702388
Fax: 49.221.4705170
ute.deichmann@uni-koeln.de
Doctorate
20c, Europe, Biological Sciences;
20c, Europe, Physical Sciences;
20c, Institutions

Deken, Jean M.
854 Jordan Ave Apt M
Los Altos, CA 94022-1422
Phone: 1.650.926.3091
Fax: 1.650.926.5371
jmdeken@slac.stanford.edu

Dekker, Elly
Utrecht University Frw
Heidelberglaan 2
Postbus 00115
Utrecht 3508 TC
The Netherlands
Phone: 31302532086
Fax: 313.481.5406
Doctorate, 1976
14c-16c, Europe, Astronomical
Sciences; 14c-16c, Europe,
Instruments and Techniques;
14c-16c, Europe, Education;
14c-16c, Europe, Mathematics

DeKosky, Robert K.
University of Kansas
History Department
Wescoe Hall
Lawrence, KS 66045-0001
Phone: 1.785.864.9462
dekosky@falcon.cc.ukans.edu

Doctorate, 1972
19c, Europe, Physical Sciences;
19c, Europe, Social Sciences;
19c, Europe, Institutions

Delisle, Richard G.
University of Witwatersrand
Dept of Anatomy
7 York Rd Parktown 2193
Johannesburg
South Africa
Fax: 1.514.381.0092
c.grondine@sympatico.ca
Doctorate, 1998
20c, Natural and Human
History; 19c, Earth Sciences;
20c, Biological Sciences

Deltete, Robert J.
Seattle University
Philosophy Department
900 Broadway
Seattle, WA 98122-4340
Phone: 1.206.296.5462
Fax: 1.206.296.5997
rdeltete@seattleu.edu
Doctorate
19c, Europe, Philosophy
and Philosophy of Science;
19c, Europe, Physical Sciences;
20c, Europe, Philosophy
and Philosophy of Science;
20c, Astronomical Sciences

Demaitre, Luke
University of Virginia
Humanities in Medicine
Charlottesville, VA 22908
ldemaitre@summit.net
Doctorate, 1973
5c-13c, Europe, Medical
Sciences; Medical Sciences;
14c-16c, Europe,
Medical Sciences;
5c-13c, Europe, Education

Demidov, Sergei S.
Inst. For History of Science
Staropanskii Per 1/5
Moscow 103012
Russia
Phone: 7.095.2983907
Fax: 7.095.9259911
postmaster@ihst.ru
Doctorate, 1991
18c, Europe, Mathematics;

19c, Europe, Mathematics;
20c, Europe, Mathematics;
20c, North America, Mathematics

Dennis, Michael A.
Cornell University
Dept of Science & Tech Studies
632 Clark Hall
Ithaca, NY 14853-2501
Phone: 1.607.255.6045
Fax: 1.601.255.6044
md11@cornell.edu

Des Chene, Dennis
Emory University
Philosophy, Bowden Hall
S Kilgo Circle
Atlanta, GA 30322
ddesche@emory.edu

Desmond, Adrian J.
Brackenber House
The Ridge Cold Ash
Berkshire RG18 9HY
Great Britain
Phone: 1635.87453
Fax: 1635.87453
ajdesmond@post.harvard.edu
Doctorate
19c, Europe, Biological Sciences;
19c, Europe, Earth Sciences

Dettelbach, Michael S.
Smith College
Department of History
Northampton, MA 01063
Phone: 1.413.585.3723
Fax: 1.413.585.3393
mdettelb@smith.edu
Doctorate, 1993
18c, Europe, Exploration,
Navigation and Expeditions;
17c, North America, Natural
and Human History;
19c, North America, Exploration,
Navigation and Expeditions;
South America, Earth Sciences

Devons, Samuel
Columbia University
Nevis Laboratory
PO Box 137
Irvington, NY 10533-0137
Phone: 1.914.591.28.60
Fax: 1.914.591.7080

devons@nevis1.columbia.edu
Doctorate, 1939
Physical Sciences; Education

DeVorkin, David H.
Smithsonian Institution
MRC 311 NASM 3530
Washington, DC 20560
Phone: 1.202.357.2784
Fax: 1.202.786.2947
david.devorkin@nasm.si.edu
Doctorate, 1978
20c, North America,
Astronomical Sciences;
20c, North America, Instruments
and Techniques

Dew, Douglas
7023 N Oceanshore Blvd
Palm Coast, FL 32137-2312
Phone: 1.3386.328.9686
Fax: 1.386.328.0600
beachferrari@yahoo.com

Dew, Nicholas
11, rue Sauffroy
Paris 75017
France
Phone: 33.1.4228.5425
nd230@cam.ac.uk

Dewitt, Stephen
16428 426th Way SE
North Bend, WA 98045-9046
Phone: 1.425.831.0831
dewitts@us.ibm.com

Dewsbury, Donald A.
University of Florida
Dept of Psychology
Gainesville, FL 32611-2250
Phone: 1.352.392.0601
Fax: 1.352.392.7985
dewsbury@ufl.edu
Doctorate, 1965
20c, Cognitive Sciences;
20c, Biological Sciences;
19c, Cognitive Sciences;
19c, Biological Sciences

Di Bono, Mario
Universita de Milano
Biblioteca
Via Conservatorio 7
Milano 20122
Italy
Phone: 39.02.5835.2977

Fax: 39.02.5835.2602
maria.dibono@unimi.it
Doctorate
14c-16c, Astronomical Sciences;
18c, Europe, Astronomical
Sciences; 19c, Europe,
Astronomical Sciences;
Education

Di Poppa, Francesca
344 N Craig St # 1
Pittsburgh, PA 15213-1245
Phone: 1.412.681.9438
frdst5@pitt.edu

Dias, Penha C.
Instituto de Fisica
UFRJ Cidade Universitaria
ILHA Do Fundao
Rio de Janeiro
Brazil
Phone: 021.562.7934
penha@if.ufrj.br
Doctorate, 1980
Physical Sciences

Dibattista, Liborio
Via Cassese 20
Gravina BA 70024
Italy
Phone: 390803261135
labldiba@tin.it

Dibner, David
The Dibner Fund
44 Old Ridgefield Road
PO Box 7575
Wilton, CT 06897-7575
Phone: 1.203.761.9904
Fax: 1.203.761.9989
dibnerfund@worldnet.att.net

Dick, Steven J.
US Naval Observatory
34th & Massachusetts Ave NW
Washington, DC 20392-5420
Phone: 1.202.762.0379
Fax: 1.202.762.1489
dick.steve@usno.navy.mil

Dietrich, Michael R.
Dartmouth College
Dept of Biological Sciences
Hanover, NH 03755
Phone: 1.603.646.1389
michael.dietrich@dartmouth.edu
Doctorate

*20c, North America, Biological
Sciences; 20c, Europe,
Biological Sciences*

Dijksterhuis, Fokko J.
University of Twente
WMW-6ES
PO Box 217
Enschede 7500 AE
The Netherlands
Phone: 31534893318
f.j.dijksterhuis@wmw.utwente.nl
Master's Level
*17c, Europe, Physical Sciences;
17c, Europe, Technology;
5c-13c, Europe, Social Relations
of Science*

DiLaura, David
1568 Bradley Dr
Boulder, CO 80305-7375
Phone: 1.303.492.4798
Fax: 1.303.492.7317
david.dilaura@colorado.edu

Dillon, John
Randall Museum
199 Museum Way
San Francisco, CA 94114
Phone: 1.415.554.9602
jdillon@randall.mus.ca.us
Bachelor's Level, 1967
*19c, Biological Sciences;
Pre-400 B.C.E., Instruments
and Techniques; 19c, Europe,
Biological Sciences, Biography*

Dirkse, Erika J.
Univ Of Minnesota Technology
Program In History of Science
Minneapolis, MN 55455
Phone: 1.612.626.8722
Fax: 1.612.626.8722
dirk0025@tc.umn.edu
Bachelor's Level
*19c, Europe, Earth Sciences;
19c, Europe; Pre-400 B.C.E.,
Europe, Earth Sciences*

Dmitriev, Igor
St Petersburg University
Mendeleev' Museum
Universiterskaia nab., 5
St Petersburg 199034
Russia
Phone: 812.3289737
igor@mdim.lgu.spb.su

Doctorate
*17c, Europe, Physical Sciences;
18c, Europe, Physical Sciences;
19c, Europe, Physical Sciences;
14c-16c, Europe, Social
Relations of Science*

Doel, Ronald E.
Oregon State University
Department of History
306 Milam Hall
Corvallis, OR 97331-5104
Phone: 1.541.737.3469
Fax: 1.541.737.1257
doelr@ucs.orst.edu
Doctorate, 1990
*20c, Earth Sciences;
Astronomical Sciences;
Biological Sciences;
20c, Science Policy*

Dolan, Brian
4 The Fairway
Bar Hill
Cambridge
Great Britain

Dold-Samplonius, Yvonne
Turkeniousisweg 14
Neckargemund 69151
Germany
Fax: 49.6221548312
dold@math.uni-heidelberg.de
Doctorate, 1977
*5c-13c, Mathematics;
17c, Europe, Mathematics;
20c, Europe, Mathematics*

Donahoe, Brian
27485 N Chevy Chase Rd
Mundelein, IL 60060-9573
Phone: 1.847.566.4265

Donahue, William H.
Green Lion Press
1611 Camino Cruz Blanca
Santa Fe, NM 87501-4553
Phone: 1.505.983.3675
Fax: 1.505.989.9314
donahue@greenlion.com
Doctorate, 1973
*17c, Europe, Astronomical
Sciences; 400 B.C.E.-400 C.E.,
Astronomical Sciences;
17c, Astronomical Sciences;
14c-16c, Astronomical Sciences*

Doncel, Manuel G.
Univ. Autonoma de Barcelona
Cehic, Edif. CC
Bellaterra 08193
Spain
Phone: 34.93674115
Fax: 34.936752461
manuel.g.doncel@uaab.es
Doctorate, 1967
*19c, Physical Sciences;
20c, Physical Sciences;
Europe, Physical Sciences*

Donovan, Arthur
US Merchant Marine Academy
Department of Humanities
Kings Point, NY 11024

Dorries, Matthias
IRIST-Gersulp
Université Louis Pasteur
7, rue de l'Universite
Strasbourg 67000
France
Phone: 33.3.8835.3021
doerries@mpiwg-berlin.mpg.de
Doctorate, 1989
*19c, Europe, Physical Sciences;
19c, Europe, Social Relations
of Science; 19c, Europe,
Humanistic Relations of Science;
19c, Europe, Instruments and
Techniques*

Douglas, Deborah
MIT Museum
Bldg N51 - 209
265 Massachusetts Ave
Cambridge, MA 02139-4307
Phone: 1.617.253.1766
Fax: 1.617.253.8994
ddouglas@mit.edu

Dowd, Matthew F.
University of Notre Dame
Dept of Hist & Phil of Science
346 O'Shaughnessey Hall
Notre Dame, IN 46556
Fax: 1.219.631.3985
matthew.f.dowd.11@nd.edu
Master's Level, 1998
*Astronomical Sciences;
5c-13c, Europe,
Astronomical Sciences;
5c-13c, Europe, Education*

Dowling, Thomas W., III
208 Royal Pines Dr NW Apt B
Huntsville, AL 35806-2323
Phone: 1.256.864.0265
t_w_dowling_3@hotmail.com

Downey, Greg
4122 Mineral Point Rd
Madison, WI 53705-5129
gjdowney@earthlink.net

Downey, William
346 Foxtail Drive
Eugene, OR 97405
Phone: 1.541.465.8300
billd@probel.com
Master's Level
14c-16c, Physical Sciences;
5c-13c, Physical Sciences;
17c, Biological Sciences; 18c

Downie, Scott R.
2136 Ohio St
Lawrence, KS 66046-3050
Master's Level
20c, North America, Technology,
Instruments and Techniques;
20c, North America, Social
Sciences; 20c, North America,
Philosophy and Philosophy of
Science; 20c, North America,
Technology

Downing, Kevin
Greenwood Publishing Group
532 Northridge Rd
Highlands Ranch, CO
80126-2045
Phone: 1.303.791.8243
kdowning@greenwood.com

Downing, Lisa
University of Illinois Chicago
Dept of Philosophy (M/C 267)
1421 University Hall
601 South Morgan St.
Chicago, IL 60607-7114
Phone: 1.312.413.3016
Doctorate
17c, Europe, Philosophy
and Philosophy of Science;
18c, Europe, Philosophy
and Philosophy of Science;
17c, Europe, Physical Sciences

Dreger, Alice D.
Michigan State Univ
Lyman Briggs School
Holmes Hall E-30
East Lansing, MI 48825-1107
Phone: 1.517.353.4628
Fax: 1.517.432.2758
dreger@msu.edu
Doctorate, 1995
Medical Sciences; Gender and
Science; 20c, North America,
Philosophy and Philosophy of
Science

Dritsas, Lawrence
GF2
1 Sciennes Hill Place
Edinburgh EH9 1NP
Great Britain
Phone: 131.667.3026
l.s.dritsas@sma.ed.ac.uk

Dronamraju, Krishna R.
Fdn for Genetics Research
PO Box 27701
Houston, TX 77227-7701

Dror, Otniel E.
Hebrew University of Jerusalem
History of Medicine
The Hebrew U-Hadassah School
of Medicine
PO Box 12272
Jerusalem 91120
Israel
Phone: 972.2.6757162
Fax: 972.2.6780278
otniel@md.huji.ac.il
Doctorate, 1998
Medical Sciences; 20c, North
America, Medical Sciences;
20c, North America, Medical
Sciences, Instruments and
Techniques; 19c, Europe,
Cognitive Sciences

Drucker, Thomas L.
University of Wisconsin
Dept of Math and CS
Whitewater, WI 53190
Phone: 1.262.472.5173
Doctorate, 1998
Philosophy and Philosophy of
Science; Mathematics

Duarte, Antonio G.
Faculdade de Farmacia
Av. Prof Gama Pinto
Lisboa
Portugal
Phone: 351.934.456.899
Fax: 351.217.946.470
agduarte@ff.ul.pt

Duarte, Antonio L.
Universidade de Coimbra
Departamento de Mathematica
Apartado Correos 3008
Coimbra 3000
Portugal
Phone: 351.239.791.150
Fax: 351.239.832.568
leal@mat.ue.pt

Duchesneau, Francois
Université de Montreal
Dept of Philosophie
CP 6128 SUCC Centre-ville
Montreal, PQ H3C3J7
Canada
Phone: 1.514.343.7373
Fax: 1.514.343.2098
francois.duchesneau@
umontreal.ca
Doctorate, 1971
18c, Europe, Biological Sciences;
19c, Europe, Biological Sciences;
18c, Europe, Philosophy
and Philosophy of Science;
19c, Europe, Philosophy and
Philosophy of Science

Duffin, J. M.
Queen's University
Hannah Chair
78 Barrie St
Kingston, ON K7L3N6
Canada
Phone: 1.613.533.6580
Fax: 1.613.533.6330
duffinj@post.queensu.ca
Doctorate, 1985
19c, Medical Sciences;
20c, Medical Sciences; North
America, Medical Sciences

Dukes, Thomas W.
PO Box 9
Kemptville, ON K0G1J0
Canada
Phone: 1.613.545.4080
twdukes@kos.net

Doctorate, 1965
19c, North America, Medical Sciences; 19c, North America, Biological Sciences; 19c, Africa

Dunbar, Gary S.
13 Church St
Cooperstown, NY 13326-1320
Phone: 1.607.547.2290
dunbargary@aol.com
Doctorate, 1956
Earth Sciences; Exploration, Navigation and Expeditions

Duncan, Kristen M.
University of Akron
ASEC 486
Dept of Biology
Akron, OH 44325
Fax: 1.3309235418
mdmd100@aol.com
Master's Level, 1999
Biological Sciences; Medical Sciences

Dunphy, Lynne M.
1288 SW 7th St
Boca Raton, FL 33486-8463
Phone: 1.521.297.3384

Dupre, Sven
University of Gent
Department of Philosophy
Blandijnberg 2
Gent 9000
Belgium
Phone: 32.9.264.39.41
Fax: 32.9.264.41.87
sven.dupre@rug.ac.be

Dupree, Hunter
975 Memorial Dr Apt 201
Cambridge, MA 02138-5755
Phone: 1.617.576.1522

Duque Naranjo, Camilo
1250S Miami Rd Apt 6
Fort Lauderdale, FL 33316-2066
camilo@utopia.com

Durham, Frank E.
836 Henry Clay Ave
New Orleans, LA 70118-5822
Phone: 1.504.891.1127
fed@tulane.edu

Duris, Pascal
Le Clos d'Arlac
5 Rue des Oeillets
Merignac 33700
France
Phone: 556475365
pascal.duris@wanadoo.fr
Doctorate, 1991
18c, Europe, Natural and Human History; 19c, Europe, Natural and Human History; 18c, Europe, Medical Sciences

Duschl, Richard A.
King's College London
School of Education
Waterloo Road
London SE1 9NN
Great Britain
Phone: 44.2.7848.3144
Fax: 44.2.7848.3182
richard.duschl@kcl.ac.uk
Doctorate, 1983
20c, North America, Education; 19c, Europe, Philosophy and Philosophy of Science; 18c, Philosophy and Philosophy of Science; Earth Sciences

Dusek, Rudolph V.
27 Villager Rd
Chester, NH 03036-4035
Phone: 1.603.862.3076
Fax: 1.603.862.2142
valdusek@aol.com

Dyck, David R.
Concord College
169 Riverton Avenue
Winnipeg, MB R2L2E5
Canada
Phone: 1.204.669.6503232
Fax: 1.204.663.2468
Doctorate
19c, Europe, Astronomical Sciences

Dyck, Martin
Mass. Inst. of Tech.
77 Mass Ave
MIT E38 6th Fl
Cambridge, MA 02139
Doctorate, 1956
18c, Europe, Mathematics; 20c, Humanistic Relations of Science

Dykstra, Gamelyn
University of Colorado
Dept of EPO Biology
Campus Box 334
Boulder, CO 80309
gamelyn.dykstra@colorado.edu

Eagleton, Catherine
Jesus College
Cambridge CB58BL
Great Britain
Phone: 44.20.7942.4208
katie_eagleton@yahoo.co.uk
Master's Level, 1999
5c-13c, Europe, Astronomical Sciences; 14c-16c, Europe, Astronomical Sciences; 5c-13c, Europe, Instruments and Techniques; 14c-16c, Europe, Instruments and Techniques

Eakin, Marshall C.
Vanderbilt University
Box 31-B
Nashville, TN 37235
Phone: 1.615.322.3328
Fax: 1.615.343.6002
marshall.c.eakin@vanderbilt.edu
Doctorate, 1981
20c, South America, Social Relations of Science; 19c, North America, Institutions; 18c, South America, Science Policy; 17c, North America, Humanistic Relations of Science

Eamon, William
New Mexico State University
Department of History
Box 30001 Dept 3H
Las Cruces, NM 88003
Phone: 1.505.646.4601
Fax: 1.505.646.8148
weamon@nmsu.edu
Doctorate, 1977
14c-16c, Europe, Medical Sciences; 14c-16c, Europe, Natural and Human History; 14c-16c, Medical Sciences

Easton, Joy B.
231 Heritage Pointe
Morgantown, WV 26505-2831
Phone: 1.304.599.6958
14c-16c, Europe, Mathematics; 17c, Europe, Education; Europe

Eastwood, Bruce S.
University of Kentucky
Patterson Office Tower 1715
Lexington, KY 40506-0027
Phone: 1.859.257.1859
Fax: 1.859.323.3885
bseast01@pop.uky.edu
Doctorate
5c-13c, Europe, Astronomical
Sciences; 5c-13c, Europe,
Education; 5c-13c, Europe,
Humanistic Relations of Science

Ebert, James
1512 Poole Rd
Street, MD 21154

Eddy, John H., Jr.
2356 Wood Ave
Colorado Spgs, CO 80907-6775
Phone: 1.719.593.8465
jeddy@coloradocollege.edu
Doctorate, 1977
18c, Europe, Biological Sciences;
18c, Europe, Natural and
Human History; 18c, Europe,
Philosophy and Philosophy of
Science

Eddy, M. D.
University of Durham
Dept of Philosophy
Durham DH13HN
Great Britain
Phone: 0.191.374.7641
Fax: 0.191.374.7641
m.d.eddy@durham.ac.uk

Eddy, Mark A.
157 Jacob St
Berea, OH 44017-2013
meddy@ou.edu

Edelson, Paul J.
Columbia College
630 W 168th St. Cntr for Soc
& Med, Physicians & Surgeons
New York, NY 10032
Phone: 1.212.305.6914
Fax: 1.212.650.9189
pedelson@pol.net
Doctorate, 1969
20c, North America, Medical
Sciences; 19c, North America,
Medical Sciences; 20c, North
America, Biological Sciences;
19c, Europe, Medical Sciences

Edelstein, Sidney M.
Dexter Chemical Corporation
845 Edgewater Rd
Bronx, NY 10474-4986

Eden, Trudy
University of Northern Iowa
History Dept
Cedar Falls, IA 50614-0701
Phone: 1.319.273.2933
Fax: 1.319.273.5846
trudy.eden@uni.edu

Edge, David O.
25 Gilmour Road
Edinburgh EH16 5NS
Great Britain
Phone: .44.131.6673497
Fax: 44.131.668.4008
d.edge@ed.ac.uk
Doctorate, 1959
20c, Astronomical Sciences;
Social Relations of Science

Edsall, John T.
985 Memorial Dr Apt 503
Cambridge, MA 02138-5804

Edwards, Diane D.
12695 Gardiner Rd N
Big Sandy, MT 59520-8476
Phone: 1.406.378.2634
sciwri@ttc-cmc.net
Doctorate, 2001
19c, North America, Medical
Sciences; 20c, North America,
Biological Sciences, Instruments
and Techniques; 20c, North
America, Medical Sciences;
19c, North America, Biological
Sciences, Instruments and
Techniques

Efron, Noah J.
Bar Ilan University
Graduate Program for Hist &
Phil of Science
Interdisciplinary Studies
Ramat Gan
Israel
Phone: 972.3.6048607
efron@mail.biu.ac.il

Egerton, Frank N.
Univ of Wisconsin-Parkside
History Dept
900 Wood Rd

Kenosha, WI 53141-2000
Phone: 1.262.595.2583
frank.egerton@uwp.edu

Egger, David
UMDNJ-RWJMS
675 Hoes Lane
Dept Neursci & Cell Biology
Piscataway, NJ 08854-5635
Phone: 1.732.235.4522
Fax: 1.609.921.3843
egger@umdnj.edu
Doctorate, 1962
Medical Sciences; Medical
Sciences, Instruments and
Techniques; Physical Sciences

Eghigian, Greg A.
Pennsylvania State University
Dept of History
108 Weaver Building
University Park, PA 16802-5500
Phone: 1.814.865.9022
Fax: 1.814.863.7840
gae2@psu.edu
Doctorate, 1993
20c, Europe, Medical Sciences,
Instruments and Techniques;
20c, Europe, Cognitive Sciences;
20c, Europe, Medical Sciences

Ehrhardt, George R.
Duke University
History Dept.
Carr Building
Durham, NC 27708-0719
gre120@aol.com
Doctorate, 1993
North America; Humanistic
Relations of Science;
Social Relations of Science; 20c

Eisberg, Joann
Univ. of Calif.- Santa Barbara
Department of Physics
Santa Barbara, CA 93106
eisbergjoa@aol.com
Doctorate
Astronomical Sciences;
Physical Sciences;
Gender and Science

Eisen, S.
York University
Department of History
237 Vanier College

Downsview, ON M3J1P3
Canada
Phone: 1.416.736.5123

Eisenman, Harry J.
University of Missouri-Rolla
Dept of History & Pol Science
1870 Miner Cir
Rolla, MO 65409-1260
Phone: 1.573.341.4808
Fax: 1.573.341.4871
hje@umr.edu
Doctorate, 1967
19c, Technology;
20c, Technology; Education

Ekman, Richard
The Andrew W Mellon
Foundation
Secretary
140 E 62nd St
New York, NY 10021-8187

El Bouazzati, Bennacer
Dept. Philosophie Fac Lettres
BP 1040
Rabat
Morocco
el_bouazzati@meganet.net.ma
Doctorate, 1997
5c-13c, Physical Sciences;
14c-16c, Europe, Astronomical
Sciences; 20c, Europe,
Philosophy and Philosophy
of Science; Philosophy and
Philosophy of Science

Elena, Alberto
Calle Emilio Coll, 22
Puerta 55
28224 Pozuelo de Alarcon
Madrid
Spain
Phone: 341.352.3254
alberto.elena@uam.es

Elina, Olga J.
Institute of the History of
Science & Technology
Russian Academy of Science
Staropansky per. 1/5
Moscow 103012
Russia
Phone: 7095921.8061
Fax: 7095925.9971
olgaelina@history.ihst.ru
Doctorate

20c, Europe, Biological Sciences;
20c, Institutions;
20c, Social Relations of Science

Elliott, Clark A.
MIT
E56-100 – 38 Memorial Drive
Burndy Library Dibner Inst.
Cambridge, MA 02139
Phone: 1.617.489.0822
claelliott@earthlink.net
Doctorate, 1970
North America; Education;
Social Sciences; 19c, North
America, Natural and Human
History

Elliott, John E.
Echidna Design
PO Box 10002
Winston-Salem, NC 27108
Phone: 1.336.727.9857
Fax: 1.336.727.9857
echidna@worldnet.att.net
Master's Level, 1999
20c, Europe, Physical Sciences;
19c, Biological Sciences;
Earth Sciences;
Humanistic Relations of Science

Ellis, Erik
306 NW Monroe Ave Apt 2
Corvallis, OR 97330-4732
Phone: 1.206.920.3745
ellisde@onid.orst.edu

Elshakry, Marwa
81 Cranbury Neck Road
Cranbury, NJ 08512
Phone: 1.609.430.7506
elshakry@princeton.edu

Elwick, James
IHPST Room 316
Victoria College
91 Charles St West
Toronto, ON M5S1K7
Canada
Phone: 1.416.978.7432
Fax: 1.416.978.3003
jelwick@chass.utoronto.ca

Emelianova, Larissa L.
Oceanography Museum
Naberezhnaia Bagramyana, 1
Kaliningrad 236000
Russia

Phone: 7.3.0112.436302
Fax: 7.3.0112.340211
postmaster@vitiaz.koenig.ru
Master's Level
19c, Europe;
18c, Europe, Earth Sciences;
18c, Europe, Social Sciences;
19c, Europe, Social Sciences

Emery, Guy T.
Bowdoin College
Physics Department
Brunswick, ME 04011
Phone: 1.207.725.3708
Fax: 1.207.725.3638
gemery@bowdoin.edu
Doctorate
20c, Physical Sciences;
19c, Physical Sciences

Emrich, John
George Washington University
Center for History of Recent
Science
Dept of History
Washington, DC 20052
alngthwy76@netscape.net

Enderlein, Joerg
Fuerbringerstr. 6
Berlin 10961
Germany
joerg.enderlein@chemie.
uni-regensburg.de

Endersby, Jim
Cambridge University
Dept of History & Philosophy
of Science
Free School Lane
Cambridge CB2 3RH
Great Britain
Phone: 01223.527.850
jje21@cam.ac.uk
Bachelor's Level
19c, Europe, Biological Sciences;
19c, Europe, Exploration,
Navigation and Expeditions

Endersby, Linda Eikemeier
606 Shakertown Way
Columbia, MO 65203-6822
Phone: 1.573.442.5274
Fax: 1.573.884.5579
endersbyl@missouri.edu

Enea, Horace
1240 Lisa Ln
Los Altos, CA 94024-6039
hjemea@earthlink.net

Engelmann, Hugo O.
421 W Hillcrest Dr
Dekalb, IL 60115
Phone: 1.815.758.7167
hugoe@svm.soci.niu.edu

Englehart, Stephen F.
California State Poly Univ.
Department of History
Pomona, CA 91768
Phone: 1.919.865.2859
Fax: 1.909.869.4724
sfengleharte@csupomowa.edu

Enros, Philip C.
Environment Canada
351 St Joseph Blvd
Floor 7
Ottawa, ON K1A0H3
Canada
Phone: 1.819.994.5434
Fax: 1.819.953.0550
philip.enros@ec.gc.ca
Doctorate, 1979
Science Policy; Mathematics;
Biological Sciences;
North America

Ensmenger, Nathan
University of Pennsylvania
History & Soc of Science
303 Logan Hall
249 S 36th Street
Philadelphia, PA 19104
Phone: 1.215.898.4643
nathanen@sas.upenn.edu

Epstein, Isaac
Rua Dr. Jesuino Maciel, 1249
São Paulo 04615-000
Brazil
Phone: 55.11.5533.8707

Erikson, G. E.
Erikson Biographical Institute
242 B Meeting Street
Providence, RI 02906

Erlen, Jonathon
University of Pittsburgh
200 Scaife Hall
Pittsburgh, PA 15261

Phone: 1.412.648.8927
Fax: 1.412.648.9020
erlen@pitt.edu
Doctorate, 1973
20c, North America, Medical
Sciences; 19c, North America,
Medical Sciences

Erlichson, Herman
College of Staten Island
Dept Engineering Sci & Physics
2800 Victory Blvd
Staten Island, NY 10314-6600
Phone: 1.718.982.2828
Fax: 1.718.982.2830
erlichson@postbox.csl.cuny.edu
Doctorate
17c, Physical Sciences

Eser, Uta
Provenceweg 3
Tuebingen D-72072
Germany
Phone: 49.7071.33119
Fax: 49.7071.295225
uta.eser@uni-tuebingen.de

Eshet, Dan
9 Goodrich Rd Apt 6
Jamaica Plain, MA 02130-2036
Phone: 1.617.983.8221
deshet@worldnet.att.net

Evans, Hughes
1600 7th Ave South
Midtown Building 201
Birmingham, AL 35233
Phone: 1.205.975.6503
hevans@uab.edu
Doctorate
19c, North America, Medical
Sciences; 20c, North America,
Medical Sciences

Evans, James
Univ of Puget Sound
Physics Dept
1500 N Warner St
Tacoma, WA 98416-0005
Phone: 1.253.879.3813
Fax: 1.253.879.3500
jcevans@ups.edu
Doctorate, 1983
Pre-400 B.C.E.,
Astronomical Sciences;

18c, Europe, Physical Sciences;
19c, Physical Sciences;
20c, Physical Sciences

Everett-Lane, Debra
155 7th Ave # 2
Brooklyn, NY 11215-2202
dal20@columbia.edu

Evernden, Neil
1722 5th Ave S
Lethbridge, AB T1J0W5
Canada

Eyler, John M.
University of Minnesota
Department History of Medicine
511 Diehl Hall
505 Essex St SE
Minneapolis, MN 55455-0350
Phone: 1.612.624.5921
Fax: 1.612.625.7938

Fagette, Paul H.
Arkansas State University
Department of History
PO Box 1690
State Univ., AR 72469
Phone: 1.870.972.3046
Fax: 1.870.972.2088
fagette@iit.edu
Doctorate
20c, North America, Technology,
Instruments and Techniques;
20c, North America, Medical
Sciences

Fairhurst, Stephen P.
123 W 93rd St #6E
New York, NY 10025-7585

Fairman, Julie A.
University of Pennsylvania
School of Nursing
Philadelphia, PA 19104
Phone: 1.215.898.4151
Fax: 1.215.573.7492
fairman@nursing.upenn.edu
Doctorate
20c, North America, Technology;
20c, North America, Gender and
Science; 20c, North America,
Medical Sciences

Faiz, Alexandria
9 Russett Rd
Sandy Hook, CT 06482-1480

Phone: 1.212.853.9090
afaiz@chek.com
Bachelor's Level, 1993
*Philosophy and Philosophy
of Science; Science Policy;
Social Relations of Science;
Humanistic Relations of Science*

Falla, William
St. John's UCC
2918 MacArthur Road
Whithall, PA 18052-3421
Phone: 1.610.821.8725
Doctorate, 1993
*Philosophy and Philosophy of
Science; Biological Sciences*

Fan, Fa-Ti
SUNY-Binghamton
History Department
PO Box 6000
Binghamton, NY 13902-6000
ffan@brighamton.edu
Master's Level
*Europe, Natural and Human
History; Asia, Exploration,
Navigation and Expeditions;
Asia, Natural and Human
History*

Fancher, Raymond E.
York University
Department of Psychology-Arts
4700 Keele St.
Toronto, ON M3J1P3
Canada
Phone: 1.416.736.5115
Fax: 1.416.736.5814
fancher@yorku.ca
Doctorate, 1967
*Cognitive Sciences;
Biological Sciences;
19c, Europe, Biological Sciences,
Biography; Social Sciences*

Farber, Paul L.
Oregon State University
Dept. of History
Corvallis, OR 97331
Phone: 1.541.737.1273
Fax: 1.541.737.1257
pfarber@orst.edu
Doctorate, 1970
*18c, Biological Sciences;
19c, Biological Sciences;*

*Natural and Human History;
Philosophy and Philosophy of
Science*

Farley, John
Dalhousie University
Biology Department
Halifax, NS B3H4J1
Canada
Phone: 1.902.494.3515
Fax: 1.902.494.3736
biology@dal.ca
Doctorate, 1964
20c, Medical Sciences

Fausto-Sterling, Anne
Brown University
Dept of Molecular, Cellular &
Developmental Biology
Box G-J163
Providence, RI 02912-0001
Phone: 1.401.863.2109
Fax: 1.401.863.2421
anne_fausto-sterling@brown
.edu
Doctorate
*20c, North America, Biological
Sciences; 20c, North America,
Gender and Science; 20c, North
America, Social Relations of
Science; 20c, North America,
Humanistic Relations of Science*

Fayter, Paul
York University
Natural Sci/Glendon College
2275 Bayview Avenue
Toronto, ON M4N3M6
Canada
Phone: 1.416.736.2100
Fax: 1.905.522.9900
fayter@yorku.ca
Master's Level
*20c, North America,
Humanistic Relations of Science;
19c, Europe, Biological Sciences;
19c, Europe, Astronomical
Sciences*

Featherman, David
810 7th Ave Fl 31
New York, NY 10019-5818

Federlin, Paul
24 Avenue du General DeGaulle
Strasbourg 67000
France

Fee, Elizabeth
National Library of Medicine
Chief Hist of Medicine
8600 Rockville Pike
Bethesda, MD 20894-0001
Phone: 1.301.496.5406
Fax: 1.301.402.0872
elizabeth_fee@ulm.nih.gov

Feest, Uljana
University of Pittsburgh
HPS Dept
1017 Cathedral of Learning
Pittsburgh, PA 15260-6299
Phone: 1.412.624.3181
feest@pitt.edu

Feferman, Solomon
Stanford University
Dept of Mathematics
Stanford, CA 94305-2125
Phone: 1.650.723.2439
Fax: 1.650.725.4066
sf@csli.stanford.edu
Doctorate, 1957
*20c, Philosophy and Philosophy
of Science; 20c, Mathematics;
Philosophy and Philosophy of
Science*

Feigenbaum, Lenore
Tufts University
Dept of Mathematics
Medford, MA 02155
Phone: 1.617.627.2353
Fax: 1.617.627.3966
lfeigenb@emerald.tufts.edu
Doctorate, 1981
*Mathematics; 18c, Europe,
Mathematics; Humanistic
Relations of Science; Education*

Feingold, Mordechai
90 Marion Rd.
Watertown, MA 02472
Phone: 1.540.231.8472
Fax: 1.540.231.7013
feingold@vt.edu
Doctorate, 1980
*17c, Europe, Education;
17c, Europe, Institutions;
17c, Europe, Humanistic
Relations of Science*

Feinman, Richard
SUNY
Biochem-8

450 Clarkson Avenue
Brooklyn, NY 11203
Phone: 1.718.871.7333
rfeinman@netmail.hscbklyn.edu

Feins, John
950 Larrabee St Apt 408
West Hollywood, CA
90069-3947
fidicen@earthlink.net

Feldberg, Georgiana
York Univ Ctr for Hlth Studies
214 York Lanes
4700 Keele Street
North York, ON M3J1P3
Canada
Phone: 1.416.736.5941
Fax: 1.416.736.5986
ginaf@yorku.ca

Feldhay, Rivka
Tel Aviv University
Cohen Institute
Tel Aviv 69978
Israel
Phone: 972.03.640.9198
Fax: 972.03.640.9463
fekdhay@post.tau.ac.il
Doctorate, 1987
17c, Europe, Education;
14c-16c, Europe,
Astronomical Sciences;
17c, Europe, Physical Sciences;
14c-16c, Europe, Social Sciences

Felt, Ulrik
Dept for Phil of Sci and Soc
Sensengasse 8/10
Vienna A-1090
Austria
Phone: 43.1.427747611
Fax: 43.1.42779476
ulrike.felt@univie.ac.at

Ferguson, Robert G.
Hong Kong Univ of Sci & Tech
Division of Social Science
Clear Water Bay
Kowloon
Hong Kong
Phone: 852.2358.7819
Fax: 852.2335.0014
sorf@ust.hk
Doctorate, 1996
20c, North America, Technology

Fernandez, Eliseo A.
Linda Hall Library
5109 Cherry St
Kansas City, MO 64110
Phone: 1.816.363.5020
Fax: 1.816.926.8790
fernande@lindahall.org
Master's Level, 1965
Philosophy and Philosophy of
Science; Physical Sciences;
Mathematics

Ferngren, Gary B.
Oregon State University
Dept of History
Milam 306
Corvallis, OR 97331
Phone: 1.541.737.1262
Fax: 1.541.737.1257
gferngren@orst.edu
Doctorate, 1973
400 B.C.E.-400 C.E., Europe,
Medical Sciences

Fernos, Rodrigo
Cond. El Monte Norte, Apt. 226
165 Ave Hostos
San Juan, PR 00918-4244
Phone: 1.787.763.3781
rodrigo@spiderlink.net

Ferrell, Thomas
1261 Cobble Pond Way
Vienna, VA 22182-6605
Phone: 1.703.757.9777
tom@faaconsulting.com

Fesimec, Melissa
PMB #305
416 Sandario Ave
LAREDO, TX 78041
Phone: 1.956.725.3138
Fax: 1.95.6.725.7015

Fichman, Martin
Bethune Coll / York Univ.
4700 Keele St.
North York, ON M4J1P3
Canada
Phone: 1.416.736.2100
Fax: 1.416.736.5892
mfichman@yorku.ca
Doctorate, 1969
19c, Biological Sciences;
Humanistic Relations of Science;
Social Relations of Science

Fickess, Douglas R.
Westminster College
501 Westminster Ave
Fulton, MO 65251-1299
Phone: 1.573.592.5273
Fax: 1.573.592.5217
fickesd@jaynet.wcmo.edu
Doctorate, 1962
20c, Europe, Biological Sciences;
5c-13c, North America

Fifield, Steve J.
Univ of Delaware
Dept of Biological Sciences
Newark, DE 19716
Phone: 1.302.831.2281
fifield@udel.edu
Doctorate, 1999
Biological Sciences; Education;
Social Relations of Science

Figueiredo, Fernando
Urbanizacao Quinto do Bosque
lote 130 - 2 Esq Post
Viseu 3510
Portugal
Phone: 351966251459
fernandofigueiredo@mail.tele
pac.pt
20c, Europe, Philosophy
and Philosophy of Science;
18c, Europe, Social Sciences

Filgueiras, Carlos A.
Un. Fed. Rio de Janeiro
Dept de Quimica Inorganica IQ
CP 68563
Rio de Janeiro 21945970
Brazil
Phone: 21.562.7812
Fax: 21.562.7559
calf@iq.ufrj.br
Doctorate
18c, South America;
18c, Europe; 17c, Europe

Filner, Robert
US Congress
333 F St Ste A
Chula Vista, CA 91910
Doctorate, 1973
20c, North America, Science
Policy; Social Relations of
Science; Humanistic Relations
of Science

Finan, Barbara
21 Bedford Ct
Concord, MA 01742-2624
bfinanp@aol.com

Findlen, Paula
Stanford University
Dept of History
Stanford, CA 94305
Phone: 1.650.723.9570
Fax: 1.650.725.0597
pfindlen@leland.stanford.edu
Doctorate, 1989
17c, Europe, Institutions;
14c-16c, Europe, Humanistic
Relations of Science;
Medical Sciences;
Natural and Human History

Fine, Arthur
University of Washington
Department of Philosophy
PO Box 353350
Seattle, WA 98195-3350
Phone: 1.206.543.5855
Fax: 1.206.685.8740
afine@u.washington.edu
Doctorate, 1963
Physical Sciences; Philosophy
and Philosophy of Science;
20c, Physical Sciences

Fink, Karl J.
St. Olaf College
1520 St. Olaf Avenue
Northfield, MN 55057-1098
Phone: 1.507.646.3864
Fax: 1.507.646.3732
kjfink@stolaf.edu
Doctorate
18c, Europe; 19c, Europe;
5c-13c, Europe; Europe

Finkelstein, Gabriel
University of Colorado/ Denver
Dept of History, CB182
P.O Box 173364
Denver, CO 80217
Phone: 1.303.556.4272
Fax: 1.303.556.6037
gabriel.finkelstein@
  cudenver.edu
Doctorate, 1996
19c, Europe, Exploration,
Navigation and Expeditions;

19c, Europe, Medical Sciences;
Medical Sciences, Instruments
and Techniques

Finlay, Mark R.
Armstrong Atlantic State Univ
Department of History
11935 Abercorn St
Savannah, GA 31419-1989
Phone: 1.912.921.5642
Fax: 1.912.921.5581
finlayma@mail.armstrong.edu
Doctorate
19c, North America, Technology;
19c, Europe, Biological Sciences;
19c, North America, Biological
Sciences; 19c, Physical Sciences

Finn, Bernard S.
Smithsonian Institution
NMAH
Washington, DC 20560-0631
Phone: 1.202.357.1840
Fax: 1.202.633.9273
finnb@si.edu
Doctorate
19c, North America, Physical
Sciences; 20c, North
America, Physical Sciences;
19c, North America, Technology;
20c, North America, Technology

Finocchiaro, Maurice A.
Univ of Nevada at Las Vegas
Dept of Philosophy
Box 455028
4505 S Maryland Pkwy
Las Vegas, NV 89154-5028
Phone: 1.702.895.3461
Fax: 1.702.895.1279
mauricef@unlv.edu
Doctorate, 1969
17c, Astronomical Sciences;
17c, Physical Sciences;
Social Sciences; 20c, Europe,
Philosophy and Philosophy of
Science

Fischer, Hans
Kath. Univ. Eichstaett, MGF
Ostenstr. 26
Eichtaett 85072
Germany
Phone: 49842180919
Fax: 498421932256
hans.fischer@ku-eichstaett.de

Fischer, Klaus
Universität Trier
Fb I - Philosophie
Trier 54286
Germany
Phone: 49651.2012342
Fax: 49651.2013922
fischer@uni-trier.de
Doctorate, 1977
Cognitive Sciences;
Astronomical Sciences;
Social Relations of Science

Fisher, Saul
Andrew W Mellon Foundation
140 E 62nd St
New York, NY 10021-8187
Phone: 1.212.838.8400
Fax: 1.212.223.2778
sf@mellon.org
Doctorate, 1997
20c, Philosophy and Philosophy
of Science; 17c, Philosophy and
Philosophy of Science

Fitzpatrick, Joseph F., Jr.
Tulane University
Museum of Natural History
Belle Chasse, LA 70037
Phone: 1.504.391.1338
Fax: 1.504.394.5045
jfitz@museum.tulane.edu
Doctorate, 1964
19c, Europe, Biological Sciences;
Europe, Cognitive Sciences;
18c, North America, Natural
and Human History;
20c, North America

Fleck, Robert C.
Embry Riddle Aeronautical Univ
Physical Sciences Dept
600 S Clyde Morris Blvd
Daytona Beach, FL 32114-3900
Phone: 1.386.226.6612
Fax: 1.376.226.6621
fleckr@cts.db.erau.edu
Doctorate, 1977
Astronomical Sciences;
Physical Sciences;
Humanistic Relations of Science

Fleming, Donald
Harvard University
785 Widener Library
Cambridge, MA 02138
Phone: 1.617.495.2488

Fleming, James R.
Colby College
STS Program
5881 Mayflower Hill
Waterville, ME 04901-8858
Phone: 1.207.872.3548
Fax: 1.207.872.3074
jrflemin@colby.edu
Doctorate, 1988
Earth Sciences;
Social Relations of Science

Flemming, Isabelle
2307 N James Ct
Arlington Hts, IL 60004-3088
Phone: 1.352.384.1558
iplatt@ufl.edu

Flesher, Mary M.
Smith College
Northampton, MA 01060
Fax: 1.413.585.2794
mflesher@smith.edu
Doctorate, 1986
19c, Europe, Cognitive Sciences;
19c, Europe, Medical Sciences;
19c, Europe, Natural and
Human History;
20c, North America, Natural and
Human History

Fliss, Edward
St Louis Community College at
Florissant Valley
Biology Dept
3400 Pershall Rd
Saint Louis, MO 63135-1499
Phone: 1.314.595.4392
Fax: 1.314.595.2047
efliss@stlcc.cc.mo.us

Folkerts, Menso
Lehrstuhl für Geschichte der
Naturwissenschaften
Museumsinsel 1
München D-80538
Germany
Phone: 49.89.21803252
Fax: 49.89.21803162
m.folkerts@lrz.uni-muenchen.de
Doctorate, 1967
5c-13c, Europe, Mathematics;
400 B.C.E.-400 C.E., Europe,
Astronomical Sciences;
14c-16c, Education; Biography

Folsom, Herbert R.
Iowa State University
Dept of History
603 Ross Hall
Ames, IA 50010-1202
Phone: 1.515.294.7266
Fax: 1.515.294.6390
folsomoon@aol.com
Master's Level, 1996
19c, Europe, Astronomical
Sciences; 20c, Europe,
Astronomical Sciences;
19c, Europe, Physical Sciences

Folta, Jaroslav
Kostelni 42
C2-170 78
Prague
Czech Republic

Fontaine, Philippe
Ecole Normale Superieure
Economics Dept
61 Avenue du President Wilson
Cachan 94235
France
Phone: 33147402450
fontaine@grid.ens-cachan.fr
Doctorate
18c, North America, Social
Sciences; 20c, Europe, Social
Sciences; Europe, Cognitive
Sciences; Social Sciences

Fontana, J. R.
4375 Via Glorieta
Santa Barbara, CA 93110-2113
Phone: 1.805.682.3387
jorgef@rain.org

Ford, Charles E.
St Louis University
Dept of Mathematics and
Computer Science
St Louis, MO 63103
Phone: 1.314.977.2434
fordce@slu.edu
Doctorate, 1968
20c, Europe, Mathematics

Forest, Denis
Université Lyon
III Philosophie
74 Rue Pasteur
Lyon 69007
France
denis.forest@wanadoo.fr

Forgan, S.
University of Teesside
School of Arts & Media
Middlesbrough TS1 3BA
Great Britain
Phone: 44.1642342364
Fax: 44.1642384099
s.forgan@tees.ac.uk

Forman, Paul
Smithsonian Institution
National Museum of American
History
Washington, DC 20013
Phone: 1.202.357.2820
Fax: 1.202.633.9273
formanp@nmah.si.edu
Doctorate, 1967
20c, Physical Sciences;
Social Relations of Science

Forstner, Christian J.
Dr - Franz-Schmitz-Str, 2
Bad Abbach
Germany
Phone: 49.179.3983198
Fax: 49.941.943.2038
c.j.forstner@web.de

Fortune, Nathaniel
Smith College
Department of Physics
Northampton, MA 01063-0001
Phone: 1.413.585.3980
Fax: 1.413.85.3786
nfortune@science.smith.edu

Foster, Amy E.
University of Minnesota
438 Walter Library
Minneapolis, MN 55455
fosteae@mail.auburn.edu
Master's Level, 1999
20c, North America, Technology;
Europe, Technology, Instruments
and Techniques

Fouke, Daniel C.
University of Dayton
Dept of Philosophy
300 College Park Ave
Dayton, OH 45469-1546
Phone: 1.937.229.2817
fouke@checkov.hm.udayton.edu
Doctorate, 1986
17c, Europe, Philosophy
and Philosophy of Science;

17c, Europe, Social Sciences;
17c, Europe, Astronomical
Sciences; 17c, Europe,
Humanistic Relations of Science

Fox, Robert
University of Oxford
Modern History Faculty
Broad Street
Oxford OX1 3BD
Great Britain
Phone: 01.865.277277
Fax: 01.865.277277
robert.fox@history.ox.ac.uk
Doctorate, 1967
19c, Europe, Physical Sciences;
19c, Europe, Education;
19c, Europe, Institutions

Foxworth, Walter L.
17111 Waterview Pkwy
Dallas, TX 75252-8005
Phone: 1.972.437.6100
wfoxworth@foxgal.com

Franceschetti, Donald R.
University of Memphis
Department of Physics
Campus Box 526670
Memphis, TN 38152-6670
Fax: 1.901.678.4733
dfrncsch@memphis.edu
Doctorate, 1974
Physical Sciences; Philosophy
and Philosophy of Science;
20c, Instruments and Techniques

Francis, Kevin
Mt Angel Seminary
St Benedict, OR 97373
Phone: 1.503.288.8670
kfrancis@mtangel.edu

Frangsmyr, Tore
Uppsala University
History of Science
Box 256
Uppsala 75105
Sweden
Phone: 46.18.4711579
Fax: 46.18.108046
tore.frangsmyr@idehist.uu.se
Doctorate
18c, Europe, Earth Sciences;
19c, Europe, Social Relations of
Science; Humanistic Relations
of Science

Frank, Robert G., Jr.
UCLA Sch of Med
Med Hist Div
Los Angeles, CA 90024

Franklin, Allan D.
University of Colorado
Dept of Physics
PO Box 390
Boulder, CO 80309-0390
Phone: 1.303.492.8610
Fax: 1.303.492.2998
allan.franklin@colorado.edu
Doctorate, 1965
Physical Sciences; Philosophy
and Philosophy of Science

Franzel, Thomas G.
2210 Cobbler Ct NW
Salem, OR 97304-2086
Phone: 1.503.375.3101
franzelt@open.org

Fraser, Craig G.
University of Toronto
Inst for History & Philosophy of
Science and Technology
Victoria College
Toronto, ON M5S1K7
Canada
Phone: 1.416.978.5135
Fax: 1.416.978.3003
cfraser@chass.utoronto.ca
Doctorate, 1981
Mathematics; Philosophy
and Philosophy of Science;
Physical Sciences;
20c, Astronomical Sciences

Fredette, Raymond F.
281, Chemin-Noel, RR3
Fitch Bay, PQ J1X3W4
Canada
Phone: 1.819.876.2895
fredetra@interlinx.qc.ca
Doctorate
14c-16c, Europe, Physical
Sciences; 5c-13c, Europe,
Philosophy and Philosophy
of Science; Pre-400 B.C.E.,
Biological Sciences

Freed, Libbie
1149 E Mifflin St
Madison, WI 53703-2434
ljfreed@students.wisc.edu

Freemon, Frank R.
2422 Valley Brook Rd
Nashville, TN 37215-2019

Freiburger, Dana A.
934 High Street #6
Madison, WI 53715
dafreiburger@students.wisc.edu
Master's Level, 1998
Instruments and Techniques;
Earth Sciences; Exploration,
Navigation and Expeditions

Freire, Olival, Jr.
Instituto De Fisica UFBA
Campus de Ondina
40210-340 Salvador
Bahia
Brazil
Phone: 55.71.247.20.33
Fax: 55.71.235.55.92
freirejr@ufba.br
Doctorate, 1995
20c, Physical Sciences;
20c, South America,
Physical Sciences;
20c, South America, Education

Freitag, Ruth S.
Library of Congress
10 First Street SE
Washington, DC 20540-4330
Phone: 1.202.707.6984
Fax: 1.202.707.0279
Master's Level, 1959
Astronomical Sciences;
Education

Freudenthal, Gad
CNRS
7 rue Guy Moquet B.P. #8
Villejuif 94801
France
Phone: 331.49583599
Fax: 331.49583547
freudent@msh-paris.fr
Doctorate, 1980
5c-13c, Astronomical Sciences;
17c, Physical Sciences;
Physical Sciences; Philosophy
and Philosophy of Science

Friedlander, Michael W.
Washington University
Department of Physics
1 Brookings Dr
Saint Louis, MO 63130-4899

Phone: 1.314.935.6279
Fax: 1.314.935.6219
mwf@wuphys.wustl.edu
Doctorate, 1955
*17c, Europe, Astronomical
Sciences; 20c, North America;
20c, North America, Social
Relations of Science; 20c, North
America, Physical Sciences*

Friedman, Ami
431 Lakeside Dr
Waterford, MI 48328-4039
Phone: 1.248.682.5130
friedma@walledlake.k12.mi.ud

Friedman, Daniel
St. Clare's Hospital
415 W. 51st Street
New York, NY 10019
Phone: 1.212.459.8704
Doctorate, 1967
*20c, Europe, Biological Sciences;
Europe, Philosophy and
Philosophy of Science; Europe,
Social Relations of Science;
Humanistic Relations of Science*

Friedman, Donald
159 Madison Ave # 2
New York, NY 10016-5428
dfriedman@lzatechnology.com

Friedman, Gerald M.
Rensselaer Ctr Appl Geology
15 3rd St
Troy, NY 12180-3205
Phone: 1.518.273.3247
Fax: 1.518.273.3249
gmfriedman@juno.com

Friedman, Judith E.
University of Victoria
PO Box 3045
Victoria, BC V8W3P4
Canada
Phone: 1.250.721.7382
Fax: 1.250.721.8772
jfriedma@uvic.ca
Master's Level, 1997
*20c, North America,
Medical Sciences; Europe;
17c, Europe, Institutions*

Friedman, Michael
3975 Inverness Farm Road
Bloomington, IN 47401

Phone: 1.812.331.1330
Fax: 1.812.855.3631
mlfriedm@indiana.edu

Friedman, Russell L.
Institute for Greek & Latin
Njalsgade 92
Copenhagen DK-2300
Denmark
russ@span.hum.ku.ak

Frost, Gary L.
800 Pritchard Avenue Ext # A13
Chapel Hill, NC 27516-1724
Phone: 1.919.968.4980
Fax: 1.919.968.4980
garyfrost@mindspring.com
Master's Level, 1999
*20c, North America, Technology;
19c, North America, Technology;
20c, North America, Physical
Sciences; 19c, North America,
Physical Sciences*

Fruton, Joseph S.
Yale University
School of Medicine
333 Cedar Street
New Haven, CT 06520
Phone: 1.203.785.4340
Fax: 1.203.737.4130
Doctorate, 1934
*Physical Sciences;
Biological Sciences; Institutions;
Instruments and Techniques*

Fry, Bertrand C.
Deshaw & Co.
22nd Floor, Tower 45
120 West 45th Street
New York, NY 10036
Phone: 1.212.478.0756
Fax: 1.212.845.1756
bryn_bert@juno.com
*Philosophy and Philosophy of
Science; 17c, Europe, Physical
Sciences; Social Sciences*

Fuchs, Alfred H.
Bowdoin College
6900 College Sta
Brunswick, ME 04011-8469
Phone: 1.207.725.3143
Fax: 1.207.725.3892
afuchs@bowdoin.edu

Doctorate, 1960
*19c, Cognitive Sciences;
20c, Cognitive Sciences*

Fujii, Kiyohisa
4-31-2 Eifuku Suginami Ku
Tokyo 168 0064
Japan
Phone: 81.353.767.558
suzk-fji@asahi-met.or.jp

Fujita, Yasumoto
Dainifujimiso 209
4-7-12 Sonan
Sagamihara-shi
Kanagawa 228 0812
Japan
rxm05777@nifty.ne.jp
Master's Level, 1996
*20c, Astronomical Sciences;
20c, Social Relations of Science*

Furukawa, Yasu
Tokyo Denki University
2-2 Kanda-Nishiki-cho,
Chiyoda-ku
Tokyo 1018457
Japan
Phone: 81.3.5280.3453
Fax: 044.989.0802
yfuru@cck.dendai.ac.jp
Doctorate, 1983
*20c, Europe, Physical Sciences;
20c, North America, Physical
Sciences; 19c, Asia, Institutions;
Social Sciences*

Furumoto, Laurel
Wellesley College
Dept of Psychology
106 Central St
Wellesley, MA 02481-8204
Phone: 1.781.283.3020
Fax: 1.781.283.3730
lfurumoto@wellesley.edu
Doctorate, 1967
*20c, Cognitive Sciences;
20c, Gender and Science;
20c, Social Relations of Science*

Furzer, Adrian
1, Collingwood Close
Anerly
London SE20 8JL
Great Britain
Phone: 181.659.3061
adrian.furzer@btinternet.com

Fye, W. Bruce
1533 Seasons Lane SW
Rochester, MN 55902
Phone: 1.507.536.7921
Fax: 1.507.266.0103
fye.bruce@mayo.edu
Doctorate
Medical Sciences;
18c, Medical Sciences;
19c, Medical Sciences;
20c, Medical Sciences

Fyfe, Aileen
National University of Ireland
Dept of History
Galway
Ireland
Phone: 353.91.524411
Fax: 353.91.750556
aileen.fyfe@nuigalway.ie

Gabbey, W. Alan
Barnard College
Dept of Philosophy
3009 Broadway
New York, NY 10027-6598
Phone: 1.212.854.2066
Fax: 1.212.854.7491
wag8@columbia.edu
Doctorate
17c, Philosophy and Philosophy
of Science; Philosophy and
Philosophy of Science;
Physical Sciences

Gage, Clarke L.
6 Elm St
Canton, NY 13617-1461
Phone: 1.315.386.3427
cgage@twcny.rr.com
Doctorate
Physical Sciences;
Earth Sciences

Gagnon, Robert
CIRST-UQAM
CP.8888
Montreal, PQ H3C3P8
Canada

Gaines, Brian R.
3635 Ocean View
Cobble Hill, BC V0R1L1
Canada
Phone: 1.250.743.7080
Fax: 1.250.743.7089
gaines@ucalgary.ca

Gal, Ofer
Ben-Gurion Univ of the Negev
Dept of Philosophy
POB 653
Beer-Sheva 84105
Israel
Phone: 972.3.6817344
Fax: 972.3.6817344
ofgal@bgumail.bgu.ac.il

Galbreath, Robert
1108 Briarcliff Rd
Greensboro, NC 27408-7534
Phone: 1.336.272.8501
robert_galbreath@yahoo.com

Galbreath, Ross A.
Onewhero
RD2
Tuakau
New Zealand
Phone: 64.9.2328113
onewhero@ps.gen.nz

Gale, George D., Jr.
UMKC
Dept of Philosophy
Kansas City, MO 64110
Phone: 1.816.235.2816
Fax: 1.816.235.2819
galeg@umkc.edu
Doctorate, 1971
19c, Europe, Biological Sciences;
20c, Europe, Astronomical
Sciences; 18c, Europe,
Physical Sciences

Galison, Peter L.
Harvard University
History of Science
Science Center 235
Cambridge, MA 02138-2901
Phone: 1.617.495.3544
Fax: 1.617.495.3344
gailson@fas.harvard.edu
Doctorate, 1983
20c, Europe, Social Sciences;
20c, North America, Philosophy
and Philosophy of Science;
20c, Physical Sciences;
20c, Instruments and Techniques

Galle, Karl L.
1401 County Road 258
Liberty Hill, TX 78642-6256
k.galle@ic.ac.uk

Master's Level, 1997
Europe, Astronomical Sciences;
Europe, Mathematics

Galvin, Cyril, Jr.
7728 Brandeis Way
Springfield, VA 22153-3406
Phone: 1.703.569.9187
galvincoastal@juno.com
Doctorate, 1963
19c, Europe, Earth Sciences;
20c, Earth Sciences;
19c, Europe, Biological Sciences,
Biography; Technology,
Instruments and Techniques

Gambaro, Ivana L.
Via Dei Franzone 2/7D
Genova 16145
Italy
Phone: 39.0103624181
md3199@mclink.it
Doctorate
20c, Europe, Physical Sciences;
20c, Philosophy and
Philosophy of Science;
20c, Europe, Education

Gamble, Susan
3546 79th St Apt 53
Jackson Heights, NY
11372-4839
sag24@hermes.cam.ac.uk

Garaizar Axpe, Isabel
Universidad del Pais Vasco
Apartado 644
Bilbao
Vizcaya 48080
Spain
Phone: 34.94.601.4007
Fax: 34.94.464.82.99
hcpgaaxi@lg.ehu.es
Doctorate, 1998
20c, Instruments and
Techniques; 20c, Europe,
Social Relations of Science;
20c, Europe, Education;
17c, Europe, Technology

Garay, Michael
450 W Cool Dr Apt 406
Tucson, AZ 85704-6466
Phone: 1.520.229.9062
mgaray_1973@yahoo.com

Garber, Daniel E.
University of Chicago
Department of Philosophy
Classics 17
1050 E 59th St
Chicago, IL 60637-1559
Phone: 1.773.702.7920
Fax: 1.773.702.9061
garb@midway.uchicago.edu
Doctorate, 1975
17c, Europe, Physical Sciences;
Philosophy and Philosophy
of Science; 17c, Europe,
Social Relations of Science;
Social Sciences

Garber, Elizabeth
SUNY - Stony Brook
History Dept
100 Nicholl Rd
Stony Brook, NY 11794
Phone: 1.631.632.7511
Fax: 1.631.632.7367
egarber@notes.sunysb.edu
Doctorate, 1967
18c, Europe, Physical Sciences;
19c, Europe, Mathematics;
20c, Europe, Social Sciences;
North America

Garber, Janet B.
7734 W 81st St
Playa Del Rey, CA 90293-7909
Phone: 1.310.821.1425
Fax: 1.310.822.4621
janetgarber@worldnet.att.net
Doctorate
19c, Natural and Human History;
18c, Biological Sciences;
17c, Biological Sciences;
20c, Biography

Garber, Margaret D.
Univ of California San Diego
Dept of History/Science Studies
9500 Gilman Dr
La Jolla, CA 92093-0104
Phone: 1.619.534.1996
Fax: 1.619.296.0122
mgarber@helix.ucsd.edu
17c, Europe, Physical Sciences;
17c, Europe, Social Sciences

Garcia, Gregorio D.
Calle D. no. 506
e/21y 23 Vedado
Habana
Cuba

Garcia-Blanco, Rolando
CEHCYT, Academia de Ciencias
Calle: Cuba 460 entre Amargura
y Brazil, Habana Vieja
La Habana 10100
Cuba
Doctorate
Biography; Science Policy;
Institutions; Social Relations of
Science

Garcia-Sanz, Javier
Universidad Nacional de
Educacion a Distancia
Departament Fisica Fundamental
Apartado Correos 60141
Madrid 28080
Spain
Phone: 34917293966
Fax: 34913986697
jgarcia@ccia.uned.es
Doctorate
20c, Physical Sciences;
400 B.C.E.-400 C.E., Physical
Sciences; 5c-13c, Philosophy
and Philosophy of Science

Garciadiego, Alejandro R.
Univ. Nac. Autonoma de Mexico
Depto. Matematicas, #016
Facultad de Ciencias, UNAM
Mexico 04510
Mexico
Phone: 525.622.4860
Fax: 525.593.9721
gardan@servidor.unam.mx
Doctorate, 1983
19c, Europe, Mathematics;
20c, Europe, Philosophy
and Philosophy of Science;
19c, North America, Education;
Social Sciences

Garfield, Eugene
Inst Scientific Information
3501 Market St
Philadelphia, PA 19104-3389
Phone: 1.215.243.2205
Fax: 1.215.387.1266
garfield@codex.cis.upenn.edu

Garfinkle, Robert A.
32924 Monrovia St
Union City, CA 94587-5433
Phone: 1.510.489.4779
Fax: 1.510.489.8454
ragerf@earthlink.net
Bachelor's Level, 1994
Astronomical Sciences

Gariepy, Thomas P.
Stonehill College
320 Washington St
Easton, MA 02357-1135
Phone: 1.508.565.1122
Fax: 1.508.565.1446
tgariepy@stonehill.edu
Doctorate, 1990
20c, North America, Medical
Sciences; 19c, North America,
Medical Sciences; 19c, Europe,
Medical Sciences

Garmon, Lucille B.
State Univ of West Georgia
Department of Chemistry
Carrollton, GA 30118-0001
Phone: 1.770.836.4547
Fax: 1.770.836.4656
lgarmon@westga.edu

Garrett, Norman L.
804 N Citrus Ave
Los Angeles, CA 90038-3404
Phone: 1.213.464.6083
Doctorate, 1971
20c, North America, Cognitive
Sciences

Gascoigne, John
University of New South Wales
School of History
Sydney 2114
Australia
Phone: 61.2.9385.2341
j.gascoigne@unsw.edu.au
Doctorate, 1981
18c, Europe, Exploration,
Navigation and Expeditions;
Europe, Education; Institutions

Gaskell, Roger
17 Ramsey Road
Cambs
Warboys PE28 2RW
Great Britain
Phone: 1487.823059
Fax: 1487.82307

roger@rogergaskell.com
Bachelor's Level, 1972
17c, Europe, Education; 18c

Gasman, Daniel
John Jay College CUNY
History Department
445 W 59th St
New York, NY 10019-1199
Phone: 1.212.237.8821
dgasman@jjay.cuny.edu
Doctorate
*20c, Europe, Humanistic
Relations of Science*

Gast, Richard D.
Deere & Company
1 John Deere Place
Moline, IL 61265-8098
Phone: 1.309.765.4715
Fax: 1.309.765.9860
rg28562@deere.com
Master's Level, 1977
*18c, Europe, Physical Sciences;
Philosophy and Philosophy of
Science; 17c, Physical Sciences;
19c, Physical Sciences*

Gatensby, Anne G.
York University
Soc & Pol Thought, S7224 Ross
4700 Keele St
North York, ON M3J1P3
Canada
Phone: 1.514.848.2509
mdsh@musica.mcgill.ca
Doctorate, 1999
*20c, Philosophy and
Philosophy of Science;
20c, Gender and Science;
20c, Social Relations of Science;
20c, Biological Sciences*

Gates, Barbara T.
University of Delaware
Dept of English
Newark, DE 19716
Phone: 1.302.831.5227
Fax: 1.302.831.1586
bgates@udel.edu
Doctorate
*19c, Europe, Natural and
Human History; 19c, Europe,
Humanistic Relations of Science;
20c, Biological Sciences;
Exploration, Navigation and
Expeditions*

Gates, Robert V.
PO Box 352
Dahlgren, VA 22448-0352
Phone: 1.540.663.3068
Fax: 1.540.653.8588
robgates@crosslink.net

Gatlin, Stephen
Churchill Court Condominiums
237 Montgomery Ave Apt 1F
Haverford, PA 19041-1849
Phone: 1.610.896.6242
sgatlin@eastern.edu

Gaukroger, Stephen
University of Sydney
Philosophy, A14
Sydney 2006
Australia
Phone: 61.2.93512477
Fax: .93516601
stephen.gaukroger@
  philosophy.usyd.edu.au
Doctorate, 1977
*17c, Physical Sciences;
17c, Philosophy and Philosophy
of Science; 17c, Mathematics*

Gautero, Jean Luc
Université de Nice
Faculte des Lettres
98 bd E. Herriot; BP 209
Nice Cedex 3 06204
France
Phone: 33.493.265434
Fax: 33.607.371561
jlg@fairesuivre.com

Gautier Dalche, Patrick
Directeur de Recherche
103 Bd de Magenta
Paris 75010
France
Phone: 0.145960439

Gauvin, Jean F.
Dept of History and Science
Science Ctr 235
1 Oxford St
Cambridge, MA 02138-2901
Phone: 1.781.393.8867
gauvin@fas.harvard.edu

Gavroglu, Kostas
University of Athens
History & Phil of Science Dept
John Kennedy 37 Ilisia

Athens 16121
Greece
Phone: 30109233578
Fax: 30107275530
kgavro@cc.uoa.gr

Gay, Hannah
Simon Fraser University
History Dept
Burnaby, BC V5A1S6
Canada
Phone: 1.604.291.4379
Fax: 1.604.291.5837
hgay@sfu.ca
Doctorate
19c, Europe, Physical Sciences

Gayon, Jean
Université Paris 7
Case Courrier 7001
Paris 75251
France
Phone: 33.1.44275795
gayon@noos.fr
Doctorate, 1989
*Philosophy and Philosophy
of Science; 19c, Europe;
20c, Europe*

Gearhart, Clayton A.
St Johns University
Department of Physics
PO Box 7055
Collegeville, MN 56321-7055
Phone: 1.320.363.3184
Fax: 1.320.363.3202
cgearhart@csbsju.edu
Doctorate, 1979
*Physical Sciences;
Astronomical Sciences*

Geijo-Barrientos, Emilio
Universidad Miguel Hernandez
Dpt de Fisiologia Fac Med
Campus de San Juan
San Juan 03550
Spain
emilio.geijo@umh.es

George, Charles R. P.
Concord Hospital
Hospital Road
Concord NSW 2139
Australia
Phone: 61.2.97675000
Fax: 61.2.94493212
georgc@crgmail.crg.

cs.nsw.gov.au
Doctorate, 1966
Medical Sciences; Philosophy and Philosophy of Science

Gerbino, Anthony
Columbia University
Dept Art Hist & Archaeology
826 Schermerhorn Hall
New York, NY 10027
ag266@columbia.edu
Doctorate, 2001
17c, Europe, Technology, Instruments and Techniques; Humanistic Relations of Science; Mathematics

Gerhart, James B.
University of Washington
Phys/Astr Building
Box 351560
Seattle, WA 98195
Phone: 1.206.616.2785
Fax: 1.206.685.0636
gerhart@phys.washington.edu
Doctorate, 1954

Gerson, Elihu
Tremont Research Institute
458 29th Street
San Francisco, CA 94131-2311
Fax: 1.415.648.7660
gerson@ieee.org
Doctorate, 1998
20c, Social Relations of Science; Natural and Human History; Biological Sciences

Geyer, Peter
PO Box 1440
Warrnambool, VIC 3280
Australia
Phone: 61.3.5562.9033
Fax: 61.3.5562.9033
alchymia@ozemail.com.au

Ghamari-Tabrizi, Sharon M.
1050 Austin Ave, NE
Atlanta, GA 30307
Phone: 1.404.525.2538
sghamari@gsu.edu
Doctorate, 1993
20c, North America, Social Relations of Science; Social Sciences; Humanistic Relations of Science; 20c, North America

Ghiselin, Michael T.
California Academy of Science
Center for the History and
Philosophy of Science
Golden Gate Park
San Francisco, CA 94118-4599
Phone: 1.415.750.7084
Fax: 1.415.750.7090
mghiselin@calacademy.org
Doctorate
19c, Europe, Biological Sciences, Biography; Biological Sciences; Philosophy and Philosophy of Science

Giard, Luce
9 Rue Eugene-Gibez
Paris 75015
France
Phone: 1.453.8591
Fax: 1.4533.8591

Gibbs, Fred
1430 Mound St Apt F1
Madison, WI 53711-2239

Gibbs, Jake W.
Lexington Comm Coll
214 Moloney Building
Cooper Dr.
Lexington, KY 40506-0235
Phone: 1.606.257.4872
jwgibboo@pop.uky.edu
Master's Level, 1983
Astronomical Sciences; 19c, Europe, Biological Sciences, Biography; Exploration, Navigation and Expeditions

Gieryn, Tom
Indiana University
Dept of Sociology
754 Ballantine Hall
Bloomington, IN 47401-5022
Phone: 1.812.855.2950
Fax: 1.812.855.0781
gieryn@indiana.edu
Doctorate, 1980
19c, Europe, Social Relations of Science; 20c, North America, Social Relations of Science; 20c, North America; 20c, Europe, Biological Sciences

Giglioni, Guido M.
Johns Hopkins University
Dept of History of Science

Charles Street
Baltimore, MD 21218
gmg12@jhunix.hcf.jhu.edu
17c, Europe, Medical Sciences; 17c, Europe, Philosophy and Philosophy of Science; 14c-16c, Europe, Philosophy and Philosophy of Science; 17c, Europe, Biological Sciences

Gilkeson, John S.
Arizona State University
Arts and Sciences
ASU West 3051
PO Box 37100
Phoenix, AZ 85069-7100
Phone: 1.602.543.6069
Fax: 1.602.543.6004
john.gilkesonjr@asu.edu
Doctorate, 1981
20c, North America, Social Sciences; 20c, North America, Natural and Human History; 20c, North America, Earth Sciences

Gillen, H. William
500 Sandcastle Ct
Wilmington, NC 28405-8386
Phone: 1.910.392.7477
5c-13c, Medical Sciences

Gillespie, Neal C.
1049 Clifton Rd NE
Atlanta, GA 30307-1227
Phone: 1.404.377.3562
Doctorate
19c, Europe, Biological Sciences; 19c, Europe, Natural and Human History

Gillmor, C. S.
Wesleyan University
Dept of History
238 Church St
Middletown, CT 06459-0002
Phone: 1.860.685.2378
Fax: 1.860.685.2781
sgillmore@wesleyan.edu
Doctorate, 1968
Physical Sciences; Technology, Instruments and Techniques; Earth Sciences; Technology

Gimmel, Millie
2393 Brandon Ct
Bloomington, IN 47401-4513
Phone: 1.812.339.6957
mgimmel@indiana.edu

Gingerich, Owen
Harvard-Smithsonian
Center for Astrophysics
60 Garden St
Cambridge, MA 02138
Phone: 1.617.495.7216
Fax: 1.617.496.7564
ginger@cfa.harvard.edu
Doctorate, 1962
Astronomical Sciences;
14c-16c, Instruments
and Techniques;
14c-16c, Astronomical Sciences

Gingras, Yves
University du Quebec Montreal
CP 8888
SVCC Centre-Ville
Montreal, PQ H3C3P8
Canada
Phone: 1.514.987.3000
Fax: 1.514.987.7726
gingras.yves@uqam.ca
Doctorate
20c, North America, Physical
Sciences; 19c, Europe, Physical
Sciences; 20c, North America,
Social Relations of Science;
20c, North America, Institutions

Giordano, Raymond V.
The Antiquarian Scientist
PO Box 448
Southampton, MA 01073-0448
Phone: 1.413.529.2731
Fax: 1.413.529.0907
antiqsci@ma.ultranet.com
Bachelor's Level, 1970
Instruments and Techniques;
Physical Sciences; Technology

Giron Sierra, Alvaro
Fernandez de la Hoz 3-8
Madrid
Spain
Phone: 34914466783
ibarrangelua@inincia.es

Giuliani, Giuseppe
Univ degli Studi di Pavia
Dipartimento di Fisica Volta

Via Bassi 6
Pavia 27100
Italy
Phone: 39.08.252.4805
Fax: 39.038.250.7563
giuliani@fisicaviolta.unipv.it

Givelber, Harry M.
Geneva General Hospital
196 North St
Geneva, NY 14456
Phone: 1.315.787.4568
Fax: 1.315.787.4557
shgivelb@epix.net
Doctorate, 1964
19c, Europe, Biological Sciences,
Biography; Medical Sciences

Givens, Douglas R.
Documents Archaeology, Inc.
Research Office
5029 Country Club Dr
High Ridge, MO 63049-3509
Phone: 1.314.954.4817
Fax: 1.314.376.4336
documents@primary.net.us
Doctorate, 1986
20c, Africa, Natural and Human
History

Givens, Jean A.
University of Connecticut
Dept of Art & Art History
Unit 1099
Storrs Mansfield, CT
06269-0001
Phone: 1.860.488.3932
jean.givens@uconn.edu

Giventer, Edwin B.
Brooklyn College, CUNY
Brooklyn, NY 11210
Doctorate, 1953
20c, Philosophy and
Philosophy of Science;
20c, Cognitive Sciences;
20c, Social Relations of Science

Gjerloff, Anne K.
University of Copenhagen
Department of History
Njalsgade 102
Copenhagen 2300 S
Denmark
Phone: 45.3526.0450
akg@hum.ku.dk

Gjertsen, Derek
1 Ash Grove
Formby Merseyside
Great Britain
Phone: 1704.876441
dgj@tinyworld.co.uk

Glas, Eduard
Delft University
Department of Mathematics
Mekelweg 4
Delft 2628CD
The Netherlands
Phone: 31.152782541
e.glas@its.tudelft.nl
Doctorate
20c, Mathematics;
19c, Social Relations of Science

Glaze, Scott
PO Box 9040
Fort Wayne, IN 46899-9040
Phone: 1.260.747.4154
Fax: 1.260.747.0398

Gleason, Mary Louise
54 Riverside Dr Apt 3D
New York, NY 10024-6551
Phone: 1.212.595.6084
Fax: 1.212.504.3054
marylougleason@cs.com
Doctorate, 1978
17c, North America, Institutions;
18c, Europe, Institutions;
19c, Europe, Institutions

Gliboff, Sander J.
1241 Elmwood Ave Apt 2A
Evanston, IL 60202-1244
Phone: 1.847.425.9480
gl@norwestern.edu
Master's Level, 1997
20c, Europe, Biological Sciences;
19c, Europe, Biological Sciences;
Biological Sciences

Glick, Thomas F.
Boston University
Department of History
226 Bay State Rd
Boston, MA 02215-1403
Phone: 1.617.353.8314
Fax: 1.617.353.2556
tglick@bu.edu
Doctorate, 1968
19c, South America,

Biological Sciences;
5c-13c, Europe, Technology;
5c-13c, Technology

Glicksman, Maurice
Brown University
Division of Engineering
Box D
Providence, RI 02912-9104
Phone: 1.401.863.1409
Fax: 1.401.863.9107
maurice_glicksman@brown.edu
Doctorate, 1954
20c, Asia, Science Policy

Glitz, Marie
Arizona State University
Department of Biology
Journal of History of Biology
PO Box 871501
Tempe, AZ 85284-3181
Phone: 1.480.491.0601
Fax: 1.480.965.6684
glitz@asu.edu

Goddu, Andre L.
Stonehill College
320 Washington St
North Easton, MA 02357-5620
Phone: 1.508.565.1280
Fax: 1.508.565.1444
agoddu@stonehill.edu
Doctorate, 1979
5c-13c, Europe, Astronomical
Sciences; 14c-16c, Europe,
Astronomical Sciences;
5c-13c, Europe, Physical
Sciences; 14c-16c, Europe,
Physical Sciences

Godin, Benoit
INRS
3465, rue Durocher
Montreal, PQ H2X2C6
Canada
Phone: 1.514.499.4074
Fax: 1.514.499.4065
benoit_godin@
  inrs-urb.uquebec.ca
Doctorate
Social Sciences; Philosophy and
Philosophy of Science

Goehring, J. B.
Washington & Lee Univ
Chemistry Department
Science Center

Lexington, VA 24450-0303
goehringb@wlu.edu
Doctorate, 1962
Physical Sciences

Golan, Tal
Ben Gurion Univ of the Negev
Ben-Gurion Research Center
Sede-Boker Cmp 84990
Israel
Phone: 972.8.6596972
talgolan@bgumail.bgu.ac.il
Doctorate, 1997

Golden, William T.
500 5th Ave Fl 50
New York, NY 10110-5099
Fax: 1.212.840.6774

Goldman, Joanne M.
Univ of Northern Iowa
Dept of History
Seerley 340
Cedar Falls, IA 50614
Phone: 1.319.273.5908
Fax: 1.515.233.0837
goldman@csbs.csbs.uni.edu
Doctorate, 1988
20c, North America, Science
Policy; 20c, North America,
Institutions; Europe, Institutions;
Social Relations of Science

Goldman, Steven L.
Lehigh University
Philosophy Department
15 University Dr
Bethlehem, PA 18015-3057
Phone: 1.610.758.3773
Fax: 1.610.758.3790
slg2@lehigh.edu
Doctorate, 1971
20c, North America, Technology;
20c, North America, Technology,
Instruments and Techniques;
Philosophy and Philosophy of
Science; Social Relations of
Science

Goldstein, Bernard R.
University of Pittsburgh
RS-2604CL
Pittsburgh, PA 15260
Fax: 1.412.624.5994
brg@pitt.edu
Doctorate, 1963
5c-13c, Asia, Astronomical

Sciences; 14c-16c, Europe,
Astronomical Sciences;
5c-13c, Astronomical Sciences

Goldstein, Daniel
105 El Cajon Ave
Davis, CA 95616-0433
dgoldstein@ucdavis.edu

Goldwhite, Harold
1704 Oak St
South Pasadena, CA 91030-4717
Phone: 1.626.441.1955
hgoldwh@calstatela.edu
Doctorate, 1956
18c, Europe, Physical Sciences;
20c, North America, Physical
Sciences

Golinski, Jan V.
University of New Hampshire
History Department
Horton Social Science Center
Durham, NH 03824-3586
Phone: 1.603.862.3789
Fax: 1.603.862.0178
jan.golinski@unh.edu
Doctorate, 1984
18c, Europe, Physical Sciences;
18c, Europe, Earth Sciences;
18c, Europe, Social Relations of
Science; Social Sciences

Golland, Louise A.
Univ of Chicago
Networking, Telecom, Comp
Serv
1155 E. 60th St.
Chicago, IL 60637
Phone: 1.773.702.7613
Fax: 1.773.702.0559
l-golland@uchicago.edu
Doctorate, 1991
20c, Europe, Mathematics;
20c, North America,
Mathematics; 19c, Europe,
Mathematics; 18c, Europe,
Astronomical Sciences

Gomory, Ralph E.
Sloan Foundation
630 5th Ave
New York, NY 10111-0100

Gonzalez, Elsa L.
5532 S Kenwood Ave Apt 411
Chicago, IL 60637-1784

Phone: 1.773.493.9762
Doctorate
18c, Europe, Biological Sciences;
17c, Europe, Physical Sciences;
20c, Europe, Physical Sciences;
14c-16c, Astronomical Sciences

Good, Gregory A.
West Virginia University
History Dept
Woodburn Hall
Morgantown, WV 26506-6303
Phone: 1.304.293.2421
Fax: 1.304.293.3616
ggood@wvu.edu
Doctorate, 1982
20c, Earth Sciences; 19c; 18c

Gooday, Graeme J.
University of Leeds
Division of HPS
School of Philosophy
Woodhouse Lane
Leeds LS2 9JT
Great Britain
Phone: 44.11.3233.3274
Fax: 44.11.3233.3265
g.j.n.gooday@leeds.ac.uk
Doctorate, 1989
19c, Technology,
Instruments and Techniques;
19c, Physical Sciences;
19c, Institutions;
19c, Instruments and Techniques

Goodfellow, Sarah
Pennsylvania State University
108 Weaver Building
History Department
University Park, PA 16802
sxg205@psu.edu
Doctorate
19c, Europe, Medical Sciences;
20c, North America,
Medical Sciences;
18c, Europe, Medical Sciences

Gooding, David C.
University of Bath
Science Studies Centre
Department of Psychology
Claverton Down
Bath BA2 7AY
Great Britain
Phone: 1225.826335
Fax: 1225.826752
hssdcg@bath.ac.uk

Doctorate
North America; Philosophy
and Philosophy of Science;
Humanistic Relations of Science;
Cognitive Sciences

Goodman, Jordan
Manchester School of
Managment
PO Box 88
Manchester M60 1QD
Great Britain
Phone: 441612003408
Fax: 441612003505
gdmnjrd@aol.com
Doctorate, 1977
19c, Europe, Biological
Sciences; 20c, North America,
Medical Sciences;
19c, Europe, Physical Sciences;
19c, Europe, Medical Sciences

Goodman, Kenneth W.
University of Miami
School of Medicine
PO Box 016960 - M-825
Miami, FL 33101-6960
Phone: 1.305.243.5723
Fax: 1.305.243.3549
kgoodman@miami.edu
Doctorate, 1991
400 B.C.E.-400 C.E.,
Astronomical Sciences;
Biological Sciences; Philosophy
and Philosophy of Science

Goodstein, Judith R.
California Institute of Tech
Institute Archives
MS 015A-74
Pasadena, CA 91125
Phone: 1.626.395.2700
Fax: 1.626.577.1296
jrg@its.caltech.edu
Doctorate
20c, North America;
20c, Medical Sciences;
19c, Medical Sciences;
Mathematics

Gordin, Michael D.
Princeton University
Department of History
Dickinson Hall
Princeton, NJ 08544
Phone: 1.617.495.4045
Fax: 1.617.495.3344

mgordin@fas.harvard.edu
Doctorate, 2001
19c, Europe, Physical Sciences;
18c, Europe, Institutions

Gorelik, Gennady
Boston University
Philosophy & Hist. of Science
745 Commonwealth Avenue
Boston, MA 02215
gorelik@bu.edu
Doctorate, 1979
20c, Europe, Physical Sciences;
20c, Europe, Humanistic
Relations of Science;
20c, Europe, Social Relations of
Science

Gortler, Leon B.
Brooklyn College
Department of Chemistry
2900 Bedford Ave.
Brooklyn, NY 11210-2889
Fax: 1.718.951.4607
lgortler@brooklyn.cuny.edu
Doctorate
20c, North America, Physical
Sciences; Europe

Goss, Andrew
1211 Vista Dr
Socorro, NM 87801-4445
Phone: 1.505.835.1908

Gossel, Patricia L.
Smithsonian Institution
Nat. Musuem of American Hist
Room 5128, MRC 636
Washington, DC 20560-0636
Phone: 1.202.768.2669
Fax: 1.202.357.1631
gosselp@si.edu
Doctorate, 1989
19c, North America, Biological
Sciences, Instruments and
Techniques; 20c, North
America, Instruments and
Techniques; 20c, North America,
Medical Sciences

Gosselin, Edward A.
California State University
Dept of History
1250 N Bellflower Blvd
Long Beach, CA 90840-1601
Phone: 1.562.985.2408
Fax: 1.562.985.5431

gosselin@csulb.edu
Doctorate, 1973
14c-16c, Europe, Astronomical Sciences; 14c-16c, Astronomical Sciences

Gotz, Martin
Theoretical Astrophysics Center
Juliane Maries Vej 30
Copenhagen DK2100
Denmark
Phone: 45.3532.5903
Fax: 35.3232.5910
gotz@tac.dk
Master's Level, 1995
Astronomical Sciences

Gough, Jerry B.
Washington State University
Department of History
Pullman, WA 99164-4030
Phone: 1.509.335.3469
Fax: 1.509.335.4171
gough@wsu.edu
Doctorate
18c, Europe, Physical Sciences; 18c, Europe, Technology

Goulding, Robert
76 Alexander St
Princeton, NJ 08540-5112
Phone: 1.609.430.1091
goulding@princeton.edu

Gouvea, Fernando Q.
Colby College
Dept of Math
Mayflower Hill 5836
Waterville, ME 04901
Phone: 1.207.872.3278
Fax: 1.207.872.3801
fqgouvea@colby.edu
Doctorate, 1987
Mathematics; Physical Sciences

Goux, J-Michel
Le Caillou
Lalandusse 47330
France
Phone: 33.0553.3681
jeanmichelgoux@aol.com
Doctorate
19c, Biological Sciences

Grabiner, Judith V.
Pitzer College
1050 N. Mills

Claremont, CA 91711-6101
Phone: 1.909.607.3160
Fax: 1.909.621.8481
judith_grabiner@pitzer.edu
Doctorate, 1966
18c, Mathematics; 19c, Mathematics; 18c, Europe, Mathematics

Graham, Katherine
1309-B South Fifth Ave.
Bozeman, MT 59715
Phone: 1.406.580.9705
katherine1graham@aol.com

Graham, Loren R.
Harvard University
Coolidge Hall 210
1737 Cambridge St
Cambridge, MA 02138
Phone: 1.617.491.1616
Fax: 1.617.491.1616
lrg@mit.edu
Doctorate, 1964
20c, Europe, Physical Sciences; 20c, Europe, Biological Sciences

Granada, Miguel A.
Universidad de Barcelona
Baldiri Reixac S/N
Barcelona 08028
Spain
Phone: 34.93.333.3466
Fax: 34.93.410.2856
granada@trivium.gh.ub.es
Doctorate, 1978
14c-16c, Europe, Philosophy and Philosophy of Science; Astronomical Sciences; Humanistic Relations of Science

Grande, Darby
2708 W Prairie Creek Dr
Richardson, TX 75080-2025

Grande, Jan
Odenseveien 5
Trondheim N-7020
Norway
Phone: 4773598715
Fax: 4773591327
jan.grande@hf.ntnu.no
19c, Europe, Medical Sciences; 19c, Biological Sciences, Instruments and Techniques; 19c, Mathematics; Philosophy and Philosophy of Science

Grandin, Karl O.
Apelgatan 15A
Uppsala SE-75435
Sweden
Phone: 46.18.24.48.66
Fax: 46.8.6739598
karlg@kva.se
Master's Level, 1992
20c, Europe, Physical Sciences; 19c, Europe, Science Policy; Humanistic Relations of Science; Social Relations of Science

Grange, Kathleen M.
3828 Pine Ave
Long Beach, CA 90807-3234
Phone: 1.562.424.0900
Europe, Humanistic Relations of Science

Grant, Edward
Indiana University
History & Philosophy of Science
130 Goodbody Hall
Bloomington, IN 47405
Phone: 1.812.855.3622
Fax: 1.812.855.3631
grant@indiana.edu
5c-13c, Europe

Grapi, Pere
Calle Latorre 62, 3-3
Sabadell 08201
Spain
Phone: 93.726.2058
pgrapi@pie.xtec.es

Grau, Kevin
612 E 13th St
Indianapolis, IN 46202-2732
Phone: 1.317.955.9520
ktgrau@indiana.edu

Graubard, Stephen R.
Norton's Woods
Editor Daedalus
136 Irving St
Cambridge, MA 02138-1929

Graves Stuart, Stella L.
908 S Lahoma Ave
Norman, OK 73069-4559
Phone: 1.405.325.2213
Fax: 1.405.325.2363
sstuart@ou.edu

Green, Daniel W. E.
Smithsonian Observatory
60 Garden St
Cambridge, MA 02138-1500

Green, Judy
Mathematics Department
Marymount University
Arlington, VA 22207-4299
Phone: 1.703.284.1678
Fax: 1.703.284.3859
judy.green@marymount.edu
Doctorate
*Philosophy and Philosophy
of Science; North America,
Mathematics; 19c; 20c*

Green, Monica M.
Duke University
Department of History
226 Carr Bldg
PO Box 90719
Durham, NC 27708-0719
Phone: 1.919.684.2439
Fax: 1.919.681.7670
mhgreen@acpub.duke.edu
Doctorate, 1985
*5c-13c, Europe, Medical
Sciences; 5c-13c, Europe,
Gender and Science*

Green Musselman, Elizabeth
Southwestern University
1001 E University Ave
Georgetown, TX 78626-6100
Phone: 1.512.863.1595
Fax: 1.512.863.1535
greenmue@southwestern.edu
Doctorate, 1998
*Earth Sciences; North America;
Gender and Science; Africa*

Greenberg, John
55 Boulevard Bara
Palaiseau 91120
France
Phone: 1133160106474
jgreen8032@aol.com
Doctorate, 1979
*18c, Europe, Earth Sciences;
18c, Europe, Mathematics;
18c, Europe, Institutions*

Greenblatt, Samuel H.
Brown University Neurosurgery
Memorial Hospital
111 Brewster St

Pawtucket, RI 02860-4400
Phone: 1.401.729.2784
Fax: 1.401.729.2781
samuel_greenblatt@brown.edu
Doctorate
*19c, Europe, Medical Sciences;
19c, Europe, Biological Sciences;
20c, North America, Medical
Sciences*

Greene, John C.
Canterbury Woods B215
651 Sinex Avenue
Pacific Grove, CA 93950
Phone: 1.831.657.4298
johngreeneca@earthlink.net

Greene, Mott T.
University of Puget Sound
153 Wyatt Hall
Tacoma, WA 98416-1061
Phone: 1.253.879.3782
Fax: 1.253.879.3500
greene@ups.edu
Doctorate
20c, Earth Sciences

Greenspan, Nancy
7201 Glenbrook Rd
Bethesda, MD 20814-1242
Phone: 1.301.657.8745
Fax: 1.301.657.8745
Master's Level
20c, Europe, Physical Sciences

Greenway, John L.
University of Kentucky
1153 Patterson Tower
Honors Program
Lexington, KY 40506-0027
Phone: 1.606.257.6973
engjlg@pop.uky.edu
Doctorate
*19c, Europe;
Europe, Medical Sciences*

Gregorio, Mario Di
Univ Degli Studi L'Aquila
Dept di Culture Comparate
Via Camponeschi 2
L'Aquila 67100
Italy
mario555@uja.net

Gregory, Frederick
University of Florida
Program in History of Science

Department of History
PO Box 117320
Gainesville, FL 32611-7320
Phone: 1.352.392.0271
Fax: 1.352.392.6927
fgregory@ufl.edu
Doctorate, 1973
*19c, Social Relations of
Science; North America;
18c, Biological Sciences*

Greiner, Allen K.
Kansas Univ Medical Center
Dept of Hist & Phil of Med
3901 Rainbow Blvd
Kansas City, KS 66160-0004
Phone: 1.913.588.7098
Fax: 1.913.588.7060
agreiner@kumc.edu
*19c, Europe, Medical Sciences;
20c, North America, Medical
Sciences, Instruments and
Techniques; 19c, North America,
Medical Sciences; 20c, North
America, Medical Sciences*

Grier, David A.
204 11th St SE
Washington, DC 20003-2101
Phone: 1.202.546.8231
grier@gwu.edu

Griesemer, James R.
University of California-Davis
Dept of Philosophy
SS/H 1238
1 Shields Ave
Davis, CA 95616-8673
Phone: 1.530.752.1068
Fax: 1.530.752.8964
jrgriesemer@ucdavis.edu
Doctorate, 1983
*Biological Sciences;
Natural and Human History*

Griffith, Robert
West Virginia University
School of Pharmacy
PO Box 9530
Morgantown, WV 26506-9530
Phone: 1.304.599.6384
Fax: 1.304.293.2576
rgriffit@wvu.edu

Griffiths, Paul E.
University of Pittsburgh
Dept of Hist & Phil of Science

1017 Cathedral of Learning
Pittsburgh, PA 15260-6299
Phone: 1.412.624.5879
Fax: 1.412.624.6825
pauleg@pitt.edu
Doctorate, 1989
Biological Sciences;
Cognitive Sciences; Philosophy
and Philosophy of Science

Grilli, Martha
via L. da Vinci, 2
Ghezzano 56010
Italy

Grinevald, Jacques
IUED University of Geneva
24 Rue Rothschild
Geneva CH-1211
Switzerland
Phone: 41.22.906594
Fax: 41.22.9065947
Doctorate, 1979
18c, Europe, Biological Sciences;
19c, Europe; Europe, Biological
Sciences; 20c, North America,
Physical Sciences

Groeben, Christiane
Stazione Zoological Anton Dohr
Villa Comunale
Naples 80121
Italy
Phone: 39.081.583.3274
Fax: 39.081.764.1355
groeben@alpha.szn.it
Doctorate, 1990
19c, Institutions; 19c, Biography

Groessing, Gerhard
Austrian Institute for Nonlinear
Studies
Parkgasse 9
Vienna 1030
Austria
Phone: 43.1715.0177
Fax: 43.1715.0177
ains@teleweb.at
14c-16c, Astronomical
Sciences; 20c, Philosophy
and Philosophy of Science;
20c, Physical Sciences;
20c, Astronomical Sciences,
Philosophy and Philosophy of
Science

Gronim, Sara S.
New Jersey Inst of Technology
Dept of History
Newark, NJ 08903
Phone: 1.973.596.5627
sgronim@erols.com
Doctorate, 1999
18c, North America, Biological
Sciences; 18c, North America,
Medical Sciences; 18c, North
America, Natural and Human
History; 18c, North America,
Astronomical Sciences

Groppi, Susan M.
3937 Ruby St
Oakland, CA 94609-2719
Phone: 1.510.658.8964
groppi@socrates.berkeley.edu

Gross, Dennis M.
Merck Research Labs
WP42-300
PO Box 4
West Point, PA 19486
Phone: 1.215.652.6819
Fax: 1.215.699.6678
gross@merck.com
Doctorate, 1974
Technology; Biological Sciences;
Natural and Human History;
Philosophy and Philosophy of
Science

Gross, Marvin A.
3900 Chestnut St Apt 818
Philadelphia, PA 19104-3123
Phone: 1.215.386.5473
mgross7369@aol.com

Gruber, Jacob W.
Temple University
Broad & Montgomery
Philadelphia, PA 19122
jakegruber@cs.com
Doctorate
19c, Europe, Biological Sciences;
Pre-400 B.C.E., North America,
Natural and Human History;
Europe, Social Sciences;
Natural and Human History

Gruender, C. D.
2403 Miranda Ave
Tallahassee, FL 32304-1324
Phone: 1.850.576.0046
Fax: 1.850.644.3832

gruender@mailer.fsu.edu
Doctorate, 1957
Philosophy and Philosophy of
Science; Education

Grunden, Walter E.
Bowling Green State University
Dept of History
Williams Hall
Bowling Green, OH 43403
Phone: 1.419.372.8639
wgrund@bgnet.bgsu.edu
Doctorate, 1998
20c, Asia, Science Policy;
20c, Asia, Technology;
20c, Asia, Physical Sciences;
20c, North America

Guedes, Manuel V.
Rua da Graciosa, 57-Ric-D
4050 Porto
Portugal
mvg@mail.telepac.pt

Guerrini, Anita
University of California
History Dept
Santa Barbara, CA 93106
Fax: 1.805.893.8686
guerrini@history.ucsb.edu
Doctorate, 1983
18c, Europe, Medical Sciences;
17c, Europe, Medical Sciences;
18c, Europe, Biological Sciences

Guillen, Reynal R.
14440 Dickens St
Sherman Oaks, CA 91423-4059
Phone: 1.818.501.4834
rguillen@ucla.edu
Doctorate, 1999
20c, North America;
20c, North America,
Social Relations of Science;
20c, Gender and Science

Guillermo Ranea, Alberto
Universidad Torcuato di Tella
Minones 2159
Buenos Aires 1428
Argentina
Phone: 54.11.47840080
Fax: 54.11.47840089
granea@utdu.edu
Doctorate
Philosophy and Philosophy of

Science; Humanistic Relations of Science; 17c, Europe; 14c-16c, Europe

Guise-Richardson, Cai
1207 Marston
Ames, IA 50010
caiguise@iastate.edu

Gullberg, Steven
801 Sprinters Row Dr
Florissant, MO 63034-3371
Phone: 1.314.831.7388
gullberg439@earthlink.net

Gumienny, Kevin P.
146 Lenox Ct
Piscataway, NJ 08854-3167

Gundlach, Horst
Universitat Passau
Institut für Geschichte der
Psychologie
Passau D-94030
Germany
Phone: 851.56098611
Fax: 851.56098612
gundlach@uni-passau.de
Doctorate
*Cognitive Sciences;
Medical Sciences, Instruments
and Techniques; Instruments and
Techniques; Institutions*

Gunn, Jennifer L.
University of Minnesota
Program in History of Medicine
420 Delaware St SE
Minneapolis, MN 55455-0392
Phone: 1.612.624.1909
Fax: 1.612.625.7938
gunnx005@tc.umn.edu

Gunnoe, Charles D., Jr.
Aquinas College
1607 Robinson Road
Grand Rapids, MI 49506
Phone: 1.616.459.8281
Fax: 1.616.245.0797
gunnocha@aquinas.edu
Doctorate, 1998
*14c-16c, Europe, Medical
Sciences*

Guntau, Martin
Ernst Alban Gesellschaft
Wokrenter Str. 40

Rostock D-18055
Germany
Phone: 381.4907511
Fax: 381.4907511
guntau@metroner.de
Doctorate, 1964
*18c, Europe, Earth Sciences;
19c, Europe, Institutions;
19c, Europe, Social Relations of
Science*

Gur, Bekir S.
925 E Magnolia Dr Apt F7
Tallahassee, FL 32301-6604
Phone: 1.850.574.8541
bbg6746@garnet.acns.fsu.edu

Guralnick, Stanley
921 15th St
Boulder, CO 80302-7311
Phone: 1.212.799.4813
Fax: 1.212.799.4813

Gutzwiller, Martin C.
370 Riverside Dr
New York, NY 10025-2179
moongutz@aol.com

Guzon, Jose L.
Inspectoria Salesiana
Avda Antibioticos, 126
Leon 24080
Spain
Phone: 947.205665
Fax: 947.205665
dtormacionsm@planalk.es
*20c, Europe, Astronomical
Sciences, Philosophy and
Philosophy of Science;
14c-16c, North America,
Physical Sciences; 17c, Europe,
Philosophy and Philosophy of
Science; 400 B.C.E.-400 C.E.,
Europe, Astronomical Sciences*

Haake, Paul
Wesleyan University
MB & B
Middletown, CT 06459
phaake@wesleyan.edu
Doctorate, 1960
*19c, Physical Sciences;
Medical Sciences;
Biological Sciences*

Haas, John W., Jr.
Gordon College
Wenham, MA 01984
haasj@mediaone.net
Doctorate, 1957
*19c, Europe, Social Relations
of Science; 20c, Europe,
Humanistic Relations of
Science; 20c, Instruments and
Techniques; Biological Sciences,
Instruments and Techniques*

Haase, Wolfgang G.
Boston University
Inst Classical Tradition
745 Commonwealth Ave
Boston, MA 02215-1401
Phone: 1.617.353.7378
Fax: 1.617.566.4033
Doctorate, 1971
*400 B.C.E.-400 C.E., Philosophy
and Philosophy of Science;
400 B.C.E.-400 C.E., Social
Sciences; 14c-16c, Philosophy
and Philosophy of Science;
14c-16c, Social Sciences*

Haave, Neil C.
Augustana University College
Biology, Chemistry &
Geography
4901- 46th Avenue
Camrose, AB T4V2R3
Canada
Phone: 1.780.679.1506
Fax: 1.780.679.1590
haavn@augustana.ab.ca
Doctorate, 1991
Biological Sciences

Habraken, Clarisse L.
Univ of Leiden
Gorlaeus Laboratory
PO Box 9502
Leiden 2300 RA
The Netherlands
Phone: 31.71.5274291
Fax: 31.71.5274537
habraken@chem.leidenuniv.nl
Doctorate
*20c, North America,
Physical Sciences;
20c, Europe, Education;
20c, Europe, Physical Sciences;
20c, North America, Education*

Hacker, Barton C.
150 12th St NE
Washington, DC 20002
Phone: 1.202.544.3084
Fax: 1.202.357.1553
hackerb@nmah.si.edu
Doctorate, 1968
*Technology, Instruments and
Techniques; Technology*

Hafer, Andreas
Werderstr 46
Schorndorf 73614
Germany
andreas.hafer@gmx.net
Doctorate
*17c, Mathematics;
19c, Technology;
17c, Philosophy and Philosophy
of Science*

Hagan, William J., Jr.
The College of Saint Rose
School of Math & Sciences
432 Western Ave
Albany, NY 12203-1490
Phone: 1.518.454.5225
Fax: 1.518.458.5446
haganw@mail.strose.edu
Doctorate, 1985
Education; Physical Sciences

Hagen, Joel B.
Radford University
Biology Department
Box 6931
Radford, VA 24142
Phone: 1.540.951.0006
Fax: 1.540.831.5129
jhagen@radford.edu

Hagner, Michael
Max Planck Institute for the
History of Science
Wilhelmstrasse 44
Berlin D-10117
Germany
Phone: 49.30.22667163
Fax: 49.30.22667169
hagner@mpiwg-berlin.mpg.de
Doctorate, 1987
*19c, Europe, Medical Sciences;
Europe, Social Sciences;
19c, Europe, Medical Sciences,
Instruments and Techniques;
19c, Europe, Natural and
Human History*

Hahm, David E.
Ohio State University
Dept of Greek & Latin
414 University Hall
210 N Oval Mall
Columbus, OH 43210-1319
Phone: 1.614.292.2724
Fax: 1.614.292.7835
hahm.1@osu.edu
Doctorate, 1966
*400 B.C.E.-400 C.E.,
Astronomical Sciences;
400 B.C.E.-400 C.E.,
Philosophy and Philosophy of
Science; 400 B.C.E.-400 C.E.,
Humanistic Relations of Science*

Hahn, Nan L.
322 Second St
Dunellen, NJ 08812-1116
Phone: 1.732.752.5841
nanhan@compuserve.com
Doctorate, 1972
*5c-13c, Europe, Astronomical
Sciences; 5c-13c, Europe,
Mathematics; 5c-13c, Europe,
Natural and Human History;
5c-13c, Europe, Technology*

Hahn, Roger
Univ of California - Berkeley
Dept of History #2550
Berkeley, CA 94720-2550
Phone: 1.510.642.5199
Fax: 1.510.643.5323
rhahn@socrates.berkeley.edu
Doctorate, 1962
*18c, Europe, Physical Sciences;
18c, Europe, Astronomical
Sciences;
17c, Europe, Institutions;
19c, Europe, Education*

Haigh, Elizabeth V.
St Mary's University
Dept of History
Robie Street
Halifax, NS B3H3C3
Canada
Phone: 1.902.420.5762
Fax: 1.902.420.5414
elizabeth.haigh@stmarys.ca
Doctorate, 1973
*20c, Europe, Natural and
Human History; 20c, Europe,
Institutions; 19c, North America,
Natural and Human History*

Hakata, Toshio
331-8-217 Toyoshiki
Chiba-Ken
Kashiwa-shi
Japan
Phone: 81471447270
thakata@typhoon.mc-jma.ac.jp

Hall, Bert S.
University of Toronto
316 Victoria College
91 Charles St W
Toronto, ON M5S1K7
Canada
Phone: 1.416.924.2724
Fax: 1.416.978.3003
bert.hall@utoronto.ca
Doctorate
*5c-13c, Europe, Technology;
17c, Europe, Technology;
14c-16c, Europe, Technology;
14c-16c, Europe, Humanistic
Relations of Science*

Hall, Brian K.
Dalhousie University
Dept of Biology
Halifax, NS B3H4J1
Canada
Phone: 1.902.494.3522
Fax: 1.902.494.3736
bkh@is.dal.ca
Doctorate
19c, Europe, Biological Sciences

Hall, Karl P.
Dibner Institute, MIT
E56-100
38 Memorial Dr
Cambridge, MA 02142-1347
Phone: 1.617.452.4508
khall@dibinst.mit.edu
Bachelor's Level, 1999
*20c, Physical Sciences;
20c, Europe, Physical Sciences;
20c, Europe, Institutions;
20c, Europe, Social Relations of
Science*

Hall, Robert D.
Rabat American School
BP 120
Rabat
Morocco
Phone: 212.37.67.14.76
Fax: 212.37.69.09.63
bhall@ras.edu.ac.ma

Doctorate, 1998
20c, North America,
Astronomical Sciences;
19c, Africa, Education;
Europe; Asia

Hall, Robert E.
28 Balmoral Court
Belfast BT97GR
Northern Ireland
Phone: 44.1232660496
r.hall@qub.ac.uk
Master's Level, 1968
5c-13c; 5c-13c, Europe;
400 B.C.E.-400 C.E.;
Physical Sciences

Hallyn, Fernand H.
University of Ghent
Blandijnberg 2
Ghent B-9000
Belgium
Phone: 329.3747227
Fax: 92644174
fernand.hallyn@rug.ac.be
Doctorate, 1974
17c, Europe, Humanistic
Relations of Science;
18c, Astronomical Sciences;
14c-16c

Halpern, Paul
University of the Sciences in
Philadelphia
600 S 43rd St
Philadelphia, PA 19104-4495
Phone: 1.215.596.8913
Fax: 1.215.895.1112
p.halper@usip.edu

Hamblin, Jacob D.
777 K Madrona Walk
Goleta, CA 93117-3520
Phone: 1.805.971.1735
jdh0@umail.ucsb.edu
Doctorate, 2001
20c, Earth Sciences;
20c, Biological Sciences;
20c, Science Policy;
20c, North America

Hamerla, Ralph R.
University of Oklahoma
Honors College
1300 Asp Ave
Norman, OK 73019-6061
Phone: 1.405.325.9286

Master's Level
19c, North America, Physical
Sciences

Hamilton, Kelly
6409 Long Lake Rd
Berrien Springs, MI 49103-9613

Hamin, Mark
Univ of Massachusettes Amherst
109 Hills N 111 Infirmary Way
75 State St
Amherst, MA 01002-1116
Phone: 1.413.545.2255
Fax: 1.413.545.1772
mhamin@larp.umass.edu

Hamlin, Christopher S.
Univ of Notre Dame
Dept of History
Notre Dame, IN 46556
Phone: 1.219.631.5092
Fax: 1.219.631.8209
hamlin.1@nd.edu
Doctorate, 1982
19c, Europe, Medical Sciences;
18c, Europe, Biological Sciences;
19c, Europe, Biological Sciences;
19c, Europe, Technology

Hamm, Ernst
York University
Science & Technology Studies
Atkinson Fac
4700 Keel St
Toronto, ON M3J1P3
Canada
Phone: 1.416.736.2100
Fax: 1.416.736.5188
ehamm@yorku.ca
Doctorate, 1991
Europe, Earth Sciences;
Humanistic Relations of Science;
Social Relations of Science;
Natural and Human History

Hammond, Babi A.
University of Pennsylvania
Hist & Sociology of Sci Dept
Logan Hall
Philadelphia, PA 19104-6304
Phone: 1.215.898.4643
babih@sas.upenn.edu

Hammonds, Evelynn
MIT
Program in Science

E51-296A
77 Massachusetts Ave
Cambridge, MA 02139-4307
Phone: 1.617.253.8780
Fax: 1.617.258.8118
eveham@mit.edu

Hamor, Glenn H.
6519 W 87th St
Los Angeles, CA 90045-3716

Hancock, James
1150 W Covered Bridge Rd
Columbia, MO 65203-9571
Phone: 1.573.874.0120
jrha@mindspring.com

Hankins, Thomas L.
University of Washington
Dept of History D-20
Box 353560
Seattle, WA 98195-3560
Phone: 1.206.543.5790
Fax: 1.206.543.9451
hankins@u.washington.edu
Doctorate, 1964
Biography; Mathematics;
Physical Sciences;
Instruments and Techniques

Hanley, James
University of Winnipeg
515 Portage Ave.
Winnipeg, MB R3B2E9
Canada
Phone: 1.204.786.9005
j.hanley@uwinnipeg.ca

Hannaway, Caroline C.
NIH Historical Office
Building 31 Room 2b09
Bethesda, MD 20892
Phone: 1.301.496.6610
Fax: 1.301.402.1434
hannawac@od.nih.gov
Doctorate, 1974
20c, North America,
Medical Sciences;
18c, Europe, Medical Sciences

Hannaway, Owen
316 Suffolk Rd
Baltimore, MD 21218-2521
Phone: 1.410.366.5633
Doctorate, 1965
14c-16c, Physical Sciences;
Humanistic Relations of Science

Hansen, Bert
Baruch College
Box A1610, Dept. of History
17 Lexington Avenue
New York, NY 10010
Phone: 1.212.675.2040
Fax: 1.212.675.2040
bert_hansen@baruch.cuny.edu
Doctorate, 1974
19c, North America, Medical Sciences; 20c, North America, Medical Sciences; Gender and Science; North America

Hansen, Lee Ann
Calif State Univ Fullerton
Department of Liberal Studies
PO Box 6868
Fullerton, CA 92834
Phone: 1.714.826.7769
Fax: 1.714.278.5820
lhansen@fullerton.edu
Doctorate, 1985
18c, Europe, Medical Sciences, Instruments and Techniques; 19c, Europe, Social Relations of Science; 19c, Europe, Philosophy and Philosophy of Science; 19c, Europe, Biological Sciences

Hanson, Elizabeth A.
661 Oak Tree Rd
Palisades, NY 10964-1535
Phone: 1.212.327.7123
Fax: 1.212.327.8123
hansone@mail.rockefeller.edu
Doctorate, 1996
20c, North America, Biological Sciences; 20c, North America, Natural and Human History

Hanson, Marta E.
Univ of Calif, San Diego
9520 Gilman Dr.
Dept of History 0104
San Diego, CA 92093-0104
Phone: 1.619.822.0586
Fax: 1.619.534.7382
mehanson@ucsd.edu
Doctorate
18c, Asia, Medical Sciences; 19c, Asia, Medical Sciences; 20c, Social Relations of Science

Hara, Sumio
Sakaine 3-13-6
Chuiba-ken
Kashiwa-shi 277-0053
Japan
Phone: 0471.74.5858
su-hara@allnet.ne.jp

Hard, Mikael S.
Tech Univ Schloss
Dept of History
Darmstadt D-64283
Germany
Phone: 49.6151.163097
Fax: 49.6151.163992
hard@its.tu.darmstadt.de

Harden, Victoria A.
National Institute of Health
Bldg. 31 Rm 5B38 MSC 2092
31 Center Drive
Bethesda, MD 20892-2092
Phone: 1.301.496.6610
Fax: 1.301.402.1434
vharden@att.net
Doctorate, 1983
20c, North America, Instruments and Techniques; Pre-400 B.C.E., North America, Science Policy; 20c, North America, Medical Sciences; 20c, North America, Institutions

Haring, Kristen A.
Harvard University
History of Science
Science Ctr 235, 1 Oxford St
Cambridge, MA 02138-3800
Phone: 1.617.495.8758
Fax: 1.617.495.3344
haring@fas.harvard.edu
Master's Level
20c, North America, Technology; Instruments and Techniques; Mathematics

Harkness, Jon M.
Cornell University
Dept. of Sci & Tech Studies
726 University Avenue
Ithaca, NY 14850
Fax: 1.607.272.5567
jonmharkness@aol.com
Doctorate, 1996
20c, North America; 20c, North America, Medical Sciences;

20c, North America, Philosophy and Philosophy of Science; 20c, North America, Institutions

Harmon, Joseph E.
Argonne National Labortory
9700 S Cass Avenue
Chemical Technology Division
Argonne, IL 60439-4837
Phone: 1.630.252.7535
harmon@cmt.anl.gov
Master's Level, 1975
Social Relations of Science; Physical Sciences

Harper, Kristine C.
946 NW Circle Blvd # 306
Corvallis, OR 97330-1410
Phone: 1.541.926.3564
kharper@proaxis.com
Master's Level, 1985
20c, Earth Sciences; 20c, Education

Harrington, Jonathan E.
110 Kinnaird St
Cambridge, MA 02139-2914
Phone: 1.617.354.8822
torjon@ix.netcom.com
Doctorate
400 B.C.E.-400 C.E., Biological Sciences; 5c-13c, Biological Sciences; 14c-16c, Biological Sciences

Harris, Benjamin
University of New Hampshire
Conant Hall
Durham, NH 03824
Phone: 1.603.862.4107
Fax: 1.603.862.4986
bh5@cisunix.unh.edu
Doctorate, 1975
20c, North America, Cognitive Sciences; 20c, North America, Medical Sciences, Instruments and Techniques; 20c, Europe, Medical Sciences, Instruments and Techniques

Harris, Cory
3690 N Country Club Rd # 1016
Tucson, AZ 85716-1294
Phone: 1.520.323.3416
cdharris@u.arizona.edu

Harris, Patricia F.
Univ of Wisconsin Hospital
600 N. Highland Avenue
J5/ 230 CSC Dept. of Medicine
Madison, WI 53792
Phone: 1.608.256.1901
pf.harris@hosp.wisc.edu
Doctorate, 1995
North America, Medical
Sciences; Medical Sciences

Harris, Steven F.
110 Central St Apt 22
Wellesly Hills, MA 02481-5740
Phone: 1.781.431.9760
Fax: 1.617.495.3344
sjharris@fas.harvard.edu
Doctorate, 1989
17c, Europe, Physical Sciences;
Earth Sciences; Natural and
Human History; Biography

Harrison, Peter
Bond University
Philosophy Department
Gold Coast, QLD 4229
Australia
Phone: 61.7.5595.2519
Fax: 61.7.5595.2545
peter_harrison@bond.edu.au

Hars, Florian
DESY
Hamburg 22607
Germany
florian@hars.de
20c, Europe, Physical Sciences;
20c, Institutions; Social Sciences;
20c, Earth Sciences

Harvey, Joy
29 Kidder Ave
Somerville, MA 02144-2005
Phone: 1.617.628.1303
jharvey368@aol.com

Harwood, Jonathan
Univ of Manchester
CHSTM
Mathematics Tower
Manchester M13 9PL
Great Britain
Phone: 44.161.275.5923
Fax: 44.161.275.5699
jonathan.harwood@man.ac.uk
Doctorate, 1970
20c, Europe, Medical Sciences;

North America,
Biological Sciences; Europe,
Biological Sciences;
Social Sciences

Hashagen, Ulf
Deutsches Museum
Museumsjnsel 1
Muenich D-80538
Germany
Phone: 49.89.2179453
Fax: 49.89.2179239
uhashagen@
    deutsches-museum.de
Master's Level
19c, Europe, Mathematics;
20c, Europe, Mathematics;
20c, Europe, Science Policy;
Europe, Education

Hashimoto, Keizo
Kansai University
3-3-35 Yamate Cho Suita Shi
Faculty of Sociology
Suita Osaka 5648680
Japan
Phone: 81.6.6.3680746
Fax: 81.75.643.0994
keizo@ipcku.kansai-u.ac.jp
Doctorate, 1987
17c, Asia, Astronomical Sciences;
400 B.C.E.-400 C.E., Asia,
Astronomical Sciences;
20c, Asia, Science Policy;
20c, Asia, Social Relations of
Science

Hashimoto, Takehiko
University of Tokyo
Research Center for Advanced
Science and Technology
4-6-1 Komaba Meguro-ku
Tokyo 153-8904
Japan
Phone: 81.3.5452.5295
Fax: 81.3.5452.5299
hasimoto@rcast.u-tokyo.ac.jp
Doctorate, 1991
Technology, Instruments and
Techniques; Physical Sciences

Hassinger, Bill, Jr.
305 W Avondale Dr
Greensboro, NC 27403-1049
Phone: 1.336.274.4623
Fax: 1.336.274.4624

Hatch, Robert A.
University of Florida
226 Keene-Flint Hall
Gainesville, FL 32611
Phone: 1.352.392.0271
Fax: 1.352.392.6927
ufhatch@ufl.edu
Doctorate, 1978
17c; 17c, Astronomical Sciences;
17c, Physical Sciences;
Social Sciences

Hatfield, Gary C.
Univ of Pennsylvania
Dept of Philosophy
433 Logan Hall
249 S 36th St
Philadelphia, PA 19104-6304
Phone: 1.215.898.6346
hatfield@linc.cis.upenn.edu
Doctorate, 1979
Cognitive Sciences;
Physical Sciences

Hattab, Helen
813 S University Ave Apt F
Carbondale, IL 62901-2832
Phone: 1.618.529.4558
hhattab@siu.edu

Haubold, Hans J.
Vienna International Centre
Room E0945
UN Outer Space Office
PO Box 500
Vienna A-1400
Austria
Phone: 43.1.260.5778
Fax: 43.1.26060.5830
haubold@kph.tuwien.ac.at
Doctorate
20c, Astronomical Sciences;
20c, Mathematics;
20c, Physical Sciences

Hauger, James S.
AAAS
1200 New York Avenue NW
Washington, DC 20005
Phone: 1.202.326.6452
Fax: 1.202.289.4950
shauger@aaas.org
Doctorate, 1995
20c, North America,
Science Policy;
20c, North America, Technology

Haugland, Karen M.
69 Tenison Rd
Cambridge CB12DG
Great Britain
Phone: 47.22.854204
k.m.haugland@hi.vio.no

Hausman, Gary J.
110 West Stinson Street
Chapel Hill, NC 27516-2146
Phone: 1.919.942.0407
ghausman@mac.com

Hawkins, Michael
Kingston University
School of Social Science
Penrhyn Rd
Kingston KT1 2EE
Great Britain
Phone: 44.2.8547.2000
m.hawkins@kingston.ac.uk

Hawley, John K.
479 Lake Rd
Delanson, NY 12053-4305
Phone: 1.518.864.5171
jkhawley@mybizz.net

Hay, Joyce
Canada Science & Tech Museum
Library
PO Box 9724 Ottawa Terminal
Ottawa, ON K1G5A3
Canada
Phone: 1.613.991.5701
Fax: 1.613.990.3636
library@nmstc.ca

Hayes, Elizabeth
University of Notre Dame
Dept of History & Phil of Sci
346 O Shaugnessy Hall
Notre Dame, IN 46556-5639
Phone: 1.219.631.5015
elizabeth.e.hayes.39@nd.edu

Hayes, Melinda K.
Univ of Southern California
Special Collections, ISD
Doheny Memorial Library
Los Angeles, CA 90089-0182
Phone: 1.213.740.5141
Fax: 1.213.740.2343
melindah@usc.edu
Master's Level, 1982
19c, Europe, Natural and

Human History; 19c, North
America, Natural and Human
History; Education

Haynes, Douglas M.
University of California
Dept of History
Irvine, CA 92697-3275

Hays, Jo N.
Loyola University
Department of History
6525 N Sheridan Rd
Chicago, IL 60626-5385
Phone: 1.773.508.2233
Fax: 1.773.508.2153
jhays@wpo.it.luc.edu
Doctorate, 1970
19c, Medical Sciences; Medical
Sciences; Biological Sciences

Hayton, Darin
University of Notre Dame
346 O'Shaughnessy Hall
Notre Dame, IN 46556
Phone: 1.219.631.5015
darin.hayton.1@nd.edu
14c-16c, Astronomical Sciences;
Humanistic Relations of Science;
Medical Sciences;
18c, Natural and Human History

Hayward, Rhodri
University of East Anglia
Wellcome Unit Hist of Medicine
School of History
Norwich NR4 7TJ
Great Britain
Phone: 1603.593657
r.hayward@wellcome.ac.uk

Heaney, Peter J.
Pennsylvania State University
Dept of Geosciences
309 Deike Bldg
University Park, PA 16802-2712
Phone: 1.814.865.6821
Fax: 1.814.863.7823
heaney@geosc.psu.edu
Doctorate, 1989
Earth Sciences; Physical
Sciences; North America

Hedrick, Elizabeth A.
Univ. of Texas at Austin
Dept. of English
108 Parlin Hall

Austin, TX 78712-1164
Phone: 1.512.471.8705
eahedrick@mail.utexas.edu
Doctorate, 1986
Social Sciences; 17c, Europe,
Physical Sciences; Gender and
Science; Humanistic Relations
of Science

Heelan, Patrick A.
Georgetown University
Dept of Philosophy
Washington, DC 20057
Phone: 1.202.687.5222
Fax: 1.202.687.8039
heelanp@georgetown.edu
Doctorate, 1952
20c, Philosophy and
Philosophy of Science;
20c, Physical Sciences;
20c, Humanistic Relations of
Science; 20c, Cognitive Sciences

Heidarzadeh, Tofigh
490 Lucera Ct, #203
Phillips Ranch, CA 91766
Phone: 1.909.622.2424
tofigh60@hotmail.com

Heidelberger, Michael
Universität Tübingen
Philosophisches Seminar
Tübingen 72070
Germany
michael.heidelberger@
  rz.hu-berlin.de
19c, Europe, Physical Sciences;
20c, Europe, Philosophy
and Philosophy of Science;
19c, Europe, Biological Sciences;
19c, Europe, Cognitive Sciences

Heilbron, John L.
April House
Shilton
Burford OX18 4AB
Great Britain
Phone: 44.19.9384.0786
Fax: 44.19.9384.0786
john.heilbron@dial.
  appleinter.net
Doctorate, 1964
17c, Europe, Physical Sciences;
18c, North America, Institutions;
19c, Astronomical Sciences;
20c, Earth Sciences

Heinecke, Berthold
Zum Galgenberg 2
Flechtingen
Am Bahnhof D-39345
Germany
heinecke_flechtingen@
  compuserve.com

Heinrich, Inge
Astronomisches Rechen-Institut
Astronomy & Astrophysics Abst
Moenchhofstr 12-14
Heidelberg 69120
Germany

Heinzmann, Bernd
Leonhardsberg 17
Augsburg 86150
Germany
Phone: 49821516288
bernd.heinzmann@
  newsfactory.net

Helfand, William H.
2 Sutton Pl S # 3-C
New York, NY 10022-3070
Phone: 1.212.758.4158
Fax: 1.610.847.6923
whelfand@aol.com
Bachelor's Level
Medical Sciences;
Humanistic Relations of Science

Hellyar, Kenneth G.
158 East Pkwy
Rochester, NY 14617-3704
19c, Physical Sciences;
19c, Technology; Technology,
Instruments and Techniques

Helms, Douglas
398 N Edison St
Arlington, VA 22203
Phone: 1.703.525.1468
Fax: 1.202.720.6473
douglas.helms@usda.gov
Doctorate, 1977
Biological Sciences

Hempstead, Colin A.
University of Teesside
Borough Rd
Middlesbrough TS1 3BA
Great Britain
Phone: 1325483439
Fax: 441325483439
colin.hempstead@ntlworld.com

Doctorate
19c, Technology,
Instruments and Techniques;
20c, Physical Sciences;
Technology

Henderson, Diane
8910 Celia Rd
Tallahassee, FL 32305-0726
Fax: 1.850.421.6393
rdh8268@garnet.acns.fsu.edu
Doctorate, 1999
Philosophy and Philosophy
of Science; Science Policy;
20c, Instruments and Techniques

Henderson, Linda D.
University of Texas
Dept of Art & Art History
Austin, TX 78712
Phone: 1.512.471.7757
dnehl@mail.utexas.edu

Hendrick, Robert M.
St Johns Univesity
Bent Hall
8000 Utopia Parkway
Jamaica, NY 11439
Phone: 1.718.990.7396
Fax: 1.718.990.1882
henricr@stjohns.edu

Hendricker, David
Ohio University
Chemistry Department
Clippinger Labs
Athens, OH 45701
Phone: 1.740.593.3729
hendrick@ohio.edu

Henninger-Voss, Mary
1704 Kent Rd
Camp Hill, PA 17011-6016
Phone: 1.717.761.5191
mhenninger-voss@prodigy.net

Henson, Pamela M.
Smithsonian Inst Archives
Arts & Industries 2135
Washington, DC 20560-0414
Phone: 1.202.786.2735
Fax: 1.202.357.2395
hensonp@osia.si.edu
Doctorate, 1990
19c, North America, Natural
and Human History; 20c, North
America, Natural and Human

History; 19c, North America,
Biological Sciences; 19c, North
America, Gender and Science

Hentschel, Klaus
Postfach 2216
Göttingen D-37012
Germany
Phone: 49.4938.596
Fax: 49.551.377.330
Doctorate, 1989
Physical Sciences;
Astronomical Sciences;
Philosophy and Philosophy of
Science

Herbert, Sandra
University of Maryland, Balt
Department of History
1000 Hilltop Cir
Baltimore, MD 21250-0002
Phone: 1.410.455.1045
Fax: 1.410.455.1045
herbert@umbc.edu
Doctorate
19c, Europe, Natural and
Human History; Biological
Sciences; Earth Sciences

Herman, Gerald
Northeastern University
Dept of History
249 Meserve Hall
Boston, MA 02115
Phone: 1.617.373.4441
Fax: 1.617.373.2661
gherman@lynx.neu.edu
Master's Level, 1967
20c, Europe, Humanistic
Relations of Science;
17c, Europe, Humanistic
Relations of Science;
19c, Europe, Humanistic
Relations of Science;
19c, Europe, Technology

Hermann, Kenneth
313 Highland Ave
Kent, OH 44240-2522

Herr, Melody R.
Johns Hopkins University
History of Science
2715 N Charles St
Baltimore, MD 21218-4319
Fax: 1.410.576.6968
mh@press.jhu.edu

Bachelor's Level
20c, North America, Natural
and Human History;
20c, North America,
Medical Sciences;
20c, North America, Biological
Sciences; 20c, Exploration,
Navigation and Expeditions

Herries-Davies, Gordon L.
Trinity College
Dublin 2
Ireland
Doctorate, 1967
17c, Europe, Earth Sciences;
18c, Europe, Earth Sciences;
19c, Europe, Earth Sciences;
20c, Institutions

Herron, Timothy
54 Tamalpais Rd
Berkeley, CA 94708-1949
Phone: 1.510.704.9103
tjh_synechism@yahoo.com

Hessenbruch, Arne
Dibner Institute
MIT E56-100
38 Memorial Dr
Cambridge, MA 02139
Phone: 1.617.253.1332
Fax: 1.617.258.7483
arne@mit.edu
Doctorate, 1995
Physical Sciences; Medical
Sciences; Social Relations of
Science; Education

Hevly, Bruce
University of Washington
Department of History
Box 353560
Seattle, WA 98195-3560
Phone: 1.206.543.9417
Fax: 1.206.543.9451
bhevly@u.washington.edu

Hewlett, Richard G.
History Associates Inc.
5 Choke Cherry Road Suite 280
Rockville, MD 20850-4004
Phone: 1.301.670.0076
rghewlett@compuserve.com
Doctorate, 1952
20c, North America, Physical

Sciences; Europe, Technology,
Instruments and Techniques;
Social Sciences; Institutions

Hibner Koblitz, Ann
Arizona State University
Women's Studies Program
PO Box 873404
Tempe, AZ 85287-3404
Phone: 1.480.965.8483
Fax: 1.480.965.2357
koblitz@asu.edu
Doctorate, 1983
19c, Europe, Gender and Science;
20c, Gender and Science

Hickel, Erika
Can Pelat
Serralongue 66230
France
Phone: 33.46.839.6334
Doctorate, 1963
14c-16c, Europe, Medical
Sciences; 17c, Medical Sciences;
Physical Sciences

Hiebert, Erwin N.
Harvard University
Widener Library 172
Cambridge, MA 02138
Phone: 1.617.495.0325
Fax: 1.617.495.3344
ehiebert@fas.harvard.edu
Doctorate, 1954
19c, Europe, Physical Sciences;
19c, Europe, Humanistic
Relations of Science

Higby, Gregory
AIHP
Rennebohm Hall
777 Highland Ave
Madison, WI 53705-2222
Phone: 1.608.256.0065
gjh@pharmacy.wisc.edu

Higdon, Mark
1200 Barton Hills Dr Apt 106
Austin, TX 78704-1901

Hijioka, Yoshito
3-13-21-304 Kita-Karasuyama
Stagayaku-ku
Tokyo 157-0061
Japan
fwin1595@mb.infoweb.ne.jp

Master's Level
17c, Europe, Physical Sciences;
18c

Hildebrand, Reinhard
Institüt für Anatomic Studien
Vesalius 2-4
Münster D-48149
Germany
Phone: 49.251.8355.23013
Fax: 49.251.835.2369
hildebra@uni-muenster.de

Hildebrecht, Douglas
Roelof Hartstraat 17-I I I
Amsterdam 1071 VG
The Netherlands
Phone: 31.2.0683.4695
dough@planet.nl

Hilfstein, Erna K.
Graduate School of CUNY
32 W. 42nd St Rm 1548
New York, NY 10036-8099
Phone: 1.212.817.8383
Doctorate, 1978
400 B.C.E.-400 C.E., Europe,
Astronomical Sciences;
5c-13c, Europe, Astronomical
Sciences; 14c-16c, Europe,
Astronomical Sciences;
400 B.C.E.-400 C.E., Europe

Hilgeman, Cecilia
11049 S Green Bay Ave
Chicago, IL 60617-6904

Hill, Benjamin
Dept of Philosophy
269 EPB
Iowa City, IA 52242
Phone: 1.319.341.5856
benjamin-hill@uiowa.edu

Hill, David K.
3816 45th Street Ct
Rock Island, IL 61201-7136

Hillier, Anna
18 Spring St
Lexington, MA 02421-7958

Hilts, Victor L.
Univ. of Wisconsin - Madison
Dept. of History of Science
Madison, WI 53706
Phone: 1.608.252.1406

Himrod, David K.
United Library
2121 Sheridan Rd
Evanston, IL 60201
Phone: 1.847.866.3910
dhimrod@garrett.edu
Doctorate, 1977
17c, North America, Humanistic
Relations of Science;
19c, Europe, Social Relations of
Science; 20c

Hine, William L.
York University
Atkinson College
4700 Keele St
Toronto, ON M3J1P3
Canada
wlhine@yorku.ca
Doctorate, 1967
17c, Physical Sciences;
14c-16c, Humanistic Relations
of Science; 17c, Education;
17c, Philosophy and Philosophy
of Science

Hinz, James A.
435 Riverview Rd
Swarthmore, PA 19081-1223

Hiromasa, Naohiko
Research Inst of Civilization
1117 Kitakaname
Hiratsuka
Kanagawa 259-1292
Japan
Phone: 46.358.1211
Fax: 46.359.4047
naohiko@keyaki.cc.u-tokai.ac.jp
Bachelor's Level
19c, Asia, Philosophy and
Philosophy of Science;
19c, Europe, Physical Sciences

Hirsh, Richard F.
Virginia Tech
Dept of History
Mail Code 0117
Blacksburg, VA 24061
Phone: 1.540.231.5331
Fax: 1.540.231.8724
richards@vt.edu
Doctorate, 1979
20c, North America, Physical
Sciences; 20c, Technology

Hirshfeld, Alan
University of Massachusetts
Physics Dept
285 Old Westport Rd
N Dartmouth, MA 02747-2300
Phone: 1.508.999.8715
Fax: 1.617.244.9303
ahirshfeld@umassd.edu
Doctorate, 1978
Astronomical Sciences;
Technology

Hirukawa, Masakazu
Ida-Cho, Okazaki-Shi
27-60 Higashi-Konba
Aichiken 444 MZ
Japan

Hiskes, Anne L.
Univ of Connecticut
Dept of Philosophy
344 Mansfield Rd
Storrs, CT 06269-2054
Phone: 1.860.486.3676
Fax: 1.860.486.0387
anne.hiskes@uconn.edu
Doctorate, 1981
Astronomical Sciences,
Philosophy and Philosophy
of Science; Philosophy
and Philosophy of Science;
Gender and Science

Hobart, Willis
NMFS Scientific Publications
7600 Sand Point Way NE
Seattle, WA 98115-6349
Phone: 1.206.526.6107
Fax: 1.206.526.4456
scientific.publications@noaa.gov

Hochheiser, Sheldon H.
A T & T
Room 2317G2
295 N Maple Avenue
Basking Ridge, NJ 07920
Fax: 1.908.221.3510
hochheiser@att.com
Doctorate, 1982
North America, Technology;
North America

Hocking, Richard
PO Box 30
Madison, NH 03849-0030
Phone: 1.603.367.4369

Hoddeson, Lillian
University of Illinois
Dept of History
309 Gregory Hall
Urbana, IL 61801
Phone: 1.217.244.8412
Fax: 1.217.344.5499
hoddeson@uiuc.edu
Doctorate
Physical Sciences; Technology

Hodges, Herbert J.
1725 E 11th St
Davenport, IA 52803-3904
Phone: 1.319.323.7701
Doctorate
400 B.C.E.-400 C.E., North
America, Natural and Human
History; 5c-13c, North
America, Biological Sciences;
14c-16c, Asia, Humanistic
Relations of Science

Hodges, Linda C.
4 Branchwood Ct
Lawrenceville, NJ 08648-1057
Phone: 1.404.471.6382
lhodges@agnesscott.edu

Hoecker-Drysdale, Susan
Concordia University
7141 Sherbrooke Street West
L-VE 223
Montreal, PQ H4B1R6
Canada
Phone: 1.514.848.2160
Fax: 1.514.848.4548
hoecker@alcor.concordia.ca
Doctorate
19c, Europe, Social Sciences;
19c, Europe, Gender and
Science; 19c, North America,
Social Sciences

Hofer, Veronika
Institute Vienna Circle
Museumsstrae 5/2/19
Vienna A-1070
Austria
Phone: 43.1.526.1005
Fax: 43.1.524.88.59
veronika.hofer@eunet.at

Hoffer, Peter T.
Univ of Sci in Philadelphia
600 S 43rd St
Philadelphia, PA 19104-4495

Phone: 1.215.596.8905
Fax: 1.610.259.5244
p.hoffer@usip.edu
Doctorate, 1975
20c, Europe, Cognitive Sciences

Hoffman, Joyce M.
83 Grozier Rd
Cambridge, MA 02138-3314
Doctorate
Cognitive Sciences;
Medical Sciences, Instruments
and Techniques

Hoffmann, Christopher
Christburger Strasse 38
Berlin 10405
Germany
Phone: 49.30.4405.6911
hoffmann@
  mpiwg-berlin.mpg.de

Hoffmann, Dieter
Am Falkplatz 4
Berlin D-10437
Germany
Phone: 49.30.449.6983
Fax: 49.30.2667299
dh@mpiwg-berlin.mpg.de

Hoffmann, Kathryn
University of Hawaii - Manoa
Dept LLEA - Moore 483
1890 East West Rd
Honolulu, HI 96822-2318
Phone: 1.308.956.4170
Fax: 1.808.956.9536
hoffmann@hawaii.edu

Hoffmann, Roald
Cornell University
Baker Laboratory
Department of Chemistry
Ithaca, NY 14853
Phone: 1.607.255.3419
Fax: 1.607.255.5707
rh34@cornell.edu

Hofmann, James R.
Calf State Univ-Fullerton
Philosophy Department
PO Box 6868
Fullerton, CA 92834
Phone: 1.714.278.7049
jhofmann@fullerton.edu

Hogan, Edward R.
East Stroudsburg University
E Stroudsburg, PA 18501
Phone: 1.570.422.3445
ehogan@esu.edu

Hogendijk, Jan P.
University of Utrecht
Dept. of Mathematics
PO Box 80010
Utrecht 3508 TA
The Netherlands
Phone: 31.30.253.3697
Fax: 31.30.251.8394
hogend@math.uu.nl
Doctorate, 1983
5c-13c, Mathematics;
400 B.C.E.-400 C.E.,
Europe, Mathematics;
5c-13c, Astronomical Sciences;
5c-13c, Asia, Astronomical
Sciences

Hogwood, B. M.
119 Crabble Hill
Dover Kent CT17 0SB
Great Britain
Phone: 44.1304.826.188

Holbrow, Charles H.
7115 Spring Hill Rd.
Hamilton, NY 13346
Phone: 1.315.228.7206
Fax: 1.315.228.7187
cholbrow@colgate.edu

Hollinger, David A.
University of California
Department of History
Berkeley, CA 94720-2550
Phone: 1.510.642.1808
Fax: 1.510.643.5323
davidhol@socrates.berkeley.edu
Doctorate, 1970
20c, North America, Social
Sciences; 20c, North America,
Humanistic Relations of Science;
20c, North America, Philosophy
and Philosophy of Science

Hollis, Corey
11820 Dorothy St.
Los Angeles, CA 90049-5406
Phone: 1.310.826.1411
chollis@ucla.edu

Holmberg, Gustav
Kung Oscars vag 5A
Lund S-222 40
Sweden
Phone: 046.15.22.93
gustav.holmberg@kult.lu.se

Holmes, F. L.
Yale Univ Sch of Medicine
Sec of History of Medicine
333 Cedar St
New Haven, CT 06510-3289

Holmfeld, John D.
212 Pratt St
Fredericksburg, VA 22405-2541

Holt, Dale Lynn
Mississippi State University
Dept of Philosophy & Religion
PO Box JS
Mississipp St, MS 39762-5833
dlh4@ra.msstate.edu

Holt, Frederick R.
2542 Clay St
San Francisco, CA 94115-1811

Holton, Gerald
Harvard University
358 Jefferson Physics Lab
Cambridge, MA 02138
Phone: 1.617.495.4474
Fax: 1.617.495.0416
holton@physics.harvard.edu
Doctorate, 1948
Humanistic Relations of
Science; Physical Sciences;
20c, Physical Sciences;
Education

Holzmann, Bruno
Danzieger Strasse 9
Hanau Am Main
Hessen D-63454
Germany
Phone: 49.69.707.5477

Homburg, Ernst
Universiteit Maastricht
Department of History
PO Box 616
Maastricht 6200MD
The Netherlands
Phone: 31433883314
Fax: 31433884816
e.homburg@history.unimaas.nl

Doctorate, 1993
20c, Europe, Physical Sciences;
19c, Europe, Technology;
18c, Europe, Social Relations of
Science; Education

Home, Roderick W.
University of Melbourne
HPS Department
Parkville VIC 3010
Australia
Phone: 61.3.8344.7037
Fax: 61.3.8344.7959
home@unimelb.edu.au
Doctorate, 1967
18c, Physical Sciences;
Australia and Oceania,
Physical Sciences; Australia and
Oceania, Institutions

Hon, Giora
University of Haifa
Department of Philosophy
Mount Carmel
Haifa 31905
Israel
Phone: 972.4.824.0611
Fax: 972.4.824.9735
hon@research.haifa.ac.il
Doctorate, 1985
Philosophy and Philosophy of
Science; Physical Sciences;
17c, Europe, Astronomical
Sciences

Hong, Sungook
University of Toronto
IHPST
73 Queens Park Cr E
Toronto, ON M5S1K7
Canada
Phone: 1.416.946.5024
Fax: 1.416.978.3003
sungook@chass.utoronto.ca
Doctorate, 1994
19c, Europe, Physical Sciences;
Social Sciences;
19c, Europe, Technology;
Social Relations of Science

Hook, Ernest B.
University of California
MCH-MC7360
Warren Hall
Berkeley, CA 94720-7360
Phone: 1.510.642.4490
Fax: 1.510.643.9588

ebhook@socrates.berkeley.edu
Doctorate
14c-16c, Europe, Medical
Sciences; Social Sciences;
Philosophy and Philosophy of
Science; Biological Sciences

Hooper, Judith
2135 E Hampton St
Tucson, AZ 85719-3811
Phone: 1.520.795.1153
sulphur@qwest.net

Hooper, Meredith
4 Western Rd
Fortis Green
London N2 9HX
Great Britain
Phone: 020.8883.7811
Fax: 020.8883.1335
meredith@hooper.demon.co.uk

Hooper, Wallace
7235 Summit St
Kansas City, MO 64114-1233
Phone: 1.816.523.5680
whooper@indiana.edu

Hopkins, Daniel P.
University of Missouri-Kansas
5100 Rockhill Rd
Kansas City, MO 64110
Phone: 1.816.235.2973
Fax: 1.816.822.0555
hopkinsd@umkc.edu
Doctorate, 1987
18c, North America,
Earth Sciences;
19c, Africa, Earth Sciences

Hoppe, Brigitte B.
Ludwig-Maximilians Univ.
Inst for Hist of Nat Science
Deutsches Museum,
Museumsinsel
München D-80306
Germany
Phone: 49.89.2180.3252
Fax: 49.89.2180.3162
ug301ah@sunmail.
lrz-muechen.de
Doctorate
17c, Europe, Biological Sciences;
19c, Europe, Physical Sciences;
20c, South America,

Medical Sciences;
14c-16c, Asia, Exploration,
Navigation and Expeditions

Hopper, David H.
Macalester College
1600 Grand Ave
Saint Paul, MN 55105
hopper@macalester.edu
Doctorate, 1959
14c-16c, Europe, Technology;
14c-16c, Europe, Philosophy
and Philosophy of Science;
Europe, Philosophy and
Philosophy of Science;
Biological Sciences

Hoptroff, Georgina
635 Lincoln Ave Apt 3
Saint Paul, MN 55105-3545
Phone: 1.612.521.5909
georginahoptroff@hotmail.com

Hopwood, Nick D.
University of Cambridge
Hist & Philosophy of Science
Free School Lane
Cambridge CB2 3RH
Great Britain
Phone: 44.1223.334.542
Fax: 44.1223.334.554
ndh12@cam.ac.uk
Doctorate
19c, Biological Sciences;
20c, Biological Sciences;
19c, Medical Sciences;
20c, Medical Sciences

Hormigon, Mariano
Universidad de Zaragoza
Seminario de Historia de la
Ciencia-Facultad de Ciencias-
Ciudad Universitaria
Zaragoza E-50009
Spain
Phone: 34.976.761.119
Fax: 34.976.761.125
hormigon@posta.unizar.es
Doctorate, 1982
19c, Mathematics;
19c, Social Relations of Science;
19c, Humanistic Relations of
Science; 19c, Europe

Horn, David G.
291 S Cassingham Rd
Bexley, OH 43209-1804

Phone: 1.614.292.2559
Fax: 1.614.292.6707
horn.5@osu.edu

Hornix, W. J.
Rijksstraatweg 46
Ubbergen 6574 AE
The Netherlands
Phone: 31.24.3656460
wjh@sci.kum.nl

Horrocks, Thomas A.
Harvard University
Countway Library of Medicine
10 Shattuck St
Boston, MA 02115-6011
Phone: 1.617.432.4142
Fax: 1.617.432.0693
thomas_horrocks@
hms.harvard.edu
Master's Level, 1984
*19c, North America, Medical
Sciences; 18c, North America,
Medical Sciences; 19c, North
America; 19c, North America,
Education*

Hossfeld, Uwe
Institut für Geschichte der
Medizin, Naturwissenschaft und
Technik
Ernst-Haeckel-Haus
Berggasse 7
Jena 07745
Germany
Phone: 43.3641.949.505
Fax: 43.3641.949.502
b7houw@nds.rz.uni-jena.de

Houwaart, Eddy
Vrije Universiteit
Deptartment of Medical History
Van Der Boechorststraat 7
1081 BT Amsterdam
Amsterdam 1081 BT
The Netherlands
Phone: 31.20.444.8217
Fax: 31.20.444.8258
e.houwaart.medhistory@
med.vu.nl
Doctorate
*20c, Europe, Medical Sciences;
19c, Europe, Medical Sciences;
Medical Sciences*

Howard, Don
University of Notre Dame
Dept. of Philosophy
100 Malloy Hall
Notre Dame, IN 46556
Phone: 1.574.631.7547
Fax: 1.574.631.3985
don.a.howard.43@nd.edu
Doctorate, 1978
*Philosophy and Philosophy of
Science; Physical Sciences;
20c, Physical Sciences;
Astronomical Sciences,
Philosophy and Philosophy of
Science*

Howard, John N.
25 Woodcliff Road
Newton Highland, MA 02461
Phone: 1.617.332.1743
howards@gis.net
Doctorate, 1954
19c, Europe, Physical Sciences

Howard, Nicole
Indiana Univeristy
Goodbody Hall, 130
1011 E 3rd St
Bloomington, IN 47405-7005
Phone: 1.812.855.3976
nichowar@indiana.edu

Howell, Joel D.
University of Michigan
6312 Medical Science 1
1150 W Medical Center Dr
Ann Arbor, MI 48109-0726
Phone: 1.734.647.4844
Fax: 1.734.647.3301
jhowell@umich.edu
Doctorate
*19c, Medical Sciences;
20c, Medical Sciences*

Howell, Kenneth J.
John Henry Newman
Inst of Catholic Thought
1007 1/2 S Wright St
Champaign, IL 61820
Phone: 1.217.384.5961
Fax: 1.217.384.5974
khowell@newmancenter.com

Hoyningen-Huene, Paul
Universität Hannover
Ze Wissenschaftstheorie
Oeltzenstr. 9

Hannover D-30169
Germany
Phone: 49.511.7624801
Fax: 41.1.3413163
hoyningen@uni-hannover.de
Doctorate, 1975
*Philosophy and Philosophy of
Science*

Hoyrup, Jens
Roskilde University
Box 260
Roskilde DK-4000
Denmark
Phone: 45.4674.2527
Fax: 45.4674.3012
jensh@ruc.dk
Master's Level, 1969
*Pre-400 B.C.E., Mathematics;
5c-13c, Mathematics;
400 B.C.E.-400 C.E.,
Mathematics;
14c-16c, Mathematics*

Hoyt, Diana P.
16960 Teagues Point Rd
Hughesville, MD 20637-2848
Phone: 1.301.932.5709
Fax: 1.301.274.4024
dhoyt@crosslink.net

Hsia, Florence
1380 E Hyde Park Blvd Apt 102
Chicago, IL 60615-2989
Phone: 1.773.548.8659
fchsia1@attglobal.net

Hu, Danian
Yale Univ, Hist of Med & Sci
L130 Sterling Hall of Medicine
333 Cedar Street
New Haven, CT 06520-8015
Phone: 1.203.432.9353
Fax: 1.203.737.4130
danianhu@yahoo.com
Master's Level, 1991
*20c, Asia, Physical Sciences;
19c, Asia, Physical Sciences;
20c, Europe, Physical Sciences*

Huang, Hsing T.
4800 Fillmore Ave Apt 854
Alexandria, VA 22311-5072
hthuang@juno.com
Doctorate

*400 B.C.E.-400 C.E., Asia,
Technology; 5c-13c, Asia,
Biological Sciences*

Huang, Huei-Hsin
National Taitung Teachers Coll
684 Sec 1, Chunghua Road
Taitung 950
Taiwan
Phone: 886.89.34918
Fax: 886.89334798
huhs@cc.ntttc.edu.tw
Doctorate, 1995
*Philosophy and Philosophy of
Science; Education;
Social Relations of Science*

Hufbauer, Karl G.
3319 37th Ave So.
Seattle, WA 98144-7015
Phone: 1.206.725.2277
Fax: 1.206.725.2277
hufbauer@uci.edu
Doctorate, 1970
*20c, North America,
Astronomical Sciences;
20c, Europe, Astronomical
Sciences*

Huff, Toby E.
Univ. of Mass. - Dartmouth
285 Old Westpat Road
N. Dartmouth, MA 02747
Phone: 1.508.999.8405
Fax: 1.508.999.8808
thuff@umassd.edu
Doctorate
*5c-13c, Social Relations of
Science; Humanistic Relations
of Science; Institutions*

Hughes, Alan
Oxford University Press
Associate Editor, OED
Great Clarendon Street
Oxford OX2 6DP
Great Britain
hughesa@oup.co.uk
Master's Level

Hughes, Brad
425 El Camino Del Mar
Laguna Beach, CA 92651-2550
Phone: 1.949.497.0857
ebh@cox.net

Hughes, Jeff A.
University of Manchester
CHSTM
Mathematics Tower
Manchester MI3 9PL
Great Britain
Phone: 44.0161.275.5857
Fax: 44.0161.275.5699
hughes@fs4.ma.man.ac.uk
Doctorate, 1993
*20c, Physical Sciences;
Social Sciences; Social Relations
of Science; Education*

Hughes, Philip J.
110 Farm Court
Salt Spring Isl, BC V8K1H7
Canada
Phone: 1.250.537.0843
Fax: 1.250.537.0873
pj@snark.org
*20c, Instruments and
Techniques; Astronomical
Sciences; Biological Sciences;
Philosophy and Philosophy of
Science*

Hughes, Sally S.
University of California
ROHO
486 Library
Berkeley, CA 94720-6000
Phone: 1.510.642.7395
Fax: 1.510.642.7589
shughes@library.berkeley.edu
Doctorate
*20c, North America;
20c, North America, Technology;
20c, North America, Medical
Sciences*

Hughes, Thomas P.
8330 Millman St
Philadelphia, PA 19118-3925
Phone: 1.215.248.0327
thughes@525.upenn.edu

Hugonnard-Roche, Henri A.
Sorbonne
EPHE IVE Section
45-47 Rue des Ecoles
Paris 75005
France
henri.hugonnard-r@wanadoo.fr
Doctorate

*Pre-400 B.C.E., Philosophy
and Philosophy of Science;
5c-13c, Physical Sciences*

Hull, David L.
Northwestern University
Dept of Philosophy
1818 Hinman Ave
Evanston, IL 60208-1315
Phone: 1.847.491.3656
Fax: 1.847.491.2547
d-hull@northwestern.edu
Doctorate, 1964
*19c, Europe, Biological Sciences;
20c, North America, Biological
Sciences; 20c, North America,
Philosophy and Philosophy of
Science*

Hume, Bradley D.
University of Dayton
Department of History
300 College Park Ave
Dayton, OH 45469-0002
Phone: 1.937.229.3381
Fax: 1.937.229.4400
hume@udayton.edu

Humphreys, Margaret E.
Duke University
Department of History
PO Box 90719
Durham, NC 27708-0719
Phone: 1.919.684.2285
Fax: 1.919.681.7670
meh@acpub.duke.edu
Doctorate, 1983
*19c, North America, Medical
Sciences; 20c, North America,
Medical Sciences*

Hunt, Bruce J.
Univ of Texas
Dept of History
Austin, TX 78712
Phone: 1.512.232.6109
Fax: 1.512.475.7222
bjhunt@mail.utexas.edu
Doctorate, 1984
*19c, Europe, Physical Sciences;
19c, Technology*

Hunt, James C.
Prince Georges College
301 Largo Rd
Largo, MD 20774-2199
Phone: 1.301.322.0429

Fax: 1.301.386.7529
huntjc@pg.cc.md.us
Master's Level, 1996
Astronomical Sciences;
20c, Astronomical Sciences;
Philosophy and Philosophy of
Science; Physical Sciences

Hunter, Andrew M.
Bernard Quaritch Ltd
5-8 Lower John Street
Golden Square
London W1F 9AU
Great Britain
Phone: 44.12.0734.2983
Fax: 44.17.1437.0967
rarebooks@quaritch.com
Education; Medical Sciences;
14c-16c, Instruments and
Techniques; Asia

Hunter, Graeme
University of Western Ontario
School of Dentistry
London, ON N6A5C1
Canada
Phone: 1.519.472.5742
Fax: 1.519.850.2316
gkhunter@julian.uwo.ca

Hunter, Melanie J.
2205 W St Charles Ave
Phoenix, AZ 85041
Phone: 1.602.268.9744
tweetie@dotplanet.com
Bachelor's Level, 1999
Biological Sciences;
Natural and Human History;
Social Relations of Science

Hunter, Patti W.
Westmont College
Dept of Mathematics
955 La Paz Rd
Santa Barbara, CA 93108-1099
Phone: 1.805.565.6000
phunter@westmont.edu
Doctorate, 1997
19c, North America,
Mathematics; 20c

Huntley, Frances
1103 B Blackwood St
Regina, SK S4X3K2
Canada
Phone: 1.306.569.0340
f.huntley@accesscomm.ca

Hutchison, Keith R.
University of Melbourne
HPS Dept
Parkville VIC 3010
Australia
Phone: 61.3.8344.7571
Fax: 61.3.8344.7959
k.hutchison@unimelb.edu.au
Doctorate, 1977
17c, Europe, Astronomical
Sciences; 19c, Physical
Sciences; 400 B.C.E.-400 C.E.,
Social Relations of Science;
14c-16c, Physical Sciences

Ibanez, Itsaso
Universidad del Pais Vasco
c/o Maria Diaz de Haro, 68
Profesora de Navegacion
E.T.S. De Nautica y M Navales
Portugalete 48920
Spain
Phone: 94.495.1261
Fax: 94.495.1400
cnpibfei@lg.ehu.es

Ilerbaig, Juan F.
5815 SW 48th Street
Miami, FL 33155
juan.f.ilerbaig-2@tc.umn.edu
Doctorate, 2002
20c, North America, Biological
Sciences; 19c, North America,
Natural and Human History;
Earth Sciences; Biological
Sciences

Imhausen, Annette
Dibner Institute
MIT E56-100
38 Memorial Dr
Cambridge, MA 02142-1347
Phone: 1.617.491.0915
aimhausen@dibinst.mit.edu

Ince, Simon
University of Arizona
Dept of Hydrology
Building 11
Tucson, AZ 85721
Fax: 1.520.621.1422
Doctorate, 1953
19c, Europe, Technology,
Instruments and Techniques;
18c, Europe, Education;
20c, Europe, Institutions;
17c, Technology

Ingolfsson, Palmi
Hraunbaer 128
Reykjavik 110
Iceland

Inman, Henry F.
2016 Park Ave # A
Richmond, VA 23220-2712
Phone: 1.804.353.0817

Innis, Nancy K.
University of Western Ontario
Psychology Department
London, ON N6A5C2
Canada
Phone: 1.519.661.3686
Fax: 1.519.661.3961
ninnis@uwo.ca
20c, North America, Cognitive
Sciences; 20c, Europe,
Cognitive Sciences

Ino, Shuji
Tokyo Kasie Gakuim
22 Sanbanchyo Chiyodaku
Tokyo 102
Japan
Phone: 81.03.3262.2256
Fax: 0462.69.8210
shujiino@ma4.justnet.ne.jp
Bachelor's Level
20c, Asia, Social Relations
of Science; South America,
Philosophy and Philosophy of
Science

Inoue, Takuya
1-3-2-2204 Toyosu
Koutou-ku
Tokyo 135-0061
Japan
Phone: 81.3.5547.2688
takuyai@kokugakuin.ac.jp

Ione, Amy
PO Box 12748
Berkeley, CA 94712-3748
Phone: 1.510.548.2052
Fax: 1.510.548.2054
ione@diatrope.com
Master's Level, 1995

Ipina, Lynne
University of Wyoming
Dept of Mathmatics
Laramie, WY 82071-3036

Phone: 1.307.766.2318
Fax: 1.307.766.6838
ipina@uwyo.edu

Ishikawa, Chigusa
51 Flexner Ln
Princeton, NJ 08540-4951
Phone: 1.609.279.2813

Isobe, Takashi
Smithsonian Astrophy. Obser.
60 Garden St MS-34
Cambridge, MA 02138-1516
Phone: 1.617.496.7335
tisobe@cfa.harvard.edu
Doctorate, 1989
20c, Astronomical Sciences

Israel, Paul B.
Rutgers University
Thomas A. Edison Papers
44 Road 3
Piscataway, NJ 08854-8049
Phone: 1.732.932.8511
Fax: 1.732.445.8512
pisrael@rci.rutgers.edu
Doctorate, 1989
19c, North America, Technology;
20c, North America, Technology;
19c, North America; 19c, North
America, Biological Sciences

Itagaki, Ryoichi
Tokai University
1117 Kita-kaname
Hiratsuka 259-1292
Japan
Phone: 81.04.63.58.1211
Fax: 81.04.63.59.4047
ita@keyaki.cc.u-tokai.ac.jp
Master's Level, 1981
20c, Europe, Physical Sciences

Ito, Kenji
Harvard University
Dept. of Hist. of Science
Science Center 235
Cambridge, MA 02138
Phone: 1.617.495.8259
Fax: 1.617.495.3344
kenjiito@fas.harvard.edu
Master's Level, 1991
20c, Asia, Physical Sciences

Ivanov, K. V.
Tula State University
Department Theoretical Physics

Pr Lenina 125
Tula 300026
Russia

Iverson, Lara
11 Staples Point Rd
Freeport, ME 04032-6003
Phone: 1.207.865.3905
laraji@yahoo.com

Iverson, Margot
3208 Bryant Ave S #2
Minneapolis, MN 55408
mliverson@mindspring.com

Iwata, Toyoto
#602 6-19-3 Kohyo-Dai
Inagi-Shi
Tokyo
Japan

Jackson, Ian
1540 Walnut St
Berkeley, CA 94709-1513
Phone: 1.510.548.1431
Fax: 1.510.548.5766
Bachelor's Level

Jackson, Jerome A.
Whitaker Center
College of Arts and Sciences
Florida Gulf Coast University
10501 Fgcu Blvd S
Fort Myers, FL 33965-6565
Phone: 1.941.590.7177
Fax: 1.941.590.7200
picus@fgcu.edu

Jackson, John P., Jr.
University of Colorado
Dept of Ethnic Studies
30 Ketchum
Boulder, CO 80309-0339
Fax: 1.303.492.7799
john.p.jackson@colorado.edu
Doctorate, 1997
20c, North America; 20c, North
America, Biological Sciences;
20c, North America, Natural
and Human History; 20c, North
America, Cognitive Sciences

Jacob, Margaret C.
10785 Weybrun Av
Los Angeles, CA 90024
Phone: 1.310.234.1139
Fax: 1.310.234.1274

mjacob@history.ucla.edu
Doctorate, 1968
18c, Europe, Humanistic
Relations of Science;
17c, Europe, Physical Sciences;
Europe, Philosophy and
Philosophy of Science;
Education

Jacyna, Stephen
Wellcome Institute for the
History of Medicine
183 Euston Road
London NW1 2BE
Great Britain

James, Frank A.
Royal Institution
21 Albemarle Street
London WIS 4BS
Great Britain
Phone: 020.7670.2924
Fax: 020.7629.3569
fjames@ri.ac.uk
Doctorate, 1981
19c, Europe, Physical Sciences;
Social Relations of Science

James, Jeremaih
History of Science Dept
235 Science Center
Cambridge, MA 02138
Phone: 1.781.684.8320
jjames@fas.harvard.edu

James, K. A.
22420 Dogwood Ln
Woodway, WA 98020-6122
Phone: 1.44.1865.209225
kjames3108@aol.com

Jansen, Sarah
Harvard University
Dept of History of Science
Science Center 235
Cambridge, MA 02138
Phone: 1.617.495.2298
Fax: 1.617.495.3344
jansen@fas.harvard.edu
Doctorate, 1997

Janssen, Alex
Martin Garcia 1539
Montevideo 11800
Uruguay

Phone: 598.2200.4239
Fax: 598.2208.8434
alejanss@adinet.com.uy

Janssen, Michel H.
University of Minnesota
Tate Laboratory of Physics
116 Church Street SE
Minneapolis, MN 55455
Phone: 1.612.624.5880
Fax: 1.612.624.5880
janss011@tc.umn.edu
Doctorate, 1995
20c, Physical Sciences;
Philosophy and Philosophy of
Science

Jason, Gary J.
Argonetics, Inc
875 Avenida Acapulco
San Clemente, CA 92672
Phone: 1.888.231.8183
drgaryjason@internetconnect.net
Doctorate, 1982

Jensen, Derek
4875 Cole St Apt 66
San Diego, CA 92117-1866

Jervis, Jane L.
Evergreen State College
Olympia, WA 98505
Phone: 1.206.773.3503
jervisi@earthlink.net
Doctorate, 1978

Jesseph, Douglas M.
North Carolina State Univ
PO Box 8103
Raleigh, NC 27606
Phone: 1.919.515.6337
doug-jesseph@ncsu.edu

Jewett, Andrew
717 22nd St. #4
Sacramento, CA 95816
Phone: 1.916.442.4441
ajewett@socrates.berkeley.edu

Jha, S. R.
Harvard University
PERC, 511 Larsen Hall
Appian Way
Cambridge, MA 02138
Phone: 1.617.495.9084
Fax: 1.617.495.5908
stefania_jha@harvard.edu

Doctorate
20c, Europe, Philosophy
and Philosophy of Science;
19c, Europe, Philosophy
and Philosophy of Science;
20c, North America, Philosophy
and Philosophy of Science;
Cognitive Sciences

Johnson, Alan T.
17058 Iron Mountain Dr
Poway, CA 92064-6316
Doctorate
Physical Sciences;
Astronomical Sciences;
Instruments and Techniques;
Philosophy and Philosophy of
Science

Johnson, Ann
Fordham University
Department of History
441 E Fordham Rd
Bronx, NY 14058
Phone: 1.718.817.3996
Fax: 1.718.817.4680
annj@sprintmail.com
Doctorate, 2000
Technology, Instruments and
Techniques; North America;
Instruments and Techniques

Johnson, Coates R.
University of Georgia
Department of Physics
Athens, GA 30602
Phone: 1.706.542.2485
Fax: 1.706.542.2492

Johnson, Dale M.
1949 Weybridge Ln
Reston, VA 20191-3621
Phone: 1.703.476.0581
sidney68@erols.com

Johnson, Jeffrey A.
Villanova University
History Department
800 E Lancaster Ave
Villanova, PA 19085-1478
Phone: 1.610.519.7404
Fax: 1.610.519.4450
jeffrey.johnson@villanova.edu
Doctorate
19c, Europe, Physical Sciences;
20c, Europe, Physical Sciences;

19c, Europe, Social Relations
of Science; 20c, Europe, Social
Relations of Science

Johnson, Stephen B.
University of North Dakota
Dept. of Space Studies
Box 9008
Grand Forks, ND 58202
Phone: 1.701.777.2480
sjohnson@space.edu
Doctorate, 1997
20c, North America,
Cognitive Sciences;
18c, Europe, Technology;
19c, Physical Sciences;
Social Sciences

Johnston, A. Sidney
51 Quaboag Rd
Acton, MA 01720-2404
Phone: 1.978.635.0293
sidj@ma.ultranet.com

Johnston, Charles
12426 Moceri Drive
Grand Blanc, MI 48439
Phone: 1.810.694.3921
xwhyz@aol.com

Johnston, Timothy D.
University of North Carolina
Psychology Dept
PO Box 26164
Greensboro, NC 27402-6164
Phone: 1.336.256.0021
johnston@uncg.edu

Jolly, James
761 SW 15th St
Corvallis, OR 97333-4132
Phone: 1.541.754.9573
jollych@ucs.orst.edu

Jones, Alexander
University of Toronto
Classics
97 St George Street
Toronto, ON M5S2E8
Canada
Phone: 1.416.978.0483
alexander.jones@utoronto.ca

Jones, Charles V.
Ball State University
Office of Teaching and Learning
Advancement

Muncie, IN 47306-0205
Phone: 1.765.285.1763
Fax: 1.765.285.2669
cvjones@bsu.edu
Doctorate, 1979
400 B.C.E.-400 C.E.,
Mathematics; 14c-16c, Europe,
Mathematics; Cognitive Sciences;
20c, Instruments and Techniques

Jones, Daniel P.
Nat Endwmnt for the Humanities
1100 Pennsylvania Ave NW
MS318
Washington, DC 20506
Phone: 1.202.606.8217
Fax: 1.202.606.8394
djones@neh.gov
Doctorate, 1969
20c, North America, Physical
Sciences; 20c, North America,
Medical Sciences; 20c, North
America, Social Relations of
Science; 19c, North America,
Technology

Jones, Greta J.
University of Ulster
Shore Road
Jordanstown BT37 0QB
Northern Ireland
Phone: 1232.365131
gj.jones@ulst.ac.uk
Doctorate, 1974
19c, Europe, Biological Sciences,
Biography; Biological Sciences;
Medical Sciences

Jones, Jessica
137 E 17th St Apt 307
Minneapolis, MN 55403-3875
Phone: 1.612.872.4381
jone0732@umn.edu

Jones, Josh
17 Braxton Dr
Belle Mead, NJ 08502-4602
Phone: 1.908.359.1411
jajones00@post.harvard.edu
Physical Sciences;
Medical Sciences

Jones, Matthew L.
Columbia University
Department of History
523 Fayerweather Hall
New York, NY 10023

Phone: 1.212.854.2421
mj340@columbia.edu
Master's Level
17c, Europe, Mathematics;
18c, Europe; Philosophy and
Philosophy of Science

Jones, Paul R.
University of Michigan
Dept of Chemistry
930 N. University
Ann Arbor, MI 48109-1055
Fax: 1.734.647.4865
prjones@umich.edu
Doctorate, 1956
19c, Europe, Physical Sciences;
20c, North America, Humanistic
Relations of Science;
18c, Europe, Institutions;
17c, Europe

Jones, Susan D.
University of Colorado
Dept of History
Campus Box 234
Boulder, CO 80309-0234
Phone: 1.303.492.2931
Fax: 1.303.492.1868
jonessu@colorado.edu
Doctorate, 1997
20c, North America, Medical
Sciences; 20c, North America,
Biological Sciences; 20c, North
America, Gender and Science

Jones-Imhotep, Edward
Harvard University
History of Science Dept
Science Center 235
Cambridge, MA 02138
Phone: 1.613.635.7435
Fax: 1.617.495.3344
imhotep@fas.harvard.edu

Jordan, Philip D.
Hastings College
7th & Turner
Hastings, NE 68901
Phone: 1.402.461.7345
Fax: 1.402.461.7480
pjordan@hastings.edu
Doctorate, 1970
19c, North America,
Biological Sciences;
19c, Europe, Biological Sciences

Jorgensen, Mark R.
University of Minnesota
Dept of Sociology
267 19th Ave S
Minneapolis, MN 55455
Phone: 1.651.696.1612
Bachelor's Level, 1988
20c, North America;
20c, Social Sciences;
Technology; 20c, Science Policy

Jorgenson, Susan
University of Oklahoma
2203 Blue Creek Drive
Norman, OK 73026
Phone: 1.405.573.9175
psiouxzy@ou.edu

Jorland, Gerard
CRH/EHESS 54 Bd
Raspail
Paris 75006
France
Phone: 33.01.4327.360
Fax: 33.01.4327.3860
jorland@ehess.fr

Joseph, George
Section of the History of Med
Yale Univ Sch of Medicine
PO Box 200779
New Haven, CT 06520-0779
Phone: 1.203.789.1586
Fax: 1.203.737.4130
dgjoseph@pantheon.yale.edu
Doctorate, 2002
19c, North America, Medical
Sciences; 20c, North America,
Medical Sciences; 20c, North
America, Biological Sciences

Josset, Patrice
2 Avenue De La Gare
Villenauxe
La Grande 10370
France
Phone: 33.144736883
Fax: 33.148519270
patrice-josset@wanadoo.fr

Joven, Fernando
Estudio Teologico Agustiniano
Paseo de Filipinos 7
Valladolid 47007
Spain
Fax: 34.9.8339.7896
bestagus@adenet.es

Doctorate, 1996
Philosophy and Philosophy of
Science; Mathematics

Jun, Umeda
Yunotsu I1105 Nima-gun
Shimane 699-2511
Japan

Jungck, John R.
Beloit College
Department of Biology
700 College St
Beloit, WI 53511-5595
Phone: 1.608.363.2226
Fax: 1.608.363.2052
jungck@beloit.edu
Doctorate, 1973
19c, Biological Sciences;
20c, Biological Sciences;
20c, Science Policy;
Cognitive Sciences

Junker, Thomas
Universität Tübingen
Sigwartstr. 20
Tuebingen 72076
Germany
Phone: 49.69.96121741
thomas.junker@
 uni-tuebingen.de
Doctorate, 1989
19c, Europe, Biological Sciences;
20c, Europe, Biological Sciences;
19c, Biological Sciences;
Social Sciences

Jurado, Jaime
The Gambrinus Company
14800 San Pedro Ave 3rd Fl
San Antonio, TX 78232-3733
Phone: 1.210.490.9128
Fax: 1.210.490.9984
jaime.jurado@gambrinusco.com

Kaadan, Abdul N.
Aleppo University
History of Arab Science
PO Box 7581
Aleppo
Syria
Phone: 963.21.223.6083
Fax: 963.21.224.8035
Master's Level, 1993
5c-13c, Asia, Medical Sciences;
5c-13c, Asia, Physical Sciences

Kahl, Mark
8409-112th St Unit # 507
Edmonton, AB T6G1K6
Canada
Phone: 1.780.436.4616
mkahl@powersurfr.com
Bachelor's Level, 2000
20c, North America,
Astronomical Sciences;
20c, North America,
Physical Sciences;
North America, Philosophy and
Philosophy of Science

Kahr, Bart
University of Washington
Department of Chemistry
Bagley Hall
PO Box 351700
Seattle, WA 98195-1700
Fax: 1.206.685.8665
kahr@chem.washington.edu

Kaiser, David
MIT
Program in Sci, Tech & Society
Building E51-185
77 Massachusetts Ave
Cambridge, MA 02139
Phone: 1.617.452.3173
Fax: 1.617.258.8118
dikaiser@mit.edu
Master's Level, 2000
20c, North America,
Physical Sciences;
20c, North America, Education;
20c, Social Sciences

Kaji, Masanori
Tokyo Institute of Technology
2-12-1 Ookayama Meguro-ku
Tokyo 152-8552
Japan
Phone: 81.3.5734.2270
Fax: 81.45.544.9768
mkaji@aqu.bekkoame.ne.jp
Doctorate, 1988
19c, Europe, Physical Sciences;
20c, Europe, Physical Sciences;
19c, Asia, Physical Sciences;
20c, Asia, Physical Sciences

Kallaher, Michael J.
Washington State University
Dept of Mathematics
Pullman, WA 99164-3113
Phone: 1.509.335.3132

mkallaher@wsu.edu
Doctorate, 1967
Mathematics

Kalthoff, Mark A.
Hillsdale College
Department of History
33 College St
Hillsdale, MI 49242-1298
Phone: 1.517.437.7341
Fax: 1.517.437.3923
mak@hillsdale.edu
Doctorate, 1998

Kang, Eugene
Suyoung-gu Manymi-1-dong
590-1 19/3
Pusan 613-131
South Korea
Phone: 82.51.510.1430
Fax: 82.51.512.0528
none2000@netian.com

Kanz, Kai T.
MUL
Konigstr 42
Lübeck 23552
Germany
Phone: 49.451.7079.9814
Fax: 49.451.7079.9899
kanz@imwg.mu-luebeck.de
Doctorate, 1996
18c, Europe, Natural and
Human History; 19c, Europe,
Biological Sciences;
18c, Europe, Technology

Kaplan, Barbara B.
Villa Julie College
Asso Dean Academic Affairs
1525 Greenspring Valley Rd
Stevenson, MD 21153
Phone: 1.410.602.6566
dea-bark@mail.vjc.edu
Doctorate
17c, Europe, Medical Sciences

Karakida, Ken-ichi
Fuji Xerox Co. Ltd.
Intelligent Devices Lab
430 Sakai, Nakai
Kanagawa 259-0157
Japan
Phone: 81.465.80.2009
Fax: 81.465.81.8961
karakida.ken-ichi@
 fujixerox.co.jp

Karathanos, Michael
1534 E 17th Pl
Tulsa, OK 74120-7219
Phone: 1.918.744.1156

Karlsson, Christer
Bjorkelundsvagen 71
Gammelstad SE-95433
Sweden
Phone: 46.90.786.5281
christer.karlsson@
  historia.umu.se

Kass, Lee B.
2127 Spencer Rd
Newfield, NY 14867
Phone: 1.607.255.4876
Fax: 1.607.255.7979
lbk7@cornell.edu

Kassler, Jamie C.
10 Wollombi Road
Northbridge NSW 2063
Australia
Phone: 61.2.9967.5755
Fax: 61.2.9967.5890
michaelk@zip.com.au
Doctorate, 1971
Humanistic Relations of Science;
Philosophy and Philosophy of
Science

Kastner, Joerg
Staatliche Bibliothek Passau
Michaeligasse 11
Passau 94032
Germany
Phone: 49.851.2712
Fax: 49.851.31704
joerg.kastner@uni-passau.de
Doctorate, 1971
17c, Europe, Medical Sciences;
18c, Europe, Natural
and Human History;
5c-13c, North America,
Philosophy and Philosophy of
Science; Social Sciences

Kato, Shigeo
University of Tokyo
3-8-1, Komaba, Meguro-ku
Tokyo 153-8902
Japan
Phone: 81.3.5454.6135
Fax: 81.422.76.3595
kato@hps.c.u-tokyo.ac.jp
Master's Level

20c, Asia, Medical Sciences,
Instruments and Techniques;
19c, Asia, Medical Sciences;
Asia, Earth Sciences

Katz, Victor J.
Univ. of Dist. of Columbia
4200 Connecticut Avenue NW
Washington, DC 20008
Phone: 1.202.274.5374
Fax: 1.301.592.0061
vkatz@udc.edu
Doctorate, 1968
Mathematics

Kauffman, George B.
1609 E. Quincy Ave
Fresno, CA 93720-2309
Phone: 1.559.323.9123
Fax: 1.559.278.4402
georgek@csufresno.edu
Doctorate, 1956
Physical Sciences; Biography

Kavey, Allison
2901 Saint Paul St Apt 2F
Baltimore, MD 21218-4124
Phone: 1.410.243.5149
akavey@mail.jhmi.edu

Kay, Charles D.
Wofford College
Department of Philosophy
429 N Church St
Spartanburg, SC 29307
Phone: 1.864.597.4583
kaycd@wofford.edu
Doctorate
Physical Sciences; Philosophy
and Philosophy of Science

Kay, Gwen E.
SUNY Oswego
Dept of History
Mahar Hall
Oswego, NY 13126
Phone: 1.315.312.3418
kay@oswego.edu
Doctorate, 1997
Pre-400 B.C.E., North America,
Medical Sciences; 20c, North
America, Medical Sciences;
20c, North America, Gender and
Science

Kaye, Joel B.
Barnard College
Department of History
3009 Broadway
New York, NY 10027-6598
Phone: 1.212.854.4350
Fax: 1.212.854.0559
jkaye@barnard.columbia.edu
Doctorate, 1991
5c-13c, Europe, Philosophy
and Philosophy of Science;
5c-13c, Europe, Humanistic
Relations of Science;
5c-13c, Europe, Technology;
5c-13c, Europe, Astronomical
Sciences

Keas, Michael N.
Oklahoma Baptist University
OBU Box 61772
500 W University St
Shawnee, OK 74804-2590
Phone: 1.405.878.2098
Fax: 1.405.878.2050
mike_keas@mail.okbu.edu
Doctorate, 1992
Biological Sciences; Social
Relations of Science; Europe;
Philosophy and Philosophy of
Science

Keegan, Robert T.
Pace University
Dept. of Psych., Marks Hall
861 Bedford Road
Pleasantville, NY 10570-2799
Phone: 1.914.773.3309
rkeegan@pace.edu
Doctorate, 1985
19c, Biological Sciences;
19c, Natural and Human History;
19c, Social Sciences

Keel, Othmar
Univ de Montreal
Dept History
CP 6128 Succ A
Montreal, PQ H3C3J7
Canada
Phone: 1.514.343.6465
Fax: 1.514.343.2483
otkeel@yahoo.fr

Keelan, Jennifer
42 Baltic Avenue
Toronto, ON M4J1S2

Canada
Phone: 1.416.466.1459
jenn.keelan@utoronto.ca

Keil, Gundolf
Inst für Geschichte De Medizin
d Universität
Oberer Neubergweg 10A
Wurzburg D-97074
Germany
Phone: 49.931796780
Fax: 49.9317967878
Doctorate
5c-13c, Europe, Medical
Sciences; 400 B.C.E.-400 C.E.,
Europe, Medical Sciences;
14c-16c, Europe, Natural
and Human History;
17c, Europe, Philosophy and
Philosophy of Science

Keirns, Carla
University of Pennsylvania
303 Logan Hall
249 S 36th Street
Philadelphia, PA 19104
Phone: 1.215.893.0504
Fax: 1.215.573.2231
ckeirns@sas.upenn.edu

Keithley, Hans
10008 Park Royal Dr
Great Falls, VA 22066-1847
Phone: 1.703.757.0032
keithley@erols.com

Keller, Evelyn F.
MIT
E51 - 171
77 Massachusetts Ave
Cambridge, MA 02139-4307
Phone: 1.617.253.8722
Fax: 1.617.258.8634
efkeller@mit.edu
Doctorate, 1963
20c, Biological Sciences;
20c, Philosophy and Philosophy
of Science; Gender and Science

Kellman, Jordan
Louisana State University
Honors College
French House, LSU
Baton Rouge, LA 70803
Phone: 1.225.388.3210
kellman@lsu.edu
Doctorate

18c, Europe, Exploration,
Navigation and Expeditions;
17c, Europe, Astronomical
Sciences; Biological Sciences;
Natural and Human History

Kellogg, Ralph H.
1400 Geary Blvd Apt 2103
San Francisco, CA 94109-9313
Phone: 1.415.346.1454
Doctorate, 1953
Medical Sciences

Kelter, Irving A.
University of St Thomas
History Department
3812 Montrose Blvd
Houston, TX 77006-4626
Phone: 1.713.525.3192
Fax: 1.713.942.3454
kelter@stthom.edu

Kennedy, Edward S.
777 Ferry Rd # D-9
Doylestown, PA 18901-2199
Doctorate, 1936
Pre-400 B.C.E., Mathematics;
Astronomical Sciences

Kerwin, Edward M.
3860 Crater Lake Ave # B
Medford, OR 97504-9741
Phone: 1.541.858.1003
Fax: 1.541.858.1091
Doctorate, 1988
14c-16c, Europe, Instruments
and Techniques; 20c, North
America, Medical Sciences

Kevles, Daniel J.
Yale University
Dept of History
PO Box 208324
New Haven, CT 06520-8324
Phone: 1.203.432.1356
Fax: 1.203.436.4624
daniel.kevles@yale.edu
Doctorate, 1964
20c, North America, Biological
Sciences; 20c, North America,
Physical Sciences; 20c, North
America, Science Policy

Khalilov, Salahadin
Azerbaijan University
84 Mirali Gashgay St
Baku 370110

Azerbaijan
Phone: 994.12.403.325
Fax: 994.12.937.773
ssx@azun.baku.az

Khatib, Bashir
Total Transport Service
Airport Industrial Park
145 Hook Creek Blvd Bldg B6A
Valley Stream, NY 11581-2299
Phone: 1.516.872.3800
Fax: 1.516.872.3840
kbs@ncc.moc.kw

Kiang, Nelson Y.
Mass. Eye & Ear Infirmary
243 Charles Street
Boston, MA 02114
Phone: 1.617.573.3745
Fax: 1.617.720.4408
bnk@epl.meei.havard.edu
Doctorate
20c, Asia, Medical Sciences;
Asia, Philosophy and
Philosophy of Science; Europe,
Medical Sciences; North
America, Cognitive Sciences

Kidd, R. Garth
University of Western Ontario
Dept of Chemistry
London, ON N6A5B7
Canada
Phone: 1.519.661.2111
Fax: 1.519.661.3022
rkidd@julian.uwo.ca
Doctorate, 1962
Physical Sciences; Philosophy
and Philosophy of Science

Kidwell, Peggy A.
Smithsonian Institution
PO Box 37012
NMAH Room 5125 MRC 671
Washington, DC 20013-7012
Phone: 1.202.357.2392
Fax: 1.202.357.1853
kidwellp@si.edu
Doctorate, 1979
20c, Astronomical Sciences;
Mathematics; Cognitive Sciences

Kiger, Robert W.
Carnegie Mellon University
Hunt Institute
5000 Forbes Ave
Pittsburgh, PA 15213-3890

Phone: 1.412.268.2434
Fax: 1.412.268.5677
rkiger@andrew.cmu.edu
Doctorate, 1972
Biological Sciences; Education

Kikuchi, Yoshiyuki
2-15-1-404 Nishi-Kameari
Katsushika-ku
Tokyo 124-0002
Japan
Phone: 81.3.5629.4335
Fax: 81.3.5629.4335
yoshik@mx6.ttcn.ne.jp

Kim, Boumsoung
Research Ctr for Sci & Tech
Program in Sci & Tech Studies
4-6-1 Komaba, Meguro-ku
Tokyo 153-8904
Japan
Phone: 81.3.5452.5297
Fax: 81.3.5454.5299
kimbs@mr.rcast.u-tokyo.ac.jp

Kim, Dong-Won
KAIST
School of Humanities & Soc Sci
373-1 Kusong-Dong, Yusong-Gu
Taejon 305-701
South Korea
Phone: 042.869.4629
Fax: 042.869.4610
dwkim3@yahoo.com
Doctorate, 1991
19c, Europe, Physical Sciences;
19c, Europe, Institutions;
20c, Asia, Education;
20c, Asia, Physical Sciences

Kim, Mi Gyung
North Carolina State Univ
Dept of History
Raleigh, NC 27695-8108
Phone: 1.919.513.2230
Fax: 1.919.515.3886
kim@social.chass.ncsu.edu
Doctorate, 1990
18c, Europe, Physical Sciences

Kim, Sang-Hyun
Harvard University
Department of the History of
Science
Science Center 235

Cambridge, MA 02138
Phone: 1.617.493.3397
kim13@fas.harvard.edu

Kim, Yung Sik
Seoul National University
Dept of Chemistry
Seoul 151-742
South Korea
Phone: 02.880.6637
Fax: 02.873.0418
kysik@plaza.snu.ac.kr

Kimler, William C.
North Carolina State Univ
Dept. of History
Box 8108
Raleigh, NC 27695-8108
Phone: 1.919.513.2238
Fax: 1.919.515.3886
kimler@ncsu.edu
Doctorate
19c, Biological Sciences;
20c, Biological Sciences;
Biological Sciences

Kimmelman, Barbara A.
95 W Stratford Ave
Lansdowne, PA 19050-1946

Kingsland, Sharon E.
Johns Hopkins Univ
3400 N. Charles St
History of Science Dept
Baltimore, MD 21218-2686
Phone: 1.410.516.7505
Fax: 1.410.516.7502
sharon@jhunix.hcf.jhu.edu
Doctorate, 1981
20c, North America,
Biological Sciences;
19c, Natural and Human
History; Biological Sciences

Kipnis, Nahum
3200 Virginia Ave S Apt 304
Minneapolis, MN 55426-3635
Phone: 1.952.936.9094
nahumk_99@yahoo.com
Doctorate, 1984
19c, Europe, Physical Sciences;
19c, Europe, Philosophy and
Philosophy of Science

Kirby, David
720 North Aurora
Ithaca, NY 14850

Phone: 1.607.255.3810
dk246@cornell.edu
Doctorate, 1996
20c, Social Relations of Science;
Pre-400 B.C.E., Philosophy and
Philosophy of Science

Kircz, Joost G.
Kircz Research Amsterdam
Prins Hendrikkade 141
Amsterdam 1011 AS
The Netherlands
Phone: 31.20.625.8188
Fax: 31.20.421.1844
kircz@wins.uva.nl
Doctorate
Philosophy and Philosophy of
Science; Social Relations of
Science; Science Policy

Kirsanov, Vladimir S.
Russian Academy of Science
Inst For History of Science
Staropansky 1/5
Moscow 103012
Russia
Phone: 7.95.928.1307
Fax: 7.95.925.9911
vladimir@kirsanov.msk.ru
Doctorate
Mathematics;
17c, Physical Sciences;
Europe, Physical Sciences

Kisling, Vernon N., Jr.
University of Florida
PO Box 117011
Marston Library
Gainesville, FL 32611-7011
Phone: 1.352.392.2838
Fax: 1.352.392.4787
vkisling@mail.uflib.ufl.edu
Doctorate, 1980
Natural and Human History

Kita, Hiroshi
Kawasaki Univ. Med Welfare
288 Matsushima
Kurashiki 701 0193
Japan
Phone: 086.462.1111
Fax: 086.464.1109
kita@med.kawasaki-m.ac.jp
Doctorate
Medical Sciences;
Biological Sciences

Kjaergaard, Peter C.
University of Aarhus
History of Ideas Building 328
NDR Ringgade 1
Aarhus DK-8000
Denmark
Phone: 45.8942.2123
Fax: 45.8942.221
idepck@hum.nau.dk
Doctorate
19c, Physical Sciences;
Social Sciences; Philosophy and
Philosophy of Science

Klein, Judy L.
Mary Baldwin College
Dept of Economics
Staunton, VA 24401
Phone: 1.540.887.7053
Fax: 1.540.887.7137
jklein@mbc.edu

Klein, Martin J.
Yale University
Sloane Physics Laboratory
217 Prospect St., Box 208120
New Haven, CT 06520-8120
Phone: 1.203.432.3103
Fax: 1.203.432.6175
Doctorate, 1948
20c, Physical Sciences;
19c, Physical Sciences

Klein, Michael
103 Mountain View Dr
Blacksburg, VA 24060-4819
Phone: 1.540.953.0562
Fax: 1.540.231.7991
mklein@vt.edu

Klein, Ursula
Max Planck Institute for the
History of Science
Wilhelmstr. 44
Berlin 10117
Germany
Phone: 49.30.2266.7301
Fax: 49.30.2266.7299
klein@mpiwg-berlin.mpg.de
Doctorate
18c, Europe, Physical Sciences;
North America, Physical
Sciences; 19c; 20c

Kleiner, Scott A.
University of Georgia
Department of Philosophy

Athens, GA 30602
Phone: 1.706.542.2826
skleiner@arches.uga.edu
Doctorate, 1968
20c, Biological Sciences;
19c, Biological Sciences;
20c, Philosophy and Philosophy
of Science

Kleinert, Andreas
Martin Luther Universität
Fachbereich Physik
Geschichte Der Naturwissensch
Kroellwitzer Strasse 44
Halle 06120
Germany
Phone: 49.3.4555.2542
Fax: 49.3.4555.27126
kleinert@physik.uni-halle.de
Doctorate, 1974
18c, Physical Sciences;
400 B.C.E.-400 C.E., Europe,
Physical Sciences; 19c, Europe,
Physical Sciences; Technology

Kleinman, Kim J.
Webster University
c/o Philosophy Dept
470 E. Lockwood
St Louis, MO 63119
Phone: 1.314.968.7144
Fax: 1.314.968.7166
kkleinma@swbell.net
Doctorate, 1997
Institutions; Biological Sciences;
Natural and Human History

Klemm, Matthew
815 Brown St
Iowa City, IA 52245-5902
Phone: 1.319.354.0249
mklemm@jhu.edu

Kline, Ronald
Cornell University
394 Rhodes Hall
Ithaca, NY 14850
Phone: 1.607.255.4307
rrk1@cornell.edu
Doctorate, 1983
20c, North America, Technology

Knight, David M.
Durham University
Dept. of Philosophy
50 Old Elvet
Durham DH1 3HN

Great Britain
Phone: 44.191.374.7641
Fax: 44.191.374.7635
d.m.knight@durham.ac.uk
Doctorate, 1964
19c, Europe, Physical Sciences;
19c, Europe, Humanistic
Relations of Science;
19c, Europe, Natural and
Human History;
19c, Europe, Institutions

Knoespel, Kenneth J.
2322 Fairoaks Rd
Decatur, GA 30033-1223
Phone: 1.404.894.8335
Fax: 1.404.315.0520
kenneth.knoespel@
  iac.gatech.edu

Knowles, Scott
1511 West Ave
Richmond, VA 23220-3721
Phone: 1.804.355.4487
opivska@aol.com

Kobayashi, Hiroyuki
3-7-A35-104
Shinsenrihigashimachi
Toyonaka 560-0082
Japan
Phone: 06.6871.4184
kobayasi@zinbun.kyoto-u.ac.jp

Koenig, Inge
1945 Cowper St
Palo Alto, CA 94301-3808
Phone: 1.650.321.2204
inge@dunmovin.com
Doctorate
Physical Sciences; Philosophy
and Philosophy of Science;
Humanistic Relations of Science

Koerner, Stephanie
101 N Dithridge St Apt 401
Pittsburgh, PA 15213-2651
Phone: 1.412.681.6724
venice@pitt.edu

Kohjiya, Shinzo
Kyoto University
Institute for Chemical Research
Uji, Kyoto 611-0011
Japan
kohjshin@scl.kyoto-u.ac.jp

Kohler, Robert E.
University of Pennsylvania
History & Sociology of Science
303 Logan Hall
249 S. 36th Street
Philadelphia, PA 19104-6304
Phone: 1.215.898.7098
Fax: 1.215.476.5856
rkohler@sas.upenn.edu
Doctorate, 1965
Biological Sciences;
North America; Exploration,
Navigation and Expeditions;
Social Relations of Science

Kohlstedt, Sally Gregory
University of Minnesota
123 Pillsbury Hall
310 Pillsbury Ave SE
Minneapolis, MN 55455-0219
Phone: 1.612.624.9368
Fax: 1.612.625.3819
sgk@mailbox.mail.umn.edu
Doctorate, 1972
North America, Gender and
Science; Earth Sciences;
Natural and Human History;
19c, North America, Institutions

Koizumi, Kenkichiro
Bunkyo University
Faculty of Intl Studies
1100 Namegaya
Chigasaki 253
Japan
Fax: 0467.54.3722
koizumi@shonan.bunkyo.ac.jp
Doctorate
20c, Asia, Technology;
20c, Asia, Science Policy;
20c, Asia, Social Relations of
Science

Kojevnikov, Alexei
University of Georgia
Department of History
Athens, GA 30602
Phone: 1.323.692.1425
anikov@arches.uga.edu

Kolchinsky, Eduard
Russian Academy of Sciences
Institute for the History of
Science & Technology
Universitetskaya NAB 5
St Petersburg 1 199034
Russia

Phone: 328.47.12
Fax: 328.46.67
ihst@spb.org.ru
Doctorate, 1986
20c, Europe, Biological Sciences

Komarower, Patricia
Monash University
Dept of Biological Sciences
Wellington Rd
Clayton, VIC 3168
Australia
Phone: 61.3.9905.5720
Fax: 61.3.9905.4903
patricia@earth.monash.edu.au
Bachelor's Level, 1999
20c, Asia, Earth Sciences;
19c, Australia and Oceania,
Natural and Human History

Konvitz, Josef W.
O.E.C.D.
TDS
2 Rue Andre-Pascal
Paris Cedex 16 75775
France
Phone: 33.14.524.9742
Fax: 33.14.524.1668
Doctorate, 1973
17c, Europe, Earth Sciences;
18c, Europe, Social Sciences

Kopp, Carolyn
651 E 14th St #10F
New York, NY 10009
Phone: 1.212.490.9000
Fax: 1.212.916.5858
ckopp@tiaa-cref.org
Master's Level, 1977
19c, Europe, Biological Sciences

Korsmo, Fae L.
National Science Foundation
OPP Suite 755
4201 Wilson Blvd
Arlington, VA 22230-0002
Phone: 1.703.292.7431
Fax: 1.703.292.9082
fkorsmo@nsf.gov

Koshland, Daniel J., Jr.
Editor Science
1333 H St NW
Washington, DC 20005-4707

Kosits, Russ
University of New Hampshire
Dept of Psychology
Conant Hall
10 Library Way # WA1
Durham, NH 03824-3520
Phone: 1.603.659.4169
rkosits@cisunix.unh.edu

Koster, Thomas A.
6298 Colby St
Oakland, CA 94618-1283

Kowalyszyn, Pedro
1316 S Peking St
McAllen, TX 78501-1140
Phone: 1.956.686.8546
Fax: 1.956.618.1611
daedalus@hiline.net

Kox, A. J.
University of Amsterdam
Valckenierstraat 65
Amsterdam 1018XE
The Netherlands
Phone: 31.2.0664.2824
Fax: 31.2.0525.5778
kox@science.uva.nl

Kozlov, Boris
Institute for the History of
Science & Technology
Russian Academy of Science
Staropansky 1/5
Moscow 103012
Russia

Kozmetsky, George
1301 W 25th St Ste 300
Austin, TX 78705-4248

Krafft, Fritz A.
Schutzenstr 18
Weimar Lahn D-35096
Germany
Phone: 49.6421.77592
Fax: 49.6421.282.2878
krafft@mailer.uni-marburg.de
Doctorate, 1971
14c-16c, Europe, Humanistic
Relations of Science;
400 B.C.E.-400 C.E., Europe,
Astronomical Sciences;
17c, Physical Sciences;
20c, Medical Sciences

Kragh, Helge S.
Aarhus University
Building 521 Ny Munkegade
History of Science Dept.
Aarhus 8000
Denmark
Phone: 45.8942.3505
Fax: 45.8942.3510
ivhhk@ifa.au.dk
Doctorate, 1981
20c, Astronomical Sciences;
20c, Physical Sciences;
19c, Physical Sciences;
Technology

Kralingen, Klaas W.
Bronovo Hospital
Bronovolaan 5
The Hague 2597 AX
The Netherlands
Phone: 31.7.0312.4257
Fax: 31.7.0312.4425
Doctorate
Medical Sciences; Philosophy
and Philosophy of Science

Kramer, James
11275 Wingfoot Dr.
Boynton Bch, FL 33437
Phone: 1.561.736.1317

Kraus, H. P.
16 E 46th St
New York, NY 10017-2404
Phone: 1.212.687.4808
hpkraus@worldnet.att.net

Krause, David J.
Henry Ford Community College
Science Division
5101 Evergreen
Dearborn, MI 48128
Phone: 1.734.995.0785
djkrause@umich.edu
Doctorate, 1986
19c, North America, Earth
Sciences; Astronomical Sciences

Kreiling, F. C.
58 Corbin Pl
Brooklyn, NY 11235-4804

Kremer, Richard L.
Dartmouth College
Department of History
6107 Reed Hall
Hanover, NH 03755-3506

Phone: 1.603.646.2228
Fax: 1.603.646.3353
richard.kremer@dartmouth.edu
Doctorate, 1984
14c-16c, Education;
19c, Europe, Medical Sciences;
19c, Europe, Institutions

Kreuzman, Henry B.
College of Wooster
Department of Philosophy
Scovel Hall
Wooster, OH 44691
Phone: 1.330.263.2481
Fax: 1.330.263.2249
hkreuzman@acs.wooster.edu
Doctorate, 1990

Krider, E. Philip
University of Arizona
Inst of Atmospheric Physics
Tucson, AZ 85721-0081
Phone: 1.520.621.6836
Fax: 1.520.621.6833
krider@atmo.arizona.edu
Doctorate, 1969
Earth Sciences;
Physical Sciences;
Instruments and Techniques;
Social Relations of Science

Krieger, Martin H.
Univ of Southern California
School of Policy, Planning, and
Development
Los Angeles, CA 90089-0626
Phone: 1.213.740.3957
Fax: 1.213.740.1801
krieger@usc.edu
Doctorate, 1969
20c, Physical Sciences;
20c, Mathematics;
19c, Mathematics;
19c, Physical Sciences

Krige, John
Georgia Inst of Technology
School of History, Technology
and Society
Atlanta, GA 30332-0345
Phone: 1.404.894.7765
Fax: 1.404.894.0535
john.krige@hts.gatech.edu
Doctorate
20c, Europe, Physical Sciences;
20c, Europe, Astronomical

Sciences; 20c, Social Relations
of Science; 20c, Instruments and
Techniques

Krikorian, Abraham D.
PO Box 404
Port Jefferson, NY 11777-0404
Phone: 1.613.473.7016
adkrikorian@earthlink.net

Krinsky, Alan D.
Univ. of Wisconsin-Madison
Madison, WI 53706
krinsky@students.wisc.edu
Master's Level, 1995
19c, Europe, Medical Sciences;
19c, Europe, Institutions;
Social Sciences

Kroker, Kenton
781 Queen ST E, Apt 3
Toronto, ON M4M1H5
Canada
Phone: 1.416.78.1480
Fax: 1.514.398.1498
kenton.kroker@utoronto.ca

Kroll, Gary M.
Plattsburgh State University
Dept of History
Cvh 325
101 Broad St
Plattsburgh, NY 12901-2637
Phone: 1.518.564.2738
krollgm@plattsburgh.edu
Doctorate, 2000
20c, North America, Natural
and Human History; 20c, North
America, Earth Sciences;
20c, North America, Exploration,
Navigation and Expeditions;
20c, North America, Biological
Sciences

Kronick, David A.
Univ of Texas Health Sci Ctr
2830 Bee Cave St
San Antonio, TX 78284
dkronick@satx.rr.com
Doctorate
18c, Europe, Education;
18c, Europe, Physical Sciences;
18c, Europe, Natural and
Human History;
18c, Europe, Medical Sciences

Krupar, Jason
777 E 9th Ave
Denver, CO 80203-3377
Phone: 1.303.283.6789
jkrupar@hotmail.com
Doctorate, 2000
20c, North America, Physical
Sciences; 20c, North America,
Science Policy; 20c, North
America, Biological Sciences;
20c, North America, Technology

Kubbinga, Henk H.
University of Groningen
Faculty of Philosophy
A-WEG 30
Groningen NL9718CW
The Netherlands
Phone: 31.50.3636161
Fax: 31.50.5735609
h.h.kubbinga@philos.rug.nl
Doctorate, 1996
Biological Sciences;
Physical Sciences

Kucher, Michael P.
Univ. of Washington - Tacoma
IAS Box 358436
1900 Commerce Street
Tacoma, WA 98402-3100
Phone: 1.253.692.5839
Fax: 1.253.692.5718
kucher@u.washington.edu
Doctorate, 1998
5c-13c, Europe, Technology;
18c, Europe, Natural and
Human History; 5c-13c, Europe,
Biological Sciences; 20c, North
America, Biological Sciences

Kudzma, Thomas G.
Univ of Massachusetts - Lowell
Lowell, MA 01854
Master's Level
19c, Europe, Physical Sciences;
19c, Europe, Mathematics;
19c, Europe, Technology

Kuklick, Henrika
313 S 22nd St
Philadelphia, PA 19103-6505
Phone: 1.215.898.5893
hkuklick@sas.upenn.edu

Kumar, Deepak
Jawaharlal Nehru Univ
Zakir Husain Ctr for Edu Stu

School of Social Sciences
New Delhi 110067
India

Kupferberg, Eric
8 Watson St
Cambridge, MA 02139
Phone: 1.617.495.8739
edkupfer@mit.edu

Kupriyanov, Alexey
Russian Academy of Sciences
Inst of History of Sci &
Technology
Universiterskaia Nab 5
St Petersburg 1
Russia

Kuriyama, Shigehisa
Intl Research Center For
Japanese Studies
3-2 Oeyama-Cho Goryo
Nishikyo
Kyoto 61011
Japan
Phone: 81.75.335.2100
Fax: 81.75.335.2090
kuriya@nichibun.ac.jp

Kurtik, Genadii
Inst For History Sci & Techno
Staropansky 1/5
Moscow 103012
Russia

Kushner, David
604 Leonidas Ct
Raleigh, NC 27604-1977
Phone: 1.919.821.9165
kushner@bigfoot.com

Kushner, Howard I.
Emory University
Biology Dept
Rollins Bldg 2101
1510 Cliffon Rd
Atlanta, GA 30322
Phone: 1.404.727.4354
Fax: 1.404.727.2820
hkushne@emory.edu
Doctorate
20c, North America, Medical
Sciences; 19c, Europe, Medical
Sciences; 20c, North America,
Medical Sciences, Instruments

and Techniques; 20c, Europe,
Medical Sciences, Instruments
and Techniques

Kuslan, Louis I.
707 Mix Ave Apt 58
Hamden, CT 06514-2208
Phone: 1.203.288.3415
kuslan@scsu.ctstateu.edu
Doctorate, 1954
19c, North America; 19c, North
America, Physical Sciences;
19c, North America, Education

Kusukawa, Sachiko
Trinity College
Cambridge CB2 1TQ
Great Britain
Phone: 44.01.2233.3993
Fax: 44.01.2333.38564
sk111@cam.ac.uk
Doctorate
14c-16c, Europe, Medical
Sciences; 14c-16c, Europe,
Astronomical Sciences;
14c-16c, Europe, Natural
and Human History;
14c-16c, Europe, Education

Kutsko, Daniel
Jersey Village High School
7600 Solomon St
Houston, TX 77040-2199
Phone: 1.713.896.3428
Fax: 1.713.896.3438
kutsko@email.edu
Master's Level, 1989
Humanistic Relations of Science;
Education

Kuwahara-Goto, Motoko
5-1-47 Satsukigaoka
Ikeda-Shi
Osaka 563-0029
Japan
Phone: 81.727.51.7784
Fax: 81.6.6627.5331
kuwahara@andrew.ac.jp

Kuznetsova, Natalia
Russian Academy of Sciences
Inst For History Sci & Tech
Staropansky 1/5
Moscow 103012
Russia

Kwa, Chunglin
Plantage Parklaan 1
Amsterdam 1018SP
The Netherlands
Phone: 31.20.525.6593
kwa@pscw.uva.nl

La Berge, Ann F.
Virginia Tech
Program in Science and
Technology Studies
124 Lane Hall
Blacksburg, VA 24061
Phone: 1.540.231.7008
Fax: 1.540.231.7013
alaberge@vt.edu
Doctorate, 1974
19c, Europe, Medical Sciences;
19c, Europe, Biological Sciences,
Instruments and Techniques;
20c, North America,
Medical Sciences;
19c, Europe, Technology

La Duke, Jeanne
De Paul University
Department of Mathematics
2219 N Kenmore Ave
Chicago, IL 60614
Phone: 1.773.325.1342
Fax: 1.773.325.7807
jladuke@condor.depaul.edu
Doctorate, 1969
20c, North America, Mathematics

LaBar, Martin
Southern Wesleyan University
PO Box 1020
Central, SC 29630-1020

LaBarr, J. Douglas
PO Box 77
522 Route 100
Wardsboro, VT 05355
Phone: 1.802.896.9760
labarrj@indiana.edu

Labisch, Alfons
Inst History of Medicine
Universitätsstrasse GEB.23.12
Dusseldorf D40225
Germany
Phone: 49.2.6081.1394
Fax: 49.2.6081.13949
histmed@uni-duesseldorf.de
Doctorate
20c, Europe, Medical Sciences;

19c, Europe, Social Sciences;
18c, North America, Philosophy
and Philosophy of Science;
Pre-400 B.C.E., Asia,
Humanistic Relations of Science

Lachapelle, Sofie
203, des Fondateurs
Aylmer, PQ J9J1M4
Canada
Phone: 1.819.684.7012
sofie.lachapelle.1@nd.edu

Lacy, Cherilyn M.
Hartwick College
Department of History
Oneonta, NY 13820
Phone: 1.607.431.4885
lacyc@hartwick.edu
Doctorate, 1997
19c, Europe, Medical Sciences;
Social Relations of Science;
Gender and Science;
Humanistic Relations of Science

Lafortune, Keith R.
13 Pine Ridge Rd
Saco, ME 04072-2118
Phone: 1.207.282.3407
keith.r.lafortune.1@nd.edu

Lagueux, Olivier
Yale University
Yale School of Medicine
333 Cedar St., L-132 SHM
New Haven, CT 06520-8015
Phone: 1.514.336.7800
Fax: 1.514.521.1686
lolivier@mac.com
Doctorate, 2001
19c, Europe, Biological Sciences;
19c, Europe, Natural and
Human History;
18c, Europe, Biological Sciences;
18c, Europe, Exploration,
Navigation and Expeditions

Laird, W. R.
Carleton University
Dept of History
Ottawa, ON K1S5B6
Canada
Phone: 1.613.520.2600
Fax: 1.613.520.2819
wrlaird@ccs.carleton.ca
Doctorate, 1983
5c-13c, Europe; 14c-16c, Europe

Lajus, Yulia
Russian Academy of Sciences
Inst of History Science & Tech
Universiterskaia Nab 5
St Petersburg 1
Russia

Lamberti, Pedro W.
Ciudad Universitaria
FaMAF - UNC
Cordoba 5000
Argentina
Fax: 54.35.1481.8360
lamberti@fis.uncor.edu

Lan, Richard M.
48 East 57th Street 4th Fl.
New York, NY 10022-2521
Phone: 1.212.308.0018
Fax: 1.212.308.0074
rlan@martayanlan.com/
14c-16c, Medical Sciences;
14c-16c, Technology,
Instruments and Techniques;
14c-16c, Earth Sciences;
14c-16c, Instruments and
Techniques

Lange, Mark F.
718 Nancy Ln
Madison, WI 53704-1304
mflange@cgirys.net
Technology

Langer, Bernard
Ramapo College of New Jersey
School of TAS
505 Ramapo Valley Rd
Mahwah, NJ 07430-1680
Phone: 1.201.684.7716
blanger@ramapo.edu
Doctorate
17c, Europe, Philosophy
and Philosophy of Science;
18c, Europe, Humanistic
Relations of Science;
Europe, Physical Sciences;
Europe, Gender and Science

Langford, Martha W.
29-5400 Dalhousie Dr NW
Calgary, AB T3A2B4
Canada
Phone: 1.403.286.8008
mwlangford@ucalgary.ca

Langins, Janis
University of Toronto
Inst for Hist & Phil of Sci
Victoria College
91 Charles St West, Rm 310
Toronto, ON M5S1K7
Canada
Phone: 1.416.978.4950
jlangins@chass.utoronto.ca

LaRande, Thierry
Conservatoire Natl des Arts et
Musee des Arts et Metiers
292 Rue St Martin
Paris Cedex 03 75141
France
Phone: 33.1.5301.8266
Fax: 33.1.5301.8235
lalande.thierry@wanadoo.fr

Largent, Mark A.
University of Puget Sound
History Department
1500 N Warner
Tacoma, WA 98416-0033
Phone: 1.253.879.3977
mlargent@ups.edu
Doctorate, 2000
20c, North America, Biological
Sciences; 19c, North America,
Biological Sciences

Larmore, Charles E.
University of Chicago
1010 E 59th Street
Department of Philosophy
Chicago, IL 60637
clarmore@midway.uchicago.edu
Doctorate
Philosophy and Philosophy of
Science

Larrabee, Jeffrey C.
17493 Cedar Lake Cir
Northville, MI 48167-3202

Larson, Bruce
University of North Carolina
One University Heights
Economics, CPO 2110
Asheville, NC 28804-8509
Phone: 1.828.251.6562
Fax: 1.828.251.6572
blarson@unca.edu

Larson, Edward J.
University of Georgia
Department of History
Le Conte Hall
Athens, GA 30602
Phone: 1.706.542.2660
Fax: 1.706.542.2455
edlarson@uga.edu
Doctorate, 1984
Biological Sciences;
Medical Sciences;
Natural and Human History;
Exploration, Navigation and
Expeditions

Larson, James L.
2451 Ashby Ave
Berkeley, CA 94705-2034
Phone: 1.510.841.9752
Doctorate, 1965
18c, Europe, Natural and
Human History; 14c-16c,
Europe, Biological Sciences;
Europe, Biological Sciences;
Europe, Medical Sciences

Larsson, Ulf
bergsunds Strand 7, 6TR
Stockholm SE-11738
Sweden

Lashinsky, Diane
3425 NW 65th St
Seattle, WA 98117-6018
Phone: 1.206.361.4286
lashinskyd@aol.com

Lassman, Thomas C.
Chemical Heritage Foundation
315 Chestnut St
Philadelphia, PA 19106-2793
Phone: 1.215.925.222
Fax: 1.215.925.1954
thomas1@chemheritage.org

Laszlo, Pierre
Ecole Polytechnique
Laboratoire DCFI
Palaiseau Cedex 91128
France
Phone: 33.05.6572.8401
Fax: 33.05.6572.8404
clouds-rest@wanadoo.fr
Doctorate
17c, Europe, Physical Sciences;
18c, South America, Philosophy
and Philosophy of Science;

19c, Europe, Biological Sciences;
20c, Europe, Natural and
Human History

Lattis, James M.
Space Astronomy Laboratory
1150 University Ave
Madison, WI 53706-1302
Phone: 1.608.263.0360
Fax: 1.608.263.0361
lattis@sal.wisc.edu
Doctorate
14c-16c, Europe, Astronomical
Sciences; 20c, North America,
Astronomical Sciences;
17c, Europe, Humanistic
Relations of Science

Laubichler, Manfred D.
Arizona State University
Department of Biology
PO Box 871501
Tempe, AZ 85287-1501
Phone: 1.480.965.5481
manfredlaubichler@asu.edu
Doctorate, 1997
20c, Europe, Biological Sciences;
Biological Sciences;
19c, Europe, Biological Sciences

Laughran, Michelle
506 Avery Creek Pointe
Woodstock, GA 30188-2311
Phone: 1.770.516.6497
michelle@unive.it

Laumbach, Gerald D.
50 E 89th St
New York, NY 10128-1225

Launius, Roger D.
NASA Headquarters
NASA History Office Code Z
300 E St SW
Washington, DC 20546-0001
Phone: 1.202.358.0383
Fax: 1.202.358.2866
roger.launius@hq.nasa.gov
Doctorate, 1982
20c, North America,
Astronomical Sciences,
Philosophy and Philosophy of
Science; 20c, North America,
Technology; 20c, North America,
Social Sciences; 20c, North
America, Science Policy

Lavoie, Daniel J.
St Anselm College
Department of Biology
100 Saint Anselms Dr
Manchester, NH 03102-1310
Phone: 1.603.641.7161
Fax: 1.603.641.7116
dlavoie@anselm.edu
Doctorate, 1977
Medical Sciences; Cognitive
Sciences; Biological Sciences;
Philosophy and Philosophy of
Science

Lawler, Ronald G.
Brown University
Dept of Chemistry
Box H
Providence, RI 02912
ronald_lawler@brown.edu

Lawrence, Christopher
Wellcome Trust Centre
24 Eversholt St
London NW1 1AD
Great Britain
Phone: 01.7161.18559
Fax: 01.7161.18562
christopher.lawrence@uci.ac.uk
Doctorate
20c, Europe, Medical Sciences;
19c, Europe, Medical Sciences;
18c, North America, Philosophy
and Philosophy of Science;
17c, Natural and Human History

Lawrence, Susan C.
University of Iowa
History Department
Iowa City, IA 52242
susan-lawrence@uiowa.edu

Lazenby, Jill
4-120 Strachan Ave
Toronto, ON M6J3W4
Canada
Phone: 1.416.979.3546
jill.lazenby@utoronto.ca

Leader, Darian
Flat 2 - 5 Bulstrode St
London W1U 2JB
Great Britain

Leary, David E.
University of Richmond
Office Dean of Arts & Sciences

Richmond, VA 23173
Phone: 1.804.289.8416
Fax: 1.804.289.8818
dleary@richmond.edu
Doctorate, 1977
Cognitive Sciences;
Humanistic Relations of Science;
Social Relations of Science;
Social Sciences

Leavitt, Judith W.
Univ of Wisconsin Med School
1300 University Ave
1420 Medical Sciences Center
Madison, WI 53706-1532
Phone: 1.608.263.4560
Fax: 1.608.262.2327
jwleavit@facstaff.wisc.edu
Doctorate, 1975
19c, North America, Medical
Sciences; 19c, North America,
Gender and Science; 20c, North
America, Medical Sciences;
20c, North America, Gender and
Science

Lebedeva, Natalya
Russian Academy of Sciences
Institute for the History of
Science and Technology
Staropansky Per 1/5
Moscow
Russia

LeBlanc, Andre
Université du Quebec Montreal
CIRST
Case postal 8888,
succursale Centre-Ville
Montreal, PQ H3C3P8
Canada
Phone: 1.514.987.3000
Fax: 1.514.987.7726
andre.leblanc@internet.uqam.ca

Lecumberry, Carmen S.
University of Puerto Rico
Mayaguez Campus
RUM
Mayaguez, PR 00680
Phone: 1.787.832.4040
Fax: 1.787.265.3849
sitio@coqui.net
Master's Level
17c, Europe, Physical Sciences;
18c, Europe, Biological Sciences;

14c-16c, Europe, Earth Sciences;
14c-16c, Europe, Natural and
Human History

Lederer, Susan E.
Yale Univ. School of Medicine
Hist. of Med. Sect
333 Cedar
SHM L130, PO Box 208015
New Haven, CT 06520-8015
Phone: 1.203.785.4338
Fax: 1.203.737.4130
susan.lederer@yale.edu
Doctorate
North America, Medical
Sciences; North America,
Philosophy and Philosophy of
Science; 20c, North America,
Medical Sciences

Lee, Carmen
8872A Towanda St
Philadelphia, PA 19118-3628
Phone: 1.215.247.7753
raa1998@cswebmail.com
Doctorate, 1996
Education; Biological Sciences

Lee, Doogab
Seoul National University
Program in History and
Philosophy of Science
College of Natural Sciences
Seoul 151-742
South Korea
Phone: 82.2.2249.4767
Fax: 82.2.878.6544
hysdangsan@hanmail.net

Lee, Robert W.
Lassen Research
31695 Forward Rd
Manton, CA 96059-9310
Phone: 1.530.474.3966
Fax: 1.530.474.1112
rlee@lassen.com

Leegwater, Arie
Calvin College
Biochemistry Departmnt
3201 Burton St SE
Grand Rapids, MI 49546-4388
Phone: 1.616.957.6373
Fax: 1.616.957.6501
leeg@calvin.edu
Doctorate

20c, Physical Sciences;
Philosophy and Philosophy of
Science

LeFevre, Darcy
305 Chautauqua Ave
Norman, OK 73069-5503
Phone: 1.405.447.8914
dlefevre@ou.edu
Bachelor's Level, 2001

Lehner, Christopher A.
California Institute of Techno
Einstein Papers Project
363 S Hill Ave
Pasadena, CA 91125
Phone: 1.626.395.8044
lehner@albert.bu.edu
Doctorate
Physical Sciences;
20c, Physical Sciences;
Philosophy and Philosophy of
Science

Lehrach, Dirk
Brehmstr. 24
Düsseldorf 40239
Germany
Phone: 49.211.614878
Fax: 49.211.639758
dlehrach@hotmail.com
Doctorate, 1996
Biography; Social Sciences;
Philosophy and Philosophy of
Science; Physical Sciences

Leibel, Rhona
Dept of Philosophy
Metropolitan State University
St Paul, MN 55106
Phone: 1.651.772.3749
Fax: 1.651.773.7675
rhona.leibel@metrostate.edu

Leitao, Henrique
CFMC University of Lisbon
Av. Prof. GAMA Pinto 2
Lisboa 1649-003
Portugal
Phone: 351.2.1790.4876
Fax: 351.2.1785.4288
leitao@alf1.cii.fc.ul.pt
Doctorate
14c-16c, Europe,
Physical Sciences; 17c, Europe,
Mathematics; 5c-13c, Asia,

Astronomical Sciences;
Exploration, Navigation and
Expeditions

Lejbowicz, Max J.
Université Paris 1
UFR 03
3 Rue Michelet
Paris 75006
France
Phone: 33.01.53.73.71.11
Fax: 33.01.53.73.71.13
max.lejbowicz@aol.com
Master's Level, 1988
5c-13c, Europe, Philosophy
and Philosophy of Science;
14c-16c, Europe, Mathematics;
Astronomical Sciences;
Europe, Technology

Lenhoff, Howard M.
Univ. of California at Irvine
Irvine, CA 92697-2310
Phone: 1.949.824.7259
Fax: 1.949.642.7644
hmlenhof@uci.edu
Doctorate
18c, Europe, Biological Sciences

Lennox, James G.
University of Pittsburgh
Center for Philosophy of Science
817 Cathedral of Learning
Pittsburgh, PA 15260-6299
Phone: 1.412.624.1051
Fax: 1.412.624.3895
jglennox@pitt.edu
Doctorate, 1977
19c, Europe, Biological Sciences;
400 B.C.E.-400 C.E., Europe,
Philosophy and Philosophy
of Science; 19c, Europe,
Social Sciences

Lenoir, Timothy
Stanford University
History & Philosophy of Sci
Bldg 200, Rm 116
Stanford, CA 94305-2024
Phone: 1.650.723.2993
Fax: 1.650.725.0597
lenoir@stanford.edu
Doctorate

Leonard, Edward, Jr.
1435 Cloverly Ln
Jenkintown, PA 19046-1403
Phone: 1.215.886.2129
ecleonard@pol.net

Leonard, Robert J.
Univ of Quebec at Montreal
315 St. Catherine East
Dept of Economics R-5610
Montreal, PQ H2X3X2
Canada
Phone: 1.514.987.3000
Fax: 1.514.987.8494
leonard.robert@er.uqam.ca
Doctorate, 1991
20c, Social Sciences

Lepicard, Etienne E.
Jewish National & Univ Library
POB 34165
Jerusalem 91341
Israel
Fax: 972.2.0433164
etiennel@netvision.net.il
Doctorate, 1999
20c, Medical Sciences;
20c, Philosophy and Philosophy
of Science; Pre-400 B.C.E.,
Medical Sciences

Lerner, Lawrence S.
Calif State Univ-Long Beach
Dept of Physics & Astronomy
1250 Bellflower Blvd
Long Beach, CA 90840
Fax: 1.650.529.9369
lslerner@csulb.edu
Doctorate, 1962
17c, Humanistic Relations of
Science; 17c, Physical Sciences;
17c, Astronomical Sciences

Lesch, John E.
University of California
Department of History
Berkeley, CA 94720-2550
Phone: 1.510.642.5524
Fax: 1.510.643.5323
jlesch@socrates.berkeley.edu
Doctorate, 1977
20c, Europe, Medical Sciences;
20c, Europe, Biological Sciences,
Instruments and Techniques;
20c, Europe, Physical Sciences;
20c, North America, Biological
Sciences

Leslie, Stuart W.
Johns Hopkins University
Dept of History of Science
216B Ames Hall
Baltimore, MD 21218
Phone: 1.410.516.7738
Fax: 1.410.516.7502
swleslie@jhu.edu
Doctorate
20c, North America, Technology;
Asia, Science Policy;
Asia, Institutions; Asia

Lestition, Steven
328 S Stanworth Dr
Princeton, NJ 08540-3142
Phone: 1.609.258.3317
Fax: 1.609.258.2889
sol@princeton.edu

Levens, Joshua P.
500 W University Pkwy Apt
15M
Baltimore, MD 21210-3383
Phone: 1.410.243.0724
JOL12@jhunix.hcf-jhu.edu

Leventhal, Richard M.
School of American Research
PO Box 2188
Santa Fe, NM 87504
Phone: 1.505.954.7211
rml@sarsf.org
Doctorate, 1979
20c, North America, Philosophy
and Philosophy of Science;
20c, North America, Social
Sciences; 5c-13c, North
America, Social Sciences

Levere, Trevor H.
University of Toronto
IHPST Victoria College
Toronto, ON M5S1K7
Canada
Phone: 1.416.421.7261
trevor.levere@utoronto.ca

Levin, Miriam R.
Case Western Reserve Univ
Dept of History
10900 Euclid Ave
Cleveland, OH 44106-7107
Phone: 1.216.368.2624
Fax: 1.216.368.4681
mrl3@po.cwru.edu

Levine, Deborah
11 Lesley Ave
Somerville, MA 02144-2606
Phone: 1.617.905.2779
dilevine@fas.harvard.edu

Levinson, Mark
630 Giltner Ln
Edmonds, WA 98020-3001
Phone: 1.425.776.3624
Doctorate, 1964
Technology, Instruments and
Techniques; Mathematics;
Physical Sciences

Leviton, Alan E.
California Academy of Sciences
Golden Gate Park
San Francisco, CA 94118
Phone: 1.415.752.1554
aleviton@calacademy.org
Doctorate
19c, North America, Natural
and Human History; 20c, Asia,
Exploration, Navigation and
Expeditions; Earth Sciences

Levitt, Theresa
36 Dana St
Cambridge, MA 02138-4204
Phone: 1.617.497.4255
levitt@fas.harvard.edu

Lewenstein, Bruce V.
Cornell University
321 Kennedy Hall
Ithaca, NY 14853-4203
Phone: 1.607.255.8310
Fax: 1.607.254.1322
b.lewenstein@cornell.edu
Doctorate, 1987
Social Relations of Science;
Science Policy;
Humanistic Relations of Science;
Earth Sciences

Lewis, Albert C.
Indiana Univ - Indianapolis
Peirce Edition Project
425 University Blvd # CA 545
Indianapolis, IN 46202-5140
Phone: 1.317.274.1581
Fax: 1.317.274.2170
alewis2@iupui.edu
Doctorate, 1975
19c, North America,
Mathematics; 20c, North

America, Mathematics;
19c, Europe, Mathematics;
20c, North America, Education

Lewis, Dan
Huntington Library
1151 Oxford Rd
San Marino, CA 91108-1299
Phone: 1.626.405.2206
Fax: 1.626.449.5720
dlewis@huntington.org
Doctorate, 1997
Education;
Astronomical Sciences;
Humanistic Relations of Science;
Social Sciences

Lewis, Jeffrey
Ohio State University
230 W 17th Ave
Columbus, OH 43210-1361
Phone: 1.614.282.4638
lewis.317@osu.edu

Li, Shang-Jen
3F, No 18, Ln 134
Hsin-Yi Road, Sec 3
Taipei
Taiwan
Phone: 020.27080304
Fax: 020.27868834
shangli@pluto.ihp.sinica.edu.tw

Lightman, Bernard V.
York University
309 Bethune College
4700 Keele Street
Toronto, ON M3J1P3
Canada
Phone: 1.416.736.5164
Fax: 1.416.736.5892
lightman@yorku.ca
Doctorate, 1979
19c, Europe, Humanistic
Relations of Science;
19c, Europe, Biological Sciences;
19c, Europe, Gender and
Science; 19c, Europe,
Astronomical Sciences

Lilleleht, Erica
Seattle University
Dept of Psychology
900 Broadway
Seattle, WA 98122-4340
Phone: 1.206.296.5399
elillele@seattleu.edu

Limerick, Patricia N.
Univ of Colorado at Boulder
Center of the Americn West
Macky 229
Campus Box 282
Boulder, CO 80309
Phone: 1.303.492.4879
Fax: 1.303.492.1671
patricia.limerick@colorado.edu

Limoges, M. Camille
1206 Rue Lajoie
Outremont, PQ H2V1P1
Canada
Phone: 1.514.270.6107

Lindberg, David C.
University of Wisconsin
History of Science Dept
1180 Observatory Drive
Madison, WI 53706
Fax: 1.608.238.9515
dclindbe@facstaff.wisc.edu
Doctorate
*5c-13c, Europe, Physical
Sciences; 5c-13c, Europe,
Humanistic Relations of Science;
17c, Europe, Humanistic
Relations of Science*

Lindee, Susan
University of Pennsylvania
Dept of History & Soc of Sci
Suite 303 Logan Hall
249 S 36th St
Philadelphia, PA 19104-6304
Phone: 1.215.898.2271
Fax: 1.215.573.2231
mlindee@mail.sas.upenn.edu
Doctorate
*20c, North America, Biological
Sciences; 20c, North America,
Medical Sciences; 20c, North
America, Gender and Science*

Lindenmann, Jean
Obere Geerenstrasse 34
Gockhausen CH-8044
Switzerland
Phone: 41.8211270
Fax: 41.1.3820844
jean.lindenmann@
  access.unizh.ch

Lindner, Rudi Paul
University of Michigan
History Dept
Ann Arbor, MI 48109

Lindqvist, Svante
H Nobel Museum
Museum Director
PO Box 2245
Stockholm SE10316
Sweden
Phone: 46.8.5195.4281
Fax: 46.8.5195.4290
svante.lindquist@nobel.se

Lindsay, Debra
University of New Brunswick
Dept of History & Politics
PO Box 5050
St John, NB E2L4L5
Canada
Phone: 1.506.648.5757
Fax: 1.506.648.5799
dlindsay@unbsj.ca

Linhard, Frank
Inst für Geschichte der Naturw
Robert Mayer Str 1
Universität Frankfurt
Frankfurt/Main 60054
Germany
Phone: 49.69.79828397
Fax: 49.69.79823275
linhard@rz.uni-frankfurt.de
Doctorate, 1999
*20c, Physical Sciences;
Physical Sciences; Philosophy
and Philosophy of Science*

Linker, Beth
Yale Univ School of Medicine
Sect History of Medicine
Sterling Hall
PO Box 208015
New Haven, CT 06520-8015
Phone: 1.203.288.9868

Lipson, Carol S.
Syracuse University
239 HBC
Syracuse, NY 13244-1160
Phone: 1.315.443.1091
Fax: 1.315.443.1220
cslipson@syr.edu
Doctorate, 1971
*Pre-400 B.C.E., Africa, Medical
Sciences; 14c-16c, Europe,*

*Medical Sciences; 17c, North
America, Humanistic Relations
of Science; Gender and Science*

Listerud, M. B.
PO Box 937
Wolf Point, MT 59201-0937

Little, Michelle Y.
529 Taylor Ct Apt 5
Mountain View, CA 94043-3639
Phone: 1.650.960.1764
mich-little@nwu.edu
Doctorate
*Philosophy and Philosophy
of Science; 20c, Astronomical
Sciences; Biological Sciences*

Littman, Richard A.
University of Oregon
Dept of Psychology
Eugene, OR 97403
Phone: 1.541.346.4909
Fax: 1.541.346.4911
rlittman@darkwing.uoregon.edu
Doctorate
*Cognitive Sciences;
Europe, Cognitive Sciences;
Social Sciences;
Social Relations of Science*

Liu, Bing
Tsinghua University
Inst Science & Tech Studies
Sch of Humanities & Soc Sci
Beijing 100084
China

Livesey, Steven J.
University of Oklahoma
Dept. of History of Science
601 Elm St Rm 622
Norman, OK 73019-0315
Phone: 1.405.325.2213
Fax: 1.405.325.2363
slivesey@ou.edu
Doctorate, 1982
*5c-13c, Europe, Institutions;
400 B.C.E.-400 C.E.,
Europe, Education;
14c-16c, Europe, Biography*

Livingston, Julie
206 President St Apt 2
Brooklyn, NY 11231-3516
Phone: 1.718.797.9293
jliving@tulrich.com

Livingston, Katherine
1835 Phelps Pl NW
Washington, DC 20008-1849
Phone: 1.202.332.0923
kliving@attglobal.net
Master's Level, 1960
North America

Llana, James W.
215 W 88th St Apt 7H
New York, NY 10024-2354
Phone: 1.212.874.4013

Locher, Kurt H.
Dantonsschule
Wetzikon 8620
Switzerland
Phone: 41.1.9323680
Fax: 41.1.9324570
locher@tommasi.ch
Master's Level
Pre-400 B.C.E., Asia,
Astronomical Sciences;
Pre-400 B.C.E., Africa,
Humanistic Relations of Science

Loettgers, Andrea
California Institute of Tech
Div of Humanities and Soc Sci
Pasadena, CA 91125-7700
Phone: 1.626.395.4487
andreal@hss.caltech.edu
20c, Europe, Physical Sciences;
19c, North America,
Astronomical Sciences

Lofstrom, Irene
Duke University
Box 3170
Duke Medical Center
Durham, NC 27710-3170
Phone: 1.919.684.4250
Fax: 1.919.684.8034
lofst001@mc.duke.edu
Master's Level
Natural and Human History;
Earth Sciences

Logan, Gabriella B.
57 Drouin Ave
Ottawa, ON K1K2A6
Canada
glogan@magma.ca

Lohff, Brigitte
Abtl Medizingeschichte
Carl-Neuberg-Str-1

Hannover 30625
Germany
Phone: 49.05.1153.24277
Fax: 49.05.1153.25650
lohff.brigitte@mh-hannover.de
19c, Europe, Medical Sciences;
20c, Europe, Philosophy
and Philosophy of Science;
18c, Europe, Medical Sciences;
18c, Europe, Philosophy and
Philosophy of Science

Lome, Louis S.
4201 Wilson Blvd # 110-528
Arlington, VA 22203-1859
Phone: 1.703.524.8720
louis.lome@mda.osd.mil

Long, Pamela O.
3100 Connecticut Ave NW # 137
Washington, DC 20008-5100
Phone: 1.202.265.0225
Fax: 1.202.265.0902
71574.174@compuserve.com
Doctorate, 1979
14c-16c, Europe, Technology;
5c-13c, Europe, Humanistic
Relations of Science;
17c, Europe;
400 B.C.E.-400 C.E.

Longton, William H.
University of Toledo
Department of History
Toledo, OH 43606-2909
Phone: 1.419.530.2904
Fax: 1.419.530.4539
wlongto@uoft02.utoledo.edu
Doctorate, 1969
19c, North America;
18c, North America

Lopes, M. Margaret
Universidade de Campinas
Inst. de Geociencias UNICAMP
CP6152
13083-970 Campinas
São Paulo
Brazil
Phone: 55.19.3788.4571
Fax: 55.19.3289.1562
mmlopes@ige.unicamp.br
Doctorate
19c, South America, Natural
and Human History; 19c, South

America, Earth Sciences;
19c, South America, Gender and
Science

Lopez Denis, Adrian
8542 Alcott St
Los Angeles, CA 90035-3664
Phone: 1.310.852.3623
aldenis@ucla.edu

Lopez Sanchez, Juan F.
Universidad de Murcia
Paseo Alfonso XIII, 48
Cartagena 30203
Spain
Phone: 34.968.325513
Fax: 34.968.325433
juanf.lopez@upct.es
Doctorate, 1994
Astronomical Sciences; Physical
Sciences; Europe, Institutions;
18c, Exploration, Navigation
and Expeditions

Loraschi, Gian C.
Via Cernuschi 59
Varese I-21100
Italy
glorasc@tin.it

Lorenz, Philip J., Jr.
390 Onteora Ln
Sewanee, TN 37375-2639
jlorenz@sewanee.edu

Lorenzano, Pablo
Charcas 2508, 6 B
Buenos Aires 1425
Argentina
Phone: 54.11.4961.4392
Fax: 54.11.4365.7182
pablol@unq.edu.ar

Losee, John P., Jr.
Lafayette College
Dept of Philosophy
Pardee Hall
Easton, PA 18040
Doctorate, 1961
20c, Europe, Philosophy
and Philosophy of Science;
19c, North America, Philosophy
and Philosophy of Science;
18c, Philosophy and Philosophy
of Science; 17c, Philosophy and
Philosophy of Science

Louis Trudel, Jean
103-4570 Queen Mary Rd
Montreal, PQ H3W1W6
Canada
Phone: 1.514.344.5258
d364034@er.uqam.ca

Loveland, Jeff A.
Univ of Cincinnati
Romance Languages
716B Old Chemistry
Cincinnati, OH 45221-0377
Phone: 1.513.556.1843
jeff.loveland@uc.edu
Doctorate, 1994
17c, Europe, Natural and
Human History; 18c, Europe,
Natural and Human History;
18c, Europe, Philosophy and
Philosophy of Science

Low, Morris F.
University of Queensland
Asian Languages and Studies
Brisbane 4072
Australia
Phone: 61.7.3365.6935
Fax: 61.7.3365.6799
m.low@mailbox.uq.edu.au
Doctorate, 1993
20c, Asia, Physical Sciences;
20c, North America, Technology;
20c, Asia, Science Policy;
20c, Asia, Technology

Lowengard, Sarah
1080 Park Ave
New York, NY 10128-1167
sarahl@panix.com

Lowood, Henry
Stanford University Libraries
FLAC-Germanic
Green Library
Stanford, CA 94305
Phone: 1.650.723.4602
Fax: 1.650.725.1068
lowood@stanford.edu
Doctorate, 1987
Education; Technology;
18c, Europe, Institutions

Lu, Lingfeng
Univ of Sci & Tech of China
Dept of History of Science
Hefei, Anhui 230026
China

Phone: 86.551.3633864
llf@mail.ustc.edu.cn
Doctorate
18c, Asia, Astronomical Sciences;
20c, Asia, Science Policy;
20c, Asia, Social Sciences

Lucas, Arthur M.
Kings College London
James Clerk Maxwell Bldg
57 Waterloo Road
London SE1 8WA
Great Britain
Phone: 44.20.7848.3434
Fax: 44.20.7848.3430
arthur.lucas@kcl.ac.uk
Doctorate, 1972
19c, Australia and Oceania,
Biological Sciences; Education

Lucas, Harry, Jr.
Educational Advancement Fdn
PO Box 1844
Austin, TX 78767-1844
Phone: 1.512.469.3581

Luchins, Edith
53 Fordham Ct
Albany, NY 12209-1192
Phone: 1.518.276.2994
Fax: 1.518.276.4824
lunchie@rpi.edu

Lucier, Paul L. M.
Rensselaer Ploytechnic Inst.
Dept. of Science & Technology
Troy, NY 12180-3590
Phone: 1.617.253.8721
plucier@dibinst.mit.edu
Doctorate, 1994
19c, North America, Earth
Sciences; 19c, North America,
Technology; 19c, North America,
Physical Sciences; 19c, Europe,
Earth Sciences

Ludmerer, Kenneth M.
Washington University
School of Medicine
660 South Euclid, Box 8066
St Louis, MO 63110
Phone: 1.314.362.8073
Fax: 1.314.362.8015
Doctorate
Social Sciences;
Medical Sciences

Lumbreras, Angel V.
Trigal 10
Urbanizacion 'Las Perdices'
Bargas Toledo 45593
Spain
Phone: 34.925.493130
avalerol@wanadoo.es

Lunbeck, Elizabeth A.
Princeton University
129 Dickinson Hall
Dept of History
Princeton, NJ 08544
Phone: 1.609.258.5527
Fax: 1.609.258.5326
lunbeck@princeton.edu
Doctorate, 1984
20c, North America, Gender
and Science; 20c, North
America, Medical Sciences;
20c, Gender and Science

Lundgren, Anders
University of Uppsala
Dept. of Hist. of Science
Slottet Ingang AO
Uppsala S75237
Sweden
Phone: 46.1847.1158
Fax: 46.1850.4422
anders.lundgren@idehist.uu.se
Doctorate, 1979
18c, Europe, Physical Sciences;
19c, Europe, Instruments and
Techniques; Social Relations of
Science; Technology

Lustig, Abigail L.
Dibner Institute
Dibner Building, MIT E56-100
38 Memorial Dr
Cambridge, MA 02142-1347
Phone: 1.617.258.0514
Fax: 1.617.258.7483
alustig@dibinst.mit.edu

Lustig, Alice
Rockefeller University
1230 York Ave
New York, NY 10021
Phone: 1.609.683.0821

Luthy, Christopher H.
University of Nijmegen
Faculty of Philosophy
PO Box 9103
Nijmegen 6500 HD

The Netherlands
Phone: 31.24.361.5750
Fax: 31.24.361.5564
luethy@phil.kun.nl
Doctorate, 1995
17c, Europe, Physical Sciences;
14c-16c, Europe, Physical
Sciences; 17c, Europe,
Instruments and Techniques;
20c, Humanistic Relations of
Science

Lutzen, Jesper
University of Copenhagen
Department of Mathematics
Universitetsparken 5
Copenhagen 0 DK 2100
Denmark
Phone: 45.3532.0741
Fax: 45.3532.0704
lutzen@math.ku.dk
Doctorate
19c, Europe, Mathematics;
18c, Europe, Physical Sciences;
20c, North America;
17c, Europe

Lynch, John
Arizona State University
Barrett Honors College
Tempe, AZ 85287
Phone: 1.480.727.7042
john.lynch@asu.edu
Doctorate, 1993
19c, Europe, Biological Sciences;
Humanistic Relations of Science;
Europe, Biological Sciences;
North America, Biological
Sciences

Lynch, William T.
Wayne State Unversity
Interdisciplinary Studies Prog
5700 Cass Ave
2313 A/AB
Detroit, MI 48202
Phone: 1.313.577.4614
Fax: 1.313.577.8585
william.lynch@wayne.edu
Doctorate, 1996
17c, Europe, Instruments and
Techniques; Social Relations
of Science; Philosophy and
Philosophy of Science

Lynn, Michael R.
Agnes Scott College
Dept. of History
141 E College Ave
Decatur, GA 30030-3797
Phone: 1.404.471.5194
mlynn@agnesscott.edu
Doctorate, 1997
18c, Europe, Physical Sciences;
18c, Europe

Lyons, Albert S.
88 Central Park W
New York, NY 10023-5209
Phone: 1.212.787.0760
Fax: 1.212.579.3158

Maas, Ad
Quitzowstrasse 121
Berlin 10559
Germany
maas@science.uva.nl

Maas, Katherine
Vassar College
Box 28061
124 Raymond Ave
Poughkeepsie, NY 12604-0002
Phone: 1.718.451.3549

Macagno, Enzo
University of Iowa
HL 301D
Iowa Institute of Hydraulics
Iowa City, IA 52242
Phone: 1.319.335.5214
Doctorate, 1953
14c-16c, Europe, Physical
Sciences; 14c-16c, Europe,
Technology, Instruments and
Techniques; 14c-16c, Europe,
Mathematics

MacCallum, Monica M.
Univ of Melbourne
Dept of Hist & Phil of Science
Melbourne 3010
Australia
Phone: 61.3.9344.6556
Fax: 61.3.9344.7959
m.maccallum@
   hps.unimel.edu.au
Doctorate, 1959
19c, Australia and Oceania,
Natural and Human History;
Biological Sciences;
Medical Sciences

Machamer, Peter
University of Pittsburgh
1017 Cathedral of Learning
Pittsburgh, PA 15260
Phone: 1.412.624.5896
Fax: 1.412.624.6825
pkmach@pitt.edu
Doctorate, 1972
17c, Europe; Cognitive Sciences;
14c-16c; Philosophy and
Philosophy of Science

MacHenry, Trueman
York University
Dept of Mathematics
4700 Keele St
Toronto, ON M3J1P3
Canada
Phone: 1.416.736.5250
machenry@mathstat.yorku.ca

Mack, Pamela E.
Clemson University
Dept of History
Clemson, SC 29634-0527
Phone: 1.864.656.5356
Fax: 1.864.656.1015
pammack@clemson.edu
Doctorate, 1983
20c, North America, Technology;
20c, North America, Gender and
Science; 20c, North America,
Science Policy

MacKinnon, Edward M.
2045 Manzanita Dr
Oakland, CA 94611-1148
Phone: 1.510.339.2749
Fax: 1.510.339.2749
emackinnon@aol.com
Doctorate, 1960
20c, Philosophy and
Philosophy of Science;
20c, Physical Sciences;
Physical Sciences

Mackowski, Maura J.
Arizona University
PO Box 872501
Tempe, AZ 85233-2558
maura.mackowski@asu.edu
Doctorate
20c, Technology;
20c, Astronomical Sciences,
Philosophy and Philosophy

of Science; Exploration,
Navigation and Expeditions;
20c, Medical Sciences

MacLeod, Roy M.
University of Sydney
Dept of History
Glebe
Sydney 2006
Australia
Phone: 61.2.9351.2855
Fax: 61.2.9351.3918
roy.macleod@history.
  usyd.edu.au
Doctorate, 1967
19c, Europe, Natural
and Human History;
20c, Europe, Physical Sciences;
20c, Europe, Institutions;
20c, Europe, Science Policy

MacMillan, Malcolm
Deakin University
School of Psychology
Burwood 3125
Australia
Phone: 61392446846
Fax: 61392446858
m.macmillan@deakin.edu.au
Doctorate, 1992
Cognitive Sciences

MacNeil, Kevin
The Culver Academies
Box 13
1300 Academy Rd
Culver, IN 46511-1234
Phone: 1.219.842.2360
Fax: 1.219.842.8161
macneik@culver.org

MacPherson, Ryan C.
University of Notre Dame
History & Philosophy of Sci.
346 O'Shaughnessy
Notre Dame, IN 46556
Phone: 1.219.631.5015
Fax: 1.219.631.4268
ryan.macpherson.1@nd.edu
Doctorate
19c, Biological Sciences;
19c, Astronomical Sciences;
19c, Earth Sciences;
18c, Natural and Human History

Maerker, Anna
Cornell University
Dept of Science & Tech Studies
Clark Hall
Ithaca, NY 14853
akm23@cornell.edu

Maffioli, Cesare S.
Ecole Europeenne
23 Bd K. Adenauer
Luxembourg L1115
Luxembourg
Phone: 352.463.299
Fax: 352.463.295
cesare.maffioli@ci.educ.lu
Doctorate
14c-16c, Europe, Social
Relations of Science;
17c, Europe, Social Relations
of Science; 18c, Europe,
Social Relations of Science

Magner, Lois N.
1521 S.W. 58th Street
Cape Coral, FL 33914-8023
Phone: 1.941.945.6179
Fax: 1.941.945.6179
writing123@aol.com
Doctorate, 1968
Medical Sciences;
Biological Sciences

Magruder, Kerry V.
University of Oklahoma
History of Science Collections
401 W Brooks St Rm 521
Norman, OK 73019-6030
Phone: 1.405.447.6521
kvmagruder@mac.com
Master's Level, 1998
17c, Astronomical Sciences;
18c, Earth Sciences;
19c, Natural and Human
History; 14c-16c, Europe,
Medical Sciences

Mahoney, Michael S.
Princeton University
History of Science
303 Dickinson Hall
Princeton, NJ 08544-0001
Phone: 1.609.258.4157
Fax: 1.609.258.5326
mike@princeton.edu
Doctorate, 1967
Mathematics; Technology

Maienschein, Jane
Arizona State University
Philosophy Dept
Tempe, AZ 85287-2004
Phone: 1.480.965.6105
Fax: 1.480.965.0902
maienschein@asu.edu
Doctorate, 1978
19c, Biological Sciences;
Education

Maier, Clifford L.
15528 Grovedale
Roseville, MI 48066

Maier, Joseph J.
Whitman College
Dept of Philosophy
Olin Hall
Walla Walla, WA 99362
maierjj@whitman.edu
Master's Level, 1961
19c, Europe, Biological
Sciences, Biography;
19c, Europe, Biological Sciences;
19c, Biological Sciences

Maille, Bernard
Librairie Maille
3 Rue Dante
Paris 75005
France
Phone: 33.1.4325.5173
Fax: 33.1.4325.5173

Mainzer, Audrey
Princeton University
Program in History of Science
208 Dickinson Hall
Princeton, NJ 08544-1017
Phone: 1.609.258.6705
Fax: 1.609.258.5326
amainzer@princeton.edu

Maisel, Merry W.
San Diego Supercomputer
Center
UCSD0505
La Jolla, CA 92093-0505
Phone: 1.858.534.5127
Fax: 1.858.822.0504
maisel@sdsc.edu
Doctorate, 1999
20c, North America, Instruments
and Techniques

Major, John S.
144 West 27th Street, Apt 5F
New York, NY 10001
Phone: 1.212.929.7651
Fax: 1.212.924.3958
steelemajor@aol.com

Makari, George J.
Cornell Univ Medical College
Inst for History of Psychiatry
PO Box 140
New York, NY 10021-0012
Phone: 1.212.746.3091
Fax: 1.212.746.8892
gjmakari@med.cornell.edu
Doctorate, 1987
19c, Europe, Medical Sciences, Instruments and Techniques; 20c, Europe, Medical Sciences, Instruments and Techniques; 20c, North America, Medical Sciences, Instruments and Techniques

Malaquias, Isabel M.
Universidade de Aveiro
Dept de Fisica
Aveiro 3800
Portugal
imalaquias@fis.ua.pt

Malley, Marjorie C.
208 Old Dock Trail
Apex, NC 27502
Phone: 1.919.387.2922
mcmjmh@worldnet.att.net
Doctorate
19c, Physical Sciences; 20c, Physical Sciences

Malone, Robert J.
History of Science Society
Executive Director
University of Washington
Box 351330
Seattle, WA 98195-1330
Phone: 1.206.543.9366
Fax: 1.206.685.9544
hssexec@u.washington.edu
Doctorate, 1996
18c, North America, Earth Sciences; 19c, North America; 18c, North America

Malouf, Benoit
2033 Jean-Talon Est
Montreal, PQ H2E1V2

Canada
Phone: 1.514.593.7559
benoitmalouf@hotmail.com

Malpas, Constance A.
The New York Academy of Med
1216 Museum Mile
New York, NY 10029-5202
Phone: 1.212.822.7311
cmalpas@nyam.org

Mancosu, Paolo
Univ of California-Berkeley
Dept of Philosophy
Berkeley, CA 94720-2390
Phone: 1.510.642.5033
mancosu@socrates.berkeley.edu
Doctorate
Mathematics; Philosophy and Philosophy of Science

Mandelbrote, Scott H.
Peterhouse
Trumpington Street
Cambridge CB2 1RD
Great Britain
Phone: 44.12.2333.8251
scott.mandelbrote@
   all-souls.ox.ac.uk
Doctorate, 1999
17c, Europe, Humanistic Relations of Science; 18c, Europe, Humanistic Relations of Science; 18c, North America, Humanistic Relations of Science

Mane-Garzon, Fernando
Casilla De Correo 157
Montevideo 11000
Uruguay
Phone: 209.7996
tmane@facmed.edu.uy
Doctorate, 1956
19c, South America, Social Sciences; Biological Sciences

Manier, Edward
University of Notre Dame
Department of Philosophy
225 Malloy
Notre Dame, IN 46556-4619
Phone: 1.219.631.6520
Fax: 1.219.631.8209
manier.1@nd.edu

Manning, Robert J.
Davidson College
Department of Physics
Davidson, NC 28036
Phone: 1.704.894.2386
Fax: 1.704.894.2720
bomanning@davidson.edu
Doctorate
20c, Astronomical Sciences; 17c, Europe, Astronomical Sciences; Humanistic Relations of Science

Marcano, Francisco A. T.
Suite # 0073
PO Box 025307
Miami, FL 33102-5308
Fax: 1.624.25.08
pmmarcano@hotmail.com

Marchand, Nicolas
7964 Rue Foucher
Montreal, PQ H2R2L1
Canada
Phone: 1.514.270.6967
n.marchand@cup.ac.ca
Master's Level
20c, North America, Cognitive Sciences; 20c, North America, Science Policy

Marche, Jordan D.
130 N Burr Oak Ave
Oregon, WI 53575-1308
tamarche@facstaff.wisc.edu
Doctorate, 1999
20c, North America, Astronomical Sciences; Earth Sciences; Natural and Human History; Education

Marcum, James A.
Baylor University
Department of Philosophy
PO Box 97273
Waco, TX 76798-7273
Phone: 1.254.751.9932
Fax: 1.254.710.3838
james_marcum@baylor.edu
Doctorate, 1995
Philosophy and Philosophy of Science; 20c, North America, Biological Sciences; 20c, North America, Medical Sciences

Marcus, Alan I.
Iowa State University
Dept of History
635 Ross Hall
Ames, IA 50011-0001
Phone: 1.515.294.5956
Fax: 1.515.294.6390
aimarcus@iastate.edu
Doctorate, 1979
19c, North America;
20c, Europe, Technology;
Social Sciences;
Medical Sciences

Mark, Joan T.
7 Clinton St # 3
Cambridge, MA 02139-2303

Mark, Kathleen
4900 Sandia Dr
Los Alamos, NM 87544-1850
Phone: 1.505.662.5025
20c, Earth Sciences;
Natural and Human History;
19c, North America, Earth
Sciences

Markel, Howard
University of Michigan
100 Simpson Memorial Inst.
102 Observatory Dr., Box 0724
Ann Arbor, MI 48109
Phone: 1.734.647.6914
Fax: 1.734.769.1080
howard@umich.edu
Doctorate, 1994
19c, North America, Medical
Sciences; 20c, North America,
Medical Sciences

Marker, Karl
Kerschlacher Str 2d
Munich D81477
Germany

Marks, Harry M.
Johns Hopkins University
Hist. of Sci, Medicine & Tech
1900 E Monument St
Baltimore, MD 21205-2113
Phone: 1.410.955.4899
Fax: 1.410.502.6819
hmarks@jhmi.edu
Doctorate, 1987
20c, North America, Medical
Sciences; 20c, North America,

Social Relations of Science;
20c, North America, Science
Policy

Marks, Jonathan
UNC-Charlotte
Dept of Sociology and Anthro
9201 University City Boulevard
Charlotte, NC 28223-0001
Phone: 1.704.687.2519
jmarks@email.uncc.edu
Doctorate, 1984
20c, Biological Sciences;
Social Sciences; Biological
Sciences; Natural and Human
History

Marks, Lawrence E.
John B Pierce Foundation Lab
290 Congress Ave
New Haven, CT 06519-1403
Phone: 1.203.562.9901
Fax: 1.203.624.4950
marks@jbpierce.org
Doctorate, 1965
19c, Cognitive Sciences;
20c, Cognitive Sciences

Marontate, Jan L.
Acadia University
Sociology Dept
Wolfville, NS B0P1X0
Canada
Phone: 1.902.585.1432
Fax: 1.902.585.1769
jan.marontate@acadiau.ca
Doctorate, 1997
20c, Physical Sciences;
19c, Mathematics;
20c, Technology;
19c, Physical Sciences

Maroulis, George
University of Patras
Dept of Chemistry
Patras 26500
Greece
Phone: 30.6199.7142
Fax: 30.6199.7118
marou@upatras.gr
Doctorate
20c, Physical Sciences;
Physical Sciences; Philosophy
and Philosophy of Science

Marr, Alexander
New College
Oxford OX13BN
Great Britain
Phone: 44.78.8191.2821
alexander.marr@new.ox.ac.uk

Marrett, Cora B.
University of Wisonsin System
Senior V. Pres For Academic
Affairs
1220 Linden Drive
1624 Van Hise Hall
Madison, WI 53706
Phone: 1.413.545.2464
Fax: 1.413.545.2328
cmarrett@uwsa.edu
Doctorate
19c, North America, Medical
Sciences; 20c, North America,
Social Sciences

Marsden, Ben
University of Aberdeen
Cultural History Dept
Old Brewery
Aberdeen AB24 3UB
Great Britain
Phone: 44.012.4427.2637
Fax: 44.012.4427.3262
bmarsden@abdn.ac.uk
Doctorate, 1992
19c, Europe, Education;
19c, Technology;
19c, Technology, Instruments
and Techniques;
19c, Europe, Physical Sciences

Marsh, Allison
3728 Yolando Rd
Baltimore, MD 21218-2041
Phone: 1.410.662.9203
allisonmarsh@yahoo.com

Martensen, Robert C.
University of Kansas
History & Philosophy of Med
3901 Rainbow Blvd
Kansas City, KS 66160-7311
Phone: 1.913.588.7041
Fax: 1.913.588.7060
rmartens@kumc.edu
Doctorate, 1993
17c, Europe, Medical Sciences

Martin, John P.
Davis & Elkins College
Department of Chemistry
100 Campus Drive
Elkins, WV 26241
Phone: 1.304.637.1219
jmartin@dne.edu
Doctorate, 1962
19c, Biological Sciences;
Europe, Institutions;
Instruments and Techniques;
North America, Institutions

Martin, Julian J.
University of Alberta
Dept. of History & Classics
2-28 H M Tory Bldg
Edmonton, AB T6G2H4
Canada
Phone: 1.780.492.3270
Fax: 1.780.492.2878
julian.martin@ualberta.ca
Doctorate, 1988
17c, Europe, Social Sciences;
14c-16c, Europe, Medical
Sciences; Philosophy and
Philosophy of Science

Martinez, Robert M.
Quinnipiac College
Biology Department
Box 61
Mt. Carmel Ave.
Hamden, CT 06518
Phone: 1.203.281.8692
martinez@quinnipiac.edu
Doctorate, 1972
Biological Sciences;
19c, Europe, Biological Sciences,
Biography; Philosophy and
Philosophy of Science

Martins, Roberto A.
Universidade de Campinas
Caixa Postal 6059
Campinas 13081970
Brazil
Phone: 55.19.788.5516
Fax: 55.19.788.5512
rmartins@ifi.unicamp.br
Doctorate, 1987
Physical Sciences;
Biological Sciences;
Europe, Education;
South America, Education

Martinsen, Hanna E. H.
3342 Juanita Court
Mississauga, ON L5A3J6
Canada
Phone: 1.905.277.1821
prostar@interlog.com

Mason, Stephen F.
12 Hills Ave
Cambridge CB1 7XA
Great Britain
Phone: 44.012.2324.7827
jml48@cam.ac.uk
Doctorate, 1947
14c-16c, Europe,
Astronomical Sciences;
19c, Europe, Physical Sciences;
20c, Europe, Physical Sciences;
20c, Europe, Biography

Massard, Jos A.
Rue Des Romains 1A
Echternach L-6478
Luxembourg
Fax: 352.727513
jmassard@pt.lu

Massey, Walter
Morehouse College
President
830 Westview Dr SW
Atlanta, GA 30314-3773

Matchett, Karin E.
University of Minnesota
116 Church Street SE
Program in Hist of Sci & Tech
Tate Laboratory of Physics
Minneapolis, MN 55455
Phone: 1.612.626.8722
Fax: 1.612.624.4578
match001@tc.umn.edu
20c, North America, Biological
Sciences; North America,
Biological Sciences

Matsen, Herbert S.
605 South 34th Avenue
Yakima, WA 98902-3928
Phone: 1.509.457.0901
hmatsen@nwinfo.net
Doctorate, 1969
14c-16c, Europe, Philosophy
and Philosophy of Science;
14c-16c, Europe, Medical

Sciences; 5c-13c, Europe,
Philosophy and Philosophy of
Science

Matsunaga, Toshio
1-1 Manabino
Izumi Osaka 594-1198
Japan
Phone: 0721.62.3114
matunaga@andrew.ac.jp

Matsushita, Masaaki
Tokyo Inst of Psychiatry
2-1-8 Kamikitazawa
Setagaya-ku
Tokyo 157-8585
Japan
Phone: 81.3.3482.3073
Fax: 81.3.3682.3073
matsu@matsuzawa.
  hp.metro.tokyo.jp
Doctorate
Philosophy and Philosophy of
Science; 19c, Europe, Medical
Sciences; Medical Sciences,
Instruments and Techniques;
Social Relations of Science

Matta, Christina
Univ of Wisconsin - Madison
History of Science
7143 Social Sciences Bldg
1180 Observatory Dr
Madison, WI 53706-1320
cmatta@students.wisc.edu
Master's Level, 2001
20c, Gender and Science;
20c, Biological Sciences;
17c, Europe, Astronomical
Sciences; 18c, Europe,
Astronomical Sciences,
Philosophy and Philosophy of
Science

Matthews, J. Rosser
5415 Connecticut Ave, NW #640
Washington, DC 20015
Phone: 1.202.248.2379
Fax: 1.301.402.1434
matthewsjr@od.nih.gov
Doctorate, 1992
20c, North America,
Medical Sciences; 20c, Europe,
Medical Sciences; Mathematics

Matthews, Michael R.
Univ of New South Wales
School of Education Studies
Sydney NSW 2052
Australia
Phone: 61.2.9418.3665
m.matthews@unsw.edu.au
Doctorate, 1980
Philosophy and Philosophy of
Science; Education

Matthews, Steven
4615 SW 47th Way
Gainesville, FL 32608-4913
Phone: 1.352.271.5718
quidamviator@yahoo.com

Mattingly, James
Georgetown University
Philosophy Department
221 New North
Washington, DC 20057
Phone: 1.202.687.2592
jmm67@georgetown.edu
Doctorate, 2001
Physical Sciences;
Astronomical Sciences,
Philosophy and Philosophy
of Science; Mathematics;
Philosophy and Philosophy of
Science

Matvievskaya, G.
U1. Muminova 12,kv.18
Tashkent 700041
Uzbekistan

Matysiak, Angela
1513 March Dr Unit A
Grand Forks, ND 58204-4017
Phone: 1.701.594.2316
matysiak@gwu.edu

Mauskopf, Seymour H.
Duke University
Department of History
PO Box 90719
Durham, NC 27708-0719
Phone: 1.919.684.2581
Fax: 1.919.681.7670
shmaus@acpub.duke.edu
Doctorate, 1996
18c, Europe, Physical Sciences;
19c, Europe, Physical Sciences;
20c

Mauviel, Maurice
4 Rue de la Baronnie
F 14440 Douvres-la
Delivrande
France
Phone: 33.231.3748.70
mauviel-f-c-@mail.cpod.fr

May, John G.
2620 Mantilla Corte
Walnut Creek, CA 94598-3517
Phone: 1.925.937.5506
rockett88@earthlink.net

Mayer, Anna K.
Darwin College
Cambridge CB3 9EU
Great Britain
Fax: 44.12.2333.3008
akm12@cus.cam.ac.uk

Mayer, Laura L.
Via Fragante 2 Talpuente
14460 Mexico DF
Mexico
Phone: 52.5622.3560
Fax: 52.5616.2670
lmayer@servidor.unam.mx

Mayoral De Lucas, Juan V.
C Valdivias 2, 2
Toledo 45003
Spain
jvmlucas@hotmail.com

Mazzotti, Massimo
Dibner Institute
MIT E56-100
38 Memorial Dr
Cambridge, MA 02142-1347
Phone: 1.617.253.3997
mmazzotti@dibinst.mit.edu
Master's Level, 1999
Europe, Mathematics;
Social Relations of Science;
Social Sciences

McAllister, James W.
University of Leiden
Faculty of Philosophy
PO Box 9515
Leiden 2300 RA
The Netherlands
Phone: 31.71.527.2004
Fax: 31.71.527.2028
j.w.mcallister@let.leidenuniv.nl
Doctorate, 1989

Philosophy and Philosophy of
Science; Humanistic Relations of
Science; Biological Sciences

McCabe, Irene M.
Royal Inst of Great Britain
21 Albermarle St
London W1S4BS
Great Britain

McCabe, Linda C.
8878 Wisteria Way
Windsor, CA 95492-8372
Fax: 1.707.837.8757
lmccabe@sonic.net

McCalman, Janet
University of Melbourne
Ctr For Study/Health & Society
Parkville, VIC 3010
Australia
Phone: 61.3.8344.0053
Fax: 61.3.8344.7959
j.mccalman@cshs.
    unimelb.edu.au

McCarl, George W.
4287 Persimmon Woods Dr
N Charleston, SC 29420-7518
Phone: 1.843.767.3123
wmpmccar@bellsouth.net
Doctorate
14c-16c, Europe, Philosophy
and Philosophy of Science;
Philosophy and Philosophy of
Science

McClellan, James E.
Stevens Inst of Technology
Humanities Dept.
Castle Point on Hudson
Hoboken, NJ 07030
Phone: 1.201.216.5395
Fax: 1.201.216.8245
jmcclell@stevens-tech.edu
Doctorate, 1975
18c, Europe, Institutions;
Europe, Social Relations
of Science; North America,
Exploration, Navigation and
Expeditions

McClintock, Robert
106 Morningside Dr #62
New York, NY 10027-6011

Phone: 1.212.866.3368
Fax: 1.212.678.8227
rom2@columbia.edu

McCluskey, Stephen C.
West Virginia Univ
Dept of History
202 Woodburn Hall
Morgantown, WV 26506-6303
Phone: 1.304.293.2421
scmcc@wvnvm.wvnet.edu
Doctorate, 1974
5c-13c, Europe, Astronomical
Sciences; Pre-400 B.C.E., North
America, Astronomical Sciences;
5c-13c, Europe, Physical
Sciences; Pre-400 B.C.E.,
Social Relations of Science

McColl, Paul
5 Calendonia Ave
Eltham VIC 3095
Australia
Phone: 03.9439.2539
pmccoll@bigpond.net.au

McCollum, Clifford G.
6511 N Revere Dr
Kansas City, MO 64151-3989
Phone: 1.816.741.0987
cmccollumi@kc.rr.com

McConnell, Craig S.
California State University
Dept of Liberal Studies
PO Box 6868
Fullerton, CA 92834
Phone: 1.714.278.0404
Fax: 1.714.278.5820
cmcconnell@fullerton.edu
Doctorate, 2000
20c, Astronomical Sciences;
20c, Physical Sciences

McCook, Stuart G.
College of New Jersey
Dept of History
PO Box 7718
Trenton, NJ 08628-0718
Phone: 1.609.771.2258
mccook@tcnj.edu
Doctorate, 1996
19c, North America, Biological
Sciences; 20c, North America,
Biological Sciences

McCormick, George E.
4924 Torrey Pines Ct
Charlotte, NC 28226-7924
Phone: 1.704.541.8946
Fax: 1.704.541.8946
gmac@vnet.net

McCormick, Maureen A.
University of Oklahoma
Dept of the History of Science
601 Elm Ave Rm 622
Norman, OK 73019-0315
Phone: 1.405.325.2213
Fax: 1.405.325.2363
mmccorm@ou.edu
Master's Level, 2000
20c, North America,
Biological Sciences;
20c, Biological Sciences;
20c, Natural and Human History

McCray, W. Patrick
American Institute of Physics
Center for History of Physics
One Physics Ellipse
College Park, MD 20740
Phone: 1.301.209.3168
Fax: 1.301.209.0882
pmccray@aip.org

McCullough, Ernest J.
North Dakota State Univ.
1625 3rd Street South
Moorehead, MN 56560
Phone: 1.403.531.9130
Fax: 1.403.531.9136
erniejohn@canada.com
Doctorate
5c-13c, Philosophy and
Philosophy of Science;
20c, Philosophy and Philosophy
of Science; Philosophy and
Philosophy of Science

McDonald, Michael
Cornell University
Dept of Entomology
Comstock Hall
Ithaca, NY 14853-0901
mjm10@cornell.edu

McElligott, John F.
Eastern Illinois University
Dept of History
Charleston, IL 61920
Phone: 1.217.581.3039
cfjfm@eiu.edu

Doctorate
Biological Sciences;
19c, Biological Sciences;
400 B.C.E.-400 C.E.,
Astronomical Sciences

McEvoy, John G.
University of Cincinnati
Dept of Philosophy
206 McMicken Hall
Cincinnati, OH 45220-0374
Phone: 1.513.556.6337
Fax: 1.513.556.2939
john.mcevoy@uc.edu
Doctorate, 1976
18c, Europe, Physical Sciences;
Social Sciences; Philosophy
and Philosophy of Science;
Social Relations of Science

McGrath, Gary L.
714 W Euclid St
Pittsburg, KS 66762-4906
Phone: 1.620.235.4406
Fax: 1.620.235.4429

McGrath, Sylvia W.
Stephen F Austin State Univ
Box 13013
Department of History
Nacogdoches, TX 75962-0001
Phone: 1.936.468.2452
Fax: 1.936.468.2478
smcgrath@sfasu.edu
Doctorate, 1966
North America; North America,
Gender and Science;
North America, Institutions;
North America, Natural and
Human History

McGuire, James E.
11 Ellsworth Ter
Pittsburgh, PA 15213-2808
Phone: 1.412.624.5896
jemcg@pih.edu

McGuire, William J.
Yale University
Department of Psychology
PO Box 208205
New Haven, CT 06520-8205
Phone: 1.203.432.4535
Fax: 1.203.432.7172
william.mcguire@yale.edu
Doctorate

Cognitive Sciences; Philosophy and Philosophy of Science; Social Sciences

McKnight, John L.
College of William & Mary
Physics Dept
PO Box 8795
Williamsburg, VA 23187-8795
Phone: 1.757.221.3521
Fax: 1.757.221.3540
jlmckn@physics.wm.edu
Doctorate, 1957
Instruments and Techniques;
20c, Physical Sciences;
18c, Social Relations of Science

McKnight, Stephen A.
University of Florida
History Department
226 Keene-Flint Hall
Gainesville, FL 32611
Phone: 1.352.392.0271
Fax: 1.352.392.6927
smcknigh@history.ufl.edu
Doctorate, 1972
14c-16c, Europe;
14c-16c, Europe, Philosophy and Philosophy of Science;
14c-16c, Europe, Gender and Science

McLaughlin, Peter C.
Max Planck Institute for the History of Science
Wilhelmstrasse 44
Berlin 10117
Germany
Phone: 49.75.3188.2567
Fax: 49.30.2266.7299
peter.mclaughlin@
  uni-konstanz.de
Doctorate, 1986
Philosophy and Philosophy of Science;
18c, Biological Sciences;
17c, Physical Sciences

McLaughlin-Jenkins, Erin K.
815-433 Jarvis Street
Toronto, ON M4Y2G9
Canada
Phone: 1.416.736.5164
erink@yorku.ca
Doctorate
19c, Europe, Social

Relations of Science;
19c, Europe, Education;
19c, Europe, Biological Sciences

McLelland, G.
UTS
Sch of Math Sci
PO Box 123
Broadway 2007
Australia
Phone: 602.9514.2259
gordon.mclelland@uts.edu.au

McLoughlin, Lisa A.
Rensselaer Polytechnic Inst.
Dept. of Sts, Sage Lab, 5th Fl
110 8th Street
Troy, NY 12180
mcloul@rpi.edu
Master's Level, 1992
Gender and Science;
19c, Europe, Mathematics;
Philosophy and Philosophy of Science

McMahon, Susan
Max Planck Institute for the History of Science
Wilhelmstr 44
Berlin D-10117
Germany
Phone: 49.30.226.67312
Fax: 49.30.226.67299
mcmahon@
  mpiwg-berlin.mpg.de
Master's Level
17c, Europe, Natural and Human History

McMullen, Emerson T.
Georgia Southern University
PO Box 8054
Statesboro, GA 30460
Phone: 1.912.871.1873
Fax: 1.912.681.0377
etmcmullen@gsvms2.gasou.edu
Doctorate, 1989
17c, Europe;
17c, Europe, Medical Sciences;
Technology

McOuat, Gordon R.
University of King's College
Dept of Contemporary Studies
History of Sci & Tech Programm
Halifax, NS B3H2A1
Canada

Phone: 1.902.422.1271
Fax: 1.902.423.3357
gmcouat@is.dal.ca

McRae, Robert J.
2606 Raymond Pl
Billings, MT 59102-1023
Phone: 1.406.248.6284
bevbob13@aol.com
Doctorate, 1969
19c, Europe, Physical Sciences;
18c, Physical Sciences;
20c, Astronomical Sciences

McTavish, J.R.
Alcorn State University
Social Sciences
Lorman, MS 39096
Phone: 1.601.661.7728
mctavish@lorman.alcorn.edu

McVaugh, Michael R.
University of North Carolina
History Department
CB 3195
Chapel Hill, NC 27599-3195
Phone: 1.919.962.2373
Fax: 1.919.962.1403
mcvaugh@email.unc.edu
Doctorate, 1965
5c-13c, Europe,
Medical Sciences;
17c, Europe, Education;
400 B.C.E.-400 C.E., Europe

Medeiros, Djalma
Rua Santos Dumont, 208
Ap. 51
Campinas SP 13024-020
Brazil
Phone: 551932558776
djalma@correionet.com.br

Medicus, Heinrich A.
Rensselaer Polytechnic Inst.
Physics Department
Troy, NY 12180
Fax: 1.518.276.6680
medich@rpi.edu
Doctorate, 1949
20c, Europe, Physical Sciences;
20c, Physical Sciences

Mehos, Donna C.
Nieuwend Ammerdijk 79
Amsterdam LD 1025
The Netherlands

Phone: 31.20.6325287
boon.mehos@wxs.nl
Doctorate
19c, Europe, Natural and
Human History;
19c, Europe, Biological Sciences;
19c, Europe, Medical Sciences

Mehra, Jagdish
New Territory
15 Turtle Creek Mnr
Sugar Land, TX 77479-5942
Phone: 1.281.565.5655
Fax: 1.281.565.9205

Meinel, Christoph
Univ of Regensburg
Institute of Philosophy
Wissenchaftgeschicte
Regensburg D-93040
Germany
Phone: 49.941.943.3659
Fax: 49.941.943.1985
christoph.meinel@psk.
  uni-regensburg.de
Doctorate, 1977
Europe, Physical Sciences;
Instruments and Techniques;
17c; 17c, Physical Sciences

Melamed, Lawrence
Dept. of Psychology
314 Kent Hall
Kent, OH 44242
Phone: 1.330.673.2986
lmelamed@kent.edu

Meldrum, Marcia L.
4053 Irving Place #9
Culver City, CA 90232
mylnnmel@earthlink.net

Melhado, Evan M.
University of Illinois
Dept of History
309 Gregory Hall
810 S Wright St
Urbana, IL 61801-3644
Phone: 1.217.333.6175
Fax: 1.217.333.8868
melhado@uiuc.edu
Doctorate, 1977
Earth Sciences; Medical
Sciences; Science Policy

Melia, Trevor
4152 55th Way N Apt 1031
Kenneth City, FL 33709-5647
Phone: 1.208.523.3832
trevor64@juno.com

Meloni, Ronald P.
197 Chatham St
New Haven, CT 06513-3123

Meltzer, David J.
Southern Methodist University
Dept of Anthropology
Dallas, TX 75275-0001
Phone: 1.214.768.2826
Fax: 1.214.768.2906
dmeltzer@mail.smu.edu
Doctorate, 1984
19c, North America, Natural
and Human History; 19c, North
America, Earth Sciences

Mendell, Henry R.
California State University
Philosophy Dept
5151 State University Dr
Los Angeles, CA 90032-4226
Phone: 1.323.343.4178
Fax: 1.323.343.4193
hmendel@calstatea.edu
Doctorate, 1986
400 B.C.E.-400 C.E., Philosophy
and Philosophy of Science;
400 B.C.E.-400 C.E.,
Mathematics;
400 B.C.E.-400 C.E.,
Astronomical Sciences;
Philosophy and Philosophy of
Science

Mendelsohn, Everett I.
Harvard University
Science Ctr 235
Cambridge, MA 02138
Phone: 1.617.495.9967
Fax: 1.617.495.3344
emendels@fas.harvard.edu
Doctorate, 1960
19c, Biological Sciences;
20c, Social Relations of Science;
20c, Science Policy;
20c, Biological Sciences

Mendelsohn, J. Andrew
Imperial College
Ctr for the History of Science
Technology & Medicine

Sherfield Bldg
London SW72AZ
Great Britain
Phone: 4420.7594.9362
Fax: 4420.7594.9353
a.mendelsohn@ic.ac.uk
Doctorate, 1996
Biological Sciences; Medical
Sciences; Social Sciences

Menkes, Diana D.
Casa Jacaranda
Praia Da Luz
Lagos 8600
Portugal
Phone: 351.82.789.323
Bachelor's Level

Mentsin, Yulii L.
Inshenernaia ul.7, apt. 23
Moskovskaia Oblast
Sergiev Posad 141300
Russia
Doctorate, 1987
17c, Europe, Philosophy
and Philosophy of Science;
18c, Europe, Physical Sciences;
19c, North America, Science
Policy; 20c, Europe, Education

Meo, M.
Grant High School
2245 NE 36th Ave
Portland, OR 97212
Phone: 1.503.916.5160
Fax: 1.503.916.2695
mmeo@pps.k12.or.us
Master's Level
19c, Europe,
Astronomical Sciences;
19c, Europe, Mathematics;
19c, Europe, Education

Merchant, Carolyn
Univ of Calif-Berkeley
Dept of Envir Sci Policy Mngnt
135 Giannini Hall
Berkeley, CA 94720-3312
Phone: 1.510.642.0326
Fax: 1.510.643.8911
merchant@nature.berkeley.edu
Doctorate, 1967
North America, Biological
Sciences; Gender and Science;
20c, Instruments and Techniques

Meredith, Margaret O.
University of California
Department of History - 0104
9500 Gilman Drive
La Jolla, CA 92093
meredith@helix.ucsd.edu
Master's Level, 1999
*18c, North America, Natural
and Human History; 20c, Social
Sciences; 18c, North America,
Earth Sciences*

Merrill, Daniel D.
Oberlin College
Dept. of Philosophy
Oberlin, OH 44074
Fax: 1.440.775.8084
dandmerr@aol.com
Doctorate, 1962
*Philosophy and Philosophy
of Science; 19c, Europe,
Mathematics; Mathematics*

Mertens, Robert
Quinten Matsijslei 5
Mortsel 2640
Belgium
Phone: 03.2180698
metrob@ruca.ua.ac.be

Messinger, Susan F.
1400 N Lake Shore Dr Apt 11E
Chicago, IL 60610-6637
Phone: 1.312.664.0708

Metzler, Gabriele
Klenzestr 74
München D-80469
Germany
Phone: 89.20244829
gabriele.metzler@
  uni-tuebingen.de
Doctorate, 1994
*20c, Europe, Social Sciences;
20c, Europe, Physical Sciences;
20c, Europe, Technology*

Meyer, Steven
4429 Westminster Pl
Saint Louis, MO 63108-1812
Phone: 1.314.533.1531
sjmeyer@artsci.wustl.edu

Michel, John L.
5046 Aldrich Ave S
Minneapolis, MN 55419-1208
Phone: 1.612.825.0128

Master's Level
*20c, North America, Physical
Sciences; 19c, North America,
Physical Sciences*

Michler, Markwart
Ernst Putz Str 36
Bad Bruck 97769
Germany

Mickens, Ronald E.
Clark Atlanta University
233 James P. Brawley Dr.
Atlanta, GA 30314-4391
Phone: 1.404.880.6923
Fax: 1.404.880.6258
rohrs @ math.gatech.edu
Doctorate
*20c, North America, Physical
Sciences; 20c, North America,
Mathematics*

Middents, Paul W.
Olympic College
1600 Chester Avenue
Bremerton, WA 98310
Phone: 1.360.478.4676
pmiddents@bazillion.com
Master's Level, 1963
*17c, Europe,
Astronomical Sciences;
18c, Europe, Mathematics;
18c, Europe, Physical Sciences;
18c, Europe, Instruments and
Techniques*

Midis, Costa M.
Akarnanias 7
Athens 115 26
Greece

Midorikawa, Nobuyuki
2-1-5 Azuma Kashiwa-Shi
Chiba-ken 2770014
Japan
Phone: 81.0471.39.9880
midorika@ulis.ac.jp

Milam, Erika
746 W Main St Apt 201
Madison, WI 53715-1474
Phone: 1.608.260.8591
emilam@students.wisc.edu

Miles, Sara J.
Eastern University
1300 Eagle Rd

St Davids, PA 19087-3696
Phone: 1.610.341.5949
Fax: 1.610.225.5573
smiles@eastern.edu
Doctorate, 1988
*19c, Europe, Biological Sciences;
20c, North America, Humanistic
Relations of Science;
19c, Europe, Gender and Science;
20c, Gender and Science*

Millbrooke, Anne
Consulting Historian
PO Box 240
Havre, MT 59501-0240
Phone: 1.907.443.5581
anne27m@yahoo.com
Doctorate
*20c, Technology;
19c, North America, Technology;
19c, North America, Earth
Sciences; 19c, North America,
Institutions*

Miller, Arthur I.
University College London
Science & Technology Studies
Gower Street
London WC1E 6BT
Great Britain
Phone: 44.17.1391.1321
Fax: 44.17.1916.2425
a.miller@ucl.ac.uk
Doctorate, 1965
*20c, Europe, Cognitive Sciences;
20c, Europe, Philosophy
and Philosophy of Science;
20c, Europe, Physical Sciences*

Miller, Dodd W.
2503 N 83rd St
Wauwatosa, WI 53213-1026
Phone: 1.414.302.9878
anaxagoras@att.net

Miller, Donald
2862 Waverly Way
Livermore, CA 94550-1740

Miller, Jack M.
Brock University
Dept of Chemistry
500 Glenridge Ave
St Catherine, ON L2S3A1
Canada
Phone: 1.905.688.5550
Fax: 1.905.682.9020

jimiller@brocku.ca
Doctorate, 1964
Physical Sciences; Instruments
and Techniques; Technology

Miller, Jane A.
7252 Northmoor Dr
Saint Louis, MO 63105-2110
Phone: 1.314.863.7273
janemillerphd@cs.com

Millman, Arthur B.
114 Pine St
Belmont, MA 02478-2726
Phone: 1.617.489.1680
Fax: 1.617.265.7173
millmanab@aol.com

Mills, Eric L.
Dalhousie University
Dept of Oceanography
Halifax, NS B3H4J1
Canada
Phone: 1.902.494.3437
Fax: 1.902.494.3877
e.mills@dal.ca
Doctorate, 1964
19c, Earth Sciences;
20c, Earth Sciences;
19c, Natural and Human History

Millstein, Roberta L.
California State Univ Hayward
Dept of Philosophy
25800 Carlos Bee Blvd
Hayward, CA 94542-3042
Phone: 1.510.885.3546
Fax: 1.510.885.2123
rmillstein@csuhayward.edu
Doctorate, 1997
Philosophy and Philosophy of
Science; Biological Sciences

Milne, James L.
1109 Shasta Ln
Polson, MT 59860-3309
Phone: 1.416.883.3339

Mindell, David A.
MIT
77 Massachusetts Ave
Room E51-194
Cambridge, MA 02139-4307
Phone: 1.617.253.0221
Fax: 1.617.258.8118
mindell@mit.edu
Doctorate, 1996

Technology, Instruments and
Techniques; Technology;
Social Sciences

Minelli, Alessandro
Univ degli Studi di Padova
Dipartimento di Biologia
Via U. Bassi 58B
Padova I35131
Italy
Phone: 39.49.827.6303
Fax: 39.49.827.6230
almin@civ.bio.unipd.it
18c, Europe, Natural and
Human History; 19c, Europe,
Natural and Human History

Minter, Adam
23 Inner Dr
Saint Paul, MN 55116-1859

Mirowski, Philip
University of Notre Dame
400 Decio Hall
Notre Dame, IN 46556-5644
Phone: 1.219.631.7580
Fax: 1.219.631.4794
mirowski.1@nd.edu
Doctorate, 1979
20c, North America, Social
Sciences; 19c, Europe, Social
Sciences; 20c, North America,
Science Policy; 20c, North
America, Cognitive Sciences

Misa, Thomas J.
Illinois Inst. of Technology
Department of Humanities
Chicago, IL 60616
Phone: 1.312.567.7967
Fax: 1.312.567.5187
misa@iit.edu

Mitman, Gregg
University of Wisconsin
Dept of History & Medicine
1420 Medical Sciences Center
1300 University Ave
Madison, WI 53706-1510
Phone: 1.608.262.9140
Fax: 1.608.262.2327
gmitman@med.wisc.edu
Doctorate
20c, North America,
Biological Sciences;
20c, Biological Sciences;
Social Relations of Science

Miura, Nobuo
Kobe University
Cross-Cultural Studies
1-2-1 Tsurukabuto, Nada
Kobe 6578501
Japan
Phone: 81.78.803.7437
miuranob@kobe-u.ac.jp
5c-13c, Mathematics;
17c, Europe, Instruments
and Techniques;
5c-13c, Europe, Mathematics;
19c, Asia, Mathematics

Miyashiro, Akiho
14 Stonehenge Dr
Albany, NY 12203-2019
Phone: 1.518.482.0654
Doctorate, 1953
Earth Sciences; Philosophy and
Philosophy of Science

Mocholi, Christina S.
Avda. de los Pinares,
no 73,111
46012
Valencia
Spain

Moffett, John P. C.
The Needham Research Institute
East Asian History of Science
8 Sylvester Road
Cambridge CB3 9AF
Great Britain
Fax: 44.12.2336.2703
jm10019@cus.cam.ac.uk

Mole, Philip
7949 W Country Club Ln
Elmwood Park, IL 60707-3533
Phone: 1.708.456.6079
pmolejr@mostardiplattenv.com

Molinori, Bruno
R Dr reng - 12
São Paulo 01331020
Brazil
Phone: 55.11.4277.4439
Fax: 55.11.3262.06022
sportbru@aol.com.br

Mollan, R. Charles
Samton Limited
17 Pine Lawn
Newtownpark Avenue
Blackrock

Ireland
Phone: 353.289.5834
Fax: 353.289.7618
cmol@iol.ie
Doctorate
19c, Europe, Instruments
and Techniques;
19c, Europe, Biography;
19c, Europe, Physical Sciences

Molland, George
University of Aberdeen
King's College
Aberdeen AB24 3FX
Great Britain
Phone: 44.1779.841675
Fax: 44.1779.841805
molland@globalnet.co.uk
Doctorate, 1967
5c-13c, Europe, Mathematics;
14c-16c, Europe,
Physical Sciences;
17c, Europe, Philosophy
and Philosophy of Science;
400 B.C.E.-400 C.E., Asia

Molvig, Ole
Princeton University
Dept of History
207 Dickinson Hall
Princeton, NJ 08544-1174
ormolvig@princeton.edu

Monsen, Arve
Steinborgveien 36
Oslo 0678
Norway
Phone: 472.268.8546
arve.monsen@tik.uio.no

Montgomery, William M.
Mass College of Liberal Arts
Church Street
Interdisciplinary Studies Dept
North Adams, MA 01247
Phone: 1.413.662.5516
Fax: 1.413.662.5010
wmontgom@mcla.mass.edu
Doctorate, 1973
20c, North America,
Biological Sciences;
19c, Europe, Biological Sciences

Mooers, Charlotte
13 Bowdoin St
Cambridge, MA 02138-1705

Moon, Joong Yang
Unit. 1507-1401
Hugok-Maul, Ilsan3-dong
Koyang-shi
Kyong Gi-do 411736
South Korea
Phone: 82.344.914.4915
moonjym@dreamx.net
Doctorate, 1995
18c, Asia, Astronomical Sciences;
5c-13c, Asia, Biological Sciences;
5c-13c, Asia, Humanistic
Relations of Science

Moon, Suzanne M.
501 Golden Circle #203
Golden, CO 80401
smoon@mines.edu
Bachelor's Level
20c, Asia, Technology;
20c, Asia, Biological Sciences;
20c, Europe, Technology

Mooney Digrius, Dawn
PO Box 310
Pluckemin, NJ 07978-0310
Phone: 1.973.408.8129
ddigrius@drew.edu

Moore, Carl E.
Loyola University of Chicago
6525 N Sheridan Road
Chicago, IL 60626
Phone: 1.773.508.3100
Fax: 1.847.965.1614
drcemoore@aol.com
Doctorate, 1952
17c, Europe, Physical Sciences

Moore, Charles E.
5993 Klusman Ave
Alta Loma, CA 91737-2225
Phone: 1.909.944.3752
Fax: 1.909.483.4037
cemoore@alum.calberkeley.org
Master's Level, 1978
Social Relations of Science;
Philosophy and
Philosophy of Science;
Natural and Human History

Moore, Lara
Princeton University
Firestone Library
1 Washington Rd

Princeton, NJ 08544-2002
Phone: 1.609.430.2109
lmoore@princeton.edu

Mora, George
100 Shadow Farm Way # 13
Wakefield, RI 02879-3632
Phone: 1.401.783.5515

Moran, Bruce T.
Univ of Nevada
Dept of History
Reno, NV 89557-0001
Phone: 1.702.784.6677
Fax: 1.702.784.6805
moran@equinox.unr.edu
Doctorate
14c-16c, Europe, Social Relations
of Science; 14c-16c, Europe,
Physical Sciences;
17c, Europe, Medical Sciences;
17c, Europe, Physical Sciences

Morawski, Jill
Wesleyan University
Dept of Psychology
Judd Hall, 207 High Street
Middletown, CT 06459-0001
jmorawski@wesleyan.edu

Morey, Hector L.
10205 Baltimore Ave Apt 7104
College Park, MD 20740
Phone: 1.301.559.6256
bookbeetles@worldnet.att.net
Master's Level, 2004
Cognitive Sciences;
Social Sciences; North
America, Cognitive Sciences;
Medical Sciences, Instruments
and Techniques

Morf, Paul F.
University of Minnesota
Minneapolis, MN 55104-5035
morf0007@tc.umn.edu
Bachelor's Level, 1994
14c-16c, Europe, Astronomical
Sciences; 400 B.C.E.-400 C.E.,
Mathematics

Morgan, Jerry, Jr.
850 N 5th St
Baton Rouge, LA 70802-5263
Phone: 1.225.379.3228
revjerrymorgan@hotmail.com

Morgan, Mary S.
London School of Economics
Dept of Economic History
Houghton Street
London WC2A 2AE
Great Britain
Phone: 44.207.955.7081
Fax: 44.207.955.7730
m.morgan@lse.ac.uk
Doctorate, 1984
Social Sciences;
Instruments and Techniques

Morman, Edward T.
Francis C. Wood Institute for the
History of Medicine
Hist of Med
The College Phys of Philadelph
19 S 22nd St
Philadelphia, PA 19103-3001
Phone: 1.215.563.3737
Fax: 1.215.569.0356
emorman@collphyphil.org
Doctorate
19c, North America, Medical
Sciences; 20c, North America,
Medical Sciences

Morpurgo, Piero A.
PO Box 624
Vicenza 36100
Italy
Phone: 39.4.4432.0194
Fax: 39.4.4432.0194
pmorpurgo@libezo.it
Doctorate, 1988
5c-13c, Europe, Education;
5c-13c, Europe, Humanistic
Relations of Science;
14c-16c, Europe, Institutions;
5c-13c, Europe, Social Relations
of Science

Morrell, J. B.
8 Randall Place
Bradford, Yorks BD9 4AE
Great Britain
Phone: 1274.542946

Morris, Edward K.
University of Kansas
Dept of Human Development
1000 Sunnyside Ave
Lawrence, KS 66044-2133
Phone: 1.785.864.4840
Fax: 1.785.864.5202
ekm@ku.edu

Doctorate, 1976
20c, North America, Cognitive
Sciences; 20c, North America,
Philosophy and Philosophy of
Science; 20c, North America,
Social Sciences

Morris, Henry M.
I C R
PO Box 2667
El Cajon, CA 92021-0667

Morris, Susan W.
Johns Hopkins University
Ames Hall
Dept of Hist Sci Med & Tech
Baltimore, MD 21218-2318
Fax: 1.410.467.4436
Doctorate, 1999
19c, North America,
Physical Sciences;
19c, North America, Technology,
Instruments and Techniques;
19c, Europe, Social Sciences

Morrone, Juan J.
Museo de Zoologia
Fac Ciencias, UNAM
Apdo postal 70-399
Mexico DF 04510
Mexico
Phone: 52.5622.4832
Fax: 52.5622.4828
jjm@hp.fciencias.unam.mx

Morselli, Mario A.
49 Manor Woods Apts
Kennedy Dr
S. Burlington, VT 05403
Phone: 1.802.658.4631
Doctorate
20c, Europe, Physical Sciences

Morton, Alan
Science Museum
South Kensington
London SW7 2DD
Great Britain
Phone: 44.20.7942.4169
Fax: 44.20.7942.4102
a:morton@nmsi.ac.ur
Doctorate, 1982
20c, Physical Sciences;
18c, Europe, Physical Sciences

Morus, Iwan R.
Queens University
Sch of Anthropological Studies
14 University Square
Belfast BT71NN
Northern Ireland
Phone: 028.9027.3861
Fax: 028.9027.3700
i.morus@qub.ac.uk
Doctorate, 1989
19c, Europe, Physical Sciences;
19c, Technology;
Medical Sciences;
Instruments and Techniques

Mosley, Adam
Trinity College
Cambridge CB2 ITQ
Great Britain
ajm1006@cus.cam.ac.uk

Moss, Laurence S.
Babson College
Economics Dept
Mustard Hall
Babson Park, MA 02457
Phone: 1.617.728.4949
Fax: 1.617.728.4947
lmos@aol.com
Doctorate
19c, Europe, Social Sciences;
20c, North America, Social
Sciences; 20c, Europe, Social
Sciences; 19c, North America,
Social Sciences

Moy, Timothy D.
University of New Mexico
Dept of History
1104 Mesa Vista Hall
Albuquerque, NM 87131-1181
Phone: 1.505.277.7851
Fax: 1.505.277.6023
tdmoy@unm.edu
Doctorate, 1992
20c, North America;
20c, Technology;
Social Relations of Science;
Science Policy

Moyer, Albert E.
Virginia Tech
431 Major Williams Hall
Dept of Hist, Mail Code 0117
Blacksburg, VA 24061
Phone: 1.540.231.8361
Fax: 1.540.231.8724

Doctorate, 1977
19c, North America, Physical Sciences; 20c, North America, Physical Sciences; Education; Humanistic Relations of Science

Moynahan, Gregory B.
Bard College
History Dept
Rt 96
Annendale on Hudson, NY 12504

Moyzis, Joseph A., Jr.
5824 Pondview Dr
Kettering, OH 45440-2342
Phone: 1.937.434.6546
jmoyzis@aol.com

Mueller, Paul R.
University of Chicago
1126 E 59th St Room 207
Chicago, IL 60637-1621
p-mueller@uchicago.edu
Master's Level, 2001
17c, Europe, Physical Sciences; Philosophy and Philosophy of Science

Muendel, John
1948 Michigan Ave
Waukesha, WI 53188-4247
Phone: 1.262.521.2272
jmuen@juno.com
Doctorate, 1972
5c-13c, Europe, Technology

Müller-Wille, Staffan
Max Planck Institute for the History of Science
Wilhelm Str 44
Berlin D-10117
Germany
Phone: 49.30.2766.7107
Fax: 49.35.1484.6591
smuewi@mpiwg-berlin-wpg.de
Doctorate, 1997
20c, Europe, Biological Sciences; 20c, North America, Biological Sciences; 18c, Europe, Natural and Human History; 18c, Europe, Exploration, Navigation and Expeditions

Mullet, Shawn
GSAS Mail Center
Perkins 226

35 Oxford St
Cambridge, MA 02138-1956
Phone: 1.617.493.7111
smullet1@yahoo.com

Mulligan, Joseph F.
228 Canal Park Dr Apt G103
Salisbury, MD 21804
Phone: 1.410.546.4246
jmull68640@aol.com
Doctorate, 1951
19c, Europe, Physical Sciences; Physical Sciences; Humanistic Relations of Science; Biography

Multhauf, Robert P.
9 Nunan Ln
San Rafael, CA 94901-2220
Phone: 1.415.453.4002

Munns, David P.
Johns Hopkins University
Dept of Hist of Sci, Med, Tech
Ames 204
3400 N Charles St
Baltimore, MD 21218
Phone: 1.410.243.0724
dpm1@jhunix.hcf.jhu.edu
Master's Level
20c, Australia and Oceania, Astronomical Sciences; 14c-16c, Astronomical Sciences; 20c, North America, Education

Muntersbjorn, Madeline M.
University of Toledo
Department of Philosophy
Toledo, OH 43606
Phone: 1.419.530.4513
Fax: 1.419.530.6189
mmunter@uoft02.utoledo.edu
Doctorate, 1994
17c, Europe, Mathematics; 20c, North America, Biological Sciences; Philosophy and Philosophy of Science

Munz, Tania
University of Minnesota
435 Walter Library
Minneapolis, MN 55455
Fax: 1.612.934.6868
tmunz@princeton.edu
Bachelor's Level, 1995
19c, Europe, Biological Sciences;

20c, Biological Sciences; 19c, North America, Biological Sciences

Murad, Edmond
20 Kenrick Ter
Newton, MA 02458-2419
Phone: 1.781.377.3176
Fax: 1.781.377.3160
emurad@mediaone.net

Murad, Hazim B.
University of Malaya
Faculty of Science
Dept of Sci & Tech Studies
Kuala Lumpur 50603
Malaysia
Phone: 603.79674166
Fax: 603.79674396
j8hazim@umcsd.um.edu.my
Doctorate
Humanistic Relations of Science; Philosophy and Philosophy of Science

Murdoch, John E.
Harvard University
Science Ctr 235
Cambridge, MA 02138
Phone: 1.617.495.3480
Fax: 1.617.495.3344
murdoch@harvard.fas.edu

Murphy, Clarence J.
East Stroudsburg Univ
Dept of Chemistry
Stroudsburg, PA 18301-2999
Fax: 1.570.422.3709
cjmurphy@ptd.net
Doctorate, 1962

Murphy, Jane
404A Butler Ave
Princeton, NJ 08540-5648
Phone: 1.609.683.8816
jmurphy@princeton.edu

Mylott, Anne L.
Indiana University
History & Philosophy of Sci.
Goodbody Hall 130
Bloomington, IN 47405-7005
Phone: 1.812.855.3622
Fax: 1.812.855.3631
mylotta@indiana.edu
Master's Level
19c, Europe, Biological Sciences;

19c, Biological Sciences,
Instruments and Techniques;
20c, Biological Sciences

Naddeo, Barbara A.
538 W Belmont Ave
Chicago, IL 60657-4676

Nakajima, Hideto
Tokyo Institute of Technology
2-12-1, Ookayama Meguro-Ku
Tokyo 152-8552
Japan
Phone: 03.5734.3255
Fax: 03.5734.3255
nakajima@mail.me.titech.ac.jp
Doctorate
17c, Europe, Physical Sciences;
19c, Asia, Technology,
Instruments and Techniques

Nakamura, Akika
Akita University Sch of Med
Medical Information Science
Hondo 1-1-1
Akita 010-8543
Japan
Phone: 81.18.884.6096
Fax: 81.18.838.3317
nakamura@ipc.akita-u.ac.jp
Doctorate
17c, Asia, Biological Sciences;
18c, Asia, Physical Sciences;
19c, Asia, Medical Sciences;
5c-13c, Europe, Mathematics

Nakamura, Masaki
3-4-21 Toride
Torde-shi
Ibaraki 3020004
Japan
Phone: 81.297.72.0953
Fax: 81.297.72.0953
nakamura.masaki@nifty.ne.jp

Nakamura, Tuto
Ao Madani Nishi 2 Chome
1-9, Mino City
Osaka 562 0023
Japan

Nanney, David L.
University of Illinois at
Champaign-Urbana
515 Morrill Hall
505 S Goodwin Ave
Urbana, IL 61801-3707

Phone: 1.217.333.2308
d-nanney@uiuc.edu
Doctorate
20c, Biological Sciences;
20c, Biological Sciences,
Instruments and Techniques;
20c, Philosophy and Philosophy
of Science

Nappi, Carla S.
222B Marshall Ave
Princeton, NJ 08540
Phone: 1.609.252.9973
nappi@princeton.edu
Master's Level, 2000
Asia, Biological Sciences;
Europe, Biological Sciences;
Philosophy and Philosophy of
Science; Social Sciences

Nasiopoulos, Christos
9 Gymnasiarhou Dimitriadi Str
Thessaloniki 54655
Greece
Phone: 3.10415377
nasiopoulos@mail.gr
Doctorate, 2003
20c, Europe, Physical Sciences;
19c, Europe, Humanistic
Relations of Science;
20c, Europe, Social Sciences;
20c, Europe, Philosophy and
Philosophy of Science

Nathans, Jinny
1134 Massachusetts Ave
Cambridge, MA 02138-5204
Phone: 1.617.492.3735
jinnyn@mindspring.com

Navarro, Luis
Universitat de Barcelona
Dept de Fisica
Fonemental, U.B.
Diagonal 647
Barcelona 08028
Spain
Phone: 34.93.402.1152
Fax: 34.93.402.1149
lunave@ffn.ub.es
Doctorate, 1965
19c, Physical Sciences; 20c

Navarro-Brotons, Victor
Mestral, 9
Godella
Valencia 46110

Spain
Phone: 34.963.643.175
victor.navarro@uv.es

Neal, Katherine
University of Sydney
Unit for HPS
F07 Calslaw
Sydney 2006
Australia
Phone: 9664.8490
kneal@scifac.usyd.edu.au

Needell, Allan A.
402 F Street NE
Washington, DC 20002
Phone: 1.202.547.7592
Fax: 1.202.786.2947
allan.needell@nasm.si.edu

Neeley, K. A.
1940 Blue Ridge Rd
Charlottesville, VA 22903-1216
Phone: 1.434.927.6117
Fax: 1.434.924.4306
neeley@virginia.edu

Nelson, Carl W.
Carl Nelson Consulting
PO Box 18371
Washington, DC 20036-8371
Phone: 1.202.331.9757
Fax: 1.202.331.9757
carl@carl-nelson.com

Nelson, Clifford M.
US Geological Survey
950 National Center
Reston, VA 20192-0001
Phone: 1.703.648.6080
Fax: 1.703.648.6373
cnelson@usgs.gov
Doctorate, 1974
19c, North America, Earth
Sciences; 20c, Earth Sciences;
Institutions; Exploration,
Navigation and Expeditions

Nelson, G. Blair
720 Jefferson St
Baraboo, WI 53913-2317
Phone: 1.608.356.8472
gbnelson@students.wisc.edu

Neri, Janice
1015 Myrtle Way
San Diego, CA 92103-5122
jneri@uci.edu

Neri-Vela, Rolando
Col. Barrio Actipan
Las Huertas 96-107
Mexico DF CP03230
Mexico
Phone: 5.534.4961

Nersessian, Nancy J.
Georgia Inst of Technology
School of Public Policy
Atlanta, GA 30332-0345
Phone: 1.404.894.1232
Fax: 1.404.385.0504
nancyn@cc.gatech.edu
Doctorate, 1977
*19c, North America, Cognitive
Sciences; 20c, Europe,
Philosophy and Philosophy of
Science; Physical Sciences;
Education*

Neu, John
16660 County Rd T
Townsend, WI 54175
Phone: 1.715.276.2513
neu@macc.wisc.edu
Master's Level, 1959
Education

Neuenschwander, Erwin
Universitat Zurich-Irchel
Institut Für Mathematik
Winterthurerstrasse 190
Zurich 8057
Switzerland
Phone: 41.1.635.58.62
Fax: 41.1.635.57.06
Doctorate
*19c, Europe, Mathematics;
5c-13c; Social Sciences;
Philosophy and Philosophy of
Science*

Neufeld, Michael J.
304 Philadelphia Ave
Takoma Park, MD 20912
Phone: 1.301.587.0259
Fax: 1.202.786.2447
mike.neufeld@nasm.si.edu
Doctorate, 1984
*20c, Europe, Technology;
20c, North America, Technology*

Neumann, D.
Universität Halle-Wittenberg
Medizinische Fakutat
Institut für Geschichte d. Med
Magdeburger Str.27
Halle 06097
Germany

Neumann, Theresa
32 Meadow Ct
Walnut Creek, CA 94595-2626

Neushul, Peter
915 Camino Lindo
Goleta, CA 93117-4324
peter@hss.caltech.edu

Newcomb, Sally E.
13120 Two Farm Dr
Silver Spring, MD 20904-3418
Phone: 1.301.622.0177
senewcomb@earthlink.net
Master's Level
*18c, Europe, Earth Sciences;
18c, Europe, Physical Sciences;
18c, Europe*

Newell, Julie R.
Southern Polytechnic St Univ
SIS Department
1100 S Marietta Pkwy SE
Marietta, GA 30060-2896
Phone: 1.770.528.7481
Fax: 1.770.528.4949
jnewell@spsu.edu
Doctorate, 1993
*North America, Earth Sciences;
North America; North America,
Technology; North America,
Social Relations of Science*

Newsome, Daniel
420 W 24th St Apt 17F
New York, NY 10011-1326
Phone: 1.212.206.7018
dlnuzum@yahoo.com

Nicholson, Richard S.
American Association for the
Advancement of Science
1333 H St NW
Washington, DC 20005-4707

Nickles, Thomas
University of Nevada, Reno
Dept of Philosophy 102
Reno, NV 89557-0056

Phone: 1.775.337.6330
Fax: 1.702.327.5024
nickles@unr.edu
Doctorate, 1969
*Philosophy and Philosophy of
Science; Cognitive Sciences;
Social Sciences*

Nier, Keith A.
Rutgers The State Univ of NJ
Thomas A. Edison Papers
113 Van Dyck Hall 16 Seminar
New Brunswick, NJ 08901-1108
nierfam@earthlink.net
Doctorate, 1975
*Technology; Instruments and
Techniques; Social Relations
of Science; Philosophy and
Philosophy of Science*

Nieto-Galan, Agusti
CEHIC
Facultat de Ciencies Edificic
Univ. Autonoma de Barcelona
Bellaterra 08193
Spain
Phone: 34.9.3581.2966
Fax: 34.9.3581.2790
agusti.nieto@uab.es
Doctorate
*19c, Europe, Physical Sciences;
19c, Europe, Technology;
18c, Europe, Technology*

Nigatu, Tadesse
1022 McClelland St S
Maplewood, MN 55119-5981
Phone: 1.651.730.8675

Nilsson, Ingemar
Karret 3
Sjfvik S-443 74
Sweden
Phone: 46.3.024.3051
ingemar.nilsson@hum.gu.se

Noakes, Richard
Dept of Hist and Philos of Sci
Free School Lane
Cambridge CB2 3RH
Great Britain
Phone: 0113.269.7761
r.j.noakes@leeds.ac.uk

Noll, Richard
DeSales University
2755 Station Ave

Center Valley, PA 18034-9565
Phone: 1.610.282.1100
Fax: 1.610.282.1078
rn00@desales.edu

Nonnoi, Giancarlo
V. Le. Merello, 74
Cagliari I-09123
Italy
Phone: 39070282140
Fax: 39.07.0675.7291
nonnoi@unica.it
Doctorate
17c, Europe, Physical Sciences;
18c, Europe, Physical Sciences;
17c, Europe, Philosophy and
Philosophy of Science

Nooney, Kevin
University of Georgia
College of Education
Mathematics Education Dept.
105 Aderhold Hall
Athens, GA 30602
Phone: 1.706.542.4194
knooney@coe.uga.edu
Master's Level, 2001
Education; Mathematics;
Humanistic Relations of Science;
Social Relations of Science

Norberg, Arthur L.
University of Minnesota
Dept of Computer Science
200 Union Street
4-192 EE CS Building
Minneapolis, MN 55455
Phone: 1.612.625.1067
Fax: 1.612.625.0572
norberg@cs.umn.edu
Doctorate, 1974
20c, North America, Technology;
19c, North America, Technology;
20c, Europe, Technology

Nordmann, Alfred
University of South Carolina
Philosophy Department
Columbia, SC 29208-0001
Phone: 1.803.777.3739
Fax: 1.803.777.9178
anordmann@sc.edu
Doctorate, 1986
Philosophy and Philosophy of
Science; 18c, Physical Sciences;
Social Relations of Science;
Instruments and Techniques

Norman, Jeremy
PO Box 867
Novato, CA 94947-0867
Phone: 1.415.892.3181
jnorman@jnorman.com

Norris, Wilfred G.
Juniata College
Department of Physics
1700 Moore Street
Huntingdon, PA 16652
Phone: 1.814.641.3580
Fax: 1.814.641.3685
norris@juniata.edu
Doctorate, 1963
Physical Sciences

North, John D.
Philosophical Institute RUG
A-WEG 30
Groningen 9718CW
The Netherlands
Phone: 44.1865.558458
Doctorate
Astronomical Sciences

Norton, John
University of Pittsburgh
Dept of History and Philosophy
of Science
Pittsburgh, PA 15260
Phone: 1.412.624.5896
jdnorton@pitt.edu
Doctorate, 1982
20c, Physical Sciences;
Philosophy and Philosophy of
Science

Nozawa, Satoshi
210 Dai-5-Shinzanso
Shinkawa 5-8-2
Tokyo 181-0004
Japan
Phone: 81.422.42.3826
vya05535@nifty.com

Nucci, Mary L.
62 Bunnvale Rd
Califon, NJ 07830-4139
Phone: 1.908.638.5783
Fax: 1.908.638.4253
npcinc2@blast.nct

Numbers, Ronald L.
University of Wisconsin
Dept of History of Medicine
1300 University Ave

Madison, WI 53706-1510
Phone: 1.608.262.1460
Fax: 1.608.262.2327
rnumbers@med.wisc.edu
Doctorate, 1969
North America; North America,
Medical Sciences; North
America, Biological Sciences

Nummedal, Tara E.
Brown University
Department of History
Box N
Providence, RI 02912
Phone: 1.401.963.2131
Fax: 1.401.863.1040
tara_nummedal@brown.edu
Doctorate, 2001
14c-16c, Europe, Physical
Sciences; 14c-16c, Europe,
Gender and Science;
14c-16c, Europe, Social
Relations of Science

Nunes Santos, A. M.
Quinta Da Torre
Fac de Ciencias Etech Unl
Monte Capa 2825
Portugal

Nye, Mary Jo
Oregon State University
Department of History
Milam Hall 306
Corvallis, OR 97333-5104
Phone: 1.541.737.1308
Fax: 1.541.737.1257
nyem@ucs.orst.edu
Doctorate, 1970
Physical Sciences; Institutions;
Social Relations of Science

Nye, Robert A.
Oregon State University
Dept of History
Milam Hall 306
Corvallis, OR 97333-5104
Phone: 1.541.737.1310
Fax: 1.514.737.1257
nyer@ucs.orst.edu

Nyhart, Lynn K.
University of Wisconsin
History of Science
7143 Social Science Bldg
1180 Observation Dr
Madison, WI 53706-1393

Phone: 1.608.262.3970
Fax: 1.608.262.3984
lknyhart@facstaff.wisc.edu
Doctorate, 1986
19c, Europe, Biological Sciences;
Earth Sciences; 19c; 20c

O'Boyle, Cornelius
33 Harcourt Road
London N22 7XW
Great Britain
Phone: 44.20.8888.2784
cornelius@o-boyle.com

O'Brien-Weintraub, Katy
University of Chicago
Harper Memorial Box 29
1050 E 59th St
Chicago, IL 60637
Phone: 1.773.702.0226
katy@semcoop.com
Doctorate, 1987
14c-16c, Europe, Humanistic
Relations of Science;
5c-13c, Europe, Physical
Sciences

O'Connor, Amy
Bryn Mawr College
Box C-1466
Bryn Mawr, PA 19010

O'Hara, Robert J.
University of North Carolina
Department of Biology
Greensboro, NC 27402-0001
rjohara@post.harvard.edu

O'Neill, Jeffrey
3282 4th St
Boulder, CO 80304-2148
Phone: 1.303.998.1396
oneilljeff@hotmail.com

O'Neill, Ynez
UCLA
Dept of Neurobiology
UCLA School of Medicine
Los Angeles, CA 90005-1763
Phone: 1.310.825.4933
Fax: 1.323.933.6025
yvonmhi@ucla.edu
Doctorate
5c-13c, Medical Sciences

Oaks, Jeffrey
University of Indianapolis
Mathmatics Dept
1400 E Hanna Ave
Indianapolis, IN 46227-3697
Phone: 1.317.788.3454
Fax: 1.317.788.3569
oaks@uindy.edu

Ochs, Kathleen
1205 Georgetown Rd
Boulder, CO 80305-6443

Ochs, Sidney
912 Forest Blvd North Dr
Indianapolis, IN 46240
Phone: 1.317.253.2368
Fax: 1.317.274.3318
sochs@iupui.edu
Doctorate
400 B.C.E.-400 C.E.,
Europe, Medical Sciences;
5c-13c, North America, Medical
Sciences; 19c, Instruments and
Techniques; 14c-16c, Philosophy
and Philosophy of Science

Ogata, Philip H.
2580 Panorama Ave
Boulder, CO 80304-3728
Phone: 1.308.444.6662
Master's Level, 1970
Physical Sciences

Ogawa, Mariko
Mie University
1515 Edobashi
Tsu 514-8507
Japan
Phone: 81.052.231.9154
Fax: 81.052.789.1354
ogawa@human.mle-u.ac.jp
Master's Level
19c, Europe, Biological Sciences;
19c, Europe, Medical Sciences

Ogawa, Teruaki
4-5-3 Mihori
Akishima Tokyo 196 0001
Japan
Phone: 425468555
Fax: 81.4.2546.8559
ogawa@mbp.sphere.ne.jp
Bachelor's Level
5c-13c, Asia, Technology,
Instruments and Techniques;
400 B.C.E.-400 C.E., Asia,

Technology; 14c-16c, Asia,
Instruments and Techniques;
Pre-400 B.C.E.

Ogilvie, Brian W.
Univ of Mass-Amherst
Dept of History
Herter Hall, Box 33930
Amherst, MA 01003
Phone: 1.413.545.1599
Fax: 1.413.545.6137
ogilvie@history.umass.edu
Doctorate, 1997
14c-16c, Natural and Human
History; Social Sciences

Ogilvie, Marilyn B.
University of Oklahoma
History of Science Collections
401 W Brooks Rm 521
Norman, OK 73019
Phone: 1.405.325.2741
Fax: 1.405.325.7618
mogilvie@ou.edu
Doctorate
Education; Biography;
19c, Biological Sciences;
Humanistic Relations of Science

Ohno, Makoto
Aichi Prefectural University
Faculty of Foreign Studies
Kumabari
Nagakute, Aichi 4801198
Japan
Phone: 81.0561.64.1111
Fax: 81.0561.64.1107
ohno@for.aichi-pu.ac.jp
Master's Level, 1982
18c, Europe, Institutions;
18c, Europe, Social
Relations of Science;
18c, Europe, Physical Sciences

Olby, Robert C.
University of Pittsburgh
1017 Cathedral of Learning
Pittsburgh, PA 15260
Phone: 1.412.624.5881
olbyr@pitt.edu
Doctorate, 1963
20c, Biological Sciences;
19c, Europe, Biological Sciences;
20c, North America, Technology

Oldroyd, David R.
Univ of New South Wales
Sch of Science & Technology
Studies
Sydney, NSW 2052
Australia
Phone: 61.2.9449.5559
Fax: 61.2.9144.4529
d.oldroyd@unsw.edu.au
Doctorate, 1974
17c, Earth Sciences;
18c, Earth Sciences;
19c, Earth Sciences;
20c

Olesko, Kathryn M.
Georgetown University
Dept of History
Washington, DC 20057-1035
Phone: 1.202.687.8300
Fax: 1.202.687.8359
oleskok@georgetown.edu
Doctorate, 1980
18c, Europe; 19c, Europe;
20c, Europe; Social Sciences

Oliveira Carneiro, Ana Maria
Av Visconde De Valmor, 32 1 D
Lisboa 1050-240
Portugal
Phone: 351.21.793.8572

Oliver, Brian
1655 E 55th St
Chicago, IL 60615-5838
Phone: 1.773.955.5619

Oliver, Melissa A.
1819 Eutaw Pl # 2
Baltimore, MD 21217-3807
Phone: 1.916.441.5851
callipides@yahoo.com
Master's Level
14c-16c, Europe, Medical
Sciences; 17c, Europe,
Social Relations of Science;
14c-16c, Europe, Physical
Sciences

Oliver, Robert
Rice University
Dept of Biochemistry & Cell
Biology
MS-140
PO Box 1892

Houston, TX 77251-1892
Phone: 1.713.348.5617
Fax: 1.713.348.5154

Olsen, Kenneth H.
1029 187th Pl SW
Lynnwood, WA 98036-4986
Phone: 1.425.776.5244
kolsen@geophys.
  washington.edu

Olsen, Robert J.
Wabash College
Chemistry Department
PO Box 352
Crawfordsville, IN 47933-0352
Phone: 1.765.361.6211
Fax: 1.765.361.6340
olsenr@wabash.edu
Doctorate, 1976
Physical Sciences

Olson, Richard G.
Harvey Mudd College
Humanities & Social Sciences
260 W Foothill Blvd # 1257
Claremont, CA 91711-2707
Phone: 1.909.607.4476
Fax: 1.909.607.7600
olson@hmc.edu
Doctorate, 1967
20c, Europe, Humanistic
Relations of Science;
20c, Europe, Physical Sciences;
20c, Europe, Social
Relations of Science;
20c, Europe, Social Sciences

Onabe, Tomoko
Motozushicho 477
Kamigyo-ku
Kyoto 602-0938
Japan
Phone: 075.417.6785
tsalz@theia.ocn.ne.jp

Ongaro, Guiseppe
Via S Biagio 41
Padova 35100
Italy

Opitz, Donald L.
University of Minnesota
339 Appleby Hall
128 Pleasant St SE
Minneapolis, MN 55455-0434
Phone: 1.612.624.4894

opit0002@tc.umn.edu
Master's Level, 1998
19c, Europe, Gender and
Science; 20c, North America,
Medical Sciences

Ordonez, Javier
Universidad Autonoma Madrid
Campus Cantoblanco
Facultad de Filosofia
Madrid 28049
Spain
Phone: 34.9.1397.4475
Fax: 34.9.1360.4355
javier.ordonnez@uam.es
Doctorate, 1977
19c, Europe, Physical Sciences;
19c, Astronomical Sciences;
Physical Sciences;
Astronomical Sciences

Oren, Ido
University of Florida
Dept of Political Science
PO Box 117325
Gainesville, FL 32611-7352
Phone: 1.352.392.0262
Fax: 1.352.392.8127
oren@polisci.ufl.edu
Doctorate, 1992
20c, North America, Social
Sciences

Oreskes, Naomi
Univ of California-San Diego
Department of History, 0104
9500 Gilman Dr
La Jolla, CA 92093-0104
Phone: 1.858.534.4695
Fax: 1.858.534.7283
noreskes@ucsd.edu
Doctorate, 1990
Philosophy and Philosophy of
Science; Gender and Science

Orlowski, E.
125 Van Cortlandt Ave W
Bronx, NY 10463-2704

Orna, Mary V.
16 Hemlock Pl
New Rochelle, NY 10805
mvorna@chemheritage.org

Ortega, Lina
511 Fleetwood Drive
Norman, OK 73072
lortega@ou.edu

Osborne, Michael A.
University of California
Dept of History
Santa Barbara, CA 93106-9410
Phone: 1.805.893.2901
Fax: 1.805.893.8795
osborne@history.ucsb.edu
Doctorate, 1987
19c, Europe, Medical Sciences;
19c, Europe, Natural and
Human History; 20c, Europe,
Biological Sciences; Philosophy
and Philosophy of Science

Oshima, Kouji
142-71 Jigozen Hatsukaichi Shi
Hiroshima
Japan

Osler, Margaret J.
University of Calgary
Dept of History
2500 University Dr. NW
Calgary, AB T2N1N4
Canada
Phone: 1.403.220.6401
Fax: 1.403.289.8566
mjosler@ucalgary.ca
Doctorate, 1968
17c, Europe, Philosophy
and Philosophy of Science;
17c, Europe, Social Sciences;
17c, Europe, Humanistic
Relations of Science

Oster, Malcolm
The Open University
North West Region
70 Manchester Road
Chorlton-cum-Hardy M219UN
Manchester
Great Britain
Phone: 44.0161.956.6858
Fax: 44.0161.956.6811
m.r.oster@open.ac.uk

Osterbrock, Donald E.
University of California
Lick Observatory
Santa Cruz, CA 95064
Phone: 1.831.459.2605
Fax: 1.831.426.3115

don@ucolick.org
Doctorate, 1952
19c, Astronomical Sciences;
20c, Astronomical Sciences;
19c, Physical Sciences;
Institutions

Otis, Laura
Max Planck Institute for the
History of Science
Wilhelmstr 44
Berlin 10117
Germany
Phone: 49.30.2266.7174
Fax: 49.30.2267.299
otis@mpiwg-berlin.mpg.de

Outram, Dorinda
University of Rochester
Department of History
Rochester, NY 14627
Phone: 1.716.275.4097
otrm@mail.rochester.edu
Doctorate
18c, Europe, Exploration,
Navigation and Expeditions

Outten, Burnet, Jr.
Western Metal Products Co
1300 Weber St
Orlando, FL 32803-3336
westmet@kua.net
Science Policy; Education;
Philosophy and Philosophy of
Science; Institutions

Ouwendijk, George
20 Audrey Ct
Malverne, NY 11565-1010
Phone: 1.516.599.3950
g.ouwendijk@att.net

Overfield, Richard A.
11606 Spaulding St
Omaha, NE 68164-2367
Phone: 1.402.431.9784
doverfield@msn.com

Overmann, Ronald J.
20 Beaver St
San Francisco, CA 94114-1515
Phone: 1.415.522.0501
ronjim@gte.net
Doctorate, 1974

Owens, Larry
University of Massachusetts
History Department
Amherst, MA 01003
lowens@history.umass.edu

Oweyssi, Nick
2700 W W Thorne Blvd
Houston, TX 77073-3410
Phone: 1.281.318.5547
Fax: 1.281.318.7165
nick.oweyssi@nhmccd.edu

Ozkarol, Tuygun
Erensoy Gida Ve Ambalaj Sti
C. Topuzlu Cad. Tibas C Bloku
Fenervolu
Kadikoy 81030
Istanbul
Turkey
Fax: 90.216.411.2199

Pace, Seth
106 Rabaul Rd
Seaside, CA 93955-6674
Phone: 1.915.695.7615
sep98k@acu.edu

Page, Brian R.
OCE
2 Concourse Park, Suite 550
Atlanta, GA 30328
Phone: 1.770.395.3415
Fax: 1.770.395.3402
bpage@randomc.com
Master's Level, 1980
North America; 20c, North
America, Science Policy;
Technology; Mathematics

Paletta, Francisco
Rua Congonhas, 682
Apartamento 02
Belo Horizonte 30330-10
Brazil
Phone: 55.31.293.1408
chico@retiro.cjb.net

Palladino, Paolo S.
Lancaster University
Lancaster LA1 4YG
Great Britain
Phone: 44.1524.592.793
Fax: 44.1524.846.1
p.palladino@lancaster.ac.uk
Doctorate, 1989

20c, Europe, Biological Sciences;
20c, North America, Biological
Sciences; Social Sciences

Pallasch, Thomas J.
343 Helmuth Ln
Alexandria, VA 22304-8668
Phone: 1.703.461.0620
tpallasch@earthlink.net

Palm, Lodewijk C.
Inst. of History of Science
Buys Ballotlaboratorium
Princetonplein 5
Utrecht 3584 CC
The Netherlands
Phone: 31.30.253.8283
Fax: 31.30.253.6313
l.c.palm@phys.uu.nl
Master's Level
17c, Biological Sciences;
18c, Biological Sciences;
Europe, Institutions;
Europe, Education

Palmer, William P.
PO Box 41622
Casuarina, NT 0811
Australia
Phone: 889855946
Fax: 61.8.8946.6151
bill.palmer@darwin.ntu.edu.au
Master's Level
18c, Europe, Physical Sciences;
19c, Europe, Physical Sciences;
20c, Europe, Physical Sciences;
19c, Europe, Education

Palmeri, JoAnn
University of Oklahoma
History of Science Dept
601 Elm, Rm 622
Norman, OK 73019
Phone: 1.405.325.2213
palmerij@aol.com
Doctorate, 2000
20c, North America,
Astronomical Sciences;
20c, North America;
20c, North America, Humanistic
Relations of Science;
20c, North America, Education

Palter, Robert
135 Brittany Farms Rd Apt H
New Britain, CT 06053-1127
Phone: 1.860.223.8193

Doctorate
17c, Africa, Social Sciences;
18c, Europe, Humanistic
Relations of Science;
400 B.C.E.-400 C.E.,
Europe, Physical Sciences;
14c-16c, Europe, Philosophy
and Philosophy of Science

Pancaldi, Giuliano
University of Bologna
Via Zamboni 38
Bologna 40126
Italy
Phone: 39.5.1209.8331
Fax: 39.5.1675.0512
pancaldi@alma.unibo.it
Doctorate, 1993
18c, Europe, Biological Sciences;
19c, Europe, Biological Sciences;
18c, Europe, Physical Sciences

Pandora, Katherine A.
205 Palfrey Street
Watertown, MA 02472
Phone: 1.405.447.2076
Fax: 1.405.325.2363
kpandora@ou.edu
Doctorate
North America; Humanistic
Relations of Science; 20c, North
America, Social Sciences;
20c, North America, Technology

Pantalony, David
Dartmouth College
6127 Wilder Hall
Hanover, NH 03755-3528
Fax: 1.603.646.1446
david.a.pantalony@
  dartmouth.edu

Pantin, Isabelle
Université de Paris
Nanterre
France
Fax: 33.1.30.82.25.55
jpantin@dictis.com
Doctorate
14c-16c, Europe, Humanistic
Relations of Science;
14c-16c, Europe, Astronomical
Sciences; 17c, Europe,
Astronomical Sciences;
14c-16c, Education

Paoloni, Leonello
Via A. Borrelli 1/A
Palermo I-90139
Italy
Phone: 39.0.9134.4919
leopa@infcom.it
Doctorate
19c, Europe, Physical Sciences;
20c, Europe, Philosophy
and Philosophy of Science;
20c, Europe, Physical Sciences

Paradis, James G.
Mass Inst of Technology
Program in Writing and
Humanistic Studies
14E-303
Cambridge, MA 02139
Phone: 1.617.253.7392
Fax: 1.617.253.6910
jparadis@mit.edu
Doctorate, 1975
19c, Humanistic
Relations of Science;
20c, Biological Sciences;
Natural and Human History;
Philosophy and Philosophy of
Science

Paradowski, Robert
Rochester Inst of Tech
College of Liberal Arts
92 Lomb Memorial Dr
Rochester, NY 14623-5604
Phone: 1.716.475.6950
Fax: 1.716.475.7120
rxpgsh@rit.edu

Parascandola, John L.
Public Health Service
5600 Fishers Ln Rm 18-23
Rockville, MD 20857-0002
Phone: 1.301.443.5363
Fax: 1.301.443.4193
jparascandola@psc.gov
Doctorate, 1968
20c, Medical Sciences;
20c, Physical Sciences

Paris, Elizabeth
Dibner Institute
MIT E56-100
38 Memorial Dr
Cambridge, MA 02139
Fax: 1.617.495.3344

eparis@dibinst.mit.edu
Doctorate, 1999
20c, Physical Sciences

Park, Buhm Soon
Johns Hopkins University
Hist. of Sci., Med. & Tech.
3400 N. Charles Street
Baltimore, MD 21218
Phone: 1.301.496.7388
Fax: 1.301.402.1434
parkb@jhunix.hcf.jhu.edu
Master's Level, 1991
*20c, North America, Physical
Sciences*

Park, Hyungwook
Program in History of Science
College of Natural Sciences
Seoul National University
Seoul 151-747
South Korea
phpsjournal@hotmail.com

Park, Katharine
Harvard University
Dept. of Hist. of Science
Science Center 235
Cambridge, MA 02138
Phone: 1.617.495.9922
Fax: 1.617.495.3344
park28@fas.harvard.edu
Doctorate, 1981
*14c-16c, Europe, Medical
Sciences; 5c-13c, Europe,
Medical Sciences; 14c-16c,
Europe, Gender and Science;
14c-16c, Europe, Philosophy
and Philosophy of Science*

Parker, Reno
MSU- Northern
Science-Math Dept
Havre, MT 59501
Phone: 1.406.265.3598
Fax: 1.406.265.3777
parker@msun.edu

Parlee, Mary B.
MIT
Room 14E-316
77 Massachusetts Avenue
Cambridge, MA 02139
Phone: 1.619.452.2584
mparlee@mit.edu
*20c, North America, Cognitive
Sciences; 20c, North America,*

*Biological Sciences; 20c, North
America, Medical Sciences,
Instruments and Techniques;
Gender and Science*

Parshall, Karen
University of Virginia
Department of Mathematics
Kerchof Hall
Charlottesville, VA 22903-3199
Phone: 1.804.924.1411
Fax: 1.804.982.3084
khp3k@virginia.edu
Doctorate, 1982
*19c, North America, Mathematics;
20c, North America, Mathematics;
19c, Europe, Mathematics*

Pastor, Henri A.
18 Bvd Marechal Joffre
Grenoble 38000
France
Phone: 33.476.873.990
Fax: 33.473.432.265
hpastor@fr.packardbell.org
Doctorate
*19c, Europe, Physical Sciences;
400 B.C.E.-400 C.E., Europe,
Physical Sciences;
20c, Europe, Physical Sciences;
20c, Europe, Education*

Paterson, Carla
13960 66 Avenue
Surrey, BC V3W9B1
Canada
Phone: 1.604.590.1516
cpaters@interchange.ubc.ca

Patey, Douglas L.
Smith College
Northampton, MA 01063
Phone: 1.413.585.3314
Fax: 1.202.333.7174
dpatey@hotmail.com
Doctorate, 1979

Patiniotis, Manolis
Athens University
John Kennedy 37
Kesariani
Athens GR-16121
Greece
Phone: 30.1.727.5515
Fax: 30.1.935.0151
mpatinio@cc.uoa.gr
Master's Level, 1996

Social Sciences;
*18c, Europe, Physical Sciences;
18c, Europe, Institutions*

Paton, Miranda
Cornell University
Science & Technolgy Studies
630 Clark Hall
Cornell, NY 14853
mvp3@cornell.edu

Paul, Diane B.
Univ of Mass - Boston
Dept of Political Science
1716 Cambridge St., #17
Boston, MA 02125-3393
Phone: 1.617.287.6936
Fax: 1.617.287.6511
diane.paul@umb.edu
Doctorate, 1975
*20c, Biological Sciences;
19c, Biological Sciences;
20c, Science Policy;
20c, Social Relations of Science*

Paul, Robert
11-A Parkhill Rd
Halifax, NS B3P1R2
Canada
Phone: 1.902.477.0195
robert.paul@eoascientific.com

Pauly, Philip J.
34 Second Street
Brooklyn, NY 11231
Phone: 1.718.852.2656
pauly@rci.rutgers.edu
Doctorate, 1981
*19c, North America, Biological
Sciences; North America;
20c, North America, Education*

Pauly, Roger, Jr.
University of Central Arkansas
Dept of History
PO Box 4935
201 Donaghey Ave
Conway, AR 72035-5000
Phone: 1.501.408.5619
rpauley@mail.uca.edu

Pav, Peter A.
Eckerd College
4200 54th Avenue South
St Petersburg, FL 33711
Fax: 1.727.895.8304
ppav@mindspring.com

Doctorate
17c, Europe, Physical Sciences;
18c, Philosophy and Philosophy
of Science; 400 B.C.E.-400 C.E.,
Philosophy and Philosophy of
Science; 14c-16c, Humanistic
Relations of Science

Pawley, Emily
6 Hoskin Ave
Toronto, ON M5S1H8
Canada
Phone: 1.416.341.0471
emilypawley@hotmail.com

Payne, Lynda E.
University of Missouri
History Department
5121 Rockhill Road
Kansas City, MO 64110
Phone: 1.816.238.2539
Fax: 1.816.235.5723
paynel@umkc.edu
Doctorate, 1997
17c, Europe,
Medical Sciences; 18c, Europe,
Gender and Science;
14c-16c, Europe, Philosophy
and Philosophy of Science;
5c-13c, Medical Sciences

Paynter, Henry
PO Box 568
Pittsford, VT 05763-0568
Phone: 1.802.483.2155
hankusp@aol.com

Pearl, Sharrona
Harvard University
History of Science Department
Science Center 235
Cambridge, MA 02138
Phone: 1.617.493.2597
Fax: 1.617.495.3344
spearl@fas.harvard.edu

Pearlman, Norman
720 Vine St
West Lafayette, IN 47906-2612
Phone: 1.765.743.2662
np@physicspurdue.edu

Peck, Natalie
106 Castro St
Norman, OK 73069
Phone: 1.405.329.5315
npeck@ou.edu

Bachelor's Level, 2002
Natural and Human History;
Gender and Science;
Social Relations of Science;
North America

Pedersen, Stig A.
Roskilde University
Universitets Veg 1
PO Box 260
Roskilde DK-4000
Denmark
Phone: 45.4674.2265
Fax: 45.4674.3012
sap@ruc.dk
Doctorate
Philosophy and Philosophy
of Science; Mathematics;
Physical Sciences

Pemberton, S. George
University of Alberta
Dept of Earth & Atmos Sci
Edmonton, AB T6G2E3
Canada
Phone: 1.403.492.2044
Fax: 1.403.492.8380
gpembert@gpu.sru.ualberta.ca
Doctorate, 1979
19c, North America, Earth
Sciences

Pemberton, Stephen
126 Montgomery St, Apt 1D
Highland Park, NJ 08904
Phone: 1.732.729.9731
pemberton@history.rutgers.edu
Doctorate, 2001
20c, North America, Medical
Sciences

Pence, Harry E.
SUNY Oneonta
Dept of Chemistry
Oneonta, NY 13820
Phone: 1.607.436.3179
Fax: 1.607.436.2654
pencehe@oneonta.edu
Doctorate, 1968
Physical Sciences; Biological
Sciences; Education; Philosophy
and Philosophy of Science

Pendick, Daniel A.
53 Telegraph Rd
Peru, NY 12972-4110

Phone: 1.518.834.7359
danielpendick@earthlink.net
Master's Level, 1992

Perdomo, I.
c/o La Aranita N 4
38280 Tegueste
Tenerife
Spain
Phone: 34.922.54.5734
Fax: 34.922.31.7879
mperdomo@ull.es

Perdrix, John L.
PO Box 107
Wembley 6913
Australia
Phone: 618.938.74250
Fax: 618.938.73981
astral@iinet.net.au
19c, Australia and Oceania,
Astronomical Sciences;
20c, Australia and Oceania,
Astronomical Sciences

Perez, Kim
2401 Walnut St
Hays, KS 67601-3044
Phone: 1.785.628.0792
Fax: 1.785.628.4087
kperez@fhsu.edu

Perez Tamayo, Ruy
Facultad de Medicina UNAM
Apartado Postal 70641
Mexico 04510
Mexico
Phone: 5623.2651
Fax: 5761.0249
ruypte@hotmail.com
Doctorate, 1971
400 B.C.E.-400 C.E.,
Biological Sciences;
5c-13c, Biological Sciences;
5c-13c, Medical Sciences;
14c-16c, Philosophy and
Philosophy of Science

Perkins, John H.
The Evergreen State College
Lab I
Olympia, WA 98505
Phone: 1.360.867.6503
Fax: 1.360.867.5430
perkinsj@evergreen.edu
Doctorate, 1969
20c, North America, Biological

Sciences; 20c, North America,
Science Policy; 20c, North
America, Social Relations of
Science

Pernick, Martin S.
University of Michigan
1029 Tisch Hall
Ann Arbor, MI 48109
Phone: 1.734.647.4876
Fax: 1.734.647.4881
mpernick@umich.edu
Doctorate, 1979
20c, North America, Medical
Sciences; 19c, North America,
Medical Sciences; 18c, North
America, Medical Sciences

Pertzman, Steven
2823 W Queen Ln
Philadelphia, PA 19129-1031
Phone: 1.215.843.7412
sjp27@drexel.edu

Pesic, Peter
905 Trail Cross Court
Santa Fe, NM 87505
Phone: 1.505.983.3168
ppesic@mail.sjcsf.edu
Doctorate, 1975
Physical Sciences; Mathematics;
Astronomical Sciences

Peterfreund, Stuart S.
Northeastern University
Department of English
360 Huntington Ave
406 Holmes Hall
Boston, MA 02115-5096
Phone: 1.617.373.2605
Fax: 1.617.373.2509
speterfr@lynx.doc.neu.edu
Doctorate
17c, Europe, Humanistic
Relations of Science;
18c, Europe, Humanistic
Relations of Science;
19c, Europe, Humanistic
Relations of Science;
18c, Europe, Social
Relations of Science

Petersen, Greta
11227 Farm Ln
Minnetonka, MN 55305-4341
Phone: 1.612.935.3598
petel462@tc.umn.edu

Peterson, Charles J.
University of Missouri
Dept of Physics & Astronomy
223 Physics Building
Columbia, MO 65211-0001
Phone: 1.573.882.3217
Fax: 1.573.882.4195
petersonc@missouri.edu
Doctorate
20c, North America,
Astronomical Sciences

Peterson, Kristin E.
Loeb & Troper
655 Third Avenue
New York, NY 10017
Phone: 1.718.951.6103
kpeterson@loebandtroper.com
Doctorate, 1993

Peterson, Thomas F., Jr.
3060 Lander Rd
Cleveland, OH 44124-5441
Phone: 1.216.831.8077
tpeterson@alum.mit.edu

Petri, H.
Boslaan 16
Warnsve DH-7231
The Netherlands

Petrina, Ecaterina
201 1/2 Wyckoff Ave
Ithaca, NY 14850-2441
Phone: 1.607.257.6480
ep19@cornell.com

Petto, Christine
Southern Ct State University
Seabury Hall 305
501 Crescent St
New Haven, CT 06515-1355
Phone: 1.203.392.5612
Fax: 1.203.392.5670
petto@scsu.ctstateu.edu

Philips, Thomas O.
PO Box 607
West Harwich, MA 02671-0607
Phone: 1.973.267.6253
Fax: 1.973.539.1275
tophilips@nac.net
Doctorate, 1963
Physical Sciences

Phillips, Denise
107 Hammond St
Cambridge, MA 02138-1959
Phone: 1.617.441.5178
phillips@fas.harvard.edu
Doctorate
19c, Europe, Natural
and Human History;
19c, Europe, Biological Sciences

Phillips, Mona
3140 Warrington Rd
Shaker Heights, OH 44120-2429

Piccirilli, Robert
10 Buckman Dr
Newtown, PA 18940-9684
Phone: 1.215.968.2042
Fax: 1.215.968.0787
robertpiccirilli@aol.com
Bachelor's Level, 1969
20c, Instruments and
Techniques; 17c, Europe,
Astronomical Sciences,
Philosophy and Philosophy
of Science; 17c, Europe,
Mathematics

Pickering, Andrew R.
University of Illinois
Department of Sociology
702 S Wright St
Urbana, IL 61801-3631
Phone: 1.217.333.8067
Fax: 1.217.333.5225
pickerin@uiuc.edu
Doctorate, 1984
Social Relations of Science;
Technology; Social Sciences;
20c, Instruments and Techniques

Pickren, Wade
American Psychological Assoc.
Archives & Library Services
750 1st St NE
Washington, DC 20002-4242
Phone: 1.202.336.5645
Fax: 1.202.336.5643

Picolet, Guy
Centre Alexandre Koyre
Museum National d'Histoire Nat
57 Rue Cuvier
Paris Cedex 05 75231
France
Phone: 33.17.4336.7069
Fax: 33.01.4337.3449

guy.picolet@damesme.cnrs.fr
Doctorate, 1982
17c, Europe, Institutions;
17c, Europe, Education;
17c, Europe, Biography

Piel, Henry, III
523 Central Street
Holliston, MA 01746
Phone: 1.508.893.6963
Fax: 1.781.861.0731
hwpiel@post.harvard.edu

Pierson, Stuart O.
Memorial Univ
Department of History
St John's, NF A1C5S7
Canada
Phone: 1.709.737.8431
Fax: 1.709.737.2164
spierson@mun.ca
Doctorate, 1969
17c, Europe, Philosophy
and Philosophy of Science;
18c, Europe, Physical Sciences;
19c, Europe, Astronomical
Sciences; 14c-16c, Europe,
Social Sciences

Pieters, Toine
MGR. vd. Weteringstr 23
Utrecht 3581 EB
The Netherlands
Phone: 31302516487
tpieters@bio.vu.nl

Pighetti, Clelia
Universita Dipartimento Fisica
Via Paradiso 12
Ferrara 44100
Italy
Doctorate
17c, Europe, Humanistic
Relations of Science;
18c, Europe, Institutions;
19c, North America,
Social Relations of Science

Pincus, Irwin J.
610 N Roxbury Dr
Beverly Hills, CA 90210-3228

Pingree, David E.
Brown University
PO Box 1900
Providence, RI 02912-1900
Phone: 1.401.863.2101

Doctorate, 1960
Pre-400 B.C.E.,
Astronomical Sciences;
400 B.C.E.-400 C.E., Asia,
Astronomical Sciences;
5c-13c, Education;
14c-16c, Europe, Mathematics

Pinna, Giovanni
Viale Cassiodoro 1
Milano I-20145
Italy
Phone: 39.02.4801.4352
Fax: 39.02.4801.4352
jjpin@iol.it

Pinnick, Cassandra L.
Western Kentucky University
Dept of Philosophy
Bowling Green, KY 42101
Phone: 1.270.745.3136
Fax: 1.270.745.5261
pinnick@wku.edu

Pinto-Correia, Clara
Universidade Lusofona
Campo Grande
Lisbon 376 1700
Portugal
Phone: 1.351.751.5573
Fax: 1.413.585.8516
clara@ulusofona.pt
Doctorate
17c, Europe, Biological Sciences;
18c, Europe, Biological Sciences,
Instruments and Techniques;
5c-13c, Natural and Human
History; 14c-16c, Africa,
Biological Sciences

Pires, J. Chris
Dept of Agronomy
University of Wisconsin
Madison, WI 53706
Phone: 1.608.262.4424
Fax: 1.608.262.7509
jcpires@students.wisc.edu
Bachelor's Level, 2001
20c, North America,
Biological Sciences;
Natural and Human History;
20c, Instruments and Techniques

Pita, Joao Rui
Universidade de Coimbra
Faculdade de Farmacia
Rua do Norte

Coimbra 3000
Portugal
Phone: 351.239.708.870
Fax: 351.239.708.871
jnpita@ci.uc.pt
Doctorate, 1995
19c, Europe, Medical Sciences;
20c, Europe, Medical Sciences

Pitt, Joseph C.
Virginia Tech
Dept of Philosophy 0126
Blacksburg, VA 24061-0126
Phone: 1.540.231.5760
Fax: 1.540.231.6367
jcpitt@vt.edu
Doctorate, 1972
20c, North America, Philosophy
and Philosophy of Science;
20c, North America, Technology;
20c, North America, Instruments
and Techniques; 20c, North
America, Social Sciences

Pittman, James A., Jr.
5 Ridge Dr
Birmingham, AL 35213-3631
Phone: 1.205.975.4976
Fax: 1.205.975.4976
jpdoc@aol.com

Pittman, Walter E.
685 Hartford Drive
Tuscaloosa, AL 35406-3100

Platt, Harold L.
Loyola University
Dept of History
6525 N Sheridan Rd
Chicago, IL 60626-5385
Phone: 1.773.508.2237
Fax: 1.773.508.2153
hplatt@orion.it.luc.edu
Doctorate, 1974
19c, North America,
Biological Sciences;
19c, Europe, Biological Sciences;
19c, North America, Technology;
19c, Europe, Technology

Plofker, Kim L.
Brown University
History of Mathematics Dept.
PO Box 1900
Providence, RI 02912-1900
Phone: 1.401.863.1489
kim_plofker@brown.edu

Doctorate, 1995
Astronomical Sciences;
Mathematics; Asia

Pobat, Michael
2011 Headlands Cir
Reston, VA 20191-3613

Podgorny, Irina
Arqueolgia-Museo de La Plata
Paseo del Bosque 0/M
La Plata 1900
Argentina
Phone: 54.221.427.2250
Fax: 54.1.307.7145
podgorny@mail.retina.ar
Doctorate, 1994
20c, South America, Institutions;
20c, South America, Humanistic
Relations of Science;
19c, South America, Exploration,
Navigation and Expeditions;
19c, North America, Earth
Sciences

Poirier, Jean P.
26 Rue de Varenne
Paris 75007
France
Phone: 33.1.4548.6269
Fax: 33.1.4548.6249
poirierj@noos.fr
Doctorate
18c, Europe, Physical Sciences;
18c, Europe, Social Sciences

Pols, Hans
University of Sydney
Unit for History & Philosophy
of Science
Carslaw Building F07
Sydney, NSW 2006
Australia
Phone: 61.2.9351.3610
Fax: 61.2.9351.4124
h.pols@scifac.usyd.edu.au
Doctorate, 1997
19c, Europe, Medical Sciences,
Instruments and Techniques;
20c, North America,
Medical Sciences, Instruments
and Techniques;
20c, Cognitive Sciences

Popa, Tiberiu
University of Pittsburgh
Dept of Classics

1518 Cathedral of Learning
Pittsburgh, PA 15260-6299
Phone: 1.412.683.1588
Fax: 1.412.624.4419
tmpst26+@pitt.edu

Popenoe de Hatch, Marion
Universidad del Valle
Apartado Postal 82
Guatemala 01901
Guatemala
Phone: 502.364.0336
Fax: 502.364.0212
arqueolo@uvg.edu.gt
Doctorate, 1976
5c-13c, North America,
Astronomical Sciences;
Pre-400 B.C.E., Africa,
Natural and Human History;
14c-16c, Europe, Physical
Sciences; 5c-13c, North
America, Biological Sciences

Popplestone, John A.
University of Akron
225 S Main St
Akron, OH 44325-4302
Phone: 1.330.972.7285
Fax: 1.330.972.6170
jpopplestone@uakron.edu
Doctorate, 1958
North America, Cognitive
Sciences

Porter, Charlotte M.
University of Florida
Dickinson Hall
PO Box 117800
Gainesville, FL 32611-7800
Phone: 1.352.392.6763
Fax: 1.352.846.0287
cmporter@flmnh.ufl.edu
Doctorate, 1976
Natural and Human History;
North America; Biological
Sciences; Exploration,
Navigation and Expeditions

Porter, Duncan M.
Virginia Tech
Department of Biology
Blacksburg, VA 24061
Phone: 1.540.231.6768
Fax: 1.540.231.9307
duporter@vt.edu
Doctorate, 1967
19c, Europe, Biological Sciences,

Biography; Natural and Human
History; Exploration, Navigation
and Expeditions

Porter, Theodore M.
Univ of Calif - Los Angeles
Dept of History
6265 Bunche Hall
PO Box 951473
Los Angeles, CA 90095-1473
Phone: 1.310.206.2352
Fax: 1.310.206.9630
tporter@history.ucla.edu
Doctorate, 1981

Portet, Pierre
11 rue Delambre
Paris 75014
France
Phone: 609893469
delambre@cybercable.fr.

Portolano, Marlana
7219 16th Ave
Takoma Park, MD 20912-7045
Phone: 1.301.431.2386
port@tidalwave.net

Ports, Jessica
4901 Sand Dune Cir Apt 201
West Palm Beach, FL
33417-7501
Phone: 1.561.818.3496
jesssports@hotmail.com

Portuondo, Maria M.
3929 Canterbury Rd
Baltimore, MD 21218-1704
Phone: 1.410.467.7585
Fax: 1.410.467.8130
mportuondo@jhu.edu

Post, Joseph
44 Gentry Drive
Fair Haven, NJ 07704
Phone: 1.212.395.6509
Fax: 1.212.768.7569
joepost@bellatlantic.net

Potter, Elizabeth
Mills College
5000 MacArthur Blvd
Oakland, CA 94613-1000

Potthast, Thomas
University of Tübingen
Centre for Ethics in the Sciences
and the Humanities

Keplerstr. 17
Tuebingen D-72074
Germany
Phone: 608.238.1002
tpotthast@facstaff.wisc.edu

Poupard, James A.
3612 W Earlham St
Philadelphia, PA 19129-1609

Powers, John C.
Indiana University
Dept History & Philosophy
Goodbody Hall 130
Bloomington, IN 47405
Fax: 1.914.376.0987
john_powers@hotmail.com
Doctorate, 2001
Physical Sciences;
17c, Instruments and Techniques;
17c, Europe, Medical Sciences

Pratt, Herbert T.
23 Colesbery Dr
New Castle, DE 19720-3201
Phone: 1.302.328.7273
Fax: 1.302.328.3711
rtprtcche@earthlink.net
Master's Level, 1988
18c, Europe, Physical Sciences;
18c, Europe, Technology;
Europe, Biography;
North America, Education

Pratt, Joan
Univ of Northern Colorado
History Department
Greeley, CO 80639
Phone: 1.970.351.2082
Fax: 1.970.351.2199
jkpratt@unco.edu
Doctorate
20c, Europe, Biological Sciences

Press, Frank
National Acad of Sciences
President
2101 Constitution Ave NW
Washington, DC 20418-0006

Pribram, John K.
Bates College
Dept of Physics & Astronomy
Lewiston, ME 04240-6084
Phone: 1.207.786.6321
Fax: 1.207.786.8334
jpribram@bates.edu

Doctorate, 1973
19c, Physical Sciences;
20c, Physical Sciences

Princehouse, Patricia
Harvard University
235 Science Center
1 Oxford St
Cambridge, MA 02138
Phone: 1.216.368.2632
Fax: 1.216.368.0814
princeh@fas.harvard.edu
Master's Level, 1999
Biological Sciences; 20c, North
America, Earth Sciences;
Gender and Science; Philosophy
and Philosophy of Science

Principe, Lawrence M.
Johns Hopkins University
History of Science Dept
34th & Charles Streets
Baltimore, MD 21218
Phone: 1.410.516.7280
Fax: 1.410.516.8420
lmafp@jhuvms.hcf.jhu.edu
Doctorate
Physical Sciences;
17c, Physical Sciences;
18c, Physical Sciences

Printz, Deborah B.
Yeshiva Atlanta
3130 Raymond Drive
Atlanta, GA 30340
Phone: 1.770.451.5299
dprintz@mindspring.com
Doctorate, 1966
Philosophy and Philosophy of
Science

Pritchard, Sara B.
1200 Massachusetts Ave Apt 5W
Cambridge, MA 02138-5252
Phone: 1.949.675.6424
spritch@leland.stanford.edu

Proctor, George L.
2934 Bardy Rd
Santa Rosa, CA 95404-8544
Phone: 1.707.544.7015
Doctorate, 1957
20c, North America, Philosophy
and Philosophy of Science;
20c, Biological Sciences;

5c-13c, Europe, Technology;
20c, North America, Cognitive
Sciences

Provine, William B.
Cornell University
Ecology & Systematics
E139 Corson Hall
Ithaca, NY 14853-2701
Phone: 1.607.254.4264
Fax: 1.607.849.4026
wbp2@cornell.edu
20c, North America, Biological
Sciences; Asia; Europe

Pruna, Pedro M.
154 No. 326, Bajos
Reparto Nautico, Playa
Havana
Cuba

Puig-Pla, Carles
Univ Politecnica de Catalunya
Diagonal 647
(ETSEIB)
Barcelona 08028
Spain
Phone: 34.9.3401.7782
Fax: 34.9.3401.1713
puig@ma1.upc.es
Master's Level, 1994
19c, Europe, Technology;
18c, Europe, Technology;
400 B.C.E.-400 C.E., Europe,
Astronomical Sciences;
400 B.C.E.-400 C.E., Asia,
Technology

Puigdomenech, Pere
Inst de Biologia Molecular
CID-CSIC
Jordi Girona, 18
Barcelona 08034
Spain
Phone: 34.9.34.0061
Fax: 34.9.3204.5904
pprgmp@cid.csic.es
Doctorate, 1984
Europe, Biological Sciences;
20c, Biological Sciences

Pycior, Helena M.
Univ of Wisconsin-Milwaukee
History Department
PO Box 413
Milwaukee, WI 53201-0413
Phone: 1.414.229.3966

Fax: 1.414.229.2435
helena@uwm.edu
Doctorate, 1976
18c, Europe, Mathematics;
19c, Europe, Mathematics;
20c, Europe, Physical Sciences;
Gender and Science

Pyenson, Lewis R.
Center for Louisiana Studies
PO Box 40831
Lafayette, LA 70504-0831
Phone: 1.337.482.6027
loup@louisiana.edu
Doctorate, 1974
20c, Social Sciences;
20c, Education;
Physical Sciences

Qi, Minghao
143 Albany St Apt 212A
Cambridge, MA 02139-4200
Phone: 1.617.225.7249
minghaoqi@yahoo.com

Quintero, Camilo
521 W Doty St Apt 10
Madison, WI 53703-2664

Quivik, Fredric L.
633 Lexington Pkwy N
Saint Paul, MN 55104-2023
Phone: 1.651.917.0645
Fax: 1.651.917.3946
quivik@usfamily.net
Doctorate, 1998
20c, North America, Technology;
20c, North America,
Biological Sciences;
20c, North America, Technology,
Instruments and Techniques;
19c, North America, Technology

Rabin, Sheila J.
St Peter's College
Dept of History
2641 Kennedy Blvd
Jersey City, NJ 07306-5997
Phone: 1.201.915.9366
Fax: 1.201.435.3662
rabin_s@spc.edu
Doctorate
17c, Europe, Astronomical
Sciences; 14c-16c, Europe,
Astronomical Sciences

Rabson, Diane
UCAR/NCAR
PO Box 3000
Boulder, CO 80307-3000

Rader, Karen A.
Sarah Lawrence College
Science, Tech & Society
One Mead Way
Bronxville, NY 10708-5999
Phone: 1.914.395.2348
Fax: 1.914.376.0987
krader@slc.edu
Doctorate, 1995
20c, North America, Biological
Sciences; North America; Social
Relations of Science; Technology

Radick, Gregory
School of Philosophy
University of Leeds
Leeds LS2 9JT
Great Britain
Phone: 0113.233.3278
Fax: 0113.233.3265
g.m.radick@leeds.ac.uk

Rae, Ian D.
University of Melbourne
Dept. of Hist. & Phil. of Sci.
Parkville 3016
Australia
i.rae@hps.unimelb.edu.au
Doctorate, 1965
19c, Physical Sciences;
20c, Australia and Oceania,
Biological Sciences;
20c, Education; 19c, Australia
and Oceania, Technology

Raffaella, Simili
University of Bologna
Dept of Philosophy
Zamboni, 38
Bologna 40126
Italy
Phone: 39.05.1209.8356
Fax: 39.05.1209.8356
simili@philo.unibo.it

Ragep, Jamil
University of Oklahoma
Dept of History of Science
621 Physical Sciences Building
Norman, OK 73019-0001
Phone: 1.405.325.3392
Fax: 1.405.325.2363

jragep@ou.edu
Doctorate, 1982
5c-13c, Asia, Astronomical
Sciences; Asia; Pre-400 B.C.E.

Rainger, Ronald
Texas Tech University
Dept of History
Box 41013
Lubbock, TX 79409-1013
Fax: 1.806.742.1060
j3ron@ttacs.ttu.edu
Doctorate
20c, North America; 20c, North
America, Biological Sciences;
20c, North America, Earth
Sciences

Ramberg, Peter
Truman State University
Division of Science
100 E Normal St
Kirksville, MO 63501-4200
Phone: 1.660.785.4620
ramberg@truman.edu
Doctorate, 1993
19c, Europe, Physical Sciences;
Physical Sciences

Ramunni, Girolomo
CHST-MRASH - 14 Avenue
Berthelot 69007
France
Phone: 33.4.7272.6544
Fax: 33.4.7280.0008
girolomo.ramunni@
   ish-pyon.cuts.fr

Rand, Timothy
950 Massachusetts Ave Apt 207
Cambridge, MA 02139-3174
Phone: 1.617.868.9436
tdrand@attbi.com

Rankin, Alisha
Harvard Univesity
History of Science Dept
Science Center 235
Cambridge, MA 02138
rankin@fas.harvard.edu

Rappaport, Rhoda
141 Fulton Ave, Apt 810
Poughkeepsie, NY 12603
Phone: 1.845.454.6507

Rashed, Roshdi
8 Allee du Val de Bievze
Bourg-la-Reine 92340
France
Phone: 33.1.46.64.7597
Fax: 33.1.46.65.4237
rashed@paris7.jussieu.fr

Rasmussen, Nicolas
Univ of New South Wales
Science & Technology Studies
LG19 Morvan Brown Bldg
Sydney NSW 2052
Australia
nicolas.rasmussen@unsw.edu.au

Rassinier, Jean-Paul
Medical Practice
4, square Thiers
Paris 75166
France
Phone: 33.1.4505.1080
Fax: 33.1.4505.1580
jprassinier@wanadoo.fr
Doctorate
400 B.C.E.-400 C.E.,
Europe, Medical Sciences;
5c-13c, Europe, Medical
Sciences; 400 B.C.E.-400 C.E.,
Europe, Philosophy and
Philosophy of Science;
5c-13c, Europe, Philosophy and
Philosophy of Science

Rau, Erik P.
Drexel University
Dept. of History & Politics
MacAlister Hall 5010
Philadelphia, PA 19104
Phone: 1.215.895.2463
Fax: 1.302.266.9785
erau@drexel.edu
Doctorate, 1999
20c, North America, Technology;
20c, Europe, Technology;
20c, North America

Rauch, Alan
Georgia Inst of Technology
Literature, Comm, & Culture
Atlanta, GA 30332-0165
Phone: 1.404.894.7000
Fax: 1.404.894.1287
alan.rauch@lcc.gatech.edu
Doctorate, 1989
19c, Europe, Humanistic
Relations of Science;

18c, Europe, Humanistic
Relations of Science;
19c, Europe, Natural and
Human History;
19c, Europe, Institutions

Rausing, Lisbet
17 Aubrey Walk
Aubrey Road
London W8 7JJ
Great Britain
Phone: 44.207.838.7132
lr@arcticnet.com
Doctorate
18c, Europe, Biological Sciences;
19c, Europe, Biological Sciences;
Europe, Natural and Human
History; Europe

Raven, Diederick
Univ of Utrecht
Dept of Anthropology
PO Box 80.140
Utrecht 3508 TC
The Netherlands
Phone: 31555766520
Fax: 31302534666
d.raen@fss.uu.nl

Ray, Richard D.
1562 Crofton Pkwy
Crofton, MD 21114-1534
Phone: 1.301.614.6102
richard.ray@gsfc.nasa.gov

Rayner, H.L.
600 Stewart Mountain Rd
Victoria, BC V9B6J8
Canada
Phone: 1.250.391.0919
hrayner@shaw.ca

Reaves, Gibson
University of Southern Calif
MC 1342
Los Angeles, CA 90089-1342
Phone: 1.213.740.6330
Fax: 1.213.740.6342
reaves@mizar.usc.edu
Doctorate
20c, North America,
Astronomical Sciences;
14c-16c, Europe, Astronomical
Sciences

Redekop, Benjamin
Kettering University
Dept of Liberal Studies
1700 W 3rd Ave
Flint, MI 48504-4898
Phone: 1.810.762.7824
bredekop@kettering.edu

Reeds, Karen
127 Southgate Rd
New Providence, NJ 07974-1663
Phone: 1.908.464.0714
Fax: 1.908.464.6814
reeds@openix.com
Doctorate, 1975
14c-16c, Biological Sciences;
Medical Sciences;
Natural and Human History

Rees, Graham
Queen Mary Univ of London
Sch of English and Drama
London E1 4NS
Great Britain
g.c.rees@gmw.ac.uk

Reese, Garth D., Jr.
PO Box 2211
Apple Valley, CA 92307-0042
Phone: 1.909.623.1666
reesegd@earthlink.net

Reeves, Barbara J.
Virginia Tech
Interdisciplinary Studies 0227
Blacksburg, VA 24061
Phone: 1.540.231.7687
Fax: 1.540.231.7687
reeves@vt.edu

Regni, Catehrine
Dept of Biochemistry
117 Schweitzer Hall
Columbia, MO 65211-5100
Phone: 1.573.817.2576
car963@mizzou.edu

Rehbock, Philip F.
University of Hawaii
2680 Woodlawn Dr
Honolulu, HI 96822
Phone: 1.808.956.6829
Fax: 1.808.956.3467
rehbock@ifa.hawaii.edu
Doctorate, 1975
19c, Europe, Natural and
Human History;

19c, Europe, Biological Sciences;
19c, Europe, Earth Sciences;
19c, Australia and Oceania,
Exploration, Navigation and
Expeditions

Reichard, Kathryn L.
421 First Avenue
West Haven, CT 06516-3815
Phone: 1.203.937.7518
Doctorate, 1980
Physical Sciences; Instruments
and Techniques; Institutions;
Social Sciences

Reid, David A.
UCLA
Dept of History, Box 951473
6265 Bunche Hall
Los Angeles, CA 90095-1473
Phone: 1.904.620.2886
Fax: 1.904.620.1018
dreid@unf.edu
Master's Level
17c, Europe, Education;
18c, Europe, Education;
18c, Europe, Social Relations of
Science; Physical Sciences

Reidy, Michael S.
507 W Babcock St
Bozeman, MT 59715
Phone: 1.406.582.7512

Reif, Wolf E.
Institut & Museum für Geologie
und Paleontologie
Sigwartstrasse 10
Tübingen D-72076
Germany
Phone: 49.7071.297.2491
wolf-ernst.reif@
uni.tuebingen.de

Reingold, Nathan
5305 Glenwood Rd
Bethesda, MD 20814-1405
Phone: 1.301.654.4135
reingold.miles@worldnet.att.net

Reinhardt, Carsten
Arnulfsplatz 1
Regensburg D-93040
Germany
Phone: 49.941.52254
Fax: 49.941.943.1985
carsten.reinhardt@psk.

uni-regensburg.de
Doctorate, 1996
20c, Physical Sciences;
19c, Europe, Physical Sciences;
Technology

Remington, Jack
6335 E Tanuri Cir
Tucson, AZ 85750-1933
Phone: 1.520.290.9046
hjremington@aol.com

Remmert, Volker R.
Universtaet Mainz
Fachbereich Mathematik
Staudingerweg 9
Mainz 55099
Germany
remmert@mathematik.
uni-mainz.de
Doctorate, 1997
17c, Europe; 20c, Europe,
Mathematics; Social Sciences

Rensing, Susan
University of Minnesota
Program in History of Science
and Technology
167 Social Sci Bldg, West Bank
Minneapolis, MN 55455
Phone: 1.612.626.8722
ren0031@tc.umn.edu

Reszler, Michael
4515 Walnut St Apt 306A
Kansas City, MO 64111-7732
Phone: 1.816.753.6241
Fax: 1.816.234.7915
mreszler@knightridder.com

Revill, James A.
Inst. of Physics Publishing
Dirac House
Temple Back
Bristol B5I 6BE
Great Britain
Phone: 44.11.7930.1152
Fax: 44.11.7930.1186
jim.revill@iop.org
Bachelor's Level, 1971
Physical Sciences;
Astronomical Sciences;
Mathematics; Technology,
Instruments and Techniques

Rhees, David J.
140 W 48th St
Minneapolis, MN 55409-2533

Rheinberger, Hans-Jorg
Max Planck Institute for the
History of Science
Wilhelmstr 44
Berlin 10117
Germany
Phone: 49.30.2266.7161
Fax: 49.30.2266.7167
rheinbg@mpiwg-berlin.mpg.de
Doctorate, 1982
20c, Biological Sciences;
20c, Philosophy and Philosophy
of Science

Rhoads, Geoffrey B.
Digimarc Corporation
2961 SW Turner Rd
West Linn, OR 97068-9654
rhoads@digimarc.com

Rhoads, Jonathan E.
3400 Spruce St
Philadelphia, PA 19104-4204

Ribble, John C.
6200 Willers Way
Houston, TX 77057-2808
Phone: 1.716.783.6942
jribble@houston.rr.com

Ribe, Neil M.
Institut de Physique du Globe
4 Place Jussieu
Paris Cedex 05 75252
France
Phone: 33.14.4227.2479
Fax: 33.14.4227.2481
ripe@ipgp.jussieu.fr
Doctorate, 1981
17c, Europe, Physical Sciences;
400 B.C.E.-400 C.E., Europe,
Philosophy and Philosophy
of Science; Philosophy
and Philosophy of Science;
20c, Instruments and Techniques

Rice, Richard E.
PO Box 1210
Florence, MT 59833-1210
Phone: 1.540.801.8199
Fax: 1.540.568.2913
ricere@jmu.edu
Doctorate, 1982

19c, Physical Sciences;
Europe, Physical Sciences;
Social Relations of Science;
Education

Rich, Robert M.
1921 Rock St Apt 23
Mountain View, CA 94043-2553
Phone: 1.650.968.9871
Doctorate, 1966
Earth Sciences;
Biological Sciences

Richards, Joan L.
Brown University
Dept of History
Box N
142 Angell St
Providence, RI 02912-9040
Phone: 1.401.863.3246
Fax: 1.401.863.1040
Doctorate
19c, Europe, Mathematics;
19c, Europe, Philosophy
and Philosophy of Science;
18c, Mathematics;
18c, Philosophy and Philosophy
of Science

Richards, Robert J.
University of Chicago
Fishbein Center
1126 E. 59th Street
Chicago, IL 60637
Phone: 1.773.702.8348
Fax: 1.773.834.1299
r-richards@uchicago.edu
Doctorate, 1978
20c, North America,
Biological Sciences;
19c, Europe, Biological Sciences;
19c, Europe, Social Sciences;
Philosophy and Philosophy of
Science

Richardson, Alan
University of British Columbia
Department of Philosophy
1866 Main Mall - E370
Vancouver, BC V6T1Z1
Canada
Phone: 1.604.822.3967
Fax: 1.604.822.8782
alanr@interchange.ubc.ca

Richardson, Robert C.
University of Cincinnati
Dept of Philosophy
Cincinnati, OH 45221-0374
Phone: 1.513.556.6327
Fax: 1.513.556.2939
robert.richardson@uc.edu
Doctorate, 1976
19c, Europe, Biological Sciences;
20c, Europe, Cognitive Sciences;
20c, North America,
Instruments and Techniques;
Biological Sciences

Richmond, Jesse
385 Berkeley St Apt 2
Toronto, ON M5A2X8
Canada
Phone: 1.416.928.3418
j_richmond@sympatico.ca

Richmond, Marsha L.
Wayne State University
5700 Cass Ave #2307
Detroit, MI 48202
Phone: 1.313.577.6499
Fax: 1.734.930.9446
marsha.richmond@wayne.edu
Doctorate, 1986
19c, Europe, Biological Sciences;
20c, Europe, Biological Sciences;
20c, Biological Sciences

Richter, Goetz W.
5330 Brooke Ridge Dr
Atlanta, GA 30338-3127
Phone: 1.770.399.6170
Doctorate, 1948
400 B.C.E.-400 C.E.,
Europe, Medical Sciences;
400 B.C.E.-400 C.E.,
Europe, Social Sciences;
19c, Europe, Medical Sciences;
19c, North America, Education

Rickey, V. F.
United States Military Academy
Dept of Mathematical Sciences
West Point, NY 10996-1786
fred-rickey@usma.edu

Rider, Robin E.
University of Wisconsin
Dept of Special Collections
990 Memorial Library
728 State St
Madison, WI 53706-1418

Phone: 1.608.262.2809
Fax: 1.608.265.2754
rrider@library.wisc.edu
Doctorate, 1980
Social Relations of Science;
Mathematics

Ries, Chris
1735 Dexter Ave N Apt A301
Seattle, WA 98109-6219
Phone: 1.206.283.0106
chrisriess@hotmail.com

Riess, Falk
University of Oldenburg
Physics Dept
PO Box 2503
Ammerlaender Heesstr. 114
Oldenburg 26111
Germany
Phone: 49.441.798.3540
Fax: 49.441.798.3990
falk.reiss@uni-oldenburg.de

Riethmiller, Steven
VMI
Chemistry Department
Lexington, VA 24450
Phone: 1.540.464.7244
Fax: 1.540.464.7261
sr@vmi.edu
Doctorate, 1973
19c, Physical Sciences;
19c, Medical Sciences

Rigden, John S.
American Institute of Physics
1 Physics Ellipse
College Park, MD 20740-3843
Phone: 1.301.209.3124
Fax: 1.301.209.0841
jsr@aip.org
Doctorate
20c, North America, Physical
Sciences; 20c, North America;
20c, North America, Education

Riggs, Allison
298 Lakeview Dr
Morgantown, WV 26508-9254
Phone: 1.304.594.3476
ajriggs@ufl.edu

Rioja, Ana M.
Alcalde Sainz de Baranda, 23
Madrid 28009
Spain

Phone: 34.91.394.5252
Fax: 34.91.394.5252
riojan@evcmax.sim.vcm.es

Riordan, Michael
University of California
Institute for Particle Physics
Santa Cruz, CA 95064
Phone: 1.831.459.5687
Fax: 1.831.464.2406
michael@slac.stanford.edu
Doctorate, 1973
20c, Physical Sciences;
20c, North America,
Science Policy;
20c, North America, Technology;
17c, Physical Sciences

Ripley, S. Dillon
Smithsonian Institution
Rm 336
Natural History Bldg
Washington, DC 20560-0001

Riskin, Jessica G.
Stanford University
History Department
Stanford, CA 94305-2024
jriskin@stanford.edu
Doctorate, 1995
18c, Europe, Philosophy and
Philosophy of Science

Risse, Guenter B.
University of California
533 Parnassus Ave
Box 0726
San Francisco, CA 94143-0726
Phone: 1.415.476.8826
Fax: 1.415.476.9453
profgrisse@attbi.com
Doctorate, 1971
20c, North America,
Medical Sciences;
18c, Europe, Medical Sciences

Ritter, Christopher
Univ of California at Berkeley
Office for Hist of Sci & Tech
470 Stephens Hall #2350
Berkeley, CA 94720-2350
critter@socrates.berkeley.edu
Doctorate, 1999
Physical Sciences; Education

Ritvo, Harriet
Mass Inst of Technology
E51-284
History Dept
77 Massachusetts Ave
Cambridge, MA 02139-4307
Phone: 1.617.253.6960
Fax: 1.617.253.9406
hnritvo@mit.edu
Doctorate, 1975
18c, Europe, Natural and
Human History; 19c, Europe,
Natural and Human History;
18c, Europe, Biological Sciences;
19c, Europe, Biological Sciences

Roberto, Giani
Via Lucania 9
Milano Mi 20139
Italy

Roberts, David L.
12226 Valerie Ln
Laurel, MD 20708-2838
Phone: 1.301.317.7891
robertsdl@aol.com
Doctorate, 1998
North America, Mathematics;
North America, Education

Roberts, Elizabeth T.
3A Thornton St
Exeter, NH 03833-2219
liztroberts@hotmail.com
Master's Level, 1994
19c, Europe, Earth Sciences

Roberts, Jody
201 Alleghany St Apt A
Blacksburg, VA 24060-5061
Phone: 1.540.951.1904
jody@vt.edu

Roberts, Jon H.
Boston University
Dept of History
226 Baystate Rd
Boston, MA 02215
Doctorate
19c, North America, Humanistic
Relations of Science;
20c, North America, Humanistic
Relations of Science

Robertson, Laurie
5870 1st St N
Arlington, VA 22203-1102
Phone: 1.703.522.9628
lroberts@acm.org

Robinson, David K.
Truman State University
Division of Social Science
100 E. Franklin Street
Kirksville, MO 63501
Phone: 1.660.785.4321
Fax: 1.660.785.4337
drobinso@truman.edu
Doctorate, 1987
19c, Europe, Cognitive Sciences;
Europe, Education;
Europe, Cognitive Sciences

Robinson, David Z.
Carnegie Comm Sci Tech Gvmt
437 Madison Ave 27th Fl
New York, NY 10022-7001

Robinson, Gloria
Yale University
Section History of Medicine
333 Cedar St
New Haven, CT 06510-3289
Doctorate

Robinson, Joseph D.
SUNY Health Science Center
Dept. of Pharmacology
Syracuse, NY 13210
Fax: 1.434.974.9541
jdr1934@aol.com
Doctorate, 1959
Philosophy and Philosophy
of Science; Medical Sciences;
Physical Sciences

Robinson, Michael F.
177 Warrenton Ave
Hartford, CT 06105-3934
mfrobins@students.wisc.edu

Robinson, Peter
300 Rock House Dr
Liberty Hill, TX 78642-6228
Phone: 1.512.778.5902

Roca-Rosell, Antoni
ETS Enginyers Industrials
Diagonal 647
Barcelona
Spain

Phone: 34.93.401.6629
Fax: 34.93.401 1713
roca@ma1.upc.es
Doctorate, 1990

Rochberg, Francesca
Univ of California-Riverside
Dept of History
Riverside, CA 92521
Phone: 1.909.787.5401

Rocke, Alan J.
Case Western Reserve Univ
History Department
Cleveland, OH 44106
Phone: 1.216.368.2614
Fax: 1.216.368.4681
ajr@po.cwru.edu
Doctorate, 1975
19c, Europe, Physical Sciences;
Europe, Physical Sciences;
Europe, Social Sciences;
Philosophy and Philosophy of
Science

Rockefeller, David
30 Rockefeller Plz # 5600
New York, NY 10112-0002

Rodriguez, Maria A.
Av Santa Rosa
Edificio Zumaiya Apto 03
Cumana
Estado Sucre 6101
Venezuela
mrodrig@sucre.udo.edu.ve

Roe, Shirley A.
University of Connecticut
Dept of History
Storrs Manfield, CT 06269-2103
Phone: 1.860.486.3722
Fax: 1.860.486.0641
shirley.roe@uconn.edu
Doctorate, 1976
18c, Europe, Biological Sciences;
18c, Europe, Social Relations of
Science; 20c, Social Sciences

Rogers, Catherine H.
67 Regatta Rd
Canada Bay 2046
Australia
Phone: 61.2.97447416
Fax: 61.2.97440357
crogers@mpx.com.au
Master's Level, 1990

19c, Europe, Technology;
19c, Europe, Social Relations
of Science; 19c, Europe,
Instruments and Techniques;
19c, Europe, Physical Sciences

Rogers, Naomi
Yale University
History of Medicine
PO Box 208015
New Haven, CT 06520
Phone: 1.203.385.4338
Fax: 1.203.737.4130
naomi.rogers@yale.edu
Doctorate
20c, North America, Medical
Sciences; 20c, North America,
Gender and Science; 20c, North
America; 20c, North America,
Social Relations of Science

Rogers, Rosemary F.
Stanford University
Bldg 200 Rm 33
Stanford, CA 94305-2024
Phone: 1.650.725.0714
Fax: 1.650.725.0597
rrogers@leland.stanford.edu
Master's Level

Roland, Alex
Duke University
Dept of History
Box 90719
Durham, NC 27708
Phone: 1.919.684.2758
Fax: 1.919.681.7670
alex.roland@duke.edu
Doctorate, 1974
Technology

Roll-Hansen, Nils D.
Gjennomfaret 21
Oslo 0876
Norway
Phone: 47.22.23.43.42
Fax: 47.22.85.69.63
nils.rollhansen@filosofi.uio.no

Romo, Jose
Universidad de Barcelona
C/ Baldiri 1 Reixac S/N
Barcelona 08028
Spain
Phone: 93.339.65.78

Fax: 934498510
romo@cerber.mat.ub.es
Doctorate, 1991

Roof, Richard
1045 Portland Ave
Saint Paul, MN 55104-7011
Phone: 1.612.379.2956
Fax: 1.612.625.2199
richardroof@hotmail.com

Roos, Anna Mari
University of Minnesota
History Dept
265 A B Anderson Hall
10 University Dr
Duluth, MN 55812-2496
Phone: 1.218.726.7544
aroos@d.umn.edu

Rootenberg, Howard M.
B & L Rootenberg Rare Books
15422 Sutton St
Sherman Oaks, CA 91403-3808
Phone: 1.818.788.7765
Fax: 1.818.788.8839
blroot@pacificnet.net

Roque, Xavier
Universitat Autonoma Barcelona
Centre d'Estudis d'Historia de
les Ciencies Edifici CC
Bellaterra 08193
Spain
Phone: 34.9.3581.2966
Fax: 34.9.3581.2790
xavier.roque@uab.es
Doctorate, 1993
20c, Physical Sciences;
Mathematics

Rose, Anne C.
Johns Hopkins Univ
Humanities Center
Gilman 113
3400 N Charles St
Baltimore, MD 21218-2680
arose@jhu.edu

Rosenkrantz, Barbara G.
Harvard University
Science Center 235
History of Science
Cambridge, MA 02138
Phone: 1.617.496.2239
Fax: 1.617.495.3344
rosenkr@fas.harvard.edu

Doctorate
19c, North America,
Medical Sciences; 20c, North
America, Medical Sciences;
19c, Europe, Medical Sciences;
20c, Europe, Medical Sciences

Rosenstein, George M.
Franklin & Marshall College
PO Box 3003
Lancaster, PA 17604-3003
Phone: 1.717.399.7163
Fax: 1.717.358.4507
g_rosenstein@acad.fandm.edu
Doctorate, 1963
19c, North America, Mathematics

Rosenthal, Franz
80 Heloise St
Hamden, CT 06517-3422
Doctorate
5c-13c

Roske, Earl
6677 Camden Ave
San Jose, CA 95120-2142

Rosner, David
Columbia University
Hist Sch-Public Health
722 W 168th St #9FL
New York, NY 10032-2645
Phone: 1.212.304.7979
Fax: 1.212.304.7942
dr289@columbia.edu

Rosner, Lisa
Richard Stockton College
Historical Studies Program
PO Box 195
Pomona, NJ 08240-0195
Phone: 1.609.652.4434
Fax: 1.609.652.4550
rosnerl@stockton.edu
Doctorate
18c, Europe, Medical Sciences;
18c, Europe, Institutions;
17c, Europe, Institutions

Ross, Dorothy
Johns Hopkins University
Department of History
Baltimore, MD 21218
Phone: 1.410.516.8616
Fax: 1.410.516.7586
doross1@attglobal.net
Doctorate, 1965

19c, North America,
Social Sciences; 20c, Europe;
18c, Europe

Ross, Joseph T.
University of Notre Dame
102 Hesburgh Library
Notre Dame, IN 46556
Phone: 1.219.631.5835
jross@nd.edu
Master's Level, 1991
18c, Europe, Philosophy
and Philosophy of Science;
19c, Europe, Philosophy
and Philosophy of Science;
Astronomical Sciences

Ross, Karen
2807 Johnson Street NE
Minneapolis, MN 55418
Phone: 1.612.781.1643
ross0199@tc.umn.edu

Ross, Sydney
2194 Tibbits Ave
Troy, NY 12180-7015
Fax: 1.518.274.6216
rosss2@rpi.edu

Rossello' Botey, Victoria
Severo Ochoa 12-15
Valencia 46010
Spain
Phone: 34.9.6380.7781

Rossianov, Kirill
Novolesnaya 18-1-160
Moscow 103055
Russia

Rossiter, Margaret W.
Cornell University
726 University Avenue, #201
Ithaca, NY 14850
Phone: 1.607.255.2545
Fax: 1.607.255.0616
isis@cornell.edu
Doctorate, 1971
20c, North America, Gender and
Science; 19c, North America,
Gender and Science; 20c, North
America; 20c, North America,
Biological Sciences

Roth, Durell M.
902 Yaupon Valley Road
Austin, TX 78746

Phone: 1.512.328.5332
Fax: 1.512.328.9558
roth@onr.com
20c, North America, Technology;
19c, North America, Technology

Rothenberg, Marc
8533 Milford Avenue
Silver Spring, MD 20910
Phone: 1.202.357.1421
Fax: 1.202.786.2878
josephhenr@aol.com
Doctorate, 1974
North America;
19c, Astronomical Sciences;
Exploration, Navigation and
Expeditions

Rounds, Shawn P.
Minnesota Historical Society
345 Kellogg Blvd. West
St Paul, MN 55106-1906
Phone: 1.651.297.1261
shawn.rounds@mnhs.org
Master's Level, 1996
Technology; Technology,
Instruments and Techniques;
Social Sciences;
Astronomical Sciences

Rouse, Joseph T.
Wesleyan Univ
Dept of Philosophy
Middletown, CT 06459-0001
Phone: 1.860.685.3658
Fax: 1.860.685.3861
jrouse@wesleyan.edu

Routh, Donald K.
University of Miami
Psychology Annex
5665 Ponce De Leon Blvd Rm
232
Coral Gables, FL 33146-2510
Phone: 1.305.284.5222
Fax: 1.305.284.4795
drouth@miami.edu
Doctorate, 1967
20c, Cognitive Sciences;
20c, Medical Sciences,
Instruments and Techniques;
19c, Cognitive Sciences;
19c, Medical Sciences,
Instruments and Techniques

Rozwadowski, Helen M.
Georgia Inst of Technology
School of History, Technology
and Society
Atlanta, GA 30332
Phone: 1.404.894.7448
Fax: 1.404.894.0535
helenrozi@mac.com
Doctorate, 1996
20c, Earth Sciences;
19c, Biological Sciences;
19c, Exploration, Navigation
and Expeditions

Rudenberg, H. G.
15 Piper Rd Apt K208
Scarborough, ME 04074-7542
rudyone@aol.com

Rudge, David W.
Western Michigan University
Dept of Biological Sciences
3134 Wood Hall, WMU
Kalamazoo, MI 49008
Phone: 1.616.387.2779
Fax: 1.616.387.5609
david.rudge@wmich.edu
Doctorate, 1996
20c, Europe, Biological Sciences;
20c, Biological Sciences;
Philosophy and Philosophy of
Science

Rudolph, John L.
Univ of Wisconsin-Madison
Dept. of Curriculum & Inst.
225 N. Mills St.
Madison, WI 53706
Phone: 1.608.265.3431
jlrudolp@facstaff.wisc.edu
Doctorate, 1999
20c, North America

Rudwick, Martin
University of Cambridge
Department of HPS
Free School Lane
Cambridge CB2 3RH
Great Britain
Fax: 44.353.665.043
another-gnu@virgin.net
Europe, Earth Sciences;
Natural and Human History

Ruegg, Walter H.
Route de Sonchaux 36
Veytaux CH-1820

Switzerland
Phone: 41.21.963.8261
whruegg@swissonline.ch
14c-16c, Europe, Education;
400 B.C.E.-400 C.E.,
Europe, Institutions;
5c-13c, North America,
Humanistic Relations of Science;
19c, Europe, Social Sciences

Ruestow, Edward G.
University of Colorado
Dept of History
CUB 234
Boulder, CO 80309-0234
Phone: 1.303.492.6011
Fax: 1.303.492.1868
Doctorate, 1969
17c, Europe, Social Relations
of Science; 14c-16c, Europe,
Social Relations of Science;
17c, Europe, Physical Sciences;
14c-16c, Europe, Physical
Sciences

Ruffner, James A.
1347 Bedford Rd
Grosse Pt Pk, MI 48230-1117
terjar@ameritech.net

Rupke, Nicolaas A.
Inst fur Wissenschaftsgeschichte
Humboldtallee 11
Göttingen D 37073
Germany
Phone: 49.551.39.9466
Fax: 49.551.39.9748
nrupke@gwdg.de
Doctorate
19c, North America,
Biological Sciences;
20c, Europe, Biological Sciences;
Europe, Earth Sciences

Rupp, Jan C. C.
Louis Bouwmeesterlaan 141
Utrecht 3584 GG
The Netherlands
Phone: 31.030.251.4305
Fax: 31.030.251.4364
rupp@pscw.uva.nl

Ruse, Michael
Florida State University
Department of Philosophy
151 Dodd Hall
Tallahassee, FL 32306

Phone: 1.850.644.1483
Fax: 1.850.644.3832
mruse@mailer.fsu.edu

Rushton, Alan R.
Hunterdon Medical Center
1100 Wescott Dr Ste G-3
Flemington, NJ 08822-4600
Phone: 1.908.788.6468
Fax: 1.508.788.6466
Doctorate, 1977
17c, Biological Sciences;
Biological Sciences; Philosophy
and Philosophy of Science

Ruskin, Steve
8320 Bluffview Way
Colorado Spring, CO 80919
Phone: 1.719.599.3759
sruskin@nd.edu
Master's Level, 1998
19c, Europe, Astronomical
Sciences; Technology;
Exploration, Navigation and
Expeditions

Rusnock, Andrea A.
35 Bethany Rd
Wakefield, RI 02879
Phone: 1.401.874.9021
rusnock@uri.edu
Doctorate, 1990
Medical Sciences; Social
Sciences; 18c, Europe

Russell, Andrew
3765 Davidson Pl
Boulder, CO 80305-5532

Russell, Gül
Texas A & M University
Department of Humanities
Health Science Center
164 Reynolds Med Bldg
College Station, TX 77843-0001
Phone: 1.409.845.6462
Fax: 1.409.845.8634
garussell@tamu.edu
Doctorate, 1962
400 B.C.E.-400 C.E.,
Physical Sciences;
17c, Europe, Cognitive Sciences;
400 B.C.E.-400 C.E., Medical
Sciences

Russo, Arturo
Universita di Palermo
Dipartimento di Fisica e
Tecnologire Relative
Viale delle Scienze
Palermo 90128
Italy
Phone: 39.091.645.9123
Fax: 39.091.645.9106
russo@unipa.it

Rutherford, Alexandra
York University
Dept of Psychology
4700 Keele St
Toronto, ON M3J1P3
Canada
Phone: 1.416.736.5115
Fax: 1.416.736.5814
alexr@yorku.ca

Rutherford, Michael
5080 De Brebeuf
Montreal, PQ H2T3L7
Canada
Phone: 1.514.272.5095
mrutherford68@hotmail.com

Rutkin, H. Darrel
1024 Masonic Ave
San Francisco, CA 94117
Phone: 1.415.626.5009
Fax: 1.415.864.6224
Master's Level
400 B.C.E.-400 C.E., Europe,
Astronomical Sciences;
5c-13c, Europe, Astronomical
Sciences; 14c-16c, Europe,
Physical Sciences; 17c

Rynasiewicz, Robert A.
Johns Hopkins University
Dept of Philosophy
3400 N Charles Street
347 Gilman Hall
Baltimore, MD 21218
Phone: 1.410.516.7514
Fax: 1.410.516.6848
ryno@lorentz.phl.jhu.edu
Doctorate, 1981
Physical Sciences; Astronomical
Sciences, Philosophy and
Philosophy of Science;
20c, Physical Sciences

Sabra, Abdelhamid I.
Harvard University
Science Center 235
Cambridge, MA 02138
Phone: 1.617.495.3741
Fax: 1.617.495.3344
sabra@fas.harvard.edu
Doctorate
5c-13c, Asia, Physical Sciences;
17c, Asia, Physical Sciences;
Asia, Mathematics;
Astronomical Sciences

Saijo, Toshimi
Awa-High School
Kakihara, Yoshimo-Cho
Itano-Gun
Tokushima-Ken 771-1401
Japan
Master's Level
19c, Europe, Physical Sciences;
20c, Asia, Education

Saito, Ken
Dept of Human Sciences
1-1, Gakuen-cho
Sakai 599-8531
Japan
Phone: 81.722.549.626
Fax: 81.722.549.928
ksaito@hs.cias.osakafu-u.ac.jp
Doctorate
400 B.C.E.-400 C.E.,
Europe, Mathematics;
14c-16c, Europe, Mathematics

Saito, Shigeki
1-8-8 Amagawa-cho
Tsuchiurashi
Ibaraki-ken 300
Japan

Sakai, Jennifer
540 W 122nd St
New York, NY 10027-5805
Phone: 1.858.458.9391

Sakai, Shizu
Ishigaku Juntendo Univ
2-1-1 Hongo Bunkyo-Ku
Tokyo 113-8421
Japan
Phone: 81.3.3813.1592
Fax: 81.3.3813.1592
shist@med.juntendo.ac.jp

Sakurai, Ayako
Irifune 6-4-704
Urayasu 279-0012
Japan
Phone: 81.47.380.2352
sakuraia@area.c.u-tokyo.ac.jp

Sakurai, Kunitomo
Kanagawa University
Inst of Physics
3-27-1 Rokkakubashi
Kanagawa-k
Yokohama 221-8686
Japan
Phone: 81.45.481.5661
Fax: 81.45.413.7288
ksakurai@cc.kanagawa_u.ac.jp
Doctorate, 1964
20c, Physical Sciences;
Astronomical Sciences;
Astronomical Sciences,
Philosophy and
Philosophy of Science;
17c, Europe, Earth Sciences

Saldana, Juan Jose
Apartado Postal 21-023
Mexico DF 04000
Mexico
Phone: 52.5.549.0669
Fax: 52.5.544.6316

Saliba, George
Columbia University
1140 Amsterdam Ave MC3942
Middle East Lang & Culture
604 Kent Hall
New York, NY 10027
Phone: 1.212.854.4166
g.saliba@columbia.edu
Doctorate
5c-13c, Asia,
Astronomical Sciences;
14c-16c, Asia, Social Sciences;
14c-16c, Asia, Humanistic
Relations of Science;
5c-13c, Institutions

Salman, Phillips
12A, Bassett Road
London W10 6JJ
Great Britain
Phone: 44.02.8968.0151
psalman@britishlibrary.net

Salvatico, Luis
3417 D SW, 218th Terr
Gainesville, FL 32608

Samuels, Warren J.
Michigan State University
Economics
Marshall Hall
East Lansing, MI 48824
Phone: 1.517.355.1860
Fax: 1.517.432.1068
samuels@msu.edu
Doctorate, 1957
North America, Social Sciences;
Philosophy and Philosophy
of Science; North America,
Philosophy and Philosophy of
Science; Social Sciences

Sanchez, Halley D.
Univ of Puerto Rico-Mayaguez
Philosophy and Physics
PO Box 9264
Mayaguez, PR 00681-9264
Phone: 1.787.832.4040
Fax: 1.787.265.1225
h_sanchez@rumac.upr.clu.edu
Doctorate
Physical Sciences;
20c, Physical Sciences

Sanchez-Navarro, Jesus
Nava y Grimon 46, 1A
La Laguna
Tenerife 38201
Spain
Phone: 34.922.264.293
Fax: 34.922.317.878
jesannav@ull.es
Doctorate, 1985
Philosophy and Philosophy of
Science; Social Relations of
Science; Physical Sciences;
Astronomical Sciences

Sanchez-Ron, Jose M.
Calle Alonso Heredia, 31
Madrid 28028
Spain
Phone: 34.9.1397.4877
Fax: 34.9.1397.3936
josem.sanchez@uam.es

Sandler, I.
1029 NE 103rd St
Seattle, WA 98125-7521

Sandman, Alison
Dibner Institute
MIT E56-100
38 Memorial Drive
Cambridge, MA 02139
Phone: 1.617.253.0880
adsandman@yahoo.com

Santesmases, Maria J.
IESA-CSIC
Alfonso XII 18
Madrid 28029
Spain
Phone: 34.9.1521.9028
Fax: 34.9.1521.8103
mjsantesmases@iesam.csic.es
Doctorate
20c, Europe, Biological Sciences;
20c, Europe, Science Policy;
20c, North America, Biological
Sciences; 20c, North America,
Science Policy

Sapolsky, Harvey M.
MIT
E38-674
Cambridge, MA 02139
Phone: 1.617.253.5265
sapolsky@mit.edu
Doctorate, 1967
20c, North America, Science
Policy; Technology

Sapp, Jan
Univ du Quebec a Montreal
Centre Interuniv de Recherche
sur la Science et la Tech
CP 8888 Succursale Centreville
Montreal, PQ H3C3P8
Canada
sapp.jan@uqam.ca
Doctorate
Biological Sciences

Sarasohn, Lisa T.
Oregon State University
Dept of History
Corvallis, OR 97331
Phone: 1.541.737.1271
Fax: 1.541.737.1257
lsarasohn@aol.com
Doctorate, 1979
17c, Europe, Social Relations of
Science; Europe, Philosophy and
Philosophy of Science

Sargent, Rose-Mary
Merrimack College
Dept of Philosophy S7
North Andover, MA 01845
Phone: 1.978.837.5000
Fax: 1.978.837.5169
rsargent@merrimack.edu
Doctorate, 1987
17c, Europe, Philosophy
and Philosophy of Science;
Philosophy and Philosophy
of Science; Instruments and
Techniques

Saridakis, Voula
Virginia Tech
428 Major Williams
Blacksburg, VA 24061
saridakis@ameritech.net
Master's Level, 1993
17c, Astronomical Sciences;
17c, Institutions;
17c, Instruments and Techniques

Sarkar, Shahana
521 Vista Heights Rd
Richmond, CA 94805-2503
Phone: 1.510.233.7475
shahana@jhu.edu

Satter, James M.
407 7th St SE Apt 203
Minneapolis, MN 55414-1231
Phone: 1.612.378.0671
satterj@augsburgfortress.org
Bachelor's Level
Social Relations of Science;
North America; Technology

Satzinger, Helga
Zenturm für Interdisziplinare
U. Geschlechterfors tu Berlin
Sekr Tel 20-1
Ernst- Reuter-Platz 7
Berlin D-10587
Germany
Phone: 49.30.3142.6974
Fax: 49.30.3142.6988
satzing@kgw.tu.berlin.de

Sauter, Michael
2101 NW 82nd Ave
Miami, FL 33122-1508
Phone: 1.5255.5286.1780
Fax: 1.5255.5727.9897
michael.sauter@cide.edu

Savage-Smith, Emilie
Oriental Institute
Pusey Lane
Oxford OX1 2LE
Great Britain
Phone: 44.18.6527.8193
Fax: 44.18.6527.8190
emilie.savage-smith@
  orinst.ox.ac.uk

Savitt, Todd L.
East Carolina University
Dept. of Medical Humanities
School of Medicine
600 Moye Blvd
Greenville, NC 27858-4354
Phone: 1.252.816.2797
Fax: 1.252.816.2319
savitt@mail.ecu.edu
Doctorate
*19c, North America, Medical
Sciences*

Sawin, Clark T.
VAHQ (IOMI)
810 Vermont Avenue NW
Washington, DC 20420
Phone: 1.202.273.8940
Fax: 1.202.273.9090
Doctorate, 1958
Medical Sciences; Institutions

Sawyer, Richard
Newton High School
E 4th St S.
Newton, IA 50208
Phone: 1.641.792.5797
subtractman@lisco.com
Doctorate, 1990
*20c, Biological Sciences;
20c, North America, Biological
Sciences*

Schabas, Margaret L.
University of British Columbia
Dept of Philosophy
1866 Main Mail E-370
Vancouver, BC V6T1Z1
Canada
Phone: 1.604.822.2820
Fax: 1.604.822.8782
schabas@interchange.ubc.ca
Doctorate, 1983
*Social Sciences; 19c, Europe,
Philosophy and Philosophy of
Science; Technology*

Schabel, Chris
University of Cyprus
Dept of History and Archaeolgy
PO Box 20537
Nicosia CY1678
Cyprus
Fax: 357.2.756.719
schabel@ucy.ac.cy

Schafer, Elizabeth D.
PO Box 57
Loachapoka, AL 36865
Phone: 1.334.821.0580
Fax: 1.800.468.6144
edschater@reporters.net

Schafer, Wolf
SUNY - Stony Brook
Dept of History
Stony Brook, NY 11794-0001

Schagrin, Morton L.
54 Capitol St
Watertown, MA 02472-2511
Phone: 1.617.926.9916
Fax: 1.617.926.6867
schagrin@fredonia.edu
Doctorate
*Philosophy and Philosophy of
Science; Physical Sciences*

Schalick, Walton O., III
Harvard Medical School
125 Nashua Street
Department of PM & R
Boston, MA 02114
Phone: 1.314.935.5340
Fax: 1.617.576.3913
schalick@artsci.wustl.edu
Doctorate, 1997
*5c-13c, Medical Sciences;
20c, Medical Sciences*

Schechner, Sara
Harvard University
Collection of Historical
Scientific Instruments
Science Center B-6
Cambridge, MA 02138
Phone: 1.6174969542
schechn@fas.harvard.edu
Doctorate, 1988
*Astronomical Sciences;
Instruments and Techniques;
Physical Sciences;
Social Relations of Science*

Schefke, Brian
8803 Ravenna Ave NE
Seattle, WA 98115-3342
Phone: 1.206.729.8238
thelon2@yahoo.com

Scheick, William J.
University of Texas at Austin
Parlin Hall 108
English Department
Austin, TX 78712-1164
Phone: 1.512.471.8383
scheick@mail.utexas.edu
Doctorate
*17c, North America,
Biological Sciences;
18c, Europe, Biological Sciences;
19c, Biological Sciences*

Schell, Heather
123 Acorn Cir
Oxford, OH 45056-2664
Phone: 1.513.523.4509
schellh@muohio.edu

Scherer, Stefan
Hamburger Allee 46
Frankfurt am Main
Hessen D-60486
Germany
Phone: 49.69.707.91602
scherer@th.physik.
  uni-frankfurt.de

Schickore, Jutta
University of Cambridge
Dept of History and Phil of
Science
Free School Lane
Cambridge CB23RH
Great Britain
Phone: 44.12.2333.1105
js427@cam.ac.uk
Doctorate, 1996
*Philosophy and Philosophy of
Science; Biological Sciences,
Instruments and Techniques*

Schiebinger, Londa
Pennsylvania State University
Department of History
311 Weaver Building
University Park, PA 16802
Phone: 1.814.865.1367
Fax: 1.814.863.7840
lls10@psu.edu
Doctorate, 1984

Gender and Science;
18c, North America, Exploration,
Navigation and Expeditions;
18c, Europe, Medical Sciences;
18c, North America, Biological
Sciences

Schimkat, Peter
Postfach 10 35 25
Kassel 34035
Germany
schinkat.peter@vdi.de

Schleissner, Margaret R.
Rider University
2083 Lawrenceville Road
Lawrenceville, NJ 08648-3099
Phone: 1.609.895.5598
Fax: 1.609.896.5393
schleissner@rider.edu
Doctorate, 1987
5c-13c, Europe, Medical Sciences

Schling-Brodersen, Uschi
Dreissig-Morgen-Weg 11
Schriesheim D-69198
Germany
Phone: 49.6203.62825
btodetsen@writemail.com
Doctorate
19c, Biological Sciences;
Social Relations of Science;
Gender and Science;
Humanistic Relations of Science

Schloegel, Judith J.
Max Planck Institute for the
History of Science
Wilhelmstrasse 44
Berlin D-10117
Germany
Fax: 49.32.2.771.0834
114360.3616@compuserve.com
Master's Level, 1999
20c, Biological Sciences;
19c, Biological Sciences,
Instruments and Techniques;
20c, North America, Biological
Sciences; Philosophy and
Philosophy of Science

Schmaus, Warren S.
Illinois Inst of Technology
Dept of Humanities
3301 S. Dearborn
Chicago, IL 60616
Phone: 1.312.567.3473

Fax: 1.312.567.5187
schmaus@iit.edu
Doctorate
Social Sciences;
19c, Europe, Social Sciences;
20c, Europe, Social Sciences;
Philosophy and Philosophy of
Science

Schmitt, Gail K.
Princeton University
Dept of History
206 Dickinson Hall
Princeton, NJ 08544-1017
Fax: 1.609.394.5640
gschmitt@princeton.edu
Bachelor's Level, 1973
Biological Sciences; Education;
Institutions; 19c, Europe,
Biological Sciences, Biography

Schneer, Cecil J.
PO Box 181
Newfields, NH 03856-0181
Phone: 1.603.772.4597
cjs1@hopper.unh.edu
Doctorate, 1954
Earth Sciences; Physical
Sciences; 19c, North America;
20c, Institutions

Schneider, A. B.
1890 E 107th St Apt 603
Cleveland, OH 44106-2251

Schneider, Ivo H.
Uni Der Bundeswehr München
Fachb. Wissenschaftsgeschichte
Facultät Sowi
Neubiberg D-85577
Germany
Phone: 49.89.6004.3342
Fax: 49.89.6004.3344
ivo.schneider@
  unibw-muenchen.de
Doctorate, 1972
Europe, Mathematics;
20c, Social Relations of Science;
17c, Europe, Instruments and
Techniques

Schneider, Melvin S.
818 S Brentwood Blvd Apt 3A
Saint Louis, MO 63105-2551
Phone: 1.314.727.8801
msschnei@worldnet.att.net

Schneider, William H.
611 W 79th St
Indianapolis, IN 46260-3573
Phone: 1.317.274.7220
whschneiiupui.edu

Schoepflin, Rennie B.
La Sierra University
History, Politics, & Society
4700 Pierce St
Riverside, CA 92515-8247
Phone: 1.909.785.2341
Fax: 1.909.785.2215
rschoepf@lasierra.edu
Doctorate
Medical Sciences; North
America; Biological Sciences

Schoepflin, Urs
Max Planck Institute for the
History of Science
Dir of Lib
Wilhelmstr. 44
Berlin D-10117
Germany
Phone: 49.30.226.6719
Fax: 49.30.2266.7299
schoepfl@mpiwg-berlin.mpg.de

Schofield, Robt E.
44 Sycamore Rd
Princeton, NJ 08540-5323
Phone: 1.609.497.6462
Doctorate, 1955
18c, Europe, Physical Sciences;
18c, Europe, Technology;
18c, Europe, Humanistic
Relations of Science

Scholz, Erhard W.
University Wuppertal
Fachbereich 7
Gauss Str 20
Wuppertal 42097
Germany
scholz@math.uni-wuppertal.de
20c, Mathematics; 19c

Schopman, Joop
Sillgasse 6
Innsbruck A-6020
Austria
Phone: 43.512.5364.346
joop.schopman@uibk.ac.at
Doctorate, 1971
20c, Technology;
Cognitive Sciences

Schraegle, Horst, Jr.
PO Box 36 07
Osnabrueck D49026
Germany

Schuermann, Astrid
Durlacher Str. 27
Berlin 10715
Germany
Phone: 49.30.314.240
astrid.schuermann@tu-berlin.de

Schultz, David
1313 Halley Cir
Norman, OK 73069-8493
Phone: 1.405.366.0453
schultz@nssl.noaa.gov

Schuster, John
University of New South Wales
Sch of Science & Technology
Studies
Sydney, NSW 2052
Australia
Phone: 61.2.9385.2362
Fax: 61.2.9313.7984
j.a.schuster@unsw.edu.au

Schütt, Hans-Werner
Technical University Berlin
Inst of Phil & Hist of Science
Ernst-Reuter Platz 7
Berlin D10587
Germany
Phone: 49.30.314.22606
Fax: 49.30.8090.5319
hw.schuett@tu-berlin.de
Doctorate, 1967
*Physical Sciences; Philosophy
and Philosophy of Science;
14c-16c, Physical Sciences*

Schwab, Stephen
33 Parkview Dr
Tuscaloosa, AL 35401-3440
Phone: 1.205.752.6334
azteceagles@yahoo.com

Schwach, Vera
NIFU
Hegdehaugsveien 31
Oslo 0352
Norway
Phone: 47.2259.5156
Fax: 47.2259.5101
vera.schwach@nifu.no
Master's Level

*19c, Europe; 20c, Europe,
Science Policy; Europe,
Institutions; Europe, Social
Relations of Science*

Schwartz, A. Truman
Macalester College
1600 Grand Ave
Saint Paul, MN 55105-1899
Phone: 1.651.696.6271
Fax: 1.651.696.6432
schwartz@macalester.edu
Doctorate, 1963
*18c, Physical Sciences;
Education; Physical Sciences;
Humanistic Relations of Science*

Schwartz, Joel S.
College of Staten Island, CUNY
2800 Victory Boulevard
Staten Island, NY 10314
Phone: 1.718.982.3869
Fax: 1.718.982.3852
schwartz-j@postbox.
  csi.cuny.edu
Doctorate, 1970
*19c, Europe, Biological Sciences;
18c, North America, Biological
Sciences; 20c, Europe;
17c, Europe, Science Policy*

Schwartz, Rebecca P.
490 W 187th St Apt 4H
New York, NY 10033-1537
Phone: 1.212.543.3272
rpress@princeton.edu

Schwarz, Gretchen
459 W 49th St Apt 4E
New York, NY 10019-7284
Phone: 1.212.582.5610
frauline@bway.net

Schweber, Silvan S.
Brandeis University
MS 057 Dept of Physics
So Street
Waltham, MA 02254-9110
schweber@brandeis.edu
Doctorate
*20c, Physical Sciences;
20c, North America;
19c, Europe, Biological
Sciences, Biography;
20c, Biological Sciences*

Scott, Walter G.
523 N Manning
Stillwater, OK 74075-7814
Phone: 1.405.372.3381
wscott_osu@osu.net

Scratch, Lydia
Dalhousie University
Dept of History
1411 Seymour St
Halifax, NS B3H3M6
Canada
lscratch@is2.dal.ca

Scriba, Christopher J.
Inst Gesch Naturwiss
Bundesstr 55
Hamburg D-20146
Germany
Phone: 49.404.2838.2094
Fax: 49.404.283.8526
scriba@math.uni-hamburg.de
Doctorate
*17c, Europe, Mathematics;
18c, Europe, Social Sciences;
19c, Europe, Education;
14c-16c, Europe, Biography*

Scribante, Pierre
6 Ch. de la Combe Athenaz
Geneva 1285
Switzerland
Phone: 41.2.2756.2256
p.scribante@bluewin.ch
Doctorate
*20c, Philosophy and
Philosophy of Science;
Pre-400 B.C.E., Education;
20c, Physical Sciences*

Seaquist, Carl
4701 Pine St, Apt F-3
Philadelphia, PA 19143-1816
Phone: 1.215.474.2977
carlas@ccat.sas.upenn.edu

Secord, James A.
University of Cambridge
Dept of Hist & Phil of Science
Free School Lane
Cambridge CB2 3RH
Great Britain
Phone: 44.12.2333.4544
Fax: 44.12.2333.4554
jas1010@hermes.cam.ac.uk
Doctorate, 1981
19c, Europe, Biological Sciences;

19c, North America; Earth Sciences; Social Relations of Science

Seely, Bruce
Michigan Tech
Dept. of Social Sciences
Houghton, MI 49931-1295
Phone: 1.703.294.6472
Fax: 1.703.292.9068
bseely@nsf.gov

Sefrin-Weis, Heike
Oberes Daufeld 7
Daun 54550
Germany
Phone: 49.6592.2540
Fax: 49.6592.2540
urml326@hotmail.com

Segal, Ariel
11628 Le Baron Ter
Silver Spring, MD 20902-3135
Phone: 1.301.681.2976
asegal@wam.umd.edu

Segal, Howard P.
University of Maine
History Dept
5774 Stevens Hall
Orono, ME 04469-5774
Phone: 1.207.581.1920
Fax: 1.207.581.1817
howard_segal@umit.maine.edu
Doctorate, 1975
20c, North America, Biological Sciences

Segonds, A.
58 Rue de La Rochefoucauld
Paris 75009
France
Phone: 33.01.4051.2206
segonds@danof.obspm.fr

Seidel, Robert W.
University of Minnesota
151 Amundson Hall
Minneapolis, MN 55455
Phone: 1.612.722.7591
rws@tc.umn.edu

Seidel, Robert W.
University of Minnesota
103 Walter Library
117 Pleasant Street
Minneapolis, MN 55455

Phone: 1.612.624.8003
rws@tc.umn.edu
Doctorate, 1978
20c, North America, Physical Sciences; 20c, North America, Instruments and Techniques; 20c, North America, Institutions; Social Sciences

Seigman, Robert E.
5001 Columbia Rd Apt 203
Columbia, MD 21044-5643
Phone: 1.410.997.1474

Seiler, Frederick M.
13701 Winding Oak Cir, #301
Centreville, VA 20120
Phone: 1.708.815.6047
fseiler@mail.alum.rpi.edu
Master's Level
19c, Physical Sciences

Seim, David L.
2100 Greeley St
Ames, IA 50014-7020
Phone: 1.515.294.9420
dseim@iastate.edu

Seitz, Jonathan W.
Univ of Wisconsin-Madison
7143 Social Science Bldg
1180 Observatory Drive
Madison, WI 53706
Phone: 1.608.243.9348
jwseitz@students.wisc.edu
Master's Level, 1998
5c-13c, Europe;
14c-16c, Europe; 17c, Europe;
20c, North America

Selby, Roy
3619 Garrison Rd
Little Rock, AR 72223-9673
Phone: 1.501.821.2553
Doctorate
5c-13c, Asia, Medical Sciences;
Pre-400 B.C.E., Europe,
Natural and Human History;
17c, North America,
Natural and Human History;
18c, Europe, Cognitive Sciences

Selin, Helaine
Hampshire College
Amherst, MA 01002
Phone: 1.413.559.5541
Fax: 1.413.559.5419

hselin@hampshire.edu
Master's Level, 1975
Pre-400 B.C.E., Africa;
5c-13c, Asia;
Pre-400 B.C.E., Asia

Sellers, Christopher
SUNY at Stony Brook
Stony Brook, NY 11794
Phone: 1.631.632.7514
Fax: 1.516.421.8973
csellers@notes.cc.sunysb.edu
Doctorate, 1992
20c, North America, Biological Sciences; Medical Sciences

Selya, Rena E.
Harvard University
Science Center 235
Cambridge, MA 02138
Phone: 1.617.495.8758
selya@fas.harvard.edu
Master's Level
20c, Biological Sciences;
19c, Biological Sciences;
Biological Sciences

Semendeferi, Ioanna
5440 Columbus Ave
Minneapolis, MN 55417-2433
Phone: 1.612.825.2598
seme0002@umu.edu

Senechal, Majorie
Smith College
Clark Science Center
Northampton, MA 01063-0001
Phone: 1.413.585.3862
Fax: 1.413.585.3786
senechal@smith.edu
Doctorate
20c, Europe, Technology;
19c, Europe, Technology;
18c, Europe, Technology;
17c, Europe, Technology

Sengoopta, Chandak
University of Manchester
CHSTM
Mathematics Tower
Oxford Road
Manchester M139 PL
Great Britain
Phone: 44.161.275.5843
Fax: 44.161.276.5699
c.sengoopta@man.ac.uk
Doctorate, 1996

Sent, Esther M.
University of Notre Dame
Department of Economics
Notre Dame, IN 46556
Phone: 1.219.631.6979
Fax: 1.219.631.8809
sent.2@nd.edu
Doctorate, 1994
20c, Social Sciences; Social
Sciences; Philosophy and
Philosophy of Science

Sepkoski, David
University of Minnesota
Program History of Sci & Tech
435 Walter Library
Minneapolis, MN 55455
Phone: 1.612.626.8722
sepk0003@tc.umn.edu
Master's Level, 1996
17c, Europe, Mathematics;
17c, Philosophy and Philosophy
of Science; Social Sciences

Sepper, Dennis L.
University of Dallas
Philosophy Dept
1845 E Northgate Dr
Irving, TX 75062-4736
Phone: 1.972.721.5257
Fax: 1.972.721.4005
sepper@acad.udallas.edu
Doctorate, 1981
Philosophy and
Philosophy of Science;
Humanistic Relations of Science

Serjeantson, Richard
Trinity College
Trinity Street
Cambridge CB2 1TQ
Great Britain
rws1001@cam.ac.uk

Servos, John W.
Amherst College
Dept of History
Box 2254
Amherst, MA 01002-2254
Phone: 1.413.542.2035
Fax: 1.413.542.2727
jwservos@amherst.edu
Doctorate, 1979
20c, North America, Physical
Sciences; 20c, North America,

Education; 20c, North America,
Technology; 20c, North America,
Medical Sciences

Seth, Suman
Princeton University
History Department
Dickinson Hall
Princeton, NJ 08540
Phone: 1.609.430.0201
sseth@princeton.edu

Settle, Thomas B.
38 Livingston St Apt 62
Brooklyn, NY 11201-4811
Phone: 1.718.624.5438
Fax: 1.718.260.3136
tsettle@duke.poly.edu
Doctorate, 1966
17c, Europe; 5c-13c, Europe,
Technology, Instruments and
Techniques; Pre-400 B.C.E.

Sfraga, Mike P.
University of Alaska-Fairbanks
503 Gruening
Fairbanks, AK 99775
Phone: 1.907.474.7143
Fax: 1.907.474.5824
mike.sfraga@alaska.edu
Doctorate, 1997
20c, North America, Exploration,
Navigation and Expeditions;
19c, Polar Regions, Earth
Sciences; 18c, North America,
Science Policy; 17c, Biological
Sciences

Shackelford, Jole R.
University of Minnesota
146 Physics Building
Church Street
Minneapolis, MN 55455
Phone: 1.612.625.9062
shack001@tc.umn.edu
Doctorate, 1989
14c-16c, Europe, Medical
Sciences; 14c-16c, Europe,
Physical Sciences; 14c-16c,
Europe, Astronomical Sciences;
17c, Europe, Physical Sciences

Shamin, Alexei
Inst. Hist. Sci. & Technology
Staropansky Per.1-5
Moscow 103012
Russia

Shank, J. B.
Stanford University
Dept of History
Stanford, CA 94305-2024
Phone: 1.612.624.7323
jbshank@umn.edu
Doctorate, 1999
18c, Europe, Physical Sciences;
17c, Europe, Institutions;
19c, Europe, Social Relations
of Science; 14c-16c, Philosophy
and Philosophy of Science

Shank, Michael H.
Univ of Wisconsin-Madison
Social Science Bldg. 7143
1180 Observatory Dr
Madison, WI 53706-1320
Phone: 1.608.262.3972
Fax: 1.608.262.3984
mhshank@facstaff.wisc.edu
Doctorate, 1983
5c-13c, Europe, Astronomical
Sciences; 14c-16c, Europe,
Astronomical Sciences;
5c-13c, Europe, Education

Shapere, Dudley
3125 Turkey Hill Crt
Einstein Drive
Winston-Salem, NC 27106
Phone: 1.336.760.0918
dshapere@ias.edu
Doctorate, 1957
Philosophy and Philosophy of
Science; Physical Sciences;
Biological Sciences

Shapiro, Alan E.
University of Minnesota
Hist. of Science & Technology
116 Church St SE
Minneapolis, MN 55455-0149
Phone: 1.612.624.5770
Fax: 1.612.624.4578
ashapiro@physics.spa.umn.edu
Doctorate, 1970
17c, Europe, Physical Sciences;
18c, Europe, Physical Sciences;
17c, Europe, Philosophy
and Philosophy of Science;
18c, Europe, Philosophy and
Philosophy of Science

Shapiro, Barbara J.
7 Selborne Dr
Piedmont, CA 94611-3618

Phone: 1.510.482.1206
bshapiro@socrates.berkley.edu
Doctorate
17c, Europe, Humanistic Relations of Science;
17c, Europe, Institutions;
17c, Europe, Social Relations of Science; 17c, Europe, Education

Shapiro-Shapin, Carolyn G.
Grand Valley State University
Dept of History
1013 Mackinac Hall
Allendale, MI 49401
Phone: 1.616.554.4633
Fax: 1.616.895.3285
shapiroc@gvsu.edu

Sharkey, John B.
Pace University
Office of the Dean
Dyson Coll. of Arts & Sciences
One Pace Plaza
New York, NY 10038-1598
Phone: 1.212.346.1710
Fax: 1.212.346.1725
jsharkey@pace.edu
Doctorate, 1970
20c, Europe, Physical Sciences;
North America, Physical
Sciences; Biological Sciences

Shea, William R.
University Louis Pasteur
7 rue de l'Universite
Strasbourg 67000
France
Phone: 33.390.24.05.80
Fax: 33.390.24.05.81
william.shea@i
  hs-ulp.u-strasbg.fr
Doctorate, 1968
17c, Europe, Physical Sciences;
18c, Europe, Astronomical
Sciences; 19c, Europe,
Philosophy and Philosophy
of Science; 14c-16c, Europe,
Biological Sciences

Sheehan, Jonathan
928 Ballantine Rd
Bloomington, IN 47401-5054
Phone: 1.812.331.1981
josheeha@indiana.edu

Sheffield, Roy S.
Brevard College
400 N Broad St
Brevard, NC 28712-3306
Phone: 1.828.883.8292
Fax: 1.828.884.3790
scotts@brevard.edu
Doctorate, 1994
20c, Europe, Physical Sciences;
Social Sciences;
Humanistic Relations of Science;
Physical Sciences

Sheffield, Suzanne L.
York University
4700 Keele St
Toronto, ON M3J1P3
Canada
suzanne@yorku.ca
Doctorate, 1997
19c, Europe, Gender and Science;
19c, Europe, Biological Sciences

Sheinin, Rose
28 Inglewood Drive
Toronto, ON M4T1G8
Canada
Phone: 1.416.488.8687
rose.sheinin@utoronto.ca

Shen, Grace
11A Willow Dr
Hopewell Jct, NY 12533
Phone: 1.845.221.1377
gyshen@fas.harvard.edu

Shen, Yimin
3984 Northlake Creek Dr
Tucker, GA 30084-3420
Phone: 1.404.408.5596
yimins94@netzero.net

Sheng, Jia
22 John St
Ithaca, NY 14850-6351
Phone: 1.607.272.2991
sheng@lightlink.com

Shepard, Mikki
Rockefeller Foundation
Dir for Arts & Humanities
420 5th Ave
New York, NY 10018-2729

Sherbert, Donald
1104 E Mumford Dr
Urbana, IL 61801-6804
Phone: 1.217.328.6061
sherbert@math.uiuc.edu

Sheridan, Bridgette
Ctr for Gender in Organization
Simmons Graduate Sch of Mgt
409 Commonwealth Ave
Boston, MA 02215-2322
Phone: 1.617.521.3875
Fax: 1.617.521.3878
sheridan@simmons.edu
Master's Level, 2002
Europe, Gender and Science;
Europe, Medical Sciences;
Medical Sciences

Shermer, Michael
2761 Marengo Ave
Altadena, CA 91001-2204

Shields, William
311 Valeview Ct NW
Vienna, VA 22180-4150
Phone: 1.703.938.0785
highc.king@verizon.net

Shimahara, Kenzo
6-27-43 Higashi Koigakubo
Kokubunji
Tokyo 185 0014
Japan
Phone: 81.423.25.8456
Doctorate, 1974
Europe, Physical Sciences;
Asia, Physical Sciences;
Asia, Technology, Instruments
and Techniques

Shimao, Eikoh
5-14 Ohatacho
Nishinomiya 662 0836
Japan
Phone: 0798.67.2304
shimao@hccl.bai.ne.jp
Doctorate, 1966
19c, Europe, Biography;
20c, Asia, Social
Relations of Science;
Europe, Physical Sciences

Shinoda, Mariko
3-15-2-602 Takamatsu-cho
Tachikawa-shi

Japan
Phone: 81.42.529.0717
qwm00475@nifty.ne.jp

Shipley, Brian C.
Rutgers University
Thomas Edison Papers
2 Babcock Pl
West Orange, NJ 07052-5547
Phone: 1.973.243.5645
bshipley@rci.rutgers.edu
Master's Level, 1997
*North America; 19c, Natural and
Human History; Earth Sciences;
19c, Institutions*

Shklar, Gerald
Harvard University
School of Dental Medicine
188 Longwood Ave
Boston, MA 02115-5819
Phone: 1.617.432.1467
Fax: 1.617.432.1897
geraldus@hms.harvard.edu
Doctorate
*Medical Sciences;
Biological Sciences;
14c-16c, Medical Sciences*

Shoemaker, Philip S.
Rieter Corp.
PO Box 4383
Spartanburg, SC 29305
Phone: 1.864.582.5466
Fax: 1.864.948.5284
75424.2427@compuserve.com
Doctorate, 1991
*19c, North America; North
America, Astronomical Sciences;
South America*

Shore, Steve
Indiana Univ South Bend
Physics Dept
1700 Mishawaka Ave
South Bend, IN 46615-1400
Phone: 1.219.237.4401
Fax: 1.219.237.6589
sshore@paladin.iusb.edu

Shostak, Stanley
University of Pittsburgh
Dept. of Biological Sciences
Langley Hall
Pittsburgh, PA 15260
Phone: 1.412.624.4253
Fax: 1.412.624.4759

sshostkt@pitt.edu
Doctorate, 1964
Biological Sciences

Shteir, Ann B.
York University
728 Atkinson
4700 Keele Street
Toronto, ON M3J1P3
Canada
Phone: 1.416.736.2100
Fax: 1.416.736.5766
rshteir@yorku.ca
Doctorate
*18c, Europe, Humanistic
Relations of Science;
19c, Europe, Gender and
Science; 18c, Europe,
Biological Sciences;
19c, Europe, Biological Sciences*

Shutt, Graham R.
433 13th Ave E Apt 102
Seattle, WA 98102-5175
Phone: 1.206.726.9491
gshutt@u.washington.edu

Siegel, Andre
341 O'Connor Drive
Toronto, ON M4J2V4
Canada
Phone: 1.416.425.2562
andre.siegel@utoronto.ca

Siegel, Daniel M.
University of Wisconsin
7143 Social Science Bldg
1180 Observatory Drive
Madison, WI 53706-1393
Phone: 1.608.262.1406
Fax: 1.608.262.3984
dmsiegel@facstaff.wisc.edu
Doctorate, 1968
*19c, Physical Sciences;
20c, Physical Sciences;
Philosophy and Philosophy of
Science; Education*

Siegel, Joseph
2305 Brookens Circle
Urbana, IL 61801
Phone: 1.217.344.4811
joesiegel@mac.com

Siegfried, Robert
2206 Westlawn Ave
Madison, WI 53711-1952

Phone: 1.608.255.0418
Doctorate, 1952
18c, Europe, Physical Sciences

Siegmund-Schultze, Reinhard
Agder University College
Srriceboks 422
Dept of Mathematics
Kristian Sands 4604
Norway
Phone: 47.3814.1631
Fax: 47.3814.1071
reinhard-siegmund-schultze@
hia.no
Doctorate, 1987
*20c, Europe, Mathematics;
20c, North America,
Mathematics; 20c, Europe,
Social Relations of Science;
20c, North America, Social
Relations of Science*

Sierra Cuartas, Carlos E.
Apartado Aereo 95485
(Antioquia)
Medellin
Colombia
Phone: 57.4.3415151
cesierra@gerseus.
una.med.edu.co

Sigler, Fernando C.
Ave 3AB No 9208 E/92 y 94
Playa Ciudad de la
Habana
Cuba

Silliman, Robert H.
Emory University
Department of History
Bowden Hall
Atlanta, GA 30322
Phone: 1.404.727.4469
Fax: 1.404.727.4959
rsillim@emory.edu
Doctorate, 1968
*19c, North America, Earth
Sciences; 19c, Biological
Sciences; Biological Sciences*

Silverman, Timothy
3 Byland Close
Winchmore Hill
London N21 1QH
Great Britain
Phone: 44.20.88821726
tim@timsilverman.demon.co.uk

Sime, Ruth L.
Sacramemto City College
Department of Chemistry
3835 Freeport Blvd
Sacramento, CA 95822
rodsime@csus.edu
Doctorate, 1965
20c, Europe, Physical Sciences

Simha, Evelyn
Dibner Institute
Dibner Bldg., MIT E56-100
Cambridge, MA 02139
Phone: 1.617.253.8721
Fax: 1.617.253.9858
esimha@mit.edu

Simmons, Adele
Chicago Metropolis 2020
30 W Monroe St 18th Fl
Chicago, IL 60603-2495
Phone: 1.312.332.2020
Fax: 1.312.332.2626
adele.simmons@cm2020.org

Simmons, Steve
727 1/2 Smith St
Fort Collins, CO 80524-3409

Simoes, Ana I.
University of Lisbon
Departamento de Fisica
Campo Grande
c8 Piso 6
Lisboa 1749 016
Portugal
Phone: 351.21.750.0817
Fax: 351.21.750.0977
asimoes@fc.ul.pt
Doctorate
20c, Physical Sciences;
20c, Philosophy and Philosophy
of Science; 18c, Europe,
Social Relations of Science;
19c, Europe, Social Relations of
Science

Simon, Jonathan
Max Planck Institute for the
History of Science
Wilhelmstrasse 44
Berlin 10117
Germany
zxzsimon@hotmail.com

Simonsen, Judith K.
200 Riverside Dr Apt 4F
New York, NY 10025-7243
Phone: 1.212.774.7713
Master's Level
19c, Europe, Gender and
Science; 19c, North America,
Medical Sciences, Instruments
and Techniques; 17c, Europe,
Philosophy and Philosophy of
Science

Simpson, Thomas K.
Embury Apts, 82-W
Saratoga Spring, NY
12866-1324
Phone: 1.518.587.7495
spectet@nycap.rr.com
Doctorate
19c, Europe, Physical Sciences;
400 B.C.E.-400 C.E., Europe,
Philosophy and Philosophy
of Science; 18c, Europe,
Social Sciences; 17c, Africa,
Humanistic Relations of Science

Simrall, James
5406 Navajo Rd
Louisville, KY 40207-1685
Phone: 1.502.895.5929
jsimrall@uchicago.edu

Singer, Arthur
Sloan Foundation
630 5th Ave
New York, NY 10111-0100

Singleton, Jim
Science Museum Library
Imperial Col Rd/S Kensington
London SW7 5NH
Great Britain

Singleton, Rivers, Jr.
University of Delaware
Dept of Biological Sciences
Newark, DE 19716-2590
Phone: 1.302.831.1146
Fax: 1.302.831.2281
oneton@udel.edu
Doctorate
20c, North America, Biological
Sciences; 20c, North America,
Physical Sciences; 20c, North
America, Humanistic Relations
of Science; 20c, North America,
Education

Sismondo, Sergio
Queen's University
Department of Philosophy
Kingston, ON K7L3N6
Canada
Phone: 1.613.533.2182
Fax: 1.613.533.6545
sismondo@post.queensu.ca
Doctorate, 1993
20c, Biological Sciences;
Philosophy and Philosophy of
Science; Social Relations of
Science

Skabelund, D.
1218 Las Lomas Rd NE
Albuquerque, NM 87106-4526
Phone: 1.505.242.7505
dondana@aol.com

Skiff, Peter D.
Bard College
Annadale-on-Hudson, NY 12504
Phone: 1.914.758.6822
skiff@bard.edu

Slack, Nancy G.
Russell Sage College
Dept of Biology
Troy, NY 12180
Phone: 1.518.244.2288
Fax: 1.518.244.3174
slackn@sage.edu
Doctorate, 1971
20c, North America, Biological
Sciences; 19c, North America,
Biological Sciences; 19c, North
America, Gender and Science;
20c, North America, Social
Relations of Science

Slater, Leo B.
2723 Saint Paul St Apt 3
Baltimore, MD 21218-4337
Phone: 1.215.413.0690
Fax: 1.215.925.1954
leoslater@yahoo.com
Doctorate, 1997
20c, North America, Physical
Sciences; 20c, North America,
Technology

Slaton, Amy
Drexel University
Dept. of History & Politics
32nd & Chestnut Sts.
Philadelphia, PA 19104

Phone: 1.215.895.2061
Fax: 1.215.895.6614
slatonae@drexel.edu

Sleigh, Charlotte
University of Kent Canterbury
School of History
Rutherford College
Canterbury CT2 7NX
Great Britain
Phone: 1227827665
Fax: 1227827258
c.l.sleigh@ukc.ac.uk

Sloan, Norton Q.
PO Box 570
Ipswich, MA 01938-0570

Sloan, Phillip R.
University of Notre Dame
Program in Hist & Phil of Sci
346 O'Shaughnessy Hall
Notre Dame, IN 46556
Phone: 1.219.631.5221
Fax: 1.219.631.8209
phillip.r.sloan@nd.edu
Doctorate, 1970
*18c, Europe, Natural and
Human History; 19c, Europe,
Biological Sciences; 20c, North
America, Biological Sciences;
20c, Humanistic Relations of
Science*

Slotten, Hugh R.
NASA
History Office
Washington, DC 20546-0001
Phone: 1.617.625.5345
slotten@fas.harvard.edu
Doctorate, 1991
*19c, North America;
20c, North America, Technology;
19c, North America, Institutions;
20c, North America, Science
Policy*

Smeltzer, Ronald K.
Sarnoff Corp
CN5300
Princeton, NJ 08543
rksmeltzer@worldnet.att.net
Doctorate, 1970
*Instruments and Techniques;
Physical Sciences*

Smith, A. Mark
University of Missouri
Department of History
101 Read Hall
Columbia, MO 65211-7500
Phone: 1.573.882.9456
Fax: 1.573.884.5151
hissmith@showme.missouri.edu
Doctorate, 1976
*400 B.C.E.-400 C.E.,
Europe, Physical Sciences;
5c-13c, Europe,
Physical Sciences;
5c-13c, Physical Sciences;
17c, Europe, Philosophy and
Philosophy of Science*

Smith, Barbara H.
2415 Falls Dr
Chapel Hill, NC 27514-9650
Phone: 1.919.684.3970
bhsmith@duke.edu

Smith, Charles H.
Western Kentucky University
University Libraries
1 Big Red Way
Bowling Green, KY 42101-3576
Phone: 1.270.745.6079
Fax: 1.270.745.2275
charles.smith@wku.edu
Doctorate
*19c, Education;
Biological Sciences;
Natural and Human History*

Smith, Cheryl
2200 15th St
Vernon, TX 76384
Phone: 1.940.552.9538
csmith@vrjc.cc.tx.us

Smith, Crosbie W.
University of Kent
Rutherford College
Canterbury CT2 7NX
Great Britain
Phone: 44.12.2782.3791
Fax: 44.12.2782.7258
c.smith@ukc.ac.uk
Doctorate, 1975
*19c, Europe, Physical Sciences;
20c, North America,
Physical Sciences;
19c, Europe, Technology*

Smith, D. Neel
College of the Holy Cross
Classics Department
Worcester, MA 01610
Phone: 1.508.793.2621
neelsmith@yahoo.com

Smith, Dale
PO Box 5714
Bethesda, MD 20824-5714
Phone: 1.240.401.6096
dcsmith@usuhs.mil

Smith, Edward H.
444 Crowfields Dr
Asheville, NC 28803-3280
Phone: 1.828.277.0796
Fax: 1.828.274.1057
jananded@aol.com
Doctorate
*19c, North America, Biological
Sciences; 19c, North America,
Natural and Human History*

Smith, Gerard P.
Bourne Laboratory
NY Presbyterian Hospital
21 Bloomingdale Road
White Plains, NY 10605
Phone: 1.914.997.5935
Fax: 1.914.682.3793
gpsmith@mail.med.cornell.edu

Smith, Hilary
501 W. Hortter St. #C7
Philadelphia, PA 19119
Phone: 1.215.842.3955
smithhil@sas.upenn.edu

Smith, John G.
University of Leicester
History Department
University Road
Leicester LE1 7RH
Great Britain
Phone: 44.11.6270.5968
Fax: 44.11.6252.3986
sgj@leicester.ac.uk

Smith, Jonathan
Univ of Michigan - Dearborn
Humanities Dept
4901 Evergreen Road
Dearborn, MI 48128-1491
Phone: 1.313.436.9187
Fax: 1.313.593.5552
jonsmith@umich.edu

Doctorate, 1990
*19c, Europe, Humanistic
Relations of Science;
Europe, Biological Sciences;
Natural and Human History*

Smith, Laurence D.
University of Maine
Department of Psychology
5742 Little Hall
Orono, ME 04469-5742
Phone: 1.207.581.2047
Fax: 1.207.581.6128
ldsmith@maine.edu
Doctorate, 1983
*Cognitive Sciences; Philosophy
and Philosophy of Science*

Smith, Matthew
202-10305-120 Street
Edmonton, AB T5K2A5
Canada
Phone: 1.780.452.3798
mpsmith@ualberta.ca

Smith, Merritt R.
Mass Inst of Technology
STS Program
77 Massachusetts Ave
Cambridge, MA 02139-4301
Phone: 1.617.253.4008
Fax: 1.617.258.8118
roesmith@mit.edu
Doctorate, 1971
*19c, North America, Technology;
20c, North America, Technology;
19c, North America, Philosophy
and Philosophy of Science;
20c, North America, Social
Sciences*

Smith, Pamela H.
Pomona College
History Dept
551 N College Ave
Claremont, CA 91711-6337
Phone: 1.909.607.2919
Fax: 1.909.621.8574
psmith@pomona.edu
Doctorate
*Europe, Social Relations of
Science; 17c, Europe, Physical
Sciences; 14c-16c, Europe,
Physical Sciences*

Smith, Robert W.
University of Alberta
Dept of History & Classics
2-28 Tory Bldg
Edmonton, AB T6G2H4
Canada
Phone: 1.780.492.4687
Fax: 1.780.492.9125
rwsmith@ualberta.ca
Doctorate, 1979
*Astronomical Sciences;
North America;
Instruments and Techniques;
20c, Social Relations of Science*

Smith, Wade K.
1403 Wilmington Ave
Richmond, VA 23227-4427

Smith, William A., Jr.
400 3rd St
Fulton, KY 42041-1246

Smocovitis, Vassiliki B.
University of Florida
Dept History
226 Keene-Flint Hall
Gainesville, FL 32611
Phone: 1.352.392.0271
Fax: 1.352.392.6927
bsmocovi@history.ufl.edu
Doctorate
*20c, North America, Biological
Sciences*

Smoller, Laura
50 Robinwood Dr
Little Rock, AR 72227-2226

Snedegar, Keith
2125 South Shore, Apt 102
Kentwood, MI 49508-0912
Phone: 1.616.281.9371
Fax: 1.801.226.5207
snedegke@uvsc.edu
Doctorate
*Pre-400 B.C.E., Africa,
Astronomical Sciences,
Philosophy and Philosophy
of Science; 5c-13c, Europe,
Astronomical Sciences;
19c, Europe, Astronomical
Sciences; 20c, Africa,
Astronomical Sciences*

Snider, Alvin
1722 Ridgeway Dr
Iowa City, IA 52245-3232
alvin-snider@uiowa.edu

Snodgrass, S. Robert
731 W Washington Blvd
Pasadena, CA 91103-2023
Phone: 1.626.794.0869
Fax: 1.232.667.2019
rsnodgrass@chla.usc.edu

Snyder, Laura J.
St John's University
Dept of Philosophy
8000 Utopia Pkwy
Jamaica, NY 11439
Phone: 1.718.990.6378
Fax: 1.718.990.1907
snyderl@stjohns.edu
Doctorate, 1996
*Philosophy and Philosophy
of Science; 19c, Europe,
Biological Sciences;
19c, Europe, Astronomical
Sciences; 19c, Europe,
Physical Sciences*

Soderberg, David
403 W Side Dr Apt 102
Gaithersburg, MD 20878-3132

Soderqvist, Thomas
University of Copenhagen
School of Public Health
Dept of History of Medicine
Bredgade 62
Copenhagen DK-1260
Denmark
Phone: 45.3532.3815
Fax: 45.3532.3816
t.soderqvist@pubhealth.ku.dk
Doctorate, 1986
*20c, Social Sciences;
20c, Medical Sciences;
Biography*

Sokal, Michael M.
Worcester Polytechnic Inst
Dept of Humanities and Arts
100 Institute Road
Worcester, MA 01609-1614
Phone: 1.508.831.5712
Fax: 1.508.753.7240
msokal@wpi.edu

Doctorate, 1972
20c, Cognitive Sciences;
20c, North America

Sokolovskaia, Zinaida K.
Institut Istorii
Estestvoznaniia I Tekhniki
Staroponskii Per. 1/5
Moscow 103012
Russia

Solis Santos, Carlos
UNED
Senda del Rey SN
Edificio de Humanidades
Madrid 28040
Spain
Phone: 34.91.398.6992
Fax: 34.91.398.6677
csolis@fsof.uned.es
Doctorate, 1977
17c, Europe, Astronomical
Sciences; 400 B.C.E.-400 C.E.,
Astronomical Sciences;
Biological Sciences; Philosophy
and Philosophy of Science

Solomon, Harvey M.
4020 Powers Ferry Rd NW
Atlanta, GA 30342-4028
Phone: 1.404.816.5853
h4020@aol.com

Solomon, Miriam
Temple University
Philosophy Department 022-32
Philadelphia, PA 19122
Phone: 1.215.204.9629
Fax: 1.215.204.6266
msolomon@astro.temple.edu
Doctorate
Philosophy and Philosophy
of Science; 20c; Gender and
Science

Somsen, Geert J.
Maastricht University
Faculty of Arts and Sciences
PO Box 616
Maastricht 6200 MD
The Netherlands
Phone: 31.93.388.2595
Fax: 31.93.388.4816
g.somsen@history.unimaas.nl
Doctorate, 1998
19c, Europe, Science Policy;

20c, Europe, Humanistic
Relations of Science; North
America; Europe

Song, Sang-Yang
Hallym University
1 Okchon-dong
Chunchon 200-702
South Korea
Phone: 82.33.240.1200
Fax: 82.2.379.4125
DEL1200@sun.hallym.ac.kr
Master's Level
17c, Asia, Social Sciences;
18c, Europe, Humanistic
Relations of Science;
19c, Europe, Philosophy
and Philosophy of Science;
14c-16c, Asia, Social Relations
of Science

Songster, E. Elena
414 Richmond St
El Cerrito, CA 94530-3758
Phone: 1.510.527.3170
eesongster@yahoo.com

Sonntag, Otto
New York University
100 Washington Square East
Room 908
New York, NY 10003
Phone: 1.212.998.8113
otto.sonntag@nyu.edu
Doctorate
18c, Europe, Medical Sciences

Sopka, Katherine R.
Four Corners Analytic Sciences
6 Carol Rd
Marblehead, MA 01945-2118
Phone: 1.781.639.1619
jjkjsopka@cs.com
Doctorate
20c, North America,
Physical Sciences; 19c, Europe,
Physical Sciences; Education;
Social Relations of Science

Sorensen, Knut H.
Norweigian Univ of Sci & Tech
STS-Dragvoll
Trondheim N-7491
Norway
Phone: 47.735.9179
Fax: 47.735.91327
knut.sorensen@hf.tnu.no

Doctorate, 1982
20c, Europe, Technology;
20c, Europe, Gender and
Science; 20c, Europe,
Social Relations of Science;
20c, Europe, Physical Sciences

Sorensen, W. Conner
Amselweg 3
Eschbach 56357
Germany
Phone: 49.6771.599.315
Fax: 49.6771.599.316
sorensen@aol.com
Doctorate, 1985
18c, North America;
19c, Europe, Biological Sciences;
20c, North America,
Natural and Human History;
Social Relations of Science

Sotelo, Julio
Instituto de Neurologia
Insurgentes Sur #3877
Mexico City DF14269
Mexico
Phone: 525.606.4782
Fax: 525.606.2282
jsotelo@servidor.unam.mx
Medical Sciences; Institutions;
Philosophy and Philosophy of
Science

Souffrin, Pierre
Observatoire de Cote d'Azur
BP 4229
Nice Cedex 4 06304
France
Phone: 33.04.930.3512
souffrin@obs-nice.fr
5c-13c, Europe, Physical
Sciences; 14c-16c, Europe,
Physical Sciences;
17c, Europe, Physical Sciences;
400 B.C.E.-400 C.E., Europe,
Physical Sciences

Soula, Katerina
Alexandras 57
Trikala
Thessalias 4210
Greece
Phone: 30.431.28204
mvid@otenet.gr

Spanagel, David I.
Emerson College
ILAIS
100 Beacon Street
Boston, MA 02116
Phone: 1.617.824.8969
david_spanagel@emerson.edu
Doctorate, 1996
19c, North America; 19c, North America, Natural and Human History; 19c, North America, Exploration, Navigation and Expeditions; Earth Sciences

Spargo, Peter E.
PO Box 211
Rondebosch 7701
South Africa
Phone: 27.21.686.4289
Fax: 27.21.650.3342
peter@spargo.wcape.school.za
Master's Level, 1968
17c, Europe, Physical Sciences; 17c, Europe, Instruments and Techniques

Sparling, Andrew
Apt 22-G Beech Lake Apts
4800 University Dr
Durham, NC 27707-6124
aws2@duke.edu
Master's Level
14c-16c, Europe, Physical Sciences; 17c, Europe, Natural and Human History; 18c, Europe, Medical Sciences; Europe, Biological Sciences

Spary, Emma
159 Girton Road
Cambridge CB30PQ
Great Britain
Phone: 44.12.2333.3008
psw24@cam.ac.uk

Sparzani, Antonio
Universita Statale di Milano
Via Celoria 16
Milan 20133
Italy
Phone: 39.02.239.2283
Fax: 39.02.239.2480
antispar@libero.it
Doctorate
400 B.C.E.-400 C.E., Europe, Physical Sciences; 20c, Europe, Physical Sciences;

Astronomical Sciences, Philosophy and Philosophy of Science; Philosophy and Philosophy of Science

Spath, Susan B.
1615 Edith St
Berkeley, CA 94703-1306
sbspath@yahoo.com

Spear, Scott
Alameda County Public Defender
1225 Fallon Street
Oakland, CA 94612
Fax: 1.419.735.0539
rspearatty@aol.com
Doctorate, 1973
Social Sciences; Philosophy and Philosophy of Science

Spencer, Larry T.
PO Box 365
Plymouth, NH 03264-0365
Phone: 1.603.535.2322
lts@plymouth.edu

Spittler, Ernest
John Carroll University
University Heights
Cleveland, OH 44118
Fax: 1.216.397.4228
wsplittler@jcu.edu

Sponsel, Alistair
19733 La Sierra Blvd
San Antonio, TX 78256-2015
Phone: 1.210.698.5494
alistairsponsel@hotmail.com

Stafford, Barbara
University of Chicago
Department of Art History
5540 S Greenwood Ave
Chicago, IL 60637-1506
Phone: 1.773.702.0268
Fax: 1.773.702.5901
bms6@uchicago.edu
Doctorate, 1972
18c, Europe; 18c, Europe, Philosophy and Philosophy of Science

Staley, Kent W.
St. Louis University
3800 Lindell Blvd
PO Box 56907

St. Louis, MO 63156
Phone: 1.314.977.3149
Doctorate, 1997
20c, Physical Sciences; Philosophy and Philosophy of Science; Instruments and Techniques

Staley, Richard
University of Wisconsin
History of Science
7143 Social Science Bldg
1180 Observatory Dr
Madison, WI 53706-1320
Phone: 1.608.262.3978
Fax: 1.608.262.3984
rastaley@facstaff.wisc.edu
Doctorate, 1992
Physical Sciences; 20c, Physical Sciences; Instruments and Techniques; Social Sciences

Stam, Henderiku J.
University of Calgary
Department of Psychology
Calgary, AB T2N1N4
Canada
Phone: 1.403.220.5683
Fax: 1.403.289.5570
stam@ucalgary.ca
Doctorate, 1982
20c, North America, Cognitive Sciences; 20c, Europe, Cognitive Sciences

Stamhuis, Ida H.
Vrye Universiteit
Faculty of Exact Sciences
De Boelelaan 1081
Amsterdam 1081 HV
The Netherlands
Phone: 31.20.444.7983
Fax: 31.20.444.7988
stamhuis@nat.vu.nl
Doctorate, 1989
20c, Europe, Biological Sciences; 20c, Europe, Gender and Science; 19c, Europe, Mathematics; 20c, Europe, Mathematics

Stanchi, Richard R.
49 Barnes Ave
East Boston, MA 02128-1201

Stanley, Donald E.
Rutland Regional Medical Ctr.
160 Allen Street
Rutland, VT 05701
Phone: 1.207.563.1560
Fax: 1.207.563.1594
dstanley@lincoln.midcoast.com
Doctorate
20c, North America, Medical
Sciences; 20c, North America,
Philosophy and Philosophy of
Science

Stanley, Matthew
Harvard University
Department of the History of
Science
Science Center 235
Cambridge, MA 02138
Phone: 1.617.628.9213
stanley@fas.harvard.edu

Stapleford, Tom
11 Ware St Apt 19
Cambridge, MA 02138-4042
Phone: 1.617.497.5348
tstaplef@fas.harvard.edu

Stapleton, Darwin H.
Rockefeller Archive Center
15 Dayton Ave
Tarrytown, NY 10591-1519
Phone: 1.914.631.4505
Fax: 1.914.631.6017
stapled@mail.rockefeller.edu
Doctorate, 1975
19c, North America, Technology;
20c, North America, Technology;
20c, North America, Institutions

Staroverova, Zoya P.
Russian Academy of Sciences
Inst of History of Sci & Tech
Universitetskaia Nab 5
St Petersburg 1
Russia

Steedman, Ian
6 Osmondthorpe Cottages
Leeds LS9 9EQ
Great Britain
Phone: 1132493741
irs@eggconnect.net
Bachelor's Level, 1958
Europe, Technology, Instruments
and Techniques

Steen, Tomoko Y.
Library of Congress
Sci., Tech. & Business Div.
Rm. LA5224
Washington, DC 20540-4751
Phone: 1.301.468.2346
Fax: 1.781.431.0633
tste@loc.gov
Doctorate, 1996
20c, Asia, Biological Sciences

Steffens, Henry J.
University of Vermont
Dept. of History
Wheeler House
Burlington, VT 05405-0164
Phone: 1.802.656.4421
Fax: 1.802.656.8794
hsteffen@zoo.uvm.edu
Doctorate, 1968
19c, Europe

Steigerwald, Joan S.
York University
Bethune College
4700 Keele Street
Toronto, ON M3J1P3
Canada
Phone: 1.416.736.2100
Fax: 1.416.736.5892
steiger@yorku.ca
Doctorate
18c, Europe, Medical Sciences;
18c, Europe, Natural
and Human History;
Humanistic Relations of Science;
18c, Europe, Philosophy and
Philosophy of Science

Steinberg, David A.
S & A Institute
15111 Tyler Rd
Fiddletown, CA 95629-9704
Phone: 1.209.245.6178
Fax: 1.209.245.5653
dasteinberg@cdepot.net
Doctorate, 1978
20c, Physical Sciences;
Astronomical Sciences,
Philosophy and Philosophy
of Science; Pre-400 B.C.E.,
Natural and Human History

Steinbock, Ted R.
Baptist Hospital East
Radiology Department
4000 Kresge Way

Louisville, KY 40207
Phone: 1.502.897.8121
Fax: 1.502.897.8040
Doctorate, 1977
18c, North America;
14c-16c, Medical Sciences;
Medical Sciences;
Natural and Human History

Steinle, Friedrich
Max Planck Institute for the
History of Science
Wilhelmstr. 44
Berlin 10117
Germany
Phone: 49.30.22667.135
Fax: 49.30.22667.299
steinle@mpiwg-berlin.mpg.de
Doctorate, 1999
19c, Physical Sciences;
17c, Physical Sciences;
Philosophy and
Philosophy of Science;
19c, Instruments and Techniques

Stephens, Carlene E.
Smithsonian Instiution
PO Box 37012
Nat Museum of American
History
Room 5002 MRC 629
Washington, DC 20013-7012
Phone: 1.202.357.2379
Fax: 1.202.357.1853
stephenc@si.edu

Stephens, Lester D.
University of Georgia
Department of History
LeConte Hall
Athens, GA 30602-5304
Fax: 1.706.542.2053
lstephen@arches.uga.edu
Doctorate, 1964
19c, North America,
Natural and Human History;
20c, Biological Sciences;
Biological Sciences

Stephenson, Bruce
Adler Planetarium
1300 S. Lake Shore Drive
Chicago, IL 60605
Phone: 1.312.322.0820
Fax: 1.312.341.9935
bstephenson@adlernet.org
Doctorate, 1983

17c, Europe, Astronomical Sciences; 14c-16c, Europe, Astronomical Sciences; Instruments and Techniques; Astronomical Sciences

Sterling, Keir B.
US Army Combined Arms Support
3901 A Avenue
Suite 100
Fort Lee, VA 23801-1837
Phone: 1.804.734.0082
Fax: 1.804.285.9133
koos1934@cs.com
Doctorate, 1973
North America, Biological Sciences; 19c, North America, Biological Sciences; 19c, North America, Natural and Human History; Biography

Sterner, Hakan
Energiewerke Nord GMBH
Postfach 1125
Lubmin D-17507
Germany
Phone: 49.3.8354.4801
Fax: 49.3.8354.22458
hakan.sterner@ewn-gmbh.de
Master's Level, 1968
Technology, Instruments and Techniques; Natural and Human History; Technology; Instruments and Techniques

Sterrett, Susan G.
Duke University
Dept of Philosophy
201C W. Duke Bldg Box 90743
Durham, NC 27708
Phone: 1.919.660.3054
Fax: 1.919.660.3060
sterrett@duke.edu
Master's Level
Philosophy and Philosophy of Science

Stevens, Marianne P.
Univ. of Toronto - IHPST
Victoria College, Room 316
91 Charles Street West
Toronto, ON M5S1K7
Canada
Fax: 1.416.978.3003
mpstevens35@hotmail.com
Doctorate, 2000

20c, Medical Sciences; 20c, North America, Technology; 20c, Social Relations of Science

Stevens, Peter F.
Missouri Botnaical Garden
PO Box 299
Saint Louis, MO 63166-0299
Phone: 1.314.577.0861
Fax: 1.314.577.0830
peter.stevens@mobt.org
Doctorate
Biological Sciences; Natural and Human History

Stevens, Wesley M.
University of Winnipeg
Dept of History
Winnipeg, MB R3J2G2
Canada
Phone: 1.204.786.9203
Fax: 1.204.783.7981
wstevens@uwinnipeg.ca
Doctorate, 1968
5c-13c, Europe, Astronomical Sciences; 5c-13c, Europe, Mathematics; 5c-13c, Europe, Astronomical Sciences, Philosophy and Philosophy of Science

Stewart, Albert B.
1453 Corry St
Yellow Springs, OH 45387-1315
Phone: 1.937.767.9189
Doctorate
19c, North America, Physical Sciences; 20c, Asia, Philosophy and Philosophy of Science; Africa, Astronomical Sciences; Astronomical Sciences, Philosophy and Philosophy of Science

Stewart, L.
University of Saskatchewan
Dept of History
Saskatoon, SK S7N5A5
Canada
Phone: 1.306.966.5793
Fax: 1.306.966.5852
stewartl@sask.usask.ca

Stewart, Mart A.
Western Washington University
Department of History
Bellingham, WA 98225-9056

Phone: 1.360.650.3455
Fax: 1.360.650.7789
smar4@cc.wwu.edu
Doctorate
19c, North America, Biological Sciences; 20c, North America, Biological Sciences; 19c, North America, Earth Sciences

Stigler, Stephen M.
University of Chicago
Dept. of Statistics
5734 S University Ave
Chicago, IL 60637-1514
Phone: 1.773.955.9186
Fax: 1.773.702.9810
stigler@galton.uchicago.edu
Doctorate, 1967

Stiling, Rodney L.
University of Wisconsin
Integrated Liberal Studies
228 North Charter Street
Madison, WI 53715-1145
Phone: 1.206.281.2680
Fax: 1.206.281.2771
rstiling@earthlink.net
Doctorate, 1991
19c, North America, Earth Sciences; 19c, North America, Biological Sciences; 19c, North America, Social Relations of Science

Stirling, Jim
3508 Lawrence Dr
Naperville, IL 60564-4110

Stocking, George W., Jr.
University of Chicago
Dept of Anthropology
1126 E 59th St
Chicago, IL 60637-1539
Phone: 1.773.702.7702
Fax: 1.773.702.4503
g-stocking@uchicago.edu
Doctorate, 1960
20c, North America, Natural and Human History; 19c, North America, Natural and Human History; 20c, Europe, Natural and Human History; 20c, North America, Social Sciences

Stoddart, David R.
University of California
Geography Dept

Berkeley, CA 94707
Fax: 1.510.526.0274
stoddart@socrates.berkeley.edu
Doctorate, 1964
18c, Australia and Oceania,
Biological Sciences;
19c, Europe, Biological Sciences;
Australia and Oceania,
Exploration, Navigation and
Expeditions; Australia and
Oceania, Earth Sciences

Stoeltzner, Michael
University of Bielefeld
Inst for Science & Tech Studie
PO Box 10 01 31
33501 Bielefeld
Germany
Phone: 49.521.106.4661
Fax: 49.521.106.6418
stoeltzner@iwt.uni-bielefeld.de

Stolz, Michael
University of Tübingen
Department of Mathematics
Auf Der Morgenstelle 10
Tübingen D 72076
Germany
Phone: 49.70.7129.7676
Fax: 49.70.7129.4322
michael.stolz@uni-tuebingen.de

Stolzenberg, Daniel
715 60th St
Oakland, CA 94609-1421
stolzius@stanford.edu

Strachan, Graham
40 Deanewood Crescent
Etobicoke, ON M9B3B1
Canada
Phone: 1.416.622.8098
g.strachan@sympatico.ca

Stranges, Anthony N.
Texas A & M University
History Bldg
Dept of History
College Station, TX 77843-4236
Phone: 1.979.845.7151
Fax: 1.979.862.4314
a-stranges@tamu.edu
Doctorate
20c, North America, Physical
Sciences; 20c, Europe, Physical
Sciences

Strasser, Bruno
University of Geneva
CMU
Geneva 4
Switzerland
Phone: 41.22.328.9365
Fax: 41.22.702.5792
brunostr@uni2a.unige.ch

Strauss, David
Kalamazoo College
Department of History
Kalamazoo, MI 49006
Phone: 1.616.337.7055
Fax: 1.616.337.7028
strauss@kzoo.edu
Doctorate
19c, North America;
20c, Europe, Astronomical
Sciences

Strauss, Linda M.
Univ of Puget Sound
1500 N Warner
Tacoma, WA 98416
Phone: 1.253.756.8386
lstrauss@ups.edu
Doctorate, 1987
Europe, Humanistic Relations
of Science; North America,
Social Relations of Science;
Biological Sciences

Strick, James E.
Princeton University
Program in History of Science
Dickinson Hall
Princeton, NJ 08544-1017
jamesstrick@earthlink.net
Doctorate, 1997
19c, Biological Sciences,
Instruments and Techniques;
20c, Biological Sciences
; 19c, Medical Sciences;
20c, Social Relations of Science

Stromberg, Roland
7033 N Fairchild Cir
Fox Point, WI 53217-3851

Stroud, Patricia T.
613 Maplewood Rd
Wayne, PA 19087-4720

Stroup, Alice
Bard College
Department of History

Annadale on Hud, NY
12504-5000
Phone: 1.845.758.7234
Fax: 1.212.202.4901
stroup@bard.edu
Doctorate, 1978
17c, Europe, Institutions;
17c, Europe, Humanistic
Relations of Science;
17c, Europe, Astronomical
Sciences

Stuewer, Roger H.
University of Minnesota
School of Physics & Astronomy
116 Church St SE
Minneapolis, MN 55455-0149
Phone: 1.612.624.8073
Fax: 1.612.624.4578
rstruewer@physics.spa.umn.edu
Doctorate, 1968
20c, Europe, Physical Sciences;
20c, North America, Physical
Sciences

Stump, David J.
University of San Francisco
Dept of Philosophy
2130 Fulton St
San Francisco, CA 94117-1080
Phone: 1.415.422.6153
Fax: 1.415.422.2346
stumpd@usfca.edu
Doctorate, 1988
Philosophy and Philosophy of
Science; Astronomical Sciences,
Philosophy and Philosophy of
Science; 19c, Mathematics;
Social Relations of Science

Sturchio, Jeffrey L.
Merck & Co Inc
1 Merck Drive WS2A-55
Whitehouse Station, NJ
08889-0100
Phone: 1.908.423.3981
Fax: 1.908.735.1704
jeffrey_sturchio@merck.com
Doctorate, 1981
Physical Sciences; 20c, North
America; Social Relations of
Science; Technology

Sturdy, Steven
University of Edinburgh
Science Studies Unit
21 Buccleuch Place

Edinburgh EH8 9LN
Great Britain
Phone: 44.131.650.4014
Fax: 44.131.650.6886
s.sturdy@ed.ac.uk

Sudduth, William M.
Fernbank Science Center
156 Heaton Park Rd NE
Atlanta, GA 30307
Phone: 1.404.378.4311
Fax: 1.404.370.1336
mac.sudduth@fernbank.edu
Doctorate
18c, Europe, Physical Sciences;
18c, Europe, Philosophy
and Philosophy of Science;
20c, North America, Institutions;
20c, North America, Education

Sugiyama, Shigeo
Hokkaido University
Faculty of Science
Sapporo 060-0810
Japan
Phone: 011.706.4421
sugiyama@hps.sci.hokudai.ac.jp
Doctorate
19c, Physical Sciences;
20c, Physical Sciences;
Asia, Education;
Asia, Social Relations of Science

Sullivan, Gerald
2016 Westgate Dr
Bethlehem, PA 18017-7462
Phone: 1.703.960.6815
pakdjeri@earthlink.net

Sullivan, Woodruff T., III
University of Washington
Dept of Astronomy
Box 351580
Seattle, WA 98195-1580
Phone: 1.206.543.7773
Fax: 1.206.685.0403
woody@astro.washington.edu
Doctorate, 1971
20c, Astronomical Sciences;
19c, Astronomical Sciences;
20c, Physical Sciences;
19c, Physical Sciences

Sulloway, Frank J.
University of California
Dept of Psychology IPSR
4125 Tolman Hall

Berkeley, CA 94720
Phone: 1.510.642.7139
Fax: 1.510.643.9334
sulloway@uclink.berkeley.edu
Doctorate, 1978
19c, Europe, Biological Sciences,
Biography; Biological Sciences;
Cognitive Sciences; Social
Sciences

Summers, Greg P.
Orbital Sciences
1521 Century Blvd.
Germantown, MD 20874
Phone: 1.301.444.3931
Fax: 1.301.249.7020
gxsummers@worldnet.att.net
Master's Level, 1987
Philosophy and Philosophy
of Science; North America,
Exploration, Navigation and
Expeditions; North America; 20c

Summers, William C.
Yale University
333 Cedar Street
New Haven, CT 06520-8040
Fax: 1.203.785.6309
william.summers@yale.edu
Doctorate, 1967
20c, Biological Sciences;
Medical Sciences; Asia;
Education

Suppe, John
Princeton University
Department of Geosciences
Princeton, NJ 08544
Phone: 1.609.924.6519
suppe@princeton.edu

Sutphen, Molly P.
UCSF
533 Parnassus Avenue
Room 464
San Francisco, CA 94143
Phone: 1.415.476.2766
sutphen@itsa.ucsf.edu
Doctorate
20c, Asia, Biological Sciences,
Instruments and Techniques;
20c, Africa, Medical Sciences

Sutton, John
Macquarie University
Dept of Philosophy
Sydney 2109

Australia
Phone: 61.2.9850.8817
Fax: 61.2.9850.8892
jsutton@laurel.ocs.mq.edu.au
Doctorate, 1993
Cognitive Sciences; 18c, Europe,
Medical Sciences, Instruments
and Techniques; 17c, Philosophy
and Philosophy of Science

Suzuki, Takanori
Sagamioona 6-23-13-503
Sagamihara 228
Japan

Suzuki, Terumi
1-9-2 Chuo-Cho
Higashikurume
Tokyo 203 0054
Japan
Phone: 81.424.79.7979
jcome@home.ne.jp

Svantesson, Jan Olof
Thulehemsvagen 141
Lund S-22467
Sweden
Phone: 48.46.145403

Sviedrys, Romualdas
235 Gristmill Ln
Great Neck, NY 11023-1816

Swan, Patricia B.
Iowa State University
211 Beardshear
Ames, IA 50011
Phone: 1.515.294.5523
Fax: 1.515.292.3744
pswan@iastate.edu
Doctorate, 1964
20c, North America, Medical
Sciences; 20c, North America,
Biological Sciences; 20c, North
America, Physical Sciences;
20c, North America, Biography

Swan, Victor R.
Swan & Jaillet Rare Books
PO Box 161452
Austin, TX 78716-1452
Phone: 1.512.327.0626
Fax: 1.512.327.0480
rarebooks@austin.rr.com
Bachelor's Level, 1965
Astronomical Sciences;

Biological Sciences;
Earth Sciences; Natural and
Human History

Swann, John P.
Food & Drug Administration
History Office
HFC-24 Rm 1351
5600 Fishers Lane
Rockville, MD 20857
Phone: 1.301.827.3756
Fax: 1.301.827.0551
jswann@ora.fda.gov
Doctorate, 1985
20c, North America, Medical
Sciences; 20c, North America,
Science Policy; 20c, North
America, Physical Sciences

Swanson, Gavin
22 Pretoria Road
Cambridge
Great Britain
Phone: 44.1223.326223
gswanson@cambridge.org
Doctorate, 1982
Biological Sciences; Physical
Sciences; Astronomical Sciences

Swanson, Richard P.
866 Ferndale St S
Maplewood, MN 55119-5944
Phone: 1.651.739.7164
swans068@tc.umn.edu

Sweet, Victoria
1947 Alameda
Redwood City, CA 94061-3211
vsweet@itsa.ucsf.edu

Swenson, Loyd S., Jr.
1948 N Macgregor Way
Houston, TX 77023
Phone: 1.713.926.7639
Fax: 1.713.743.3216
Doctorate, 1962
19c, Europe, Physical Sciences;
20c, North America, Physical
Sciences; Europe, Technology;
20c, North America, Humanistic
Relations of Science

Swijtink, Zeno G.
Sonoma State University
Department of Philosophy
1801 E Cotati Ave
Rohnert Park, CA 94928-3609

Phone: 1.707.664.4421
Fax: 1.707.664.2505
swijtink@sonoma.edu
Doctorate, 1982
Instruments and Techniques;
Earth Sciences; Exploration,
Navigation and Expeditions;
Philosophy and Philosophy of
Science

Sy, Donna
642 Arlington Avenue
Berkeley, CA 94707
Phone: 1.510.524.9672
dsy@socrates.berkeley.edu

Sylla, Edith D.
North Carolina State Univ.
Harrelson Hall 161
Raleigh, NC 27695-8108
Phone: 1.919.513.2221
Fax: 1.919.515.3886
edith_sylla@ncsu.edu
Doctorate, 1971
5c-13c, Europe,
Physical Sciences;
5c-13c, Europe, Mathematics;
17c, Europe, Mathematics;
17c, Europe, Physical Sciences

Szabad, George
3300 Darby Rd
Haverford, PA 19041-1016

Szwaja, Lynn
The Rockefeller Foundation
1133 Avenue of the Americas
New York, NY 10036-6710

Taber, Harry W.
Wadsworth Center, NYSDH
Axelrod Institute
PO Box 22002
Albany, NY 12201-2002
Phone: 1.518.473.2760
Fax: 1.518.473.1326
taber@wadsworth.org
Doctorate, 1963
19c, Europe, Biological Sciences,
Instruments and Techniques;
19c, North America, Biological
Sciences, Instruments and
Techniques; 20c, Europe,
Biological Sciences, Instruments
and Techniques; 20c, North
America, Biological Sciences,
Instruments and Techniques

Tabery, James
University of Pittsburgh
History & Philosophy of Science
1017 Cathedral of Learning
Pittsburgh, PA 15260-6299
Phone: 1.412.422.9494
jimtabery@hotmail.com

Tachau, Katherine H.
University of Iowa
Department of History
Iowa City, IA 52242
Phone: 1.319.335.2210
Fax: 1.319.338.1308
katherine-tachau@uiowa.edu

Takahashi, Kenichi
Kyushu University
4-2-1 Ropponmatsu, Chuo-Ku
Fukuoka 810-8560
Japan
Phone: 81.92.726.4633
Fax: 81.92.726.4633
kentaka@rc.kyushu-u.ac.jp

Takarabe, Kae
Nagoya University
Furo-cho, Chikusa-ku
Nagoya 464 8601
Japan
Fax: 81.52.789.4230
takarabe@info.human.
  nagoya-u.ac.jp
Master's Level, 1996
18c, North America, Natural
and Human History; 19c, North
America, Natural and Human
History; 18c, Asia, Natural
and Human History; 19c, Asia,
Natural and Human History

Talarico, Kathryn M.
College of Staten Island
2800 Victory Blvd.
2S-109
Staten Island, NY 10314
Phone: 1.718.982.3701
Fax: 1.718.982.3712
talarico@scholar.chem.nyu.edu
Doctorate
5c-13c, Europe; Africa

Tammann, Gustav A.
Astronomisches Institut
Venusstrasse 7
Binningen CH-4102
Switzerland

Phone: 41.61.205.5454
Fax: 41.61.205.5455
tammann@ubaclu.unibas.ch
Doctorate
Astronomical Sciences

Tamny, Martin
City College
Philosophy Department
138th St. & Convent Avenue
New York, NY 10034
Phone: 1.212.650.7770
mtamny@optonline.net
Doctorate, 1976
17c, Astronomical Sciences,
Philosophy and Philosophy of
Science; 17c, Physical Sciences;
17c, Astronomical Sciences;
18c, Astronomical Sciences

Tanaka, Hiroaki
Kashii 2-chome 18-23-402
Higashi-ku
Fukuoka-shi 813-0011
Japan
Phone: 81.92.683.2543
Fax: 81.92.683.2543
tanaka.xi@nifty.com

Tandarich, John
Hey and Associates
53 W. Jackson Blvd Suite 1015
Chicago, IL 60604
Phone: 1.708.848.3716
Fax: 1.312.922.1823
chicago@heyassoc.com

Tandy, Pippa
122 East Parade
East Perth 6004
Australia
Phone: 61.8.9328.4260
Fax: 61.8.9328.4643
pippa@pc.wa.edu.au

Tanford, Charles
Tarlswood Back Lane
Easingwold York YO6 3BG
Great Britain
Phone: 44.13.4782.1029
candj@dial.pipex.com

Tang, Joyce
City University of New York
Queens College
65-30 Kissena Boulevard
Flushing, NY 11367-1597

Phone: 1.718.997.2839
Fax: 1.718.460.0439
jtang@qc.edu
Doctorate, 1991
20c, North America, Social
Sciences; Technology,
Instruments and Techniques

Tang, Paul C. L.
Cal State Univ-Long Beach
Dept of Philosophy
1250 N Bellflower Blvd
Long Beach, CA 90840-0001
Phone: 1.562.985.4343
Fax: 1.562.985.7135
pcltang@csulb.edu
Doctorate, 1982
Philosophy and Philosophy of
Science; Cognitive Sciences

Tanimoto, Tsutomu
Hosei University
Fujimi 2-17-1
Chiyoda-ku
Tokyo 1028160
Japan
Phone: 03.3264.9633
Fax: 03.3264.9663
tanimoto@i.hosei.ac.jp
Doctorate, 1990
20c, Asia, Earth Sciences;
19c, Europe, Natural and
Human History; 18c, North
America, Earth Sciences; 17c

Tarasova, Valentina
Dmitrovskii Proezd
#20 Kor 1 Apartment 95
Moscow 125206
Russia

Tarver, John R.
Academic Services
333 S Acadian Thruway
Baton Rouge, LA 70806-5022
Phone: 1.225.346.6702
Fax: 1.225.346.0932
john_tarver@baellsouth.net
Doctorate
17c, Asia, Biological Sciences;
18c, Asia, Biological Sciences;
20c, Medical Sciences;
19c, North America, Natural and
Human History

Tassava, C. James
18701 Stratford Rd Apt 331
Minnetonka, MN 55345-4064
Phone: 1.952.401.4264
c-tassava@northwestern.edu

Tatarewicz, Joseph N.
University of Maryland
1000 Hilltop Cir # AD702
Baltimore, MD 21250-0002
Phone: 1.410.455.2036
Fax: 1.410.455.1045
tatarewicz@umbc.edu
Doctorate, 1984
North America; Exploration,
Navigation and Expeditions;
Humanistic Relations of Science;
Science Policy

Tattersall, James J.
Providence College
Department of Mathematics
River Avenue
Providence, RI 02918
Phone: 1.401.865.2468
Fax: 1.401.865.1438
tat@providence.edu
Doctorate, 1971
18c, Europe, Mathematics;
18c, North America,
Mathematics; 18c, Europe,
Astronomical Sciences;
18c, North America,
Astronomical Sciences

Taub, Liba
University of Cambridge
Whipple Museum
Free School Lane
Cambridge CB2 3RH
Great Britain
Phone: 44.1223.334545
Fax: 44.1223.334554
lct1001@hermes.cam.ac.uk
Doctorate
Pre-400 B.C.E., Astronomical
Sciences; Pre-400 B.C.E.,
Earth Sciences; Pre-400 B.C.E.,
Physical Sciences; Instruments
and Techniques

Tauber, Alfred I.
Boston University
745 Commonwealth Ave
Room 506
Boston, MA 02215-1401
Phone: 1.617.353.2604

Fax: 1.617.353.6805
atauber@bu.edu
Doctorate, 1973
*20c, Europe, Philosophy
and Philosophy of Science;
19c, North America; Biological
Sciences; Philosophy and
Philosophy of Science*

Taylor, Eugene
98 Clifton St
Cambridge, MA 02140-1711

Taylor, Hunter
163 Warren St.
Brooklyn, NY 11201
Phone: 1.718.522.5469
jhctaylor@aol.com

Taylor, Kenneth L.
Univ of Oklahoma
622 Physical Sciences Building
601 Elm Ave
Norman, OK 73019-3106
Phone: 1.405.325.5416
Fax: 1.405.325.2363
ktaylor@ou.edu
Doctorate, 1968
*Earth Sciences; 18c;
Exploration, Navigation and
Expeditions; Natural and
Human History*

Taylor, Ronald C.
6100 Westchester Park Dr #1812
College Park, MD 20740-2851
Phone: 1.301.474.2836
Doctorate, 1969
19c, Earth Sciences

Tayyabkhan, Tara
University of New Hampshire
Department of Psychology
Conant Hall
Durham, NH 03824
trt@cisunix.unh.edu

Temkin, Owsei
830 W 40th St
Baltimore, MD 21211-2116
Phone: 1.410.243.6322
Doctorate, 1927
Medical Sciences

Tenner, Edward H.
4316 Hunters Glen Dr
Plainsboro, NJ 08536-3911

Phone: 1.609.716.0263
Fax: 1.609.799.9010
tenner@princeton.edu
Doctorate
*20c, Europe, Technology;
19c, North America, Biological
Sciences; 5c-13c, Europe,
Natural and Human History;
400 B.C.E.-400 C.E., Asia,
Humanistic Relations of Science*

Tercier, John
29 Mornington Terrace
3rd Floor Flat
London NW17RS
Great Britain
Phone: 020.7.387.8677
ttercier@hotmail.com

Terrall, Mary
UCLA
Dept. of History
Box 1473
Los Angeles, CA 90095-1473
Phone: 1.310.825.2013
Fax: 1.310.206.9630
terrall@history.ucla.edu
Doctorate, 1987
*18c, Europe, Physical Sciences;
18c, Europe, Gender and
Science; 18c, Europe, Biological
Sciences*

Tesdall, Eric
8133 Heatherton Ln Apt 102
Vienna, VA 22180-7420
Phone: 1.703.527.6424
tesdalle@georgetown.edu

Teslow, Tracy L.
University of Chicago
SS 205
1126 E 59th St
Chicago, IL 60637
Phone: 1.617.258.8118
tesl@midway.uchicago.edu
Master's Level, 1999
*20c, North America, Natural and
Human History; Gender and
Science; North America*

Texera, Yolanda
CENDES - UCV
Apdo Correos 6622
Caracas 1040-A
Venezuela
ytexera@telcel.net.ve

Doctorate
*19c, South America,
Biological Sciences;
20c, Biological Sciences*

Thackray, Arnold W.
Chemical Heritage Foundation
315 Chestnut St
Philadelphia, PA 19106-2702
Phone: 1.215.873.8245
Fax: 1.215.925.1954
athackray@chemheritage.org
Doctorate, 1966
*20c, North America;
Social Sciences*

Thaler, Michael M.
University of California
MU4E
San Francisco, CA 94143
Phone: 1.415.467.5892
Fax: 1.415.664.6554
mmt@itsa.ucsf.edu
Doctorate, 1958
*20c, Europe, Medical Sciences;
20c, North America, Philosophy
and Philosophy of Science;
20c, Europe, Social Relations of
Science*

Theerman, Paul
National Library of Medicine
8600 Rockville Pike
Bethesda, MD 20894
Phone: 1.301.594.0975
Fax: 1.301.402.0872
ptheerman@mediasoft.net
Doctorate, 1980
*Medical Sciences; Humanistic
Relations of Science; Social
Relations of Science; Biography*

Theunissen, Bert
Inst for History of Science
Princetonplein 5
Utrecht 3584 CC
The Netherlands
Phone: 31.30.2537218
Fax: 31.30.2536313
l.t.g.theunissen@phys.uu.nl
Doctorate
*19c, Europe, Biological Sciences;
19c, Europe, Humanistic
Relations of Science;
19c, Europe, Institutions;
19c, Europe, Social Relations of
Science*

Thibodeau, Philip
10 Kesler St
Athens, GA 30601
Phone: 1.706.542.5613
pthib@arches.uga.edu

Thibodeau, Sharon G.
5211 Wilson Ln
Bethesda, MD 20814-2407

Thiede, Walther
An der Ronne 186
Koeln 50859
Germany
Phone: 2234.70584
Doctorate
20c, Europe, Biological Sciences;
19c, Asia, Education; Biography

Thijssen, Johannes M.
Nijmegen University
Faculty of Philosophy
Erasmusplein 1
Nijmegen 6500 HD
The Netherlands
Phone: 31.24.361.5655
Fax: 31.24.361.5564
hthijssen@phil.kun.nl
Doctorate
5c-13c, Europe, Philosophy
and Philosophy of Science;
14c-16c, Europe, Philosophy
and Philosophy of Science

Thomas, Mary M.
1125 Iowa Ave W
Saint Paul, MN 55108-2241
Phone: 1.651.489.7245
thom0209@tc.umn.edu
Doctorate, 1999
Technology; Institutions;
Instruments and Techniques

Thomas, Nigel J.
86 S Sierra Madre Blvd Apt 5
Pasadena, CA 91107-4243
Phone: 1.626.578.7328
jantec@earthlink.net
Doctorate, 1987
Cognitive Sciences; Philosophy
and Philosophy of Science;
Physical Sciences

Thomas, Steven D.
70 E 10th St Apt 15D
New York, NY 10003-5116
Phone: 1.212.253.6707

Thomasset, Claude
2 Rue du 8 Mai 1945
Chaville 92370
France
Phone: 33.01.47.50.02.13

Thomaz, Manuel F.
Av. Dr. Lourenco Peixinho 175G
5E
Aveiro 3800-167
Portugal
Phone: 351.34.370.271
Fax: 351.34.424.965
mfthomaz@fis.ua.pt
Doctorate, 1968
18c, Europe, Physical Sciences;
19c, Europe, Physical Sciences;
18c, Europe, Instruments
and Techniques;
19c, Europe, Technology

Thompson, Emily
2521 Autumnwood Dr
Glenshaw, PA 15116-1801
emilyt@alumni.princeton.edu
Doctorate
20c, North America, Technology

Thompson, Michele
Southern Connecticut State
University
Dept of History
501 Crescent St
New Haven, CT 06515-1355
thompson_mc@scsu.ctstateu.edu
Doctorate, 1998
19c, Asia, Medical Sciences;
Asia, Medical Sciences

Thomson, Ron B.
Pontifical Inst of Medaev Stud
59 Queens Park Cres E
Toronto, ON M5S2C4
Canada
Phone: 1.416.926.7143
Fax: 1.416.926.7258
thomson@chass.utoronto.ca
Doctorate, 1975
5c-13c, Europe, Astronomical
Sciences

Thorpe, John
70 Olive Ave
Toronto, ON M6G1V1
Canada
Phone: 1.416.979.5195

Thurk, Jessica
1500 Chicago Ave Apt 622
Evanston, IL 60201-4434
Phone: 1.847.328.5924
j-thurk@northwestern.edu

Thurs, Daniel
719 Clark Ct
Madison, WI 53715-1421
Phone: 1.608.294.9514
dpthurs@students.wisc.edu

Thurston, Floyd
723 S Ravencrest St
Bloomington, IN 47401-4260
Phone: 1.812.331.0722
fethur@earthlink.net

Tierney, Kathleen
500 Stinson St Apt 59
Norman, OK 73072-6260
Phone: 1.405.329.6771
kftierney@ou.edu

Tighe, Janet A.
University of Pennsylvania
Logan Hall, Suite 303
249 S 36th St
Philadelphia, PA 19104-6304
Phone: 1.215.898.8400
Fax: 1.215.573.2231
jtighe@sas.upenn.edu
Doctorate, 1983
19c, North America, Medical
Sciences; 20c, North America,
Medical Sciences; 20c, North
America, Medical Sciences,
Instruments and Techniques

Tiles, Mary E.
University of Hawaii
Dept of Philosophy
2530 Dole St
Honolulu, HI 96822-2303
Phone: 1.808.956.8250
mtiles@hawaii.edu
Doctorate
Philosophy and Philosophy of
Science; 17c, Asia, Mathematics;
Technology

Timmons, William T.
Westark College
5210 Grand
Fort Smith, AR 72903

Phone: 1.501.788.7630
ttimmons@systema.westark.edu
Master's Level, 1996

Tirapicos, Luis A.
Visionarium Centro de Ciencia
Lugar de Espargo
Santa Maria de Feira 4520
Portugal
Phone: 351.256.370.626
Fax: 351.256.370.614
latirapi@aeportugal.com
Bachelor's Level, 1999
Pre-400 B.C.E., Europe,
Astronomical Sciences;
14c-16c, Europe, Astronomical
Sciences

Tjossem, Sara F.
University of Washington
Department of Zoology
Box 351800
Seattle, WA 98195-1800
Phone: 1.206.616.3312
tjossem@u.washington.edu
Doctorate, 1994
20c, North America,
Biological Sciences;
20c, Biological Sciences;
19c, North America;
20c, Gender and Science

Tobey, Ronald C.
Univ of California, Riverside
Dept of History 049
Riverside, CA 92521-0204
Phone: 1.909.787.5401
Fax: 1.909.494.4050
rtobey@horuspublications.com
Doctorate, 1969
20c, North America, Social
Sciences; 20c, North America,
Physical Sciences

Toca, Angel
Collaborator of Unit History
of Science - Faculty of Med
University of Cantabria
Torrelavega 39300
Spain
Phone: 34.9428.9064
atola@palmera.pntic.mec.es
Doctorate, 1999
20c, Europe, Physical Sciences;
19c, Europe, Physical Sciences;
19c, Europe, Technology;
20c, North America, Technology

Todd, Edmund N.
University of New Haven
Dept of History
300 Orange Ave
West Haven, CT 06516-1999
Phone: 1.203.932.7287
entodd@charger.newhaven.edu

Todes, Daniel P.
Johns Hopkins University
Inst of History of Med
1900 E Monument St
Baltimore, MD 21214
Phone: 1.410.955.7079
dtodes@jhmi.edu
Doctorate
19c, Europe, Biological Sciences;
20c, Medical Sciences;
Social Relations of Science

Togeas, James B.
Univ of Minnesota - Morris
Division of Science & Math
Morris, MN 56267
Phone: 1.320.589.6309
Fax: 1.320.589.6371
togeasjb@mrs.umn.edu
Master's Level, 1961
Physical Sciences

Toivanen, Hannes
251 10th St NW Apt 131
Atlanta, GA 30318
Phone: 1.404.874.2607
hannes.toivanen@hts.gatech.edu
Master's Level
20c, Technology

Tolstoy, I.
Knockvennie
Scotland
Castle Douglas DG7 3PA
Great Britain

Tomash, Erwin
110 S Rockingham Ave
Los Angeles, CA 90049-2514
Phone: 1.310.394.8468
etomash@gte.net
Master's Level, 1953

Tomory, Leslie
156 St. George Street
Toronto, ON M5S2G1
Canada

Phone: 1.416.598.7725
Fax: 1.416.599.3517
ltomory@yahoo.com

Toon, Elizabeth
Cornell University
Science and Tech Studies Dept
632 Clark Hall
Ithaca, NY 14853-2501
eat23@cornell.edu
Doctorate, 1998
20c, North America, Medical
Sciences; Social Relations of
Science; Gender and Science;
North America

Topham, Jonathan R.
University of Leeds
School of Philosophy
Leeds LS2 9JT
Great Britain
Phone: 44.11.3223.3383
Fax: 44.11.3233.3365
j.r.topham@leeds.ac.uk
Doctorate, 1993
19c, Europe, Humanistic
Relations of Science;
19c, Europe, Social Relations of
Science; 19c, Europe, Education

Topper, David
University of Winnipeg
Dept of History
515 Portage Ave
Winnipeg, MB R3M3J7
Canada
Phone: 1.204.786.9398
Fax: 1.204.774.4134
david.topper@ds1.uwinnipeg.ca
Doctorate, 1970
20c, Europe, Humanistic
Relations of Science; 17c, North
America, Physical Sciences;
19c, Astronomical Sciences;
18c, Astronomical Sciences

Toro, Emilio
University of Tampa
Math Dept Box 88F
401 W Kennedy Blvd
Tampa, FL 33606-1490
Phone: 1.813.253.3333
Fax: 1.813.258.7881
propro35@hotmail.com
18c, South America, Physical
Sciences

Touwaide, Alain
PO Box 25805
Washington, DC 20007-8805
atouwaide@hotmail.com

Townsend, Burke
University of Montana
Missoula, MT 59812-1038
Phone: 1.406.243.6233
Fax: 1.406.243.5313
bat@selway.umt.edu
Doctorate, 1976
Philosophy and Philosophy of Science

Townsend, Clarence W.
Lane County Health & Hum Serv
125 E 8th St
Eugene, OR 97401
Phone: 1.541.682.3686
Fax: 1.541.682.3879
cwt@efn.org
19c, Europe, Social Sciences; 20c, North America, Social Sciences; 19c, Europe, Philosophy and Philosophy of Science; 19c, Europe, Astronomical Sciences, Philosophy and Philosophy of Science

Towsley, Gary W.
SUNY Geneseo
Dept of Mathematics
1 College Cir
Geneseo, NY 14454-1302
Phone: 1.716.245.5388
towsley@uno.cc.geneseo.edu
Doctorate, 1975
5c-13c, Europe, Mathematics; Pre-400 B.C.E., Europe, Mathematics; 400 B.C.E.-400 C.E., Europe, Mathematics; 14c-16c, Europe, Mathematics

Tracy, Sarah
University of Oklahoma
Honors College
1300 Asp Ave
Norman, OK 73019-6060
Phone: 1.405.525.5534
Fax: 1.405.325.7109
swtracy@ou.edu
Doctorate, 1992
North America, Medical Sciences; North

America, Medical Sciences, Instruments and Techniques; North America, Philosophy and Philosophy of Science; North America, Gender and Science

Traver Ribes, Manuel J.
Les Sant Vicent Ferrer
Parc Salvador Castell 16
Algemesi 46680
Spain
Phone: 34.9.6242.0534
Fax: 34.9.6248.0632
manetr@wanadoo.es
Doctorate, 1996
Physical Sciences; Astronomical Sciences; Philosophy and Philosophy of Science

Travis, Anthony S.
Hebrew University
Sidney M. Edelstein Center for HPSTM
Givat Ram
Jerusalem 91904
Israel
Phone: 972.2658.5652
Fax: 972.2658.6709
travis@cc.huji.ac.il
Doctorate
19c, Europe, Technology; 19c, Europe, Physical Sciences; 20c, Europe, Physical Sciences; 20c, North America, Physical Sciences

Trepanier, Michel
INRS Urbanisation
3465 Rue Durocher
Montreal, PQ H2X2C6
Canada
Phone: 1.514.499.4057
Fax: 1.514.499.4065
michel.trepanier@
    inrs-ucs.uquebec.ca

Trepp, Anne C.
Springstr. 5
Göttingen 37077
Germany
Phone: 0551.49.56.131
Fax: 0551.49.56.170
trett@mpi-g.gwdg.de

Trickett, Susan B.
10708 Rosehaven St
Fairfax, VA 22030-2827
Phone: 1.703.691.0711
Fax: 1.703.993.1330
stricket@gmu.edu

Tricot, J. P.
Vrijheidstraat 19
Antwerpen B-2000
Belgium
Phone: 32.3237.7850
Fax: 32.3238.6664
jp.tricot@pandora.be

Trimble, Virginia
University of California
Physics Department
Irvine, CA 92697-4575
Phone: 1.949.824.6948
Fax: 1.949.824.2174
vtrimble@uci.edu
Doctorate, 1968
Astronomical Sciences; Physical Sciences

Trompoukis, Constantinos
62 Efroniou Str
Athens 161-21
Greece
Phone: 30.1.725.3417
ntek@central.ntua.fr

Truffa, Giancarlo
Via Rombon 29
Milan 20134
Italy
Phone: 39.2.215.0023
giancarlo.truffa@st.com

Tsarevsky, Nicolay V.
Carnegie Mellon University
Dept of Chemistry
Mellon Institute Box 60
4400 5th Ave
Pittsburgh, PA 15213-2617
Phone: 1.412.802.6687
Fax: 1.412.268.6897
nvt@andrew.cmu.edu

Tsukiyama, Fumiaki
Tsukiyama Clinic
8-9 Hesakayama Sakicho
Higashi-Ku
Hiroshima 732 0004
Japan
Fax: 82.229.0183

casca3@do4.enjoy.me.jp
Doctorate
19c, Europe, Medical Sciences;
Asia

Tsvetkov, Igor
St Petersburg Department
Inst of History of Science and
Tech
Universitetskaia Nab 5
St Petersburg 1
Russia

Tuan, Debbie F.
Kent State University
Department of Chemistry
Kent, OH 44242
Phone: 1.330.672.8233
Fax: 1.330.672.3816
atuan@kent.edu
Doctorate, 1961
20c, North America,
Physical Sciences;
20c, Europe, Physical Sciences;
Asia, Physical Sciences

Tucci, Pasquale
Via Bitonto 18
Milano
Italy
Phone: 39.02.805.7309
Fax: 39.02.7200.1600
pasquale.tucci@unimi.it

Tuchman, Arleen
Vanderbilt University
Department of History
Box 1652-B
Nashville, TN 37235
Phone: 1.615.322.8151
Fax: 1.615.343.6002
arleen.m.tuchman@
  vanderbilt.edu

Tucker, Linda B.
Xavier University of Lousiana
African American Studies Prog
Box 64A - 1 Drexel Drive
New Orleans, LA 70125
Phone: 1.504.485.5000
Fax: 1.504.485.7944
ltucker@xula.edu
Doctorate, 1997
19c, North America, Biological
Sciences; 20c, North America,

Biological Sciences;
19c, North America, Education;
20c, North America, Education

Tula Molina, Fernando
Universidad Nacional De
Quilme
Calle 493 Bis 2525 e/ 20 y 21
Manuel B Gonnet 1897
Argentina
Phone: 54.221.4712964
tmolina@unq.edu.ar

Tulloch, Bruce R.
Bethlehem High School
700 Delaware Ave
Delmar, NY 12054
Phone: 1.518.439.4921
brtulloch@altavista.net
Doctorate, 1981
19c, Europe, Biological Sciences;
20c, North America,
Biological Sciences;
20c, North America, Education

Tummers, Paul MJE
Philosophical Institute
Erasmusplein 1
Nymegen 6500 HD
The Netherlands
Phone: 043.321.5540
Fax: 043.388.4917
p.tummers@
  facburfdcw.unihaas.nl
Doctorate, 1984
400 B.C.E.-400 C.E.,
Philosophy and Philosophy of
Science; 5c-13c, Philosophy
and Philosophy of Science;
400 B.C.E.-400 C.E.,
Mathematics;
5c-13c, Mathematics

Turner, Frank M.
Yale Universtity
Box 208324
New Haven, CT 06520-8324
Phone: 1.203.432.1367
frank.turner@yale.edu
Master's Level, 1971
19c, Biological Sciences;
19c, Social Relations of Science;
19c

Turner, Howard R.
Youville House - Apt 711
1573 Cambridge St

Cambridge, MA 02138-4370
Phone: 1.617.497.7309
hredturner@aol.com
Bachelor's Level
5c-13c, Philosophy and
Philosophy of Science

Turner, James
Erasmus Institute
1124 Flanner Hall
Notre Dame, IN 46556-5611
Phone: 1.574.631.3434

Turner, Larry R.
Argonne National Laboratory
9700 S. Cass Ave.
Bldg 308
Argonne, IL 60439
larry@anl.gov
Doctorate, 1964
Physical Sciences;
20c, Physical Sciences

Turner, Roger
History of Science Society
Box 351330
University of Washington
Seattle, WA 98195
Phone: 1.206.543.9366
Fax: 1.206.685.9544
roger@hssonline.org

Turner, Steven
13514 Livingston Rd
Clinton, MD 20735-9403
Phone: 1.301.292.8947
Fax: 1.202.357.1631
turners@si.edu

Turner, Steven
Univ of New Brunswick
Dept of History
Fredericton, NB E3B5A3
Canada
Phone: 1.506.458.7433
Fax: 1.506.453.5068
turner@unb.ca
Doctorate, 1971
19c, Europe, Institutions;
19c, Europe, Cognitive Sciences;
20c, North America, Social
Relations of Science

Tuttle, Julianne
Indiana University
HPSC
130 Goodbody Hall

Bloomington, IN 47405
jftuttle@erols.com
Doctorate, 1999
19c, Europe, Physical Sciences;
17c, Europe, Philosophy and
Philosophy of Science

Twardy, Charles R.
Monash University, Clayton Cps
Computer Sci & Software Eng
POB 26
Clayton 3800
Australia
Phone: 61.3.9905.5056
Fax: 61.3.9905.5146
ctwardy@alumni.indiana.edu
Master's Level, 1999
Philosophy and Philosophy of
Science; Cognitive Sciences;
19c, North America, Technology;
5c-13c

Tweney, Ryan D.
Bowling Green State University
Dept of Psychology
Bowling Green, OH 43403-0228
Phone: 1.419.372.2301
Fax: 1.419.372.6013
tweney@bgnet.bgsu.edu
Doctorate, 1970
19c, Physical Sciences;
Cognitive Sciences;
Biological Sciences, Instruments
and Techniques

Twomey, Yvonne
841 Kinston Ct
Naperville, IL 60540-7133
Phone: 1.630.961.9811
ytwomey@mcs.net

Tybjerg, Karin
Darwin College
Silver St
Cambridge CB3 9EU
Great Britain
kt206@hermes.cam.ac.uk

Tyler, Noel
12104 Rohan Rd
Oklahoma City, OK 73170-4746
Phone: 1.405.692.4043
Fax: 1.405.692.5511
ntyler@flash.net

Tympas, Aristotle
PO Box 80501
Piraeus 18510
Greece
tympas@yahoo.com
Doctorate, 1999
20c, North America, Technology;
20c, North America, Technology,
Instruments and Techniques;
20c, Europe, Technology;
20c, Europe, Technology,
Instruments and Techniques

Uchida, Masao
Wako University
Machida-Shi
Tokyo 195-8585
Japan
Phone: 81.44.989.7478
Fax: 81.44.988.1435
uchidam@wako.ac.jp

Ullrich, Rebecca A.
Sandia National Laboratories
PO Box 5800
Kirtland Air Force Base
Albuquerque, NM 87185-0612
Phone: 1.505.844.1483
Fax: 1.505.284.2782
raullri@sandia.gov
Master's Level
19c, Europe, Exploration,
Navigation and Expeditions;
20c, North America

Umebayashi, Seiji
Omine 2-4-31
Kumamoto 862 0933
Japan
Phone: 81.96.365.2171
umebayas@pu-kumamoto.ac.jp

Unguru, Sabetai
Tel Aviv University
Hist. Phil. of Sci. & Ideas
Ramat-Aviv
Tel Aviv 69978
Israel
Phone: 972.3.640.9198
Fax: 972.3.640.9463
unguru@post.tau.ac.il
Doctorate, 1970
400 B.C.E.-400 C.E., Europe,
Mathematics; 5c-13c, Europe,
Physical Sciences; Social
Sciences; 5c-13c, Physical
Sciences

Usitalo, Steven
McGill University
Dept. of History
855 Sherbrooke Street West
Montreal, PQ H3A2T7
Canada
Fax: 1.514.398.8365
usitalo@hotmail.com
18c, Europe, Social Sciences;
19c, Europe, Humanistic
Relations of Science

Usselman, Melvyn C.
University of Western Ontario
Chemistry Dept
London, ON N6A5B7
Canada
Phone: 1.519.661.2111
Fax: 1.519.661.3022
usselman@uwo.ca
Doctorate, 1973
19c, Europe, Physical Sciences;
Physical Sciences; Instruments
and Techniques; Social Sciences

Vaccari, Ezio
Centro di Studio Storia Tecnic
Via Balbi 6
Genova 16126
Italy
Phone: 39.10.2099838
Fax: 39.45.565302
ezio.vaccari@lettere.unige.it
Doctorate, 1995
18c, Europe, Earth Sciences;
19c, Europe, Earth Sciences;
18c, Europe, Technology

Valderas, Jose Ma.
Prensa Cientifica S.A.
Calle Muntaner 339, Pral. 1A
Barcelona 08021
Spain
Phone: 34.9.3414.3344
Fax: 34.9.3414.5413
precisa@
  investigacionyciencia.es
Doctorate
5c-13c, Biological Sciences

Valera, Manuel
Campus de Espinardo
Dpt Hist Med/Fac Medicina
B.O. 4021
Murcia 30003
Spain

Phone: 34.68.367.180
Fax: 34.68.364.150
valera@um.es

Valerio, Vladimiro
Universita di Napoli
Via Monteoliveto 3
Napoli 80134
Italy
Fax: 39.8.1556.9536
vladimir@unina.it
Doctorate, 1971
19c, Europe, Earth Sciences;
Astronomical Sciences;
Mathematics

Valle, Ellen
University of Turku
Department of English
Turku 20014
Finland
Phone: 358.2235.3003
Fax: 358.2333.5630
ellen.valle@utu.fi

Valone, David A.
Quinnipiac College
Box 77
275 Mt Carmel Ave
Hamden, CT 06518
Phone: 1.203.582.5269
Fax: 1.203.582.3471
david.valone@quinnipiac.edu
Doctorate
Humanistic Relations of Science;
Social Sciences;
Medical Sciences

Van Berkel, Klaas
University of Groningen
Dept of History
Oude Kyk In't Jat Straat 27
Groningen 9700 AS
The Netherlands
Phone: 31.5.0363.6003
Fax: 31.5.0363.7253
k.van.berkel@let.rug.nl
Doctorate, 1983
Europe, Astronomical Sciences;
Physical Sciences; Social
Sciences; Institutions

Van Brummelen, Glen
Bennington College
1 College Dr
Bennington, VT 05201-6004
gvanbrum@bennington.edu

Van Dalen, Benno
Institut für Geschichte der
Naturwissenschaften
Postfach 111932 (FB 13)
Frankfurt/Main 60054
Germany
Fax: 49.69.7982.3275
dalen@em.uni-frankfurt.de
Doctorate, 1993
400 B.C.E.-400 C.E., Asia,
Astronomical Sciences;
5c-13c, Asia, Astronomical
Sciences; 5c-13c, Asia,
Mathematics

van Dam, Laura
222 Berkeley St
Boston, MA 02116-3748
Phone: 1.617.492.2124
laura_vandam@hmco.com

van der Heijden, Petra
Leiden Observatory
PO Box 9513
Leiden 2300 RA
The Netherlands
Phone: 31.71.527.5818
Fax: 31.71.527.5819
heigden@strw.leidenuniv.nl

Van Dongen, Jeroen
Eikenweg 48
Amsterdam 1092 CA
The Netherlands
Phone: 31.2.0525.5922
jvdongen@wins.uva.nl

Van Heiningen, Teun W.
Diepenbrocklaan 11
Losser 7582 CX
The Netherlands
Phone: 31.5.3538.8088
Doctorate
19c, Europe, Medical Sciences;
19c, Europe, Biological Sciences;
19c, Europe, Earth Sciences

Van Helden, Albert
Inst for History and Foundations
of Science
PO Box 80000
Utrecht 3508TA
The Netherlands
Phone: 31.30.253.2841
Fax: 31.30.253.6313
a.vanhelden@phys.uu.nl
Doctorate, 1970

17c, Europe, Astronomical
Sciences; 14c-16c, Europe,
Earth Sciences; 18c, Europe,
Physical Sciences; Europe,
Physical Sciences

Van Helvoort, Ton
Van Oldenbarneveldstraat 29
Elsloo 6181 BC
The Netherlands
Phone: 31.46.437.0397
tvanhelvoort@compuserve.com
Doctorate
20c, Medical Sciences;
19c, Medical Sciences;
Biological Sciences;
Physical Sciences

Van Keuren, David K.
Naval Research Laboratory
Code 5204
Washington, DC 20375-0001
Phone: 1.202.767.4263
Fax: 1.202.404.8681
dvk@ccf.nrl.navy.mil
Doctorate, 1982
20c, North America, Institutions;
20c, North America, Earth
Sciences; 19c, Europe,
Social Sciences

van Lunteren, Frans H.
v.d. Mondestraat 72
Utrecht 3515 BK
The Netherlands
Phone: 31.30.271.8046
Fax: 31.30.253.6313
f.h.vanlunteren@phys.uu.nl

van Meer, Elisabeth
Vyzkumne Centrum pro Dejiny
Ve
Legerova 61
Praha 2 120 00
Czech Republic
vamm0020@tc.umn.edu

Van Nouhuys, Tabitta
76 A Lonsdale Road
Oxford OX27ER
Great Britain
Phone: 1865.276029
tabitta.vannouhuys@
  magd.ox.ac.uk
Doctorate, 1997
14c-16c, Europe, Astronomical
Sciences; 14c-16c, Europe,

Medical Sciences;
5c-13c, Europe, Astronomical
Sciences

Van Rappard, Hans JF
Vrye Universiteit
1111 de Boelelaan
Amsterdam 1081 HU
The Netherlands
Phone: 31.20.444.8795
Fax: 31.20.444.8832
jfh.van.rappard@psy.vu.nl
Doctorate, 1976
17c, Europe, Social Sciences;
18c, Philosophy and Philosophy
of Science; 19c, Cognitive
Sciences; 20c

Van Riper, A. Bowdoin
Southern Polytechnic State Uni
Dept of Social and Inter Studies
1100 South Marietta Parkway
Marietta, GA 30062-5864
Phone: 1.770.528.7481
Fax: 1.770.509.3788
abvr@mindspring.com
Doctorate, 1990
19c, Europe, Earth Sciences;
19c, Europe, Humanistic
Relations of Science;
20c, North America,
Humanistic Relations of Science;
19c, Europe, Social Sciences

van Ronk, Suzanne
2015 Brookhaven Blvd
Norman, OK 73072-3017
Phone: 1.405.364.2229
smvanronk@aol.com

Van Rossum, Marc J. P.
IMEC
Kapeldreef 75
Leuven 8-3001
Belgium
Phone: 32.16.28.1325
Fax: 32.16.281214
vrossum@imec.be

Van Tiggelen, Brigitte R.
Université Catholique Louvain
SC/Phys/FYMA
Chemin du Cyclotron 2
Louvain La Neuv 1348
Belgium
Phone: 32.10.473.286
Fax: 32.10.472.714

vantiggelen@muse.ucl.ac.be
Doctorate, 1998
Physical Sciences; Europe,
Social Relations of Science

Vanderburgh, William L.
Wichita State University
Department of Philosophy
Campus Box 74
1845 Fairmount St
Wichita, KS 67260-0001
Phone: 1.316.978.7882
william.vanderburgh@
  wichita.edu
Doctorate, 1999
Philosophy and Philosophy of
Science; Astronomical Sciences;
Physical Sciences

Vandermeer, Jitse M.
Redeemer University College
777 Garner Road East
Ancaster, ON L9K1J4
Canada
Phone: 1.905.648.2139
Fax: 1.905.648.2134
jmvdm@redeemer.on.ca
Doctorate
Europe, Biological Sciences;
20c, Europe, Instruments and
Techniques; Europe, Philosophy
and Philosophy of Science

Vanpaemel, Geert H W.
Vanden Bemptlaan 4
Leuven B 3001
Belgium
Phone: 3216234634
Fax: 32.16.324993
geert.vanpaemel@
  arts.kuleuven.ac.be

Vanwitwsen, Anthony
3385 19th St
Boulder, CO 80304-2302
Phone: 1.303.388.9100

Vargas, Milton
Themag Eng Ltd
Rua Bela Cintra, 986
Piso 15
São Paolo 01415-906
Brazil
Phone: 55.011.255.1510
Fax: 55.011.259.2348
themag@themag.com.br
Doctorate, 1952

20c, South America, Technology,
Instruments and Techniques;
20c, South America, Earth
Sciences; Philosophy and
Philosophy of Science

Vasile, Ronald S.
Canal Corridor Association
220 S. State St. Suite 1880
Chicago, IL 60604
Phone: 1.312.427.3688
Fax: 1.312.427.6251
Master's Level, 1997
19c, North America;
20c, Biography; Exploration,
Navigation and Expeditions;
Natural and Human History

Vasishth, Ashwani
818 W 7th St 12th Fl
Los Angeles, CA 90017-3407
Phone: 1.213.236.1908
vasishth@usc.edu

Vazquez, Marcos
PO Box 450241
Laredo, TX 78045-0005
Phone: 1.956.725.3138

Vehec, Michael J.
190 Hillington Drive
Paducah, KY 42001
mvehec@indiana.edu
Master's Level, 2001

Venner, Chris
300 Yoakum Pkwy Apt 514
Alexandria, VA 22304-4053

Ventrone, C. Patrick
122 E Pleasant St
Hillsboro, OH 45133-1424
Phone: 1.937.393.1080
Technology; Cognitive Sciences;
Philosophy and Philosophy of
Science

Verbrugge, Martha H.
Bucknell University
Dept of History
229 Marts Hall
Lewisburg, PA 17837
Phone: 1.570.577.3862
Fax: 1.570.577.3760
verbrgge@bucknell.edu
Doctorate, 1978
19c, North America, Gender and

Science; 20c, North America, Gender and Science; 19c, North America, Medical Sciences; 20c, North America, Medical Sciences

Vermij, Rienk H.
Roserije 67 c
Maastricht 6228 DH
The Netherlands
Phone: 31.30.251.8773
r.vermij@history.unimaas.nl
Doctorate, 1991
14c-16c, Europe, Earth Sciences; 17c, Europe, Astronomical Sciences; 18c, Europe, Social Relations of Science

Verplancke, Marnix
St Denijslaan 103
Gent 9000
Belgium
Phone: 32.9245.0380
Fax: 32.9245.7984

Vertesi, Janet
2350 West 37th Avenue
Vancouver, BC V6M1P3
Canada
Phone: 1.604.266.4240
Fax: 1.604.266.9970
cyberlyra@yahoo.com

Vessuri, Hebe
Aptdo 47328
Caracas 1041-A
Venezuela
Phone: 58.212.992.6195
Fax: 58.212.504.1092
hvessuri@supercable.net.ve

Vetter, Jeremy
University of Pennsyvania
Logan Hall, Suite 303
249 S 36th St
Philadelphia, PA 19104-6304
Phone: 1.215.898.4643
jvetter@sas.upenn.edu
Master's Level, 2003
North America; Social Relations of Science; Exploration, Navigation and Expeditions; Technology

Vianelli, Alberto
University of Insubria
Dept. of Structural and

Functional Biology
via J.H.Dunant 3
Varese VA 21100
Italy
Phone: 39.03.3242.1408
Fax: 39.03.3242.1300
alberto.vianelli@uninsubria.it
Doctorate, 1988
20c, Biological Sciences

Vicedo, Marga
25 Irving Ter Apt 7
Cambridge, MA 02138-3019
Phone: 1.617.491.2179
vicedo@jas.harvard.edu
Doctorate, 1987
20c, North America, Biological Sciences; Philosophy and Philosophy of Science

Vicien, Pedro
Academia Nacional de Ciencias
Valvear 1711 3rd Piso
Buenos Aires
Argentina
Phone: 54.1.4433.0824
Fax: 54.1.433.0824
postmanster@vicien.cyt.edu.ar
Master's Level, 1945
Physical Sciences; Technology; Humanistic Relations of Science

Vickers, Brian
Centre for Renaissance Studies
ETH-Zentrum
Zurich CH-8092
Switzerland
vickers@english.gess.ethz.ch

Vila, Anne C.
University of Wisconsin
French & Italian Dept
1220 Linden Dr
Madison, WI 53706-1525
Phone: 1.608.262.5075
Fax: 1.608.265.3892
acvila@facstaff.wisc.edu

Villela-Gonzalez, Alicia
Calle Insurg.Sur, 3493
Edificio 11-803
14020 Tlalpan
Mexico DF
Mexico
Phone: 5.528.1639
ejeh@hp.fciencias.unam.mx

Vincent, Clare
Metropolitan Museum of Art
1000 5th Ave
New York, NY 10028-0198
Phone: 1.212.879.5500
Fax: 1.212.650.2957
Master's Level, 1963
14c-16c, Europe, Instruments and Techniques; 14c-16c, Europe, Technology

Viney, Wayne
Colorado State University
Dept of Psychology
Fort Collins, CO 80523-0001
Phone: 1.970.491.5783
vineyw@lamar.colostate.edu

Viterbo, Paula
George Washington University
Cntr for History of Recent Sci
Department of History
801 22nd St NW
Washington, DC 20052
paulaviterbo@yahoo.com
Doctorate, 1999
20c, North America, Medical Sciences; 20c, Europe, Medical Sciences; 20c, North America, Social Relations of Science; 20c, North America, Gender and Science

Vlahakis, George X.
Nat. Hellenic Res. Foundat.
Vasileos Kanstantinou 48
Athens 11635
Greece
Phone: 30.1.727.3559
gvlahakis@eie.gr
Doctorate
18c, Europe, Physical Sciences; Earth Sciences; 19c, Europe, Instruments and Techniques

Voelkel, James R.
Dibner Institute
MIT E56-100
38 Memorial Drive
Cambridge, MA 02139
Phone: 1.617.258.8033
Fax: 1.617.258.7483
jvoelkel@dibinst.mit.edu
Doctorate, 1994
Astronomical Sciences; 17c; 14c-16c

Vogel, Amber
University of North Carolina
Dept of Biology
CB#3280 - Coker Hall
Chapel Hill, NC 27599-3280

Vogel, Brant
135 N 6th St # 3R
Brooklyn, NY 11211-3201
Phone: 1.718.388.0193
brant@inch.com
Master's Level, 1994
14c-16c, Europe, Earth Sciences;
17c, Europe, Earth Sciences;
14c-16c, Europe, Humanistic
Relations of Science

Voigts, Linda E.
Univ of Missouri
English Department
Kansas City, MO 64110-2499
Phone: 1.816.235.2764
Fax: 1.816.235.1308
voigts@umkc.edu
Doctorate, 1973
5c-13c, Europe, Physical
Sciences; 5c-13c, Europe,
Astronomical Sciences;
5c-13c, Europe, Medical Sciences

Volodarski, Alexander I.
Institute of Hist of Science
Staropanskii per. 1/5
Moscow 103012
Russia
Fax: 7.095.925.9911
Doctorate, 1967
5c-13c, Asia, Mathematics;
20c, Europe, Mathematics

Von Mayrhauser, Richard T.
2616 Hillside Dr
Burlingame, CA 94010-5660
Phone: 1.724.444.5443
rtm@webpathway.com
Doctorate, 1986
20c, North America, Cognitive
Sciences; 20c, North America,
Social Sciences; 19c, North
America, Philosophy and
Philosophy of Science;
19c, North America, Technology

Von Meyenn, Karl
Gietlhausen 4A
Neuburg a.D D-86633
Germany

Phone: 49.8431.60523
Doctorate
19c, Europe, Physical Sciences;
20c, Europe, Physical Sciences;
Philosophy and Philosophy of
Science; Astronomical Sciences

von Randow, Gero
Frankfurter Allgemeine Zeitung
Weisenstasse 23
Hamurg 20255
Germany
Phone: 49.69.7591.1113
gerovonrandow@aol.com

Von Staden, Heinrich
Institute for Advanced Study
School of Historical Studies
Einstein Dr
Princeton, NJ 08540-0631
Phone: 1.609.734.8306
Fax: 1.609.951.4462
hvs@ias.edu
Doctorate
400 B.C.E.-400 C.E.,
Europe, Medical Sciences;
20c, Philosophy and Philosophy
of Science; Pre-400 B.C.E.,
Europe, Medical Sciences;
20c, Social Sciences

Voss, Diane
1310 Somerset Ave
Grosse Pt Pk, MI 48230-1031
Phone: 1.313.664.7638
Fax: 1.313.884.0727
dvoss@ccscad.edu

Vucinich, Alexander
1409 Oxford St Apt 1
Berkeley, CA 94709-1457
Phone: 1.310.848.7577

Wachelder, Joseph C.
Universiteit Maastricht
Kapoenstraat 2
Postbox 616
Maastricht 6200
The Netherlands
Phone: 31.43.388.3325
Fax: 31.43.325.9311
jo.wachelder@
   history.unimaas.nl
Doctorate
19c, Europe, Physical Sciences;
19c, Europe, Medical Sciences;

19c, Europe, Philosophy
and Philosophy of Science;
19c, Europe, Education

Waddell, Mark
3135 Abell Ave # 2
Baltimore, MD 21218-3412
Phone: 1.410.889.7278
Bachelor's Level, 2000
17c, Europe, Philosophy
and Philosophy of Science;
17c, Europe, Physical Sciences;
14c-16c, Europe, Astronomical
Sciences

Wadyko, Mike A.
128 County Rd 160
Glenwood Spring, CO
81601-9513
Phone: 1.970.285.7399
Fax: 1.970.625.0649
mjwadykog@juno.com

Waff, Craig B.
Encyclopedia Americana
Grolier Educational
6 Park Lawn Dr
Bethel, CT 06801-1042
Phone: 1.203.797.3878
Fax: 1.203.797.3428
cwaff@grolier.com
Doctorate
18c, Europe, Astronomical
Sciences; 19c, North America,
Astronomical Sciences;
Humanistic Relations of Science;
18c, Mathematics

Waggoner, Margaret A.
PO Box 306
Haydenville, MA 01039-0306
Doctorate, 1950
Physical Sciences

Wagner, Stephen C.
University of Oklahoma
Department of Philosophy
455 W Lindsey St Rm 605
Norman, OK 73019-2006
Phone: 1.405.325.6324
Fax: 1.405.325.2660
swagner@ou.edu
Master's Level, 1990
17c, Philosophy and Philosophy
of Science; Astronomical

Sciences; 400 B.C.E.-400 C.E.;
Philosophy and Philosophy of
Science

Wahrig, Bettina R.
Abteilung Geschichte der
Naturwissenschaften
Tu Baunschweig Pockelsstr. 14
Braunschweig D-38023
Germany
Phone: 49.531.391.5997
Fax: 49.531.391.5999
b.wahrig@tu-bs.de
Doctorate
18c, Europe, Medical Sciences;
19c, Europe, Medical Sciences;
18c, Europe, Social
Relations of Science;
18c, Europe, Philosophy and
Philosophy of Science

Wakefield, Andre
University of Chicago
Fishbein Center SS-205
1126 East 59th Street
Chicago, IL 60637
Phone: 1.617.258.0514
Master's Level, 1999
18c, Europe, Institutions;
18c, Europe, Physical Sciences;
18c, Europe, Technology;
18c, Europe, Social Sciences

Wald, Stephen E.
Univ of Wisconsin-Madison
7143 Social Science Bldg
1180 Observatory Dr.
Madison, WI 53706-1393
sewald@students.wisc.edu
Master's Level
19c, North America,
Humanistic Relations of Science;
20c, North America,
Humanistic Relations of Science;
17c, Europe, Humanistic
Relations of Science;
17c, Europe, Philosophy and
Philosophy of Science

Walker, Mark W.
Union College
Department of History
Schenectady, NY 12308-3163
Phone: 1.518.388.6994
Fax: 1.518.388.6422
walkerm@union.edu

Doctorate
Biological Sciences; Natural and
Human History; Europe; 20c

Wall, Byron E.
York University
Div of Natural Science
CCB Room 126
4700 Keele St
Toronto, ON M3J1P3
Canada
Phone: 1.416.467.8685
Fax: 1.416.352.5368
bwall@yorku.ca

Wallace, William A.
University of Maryland
Department of Philosophy
College Park, MD 20742-7615
Phone: 1.301.405.5711
Fax: 1.301.405.5690
wallacew@wam.umd.edu
Doctorate, 1959
14c-16c, Europe, Physical
Sciences; Philosophy and
Philosophy of Science;
5c-13c, Europe, Physical
Sciences; 400 B.C.E.-400 C.E.,
Philosophy and Philosophy of
Science

Wallis, Faith
McGill University
Dept of Social Studies of Med
3674 Peel St
Montreal, PQ H3A1X1
Canada
Phone: 1.514.398.5276
Fax: 1.514.398.1498
wallis@leacock.lan.mcgill.ca

Walloch, Karen
3160 Thorp St
Madison, WI 53714-2265
kwalloch@students.wisc.edu

Walls, Laura Dassow D.
Lafayette College
Easton, PA 18042-1781
Phone: 1.610.330.5450
Fax: 1.610.330.5606
wallsl@mail.lafayette.edu
Doctorate, 1992
19c, North America;
19c, Natural and Human History;

Humanistic Relations of Science;
Philosophy and Philosophy of
Science

Walter, Maila L.
30522 La Vue
Laguna Niguel, CA 92677-5534
Phone: 1.949.495.7778
mailawalter@home.com
Doctorate
Humanistic Relations of Science;
Social Relations of Science;
Philosophy and Philosophy of
Science

Walters, Alice N.
University of Massachusetts
Dept of History
850 Broadway St Ste 3
Lowell, MA 01854-3006
Phone: 1.978.934.4263
Fax: 1.978.934.3023
alice_walters@uml.edu
Doctorate
18c, Europe, Astronomical
Sciences; 18c, Europe,
Instruments and Techniques;
18c, Europe, Physical Sciences;
18c, Europe, Social Relations of
Science

Walton, David
University of Notre Dame
Dept of History & Philosophy
of Science
O'Shaughnessy Hall 346
Notre Dame, IN 46556
Phone: 1.219.289.1684
david.walton.10@nd.edu

Wang, Hsiu-Yun
History of Science
7143 Social Science
1180 Observatory Drive
Madison, WI 53706
hwang6@students.wisc.edu
Master's Level
19c, Asia, Medical Sciences;
20c, Asia, Medical Sciences;
19c, Asia, Gender and Science

Wang, Jessica
UCLA
Dept of History
PO Box 951473
Los Angeles, CA 90095-1473

Phone: 1.310.825.4601
Fax: 1.310.206.9630
jwang@history.ucla.edu

Wang, Shunyi
East China Normal University
Inst for the Philosophy of Sci,
Tech & History of Science
Shanghai
China

Wang, Zuoyue
California State
Polytechnic University
Dept. of History
3801 W Temple Ave
Pomona, CA 91768-2557
Phone: 1.909.869.3872
Fax: 1.909.869.4724
zywang@csupomona.edu

Ward, Gerald A.
338 Central St
Auburndale, MA 02466-2204
Phone: 1.617.969.6576
ward@shore.net

Warner, Deborah J.
Smithsonian Institution
Nat Museum of American
History
Washington, DC 20560-0636
Phone: 1.202.357.2482
warnerd@nmah.si.edu

Warnow-Blewett, Joan
1 Carolina Mdws Apt 104
Chapel Hill, NC 27514-8508
Phone: 1.919.942.9039
Fax: 1.301.209.0882
jblewett@aip.org

Watanabe, Yoshiaki
Tairadate-mura
Imadzu Sainokami
Higashitsugaru-gun
Aomori 030-1413
Japan
Phone: 81.42.342.0353
yowata@sepia.ocn.ne.jp
18c, Europe, Physical Sciences;
19c, North America, Physical
Sciences; 17c, Europe, Medical
Sciences; 14c-16c, Europe,
Philosophy and Philosophy of
Science

Watkins, Sallie A.
1081 S Lynx Dr
Pueblo, CO 81007-5032
salliewatkins@juno.com
Doctorate, 1958
20c, Europe, Physical Sciences

Watson, Neale W.
Science History Publications
PO Box 1390
Nantucket, MA 02554-1390
Phone: 1.508.228.5490
Fax: 1.508.228.7541
nww@shpusa.com
18c, North America, Medical
Sciences; 14c-16c, Europe,
Astronomical Sciences;
20c, Physical Sciences

Wear, A.
Wellcome Trust Centre for the
History of Medicine at UCL
Euston Road
24 Eversholt St
London NW1 1AD
Great Britain

Weart, Spencer R.
American Institute of Physics
Center for History of Physics
One Physics Ellipse
College Park, MD 20740-3843
Phone: 1.301.209.3174
Fax: 1.301.209.0882
sweart@aip.org
Doctorate, 1968
20c, Physical Sciences;
20c, Earth Sciences;
20c, Astronomical Sciences;
20c, Social Relations of Science

Weaver, C. S.
13931 Esworthy Rd
Germantown, MD 20874-3313
Phone: 1.301.354.4250
cweaver@alum.mit.edu

Webb, George E.
Tennessee Tech University
Dept History
Box 5064
North Dixie Ave
Cookeville, TN 38505-0001
Phone: 1.931.372.3335
gwebb@tntech.edu
Doctorate, 1978
20c, North America; 19c, North

America; 20c, North America,
Astronomical Sciences;
20c, North America, Biological
Sciences

Weber, Bruce H.
Calif State University
Dept of Chem & Biochem
800 N State College Blvd
Fullerton, CA 92834-6866
Phone: 1.714.278.3885
Fax: 1.714.278.5316
bhweber@fullerton.edu
Doctorate
Biological Sciences;
20c, Instruments and Techniques;
19c, Europe, Biological Sciences,
Biography; Philosophy and
Philosophy of Science

Weber, Jeff
Jeff Weber Rare Books
2731 Lompoc St
Los Angeles, CA 90065-5107
Phone: 1.323.344.9332
weberbks@pacbell.net
Master's Level
Education; Medical Sciences;
Astronomical Sciences;
19c, Europe, Biological
Sciences, Biography

Weber, Thomas
Lund University
Dept of Animal Ecology
Ecology Bldg
Lund S22362
Sweden
Phone: 46.46.222.3789
Fax: 46.46.222.4716
thomas.weber@zooekol.lu.se

Webster, Eleanor R.
Wellesley College
105 Central Street
Wellesley, MA 02481
ewebster@wellesley.edu
Doctorate, 1952
Physical Sciences;
20c, Europe, Physical Sciences;
20c, Europe, Biological Sciences;
20c, Europe, Education

Webster, Marjorie K.
Adler Planetarium
1300 S Lake Shore Dr
Chicago, IL 60605

Phone: 1.312.322.0594
Fax: 1.312.322.2257
Bachelor's Level, 1938
5c-13c, Europe, Instruments and
Techniques; 14c-16c, Europe,
Astronomical Sciences;
17c, Europe, Institutions

Weidenbacher, Richard L., Jr.
964 Coolidge Rd
Elizabeth, NJ 07208-1046
Phone: 1.908.965.1929
Doctorate, 1956
Philosophy and Philosophy
of Science; Medical Sciences,
Instruments and Techniques

Weidman, Nadine M.
Harvard University
Science Center 235
Cambridge, MA 02138
Phone: 1.617.495.3741
Fax: 1.617.495.3344
weidman@fas.harvard.edu
Doctorate
20c, North America, Cognitive
Sciences; 20c, North America,
Biological Sciences; 20c, North
America, Social Sciences

Weikart, Richard C.
California State University
Dept of History
Stanislaus
Turlock, CA 95382
Phone: 1.209.667.3238
rweikart@toto.csustan.edu
Doctorate, 1994
19c, Europe, Biological Sciences;
20c, Europe, Biological Sciences;
19c, Europe, Philosophy
and Philosophy of Science;
20c, Europe, Philosophy and
Philosophy of Science

Weinberg, Steven
Univ of Texas
Dept of Physics/Theory Group
Corner 26th & Speedway
RLM 5.208
Austin, TX 78712-1081
Phone: 1.512.471.4394
Fax: 1.512.471.4888
weinberg@physics.utexas.edu
Physical Sciences;
Astronomical Sciences

Weiner, Charles
Mass Institute of Technology
Bldg E51-296A
77 Mass Ave.
Cambridge, MA 02139
Phone: 1.617.253.4063
Fax: 1.617.258.8118
cweiner@mit.edu
Doctorate, 1965
20c, North America;
19c, Europe, Biological Sciences;
Asia, Physical Sciences;
Science Policy

Weiner, Dora B.
UCLA
Medical Humanities 12-138
CHS
10833 Le Conte Avenue
Los Angeles, CA 90095
Phone: 1.310.825.0599
Fax: 1.818.783.9174
dbweiner@ucla.edu
Doctorate
18c, Europe, Medical Sciences,
Instruments and Techniques;
19c, Europe, Medical Sciences;
14c-16c, Humanistic Relations
of Science; 5c-13c

Weininger, Stephen J.
Worcester Polytech Institute
Dept of Chemistry/Biochemistry
Worcester, MA 01609-2280
Phone: 1.508.831.5396
Fax: 1.508.831.5933
stevejw@wpi.edu
Doctorate
19c, Physical Sciences;
20c, Physical Sciences;
Humanistic Relations of Science;
Physical Sciences

Weinstein, Deborah
111 Longwood Ave # 3
Brookline, MA 02446-6625
Fax: 1.617.495.3344

Weinstock, Robert
Oberlin College
Department of Physics
North Professor Street
Oberlin, OH 44074
Phone: 1.440.774.4781
Fax: 1.440.775.9820
zweinsto@oberlin.net

Doctorate, 1943
17c, Europe, Physical Sciences;
18c, Europe, Mathematics

Weintraub, E. Roy
Duke University
Economics Department
DPC 90097
Durham, NC 27708-0097
Phone: 1.919.660.1838
Fax: 1.919.684.8974
erw@duke.edu
Doctorate, 1969
20c, Social Sciences

Weisel, Gary J.
Penn State University
107 Science Building
3000 Ivyside Park
Altoona, PA 16601-3760
Phone: 1.814.949.5175
Fax: 1.814.949.5011
gxw20@psu.edu
Doctorate, 1992
20c, Physical Sciences;
20c, Science Policy;
19c, Physical Sciences

Weiss, Eric A.
PO Box 537
Kailua, HI 96734-0537
Phone: 1.808.263.0630
Fax: 1.808.263.0630
eaweiss@poi.net

Weiss, Steven C.
104 W. Marshall St.
Falls Church, VA 22046
Phone: 1.703.241.2655
scweiss@attglobal.net
Doctorate, 1994
20c, North America, Science
Policy; 20c, Africa, Physical
Sciences; 20c, North America;
20c, Social Relations of Science

Wellman, Kathleen
Southern Methodist University
Dept of History
Dallas, TX 75275-0202
Phone: 1.214.768.2970
Fax: 1.214.768.2404
kwellman@mail.smu.edu
Doctorate, 1983
18c, Europe; 17c, Europe;
18c, Medical Sciences

Wellmann, Janina
Max Planck Institute for the
History of Science
Wilhelmstrasse 44
Berlin 10117
Germany
Phone: 49.30.2267.244
Fax: 49.30.2267.299
wellmann@
 mpiwg-berlin.mpg.de

Wells, Kentwood D.
Univ of Connecticut
Dept of Ecology & Evol Bio
U3043 - 75 N Eagleville Rd
Storrs, CT 06269-3043
Phone: 1.860.486.4454
Fax: 1.860.486.6364
kentus@uconnvm.uconn.edu
Doctorate, 1976
19c, Biological Sciences;
19c, Natural and Human History;
19c, Europe, Biological Sciences,
Biography

Welther, Barbara L.
Center for Astrophysics
60 Garden Street, MS-9
Cambridge, MA 02138
Phone: 1.617.495.7217
Fax: 1.617.496.7564
bwelther@cfa.harvard.edu
Master's Level, 1976
20c, North America,
Astronomical Sciences;
19c, North America, Gender and
Science; 19c, North America,
Education; 20c, North America,
Social Sciences

Wendroff, Arnold P.
544 8th St
Brooklyn, NY 11215-4201

Werdinger, Jeffrey
310 E 12th St Apt 2F
New York, NY 10003-7206

Werrett, Simon
University of Washington
History Department
Box 353560
Seattle, WA 98195-3560
Phone: 1.206.543.5790
Fax: 1.206.543.9451
srew2@hotmail.com

Westfall, Catherine L.
Argonne National Labs
Bldg 208 9700 S Cass Ave
Argonne, IL 60439-4842
cwestfall@nscl.msu.edu
Doctorate
Physical Sciences

Westman, Robert S.
Univ. of California/ San Diego
Department of History
9500 Gilman Drive
La Jolla, CA 92093-0104
Phone: 1.858.534.1996
Fax: 1.619.450.1467
rwestman@helix.ucsd.edu
Doctorate, 1971
14c-16c, Europe, Astronomical
Sciences; 14c-16c, Europe,
Social Relations of Science;
14c-16c, Europe, Social Sciences

Westwick, Peter J.
Caltech
Hss 228-77
Pasadena, CA 91125-0001
Phone: 1.626.395.4096
Fax: 1.626.793.4681
westwick@hss.caltech.edu
Doctorate, 1999
Physical Sciences;
North America;
20c, North America, Institutions

Wetmore, Karin E.
Harvard University
History of Science
Science Center 235
Cambridge, MA 02138
Fax: 1.617.731.0675
wetmore@fas.harvard.edu
Doctorate, 1991
Cognitive Sciences; Education;
Philosophy and Philosophy of
Science

Wetzell, Richard F.
German Historical Institute
1607 New Hampshire Ave NW
Washington, DC 20009-2562
Phone: 1.202.387.3355
Fax: 1.202.483.3430
r.wetzell@ghi-dc.org

Whalen, Kathleen
2407 Elendil Ln
Davis, CA 95616-3043

bwhal01@students.bbk.ac.uk
Master's Level, 1999
17c, Europe, Humanistic
Relations of Science;
17c, Europe, Biological Sciences

Wheaton, Bruce R.
Tapsha
1136 Portland Ave
Albany, CA 94706-1624
Phone: 1.510.524.3216
Fax: 1.510.524.3216
Doctorate
17c, North America,
Physical Sciences;
18c, Europe, Technology;
19c, North America,
Physical Sciences;
20c, Europe, Technology

White, Matthew A.
240 12th Pl, NE # 4
Washington, DC 20002
Phone: 1.202.786.2307
Fax: 1.202.357.3328
mattadolphus@earthlink.net

White, Paul S.
Univ of Cambridge
Univ Library
West Road
Cambridge CB3 9DR
Great Britain
Phone: 44.12.2333.3008
psw24@cam.ac.uk
Doctorate
Natural and Human History;
Biological Sciences;
19c, Europe, Biological
Sciences, Biography

Whitesell, Patricia S.
University of Michigan
Detroit Observatory
1398 E Ann St
Ann Arbor, MI 48109-2051
Phone: 1.734.763.2230
Fax: 1.734.936.4111
whitesel@umich.edu

Whitesides, John G.
222 West Ortega Apt. C
Santa Barbara, CA 93101
Phone: 1.303.963.8618
jgw@umail.ucsb.edu

Whittemore, Gilbert F.
Stalter & Kennedy, LLP
54 Canal Street, Suite 300
Boston, MA 02114
Phone: 1.617.523.8080
Fax: 1.617.441.0434
gilwhittem@aol.com
Doctorate, 1986
*20c, North America, Social
Relations of Science; 20c, North
America, Medical Sciences;
20c, North America, Technology;
20c, North America, Biological
Sciences*

Whitten, Maurice M.
11 Lincoln St
Gorham, ME 04038-1703

Wiesenfeldt, Gerhard
Zwaezengasse 6
Jena 07743
Germany
Phone: 49.3.6412.27398
Fax: 49.3.6419.4950
gwiese@mpiwg-berlin.mpg.de

Wigelsworth, Jeffrey R.
1223 13th Street East
Saskatoon, SK S7H0C3
Canada
jeffwigelsworth@hotmail.com

Wiinikka, Peter
9235 SW Brooks Bend Pl
Portland, OR 97223-7139

Wilcox, Judith
6 Margetts Rd
Monsey, NY 10952-5018
Fax: 1.845.352.1651
wilcoxfriedlander@earthlink.net

Wilders, Richard J.
North Central College
30 N. Brainard Street
PO Box 3063
Naperville, IL 60566
Phone: 1.630.637.5234
Fax: 1.630.637.5360
rjw@noctrl.edu
Doctorate
*Mathematics; Philosophy and
Philosophy of Science*

Wilkinson, Katie
276 Grandview Ave
Meadville, PA 16335-1416
Phone: 1.814.337.2564
kaw16@pitt.edu

Williams, Charles G.
415 Glen Echo Cir
Columbus, OH 43202-2419

Williams, Elizabeth A.
Oklahoma State University
Life Sciences West Rm 501
Stillwater, OK 74078-0001
Phone: 1.405.744.5680
Fax: 1.405.744.5400
williea@okstate.edu
Doctorate, 1983
*Europe, Medical Sciences;
Europe, Social
Relations of Science;
Europe, Social Sciences; Gender
and Science*

Williams, L. Pearce
207 Iradell Rd
Ithaca, NY 14850-9284
Phone: 1.607.273.9035

Williams, Mary Lou M.
Wayne State University
627 W. Alexandrine
Detroit, MI 48201
ac5964@wayne.edu
Doctorate
*Medical Sciences; Biological
Sciences; Gender and Science;
Social Sciences*

Williams, Roger L.
1701 S 17th St
Laramie, WY 82070-5406
Phone: 1.307.742.6543
Doctorate, 1951
*18c, Europe, Biological Sciences;
19c, North America, Biological
Sciences*

Williams, Thomas R.
Rice University
6100 S Main St
Houston, TX 77005-1892
trw@rice.edu
Master's Level, 1999
*19c, North America,
Astronomical Sciences;
20c, North America,*

*Astronomical Sciences;
20c, Astronomical Sciences;
Institutions*

Williford, William O.
9 Ranch Ct
Newark, DE 19711-3707

Wilsher, Kenneth
1085 Emerson St
Palo Alto, CA 94301-2417
Phone: 1.650.327.8754
ken@san-jose.tt.slb.com

Wilson, Andrew D.
11 Crescent St
Keene, NH 03431-3556
awilson@keene.edu

Wilson, Catherine W.
University of British Columbia
Dept of Philosophy
Buchanan E
Vancouver, BC V6T1Z1
Canada
Phone: 1.604.822.2520
Fax: 1.604.822.8782
catherine.wilson@ubc.ca
Doctorate, 1977
*17c, Philosophy and Philosophy
of Science; Biological Sciences,
Instruments and Techniques;
Philosophy and Philosophy of
Science; 18c, Philosophy and
Philosophy of Science*

Wilson, Curtis A.
St Johns College
PO Box 2800
Annapolis, MD 21404-2800
c.wilson@sjca.edu

Wilson, David B.
Iowa State University
Department of History
603 Ross Hall
Ames, IA 50011-1202
Phone: 1.515.294.5467
Fax: 1.515.294.6390
davidw@iastate.edu
Doctorate, 1968
*18c, Europe, Physical Sciences;
18c, Europe, Education;
Social Sciences*

Wilson, Leonard G.
History of Medicine
420 Delaware St SE
Box 506 Mayo
Minneapolis, MN 55455
Phone: 1.612.624.4416
wilso004@maroon.tc.umn.edu
Doctorate
17c, Asia, Medical Sciences;
19c, Europe, Biological Sciences;
19c, North America, Medical
Sciences

Wilson, Philip K.
Penn State College of Medicine
Milton S Hershey Medical Ctr
Dept of Humanities, H134
500 University Drive
Hershey, PA 17033-2390
Phone: 1.717.531.8779
Fax: 1.717.53.3894
pwilson@psu.edu
Doctorate, 1992
18c, Europe, Medical Sciences;
20c, North America,
Biological Sciences;
18c, Europe, Biological Sciences;
19c, North America, Earth
Sciences

Wilson, Robert A.
Pfizer Inc
235 E 42nd St
New York, NY 10017-5755

Winbauer, Mary
2100 Bryant Ave S Apt 301
Minneapolis, MN 55405-2832
Phone: 1.612.879.9727

Winnik, Herbert C.
St. Mary's College of Maryland
History Deptartment
St. Mary's City, MD 20686
Fax: 1.301.862.0450
hcwinnik@erols.com
Doctorate, 1968
20c, North America;
Biological Sciences;
Social Relations of Science

Winsor, Mary P.
University of Toronto
Victoria College
Inst Hist & Phil of Sci & Tech
73 Queen's Park Crescent E.
Toronto, ON M5S1K7

Canada
Phone: 1.416.978.6280
Fax: 1.416.978.3003
mwinsor@chass.utoronto.ca
Doctorate, 1971
Biological Sciences; Natural and
Human History; Social Sciences

Wise, M. Norton
UCLA
Dept of History
6265 Bunche Hall
Box 951473
Los Angeles, CA 90095
Phone: 1.310.825.4764
Fax: 1.310.206.9630
nortonw@history.ucla.edu
Doctorate, 1977
Europe, Physical Sciences;
Europe, Humanistic Relations
of Science; Europe, Social
Relations of Science; Philosophy
and Philosophy of Science

Wittig, Gertraude C.
3 Biscayne Dr
Edwardsville, IL 62025-2451
Phone: 1.618.656.9309
Doctorate
20c, Europe, Biological Sciences;
19c, Europe, Gender and
Science

Wojciuk, Erika
71 8th Ave Apt 2A
Brooklyn, NY 11217-3923
Phone: 1.718.398.3286

Wolf, Jacquelin H.
Ohio University
Dept of Social Medicine
College of Osteopathic Med
308 Grosvenor
Athens, OH 45701
Phone: 1.740.597.2777
Fax: 1.740.593.1730
wolfj1@ohio.edu
Doctorate, 1998
20c, Medical Sciences;
19c, Medical Sciences

Wolfe, Audra
University of Pennsylvania
Logan Hall Suite 303
249 S. 36th St.

Philadelphia, PA 19104-6304
Phone: 1.215.898.4643
awolfe@sas.upenn.edu

Wolff, Stefan
Bert-Brecht-Allee 7
München 81737
Germany
Phone: 49896706278
s.wolff@lrz.uni-muenchen.de

Wolfram, Stephen
Wolfram Research
100 Trade Centre Dr
Champaign, IL 61820-7237
wri-library@wolfram.com

Wolfson, Paul R.
West Chester University
Dept of Mathematics
West Chester, PA 19383
Phone: 1.610.436.1081
Fax: 1.610.436.2419
pwolfson@wcupa.edu
Doctorate, 1973
19c, Mathematics; Mathematics

Wood, Becky
Indiana University
History & Philosophy of Sci
Goodbody 130
1011 E. 3rd St
Bloomington, IN 47405-7005
Phone: 1.812.855.9334
beckyw@indiana.edu
Doctorate, 1993
18c, Europe, Education;
18c, Exploration, Navigation
and Expeditions; 18c, Europe,
Physical Sciences; Exploration,
Navigation and Expeditions

Wood, Paul B.
University of Victoria
Department of History
PO Box 3045
Victoria, BC V8W3P4
Canada
Phone: 1.250.721.7289
Fax: 1.250.721.8772
pbwood@uvic.ca

Woods, Emilie
25765 W Oaklane Rd
Ingleside, IL 60041
Phone: 1.847.587.8698
mavao@aol.com

Woodward, James F.
California State University
Dept of History
Fullerton, CA 92634
Phone: 1.714.278.3167
Fax: 1.714.278.2101
jwoodward@fullerton.edu
Doctorate, 1972
20c, Physical Sciences;
20c, Astronomical Sciences,
Philosophy and Philosophy of
Science

Woody, Andrea I.
University of Washington
345 Savery Hall
Box 353350
Seattle, WA 98195-3350
Phone: 1.206.685.2663
Fax: 1.206.685.8740
awoody@u.washington.edu
Doctorate, 1997
Philosophy and Philosophy of
Science; Physical Sciences;
Gender and Science

Wooley, John
107-180
4130 La Jolla Village Dr
La Jolla, CA 92037-9121

Woolf, Harry
Inst for Advanced Study
Einstein Drive
Princeton, NJ 08540

Woolf, Shirley L.
13320 Highway 99 Unit 6
Everett, WA 98204-5445
Phone: 1.425.258.3351
slwoodlf2@earthlink.net

Word, Thomas
123 Spruce Ave
Gearhart, OR 97138-4243
Phone: 1.503.738.3338
tomword@teleport.com

Wourms, John P.
Clemson University
Dept of Biological Sciences
132 Long Hall
Clemson, SC 29634-0326
Phone: 1.864.656.3598
Fax: 1.864.656.0435
wjohn@clemson.edu

Doctorate, 1966
Biological Sciences; Natural and
Human History; Earth Sciences

Wouters, Paul
NIWI-KNAW
PO Box 95110
Amsterdam 1090 HC
The Netherlands
Phone: 312.046.28654
Fax: 312.066.58013
paul.wouters@niwi.knaw.nl

Wrakberg, Urban
Royal Swedish Acad of Science
Ctr for the History of Science
Box 50005
Stockholm S-10405
Sweden
Phone: 46.8.673.9613
urban@kva.se

Wright, Jay
PO Box 381
Bradford, VT 05033-0381
Phone: 1.802.222.5286
Master's Level, 1967
Social Sciences;
Natural and Human History;
Astronomical Sciences;
Physical Sciences

Wright, Susan P.
University of Michigan
Residential Coll E Quad
Ann Arbor, MI 48109
Phone: 1.734.665.4615
Fax: 1.734.763.7712
spwright@umich.edu

Wroth, Celestina
1913 Sussex Dr
Bloomington, IN 47401
Phone: 1.812.332.1481
cewroth@indiana.edu

Wuertenberg, Jens J.
Daiserstrasse 58
München D-81371
Germany
Phone: 49.89.725.1720
jw.mue@t-online.de
Doctorate, 1988
19c, Europe, Biological Sciences;
18c, Europe, Natural and

Human History; 19c, Europe,
Instruments and Techniques;
19c, Europe, Physical Sciences

Wukovitz, Stephen G.
269 Bloom St
Bloomsburg, PA 17815-8352

Xu, Yibao
144-39, 2D Sanford Ave
Flushing, NY 11355
Phone: 1.718.939.1227
yxu2@gc.cuny.edc

Yadlowsky, Jerry
308 N 13th Ave
Manville, NJ 08835-1110
Phone: 1.908.526.0238

Yagi, Eri
Toyo University
Fac. Eng.
2100 Kujirai
Kawagoe 350 8585
Japan
Phone: 81.4.9239.1484
Fax: 81.4.9231.9807
eri@eng.toyo.ac.jp
Doctorate
19c, Europe, Physical Sciences;
20c, Asia, Physical Sciences;
20c, Gender and Science;
20c, Institutions

Yamada, Toshihiro
4-4-2-908, Takasu, Mihama-ku
Chiba 261-0004
Japan
Phone: 81.43.279.7094
Fax: 81.43.279.7094
tosmak-yamada@
  muf.biglobe.ne.jp

Yano, Michio
Kyoto Sangyo University
Dept of Cultural Studies
Kamigamo-motoyama Kita-ku
Kyoto 6038555
Japan
Phone: 81.75.705.1781
Fax: 81.75.705.1799
yanom@cc.kyoto-su.ac.jp
Doctorate, 1996
Asia, Astronomical Sciences;
Asia, Mathematics

Yarovoi, Serge V.
WFBR/UMCC, Four Biotech
377 Plantation Street
3rd Floor
Worcester, MA 01655
Phone: 1.781.681.2328
Doctorate, 1993
Biological Sciences;
Physical Sciences

Yates, Sydney
US House of Representatives
Washington, DC 20540

Yatsumimi, Toshifumi
1-22-68 Omiya
Suginami 168-0061
Japan
Phone: 81.3.5377.2027
uptijg@coral.plala.or.jp

Yavari, R.
85 Island Ave
Madison, CT 06443-3030

Ybarra-Frausto, Tomas
The Rockefeller Foundation
420 5th Ave
New York, NY 10018-2729

Yeary, David W.
3803 Partridgeberry Ct
Houston, TX 77059-4067

Yeo, Richard R.
Griffith University
School of Humanities
Kessels Rd.
Brisbane 4111
Australia
Phone: 61.7.3875.7692
Fax: 61.7.3875.7730
r.yeo@mailbox.gu.edu.au
Doctorate, 1978
18c, Europe, Humanistic
Relations of Science;
19c, Europe, Humanistic
Relations of Science;
18c, North America, Humanistic
Relations of Science;
North America, Philosophy and
Philosophy of Science

Yeung, King-Fai
PO Box 749
Daly City, CA 94017-0749
Phone: 1.650.991.9532
bfyeung@aol.com

Ymele, Jean P.
PO Box 2905
Yaounde
Cameroon

Yoder, Hatten S., Jr.
Carnegie Inst of Washington
Geophysical Laboratory
5251 Broad Branch Rd NW
Washington, DC 20015-1305
Phone: 1.202.478.8966
Fax: 1.202.478.8901
yoder@gl.ciw.edu
Doctorate, 1948
Earth Sciences;
Instruments and Techniques

Yoder, Joella G.
11720 SE 92nd St
Newcastle, WA 98056-2062
Phone: 1.425.271.4025
jgyoder@u.washington.edu
Doctorate, 1985
17c, Europe, Mathematics;
17c, Europe, Physical Sciences;
17c, Europe, Astronomical
Sciences; 17c, Europe,
Instruments and Techniques

Yokoyama, Toshiaki
Toho University
Faculty of Science
Miyama 2-2-1
Funabashi 272
Japan
Phone: 81.474.72.3517
Fax: 81.480.33.2857
yokoyama@s.sci.toho-u.ac.jp
Master's Level, 1968
19c, Europe, Biological Sciences;
19c, Europe, Cognitive Sciences

Yong, Shi
Beijing Plant 251
PO Box 261
Beijing 102101
China
Phone: 86.10.69132140
Fax: 86.10.69132500
Bachelor's Level
19c, Physical Sciences

York, William
4820 NW Kahneeta Dr
Portland, OR 97229-2107
Phone: 1.503.617.4733

Yoshimoto, Hideyuki
2-20-22 Nishiogikita
Suginami-Ku
Tokyo 167-0042
Japan
Phone: 03.3397.6785
Fax: 81.03.3335.1692
h2ysmt@t3.rim.or.jp

Young, Christian C.
1316 N Astor St
Milwaukee, WI 53202
Phone: 1.414.298.9138
cyoung@aero.net
Doctorate, 1997
20c, North America, Biological
Sciences

Young, William H.
William H. Young & Assoc.
1442 Sequoia Cir
Toms River, NJ 08753-2864
Phone: 1.732.505.3250
Fax: 1.732.505.3668
whytr@msn.com
Master's Level
20c, North America, Technology,
Instruments and Techniques;
20c, North America, Humanistic
Relations of Science;
19c, North America, Philosophy
and Philosophy of Science;
20c, North America, Social
Relations of Science

Yuan, Sun-shine
93 Woodridge Rd
Wayland, MA 01778-3624
Phone: 1.508.358.7541
Fax: 1.508.358.3487
jusanin@mediaone.net
Doctorate
19c, Asia, Physical Sciences;
Biological Sciences;
Earth Sciences;
Physical Sciences

Zacharias, Kristen L.
Albright College
Thirteenth and Bern Sts
Reading, PA 19612-5234
Phone: 1.610.921.7706

Fax: 1.610.926.6005
kristenz@alb.edu
Doctorate, 1980
19c, Biological Sciences;
Philosophy and Philosophy of
Science; Biological Sciences

Zaitsev, Evgueny
Inst für Gesch der Naturwissen
Mathematik und Technik
Bundesstrasse 55
Hamburg D-20146
Germany

Zallen, Doris T.
Virginia Tech
233 Lane Hall
Blacksburg, VA 24061-0227
Phone: 1.540.231.4216
Fax: 1.540.231.7013
dtzallen@vt.edu
Doctorate
20c, Biological Sciences;
20c, Science Policy;
20c, Europe, Medical Sciences

Zamorano, Raul
Univ Nacional de Mar del Plata
Dept de Fisica
Funes 3350
Mar del Plata
Argentina
Phone: 54.0223.475.6951
Fax: 54.0223.475.3150
dpfisica@mdp.edu.ar

Zanish-Belcher, Tanya
Iowa State University
403 Parks Library
Ames, IA 50011-2140
Phone: 1.515.294.6648
Fax: 1.515.294.5525
tzanish@iastate.edu
Master's Level, 1990
20c, North America, Gender and
Science

Zeller, Suzanne E.
Wilfrid Laurier University
Department of History
Waterloo, ON N2L3C5
Canada
Phone: 1.519.884.0710
Fax: 1.519.746.3655
szeller@wlu.ca
Doctorate
19c, North America, Social

Relations of Science; Physical
Sciences; Natural and Human
History; Biological Sciences

Zernel, John J.
Oregon State University
Distance/ Continuing Education
Corvallis, OR 97331
Phone: 1.503.393.6850
Fax: 1.541.737.2734
zernelj@ucs.orst.edu
Doctorate, 1983
19c, North America; Earth
Sciences; Biological Sciences

Ziche, Paul G.
Schelling-Kommission
Bayerische Akademie
der Wissenschaften
Marstallplatz 8
München 80539
Germany
Phone: 089.23031.231
Fax: 089.23031.100
schelling.kommission@
    lrz.badw-muenchen.de

Ziegler, Renatus J.
Verein für Krebsforschung
Kirschweg 9
Arlesheim 4144
Switzerland
Phone: 41.6.1706.7245
Fax: 41.6.1706.7200
renatus.ziegler@hiscia.ch
Doctorate, 1985
Mathematics; Philosophy
and Philosophy of Science;
Astronomical Sciences;
Physical Sciences

Zik, Yaakov
51 Hovevey Zion St
Tel Aviv 63346
Israel
Phone: 972.3.528.7707
zikya@attglobal.net
17c, Europe, Physical Sciences;
17c, Europe, Instruments and
Techniques; 17c, Europe,
Astronomical Sciences

Zingrone, Nancy L.
PO Box 41
New York, NY 10021
Phone: 1.212.396.0096
Fax: 1.212.628.1559

zingrone@parapsychology.org
Master's Level, 1977
19c, North America;
20c, Europe, Cognitive Sciences;
South America, Social Relations
of Science; Medical Sciences,
Instruments and Techniques

Zitarelli, David E.
Temple University
Department of Mathematics
1807 North Broad Street
Philadelphia, PA 19122
Phone: 1.215.787.7844
Fax: 1.215.204.6433
zit@temple.edu
Doctorate, 1970
19c, Mathematics;
20c, Mathematics;
North America, Mathematics

Zuckerman, Harriet
The Andrew W Mellon
Foundation
140 E 62nd St
New York, NY 10021-8187

Zuidervaart, Huibert J.
Bellinkstraat 29
Middelburg 4331 GV
The Netherlands
Phone: 031.118.637375
hzuidervaart@hotmail.com
18c, Europe,
Astronomical Sciences;
18c, Europe, Earth Sciences;
18c, Europe, Physical Sciences;
18c, Europe, Institutions

Zulueta, Benjamin
University of California
Department of History
Santa Barbara, CA 93106
Phone: 1.805.971.5948
Fax: 1.805.893.8795
bcz0@umail.ucsb.edu

Graduate Programs

The HSS Executive Committee approved the following definition for graduate academic programs. A graduate program in the history of science must have at least one of the following characteristics:

- A formal history of science designation
- Graduate students pursuing a degree in the history of science
- Granted a Ph.D. in the history of science in the past three years
- Full-time, core faculty members who offer graduate-level courses in the history of science

Using the above definition, 138 respondents described their institutions as graduate programs in the history of science. Entries are organized by country, and then by state for programs within the United States. Not all programs responded with complete data. Missing elements generally indicate the program did not respond completely. All entries contain some portion of the following information:

Institution name
Name of program in history of science, technology, or medicine
Primary contact person's name, title, and e-mail address
Primary e-mail address, if different from the contact person's email
Mailing address
Telephone number
Facsimile number
Journal editorial offices and ongoing serial publication projects
Special collections and research resources
Does the program offer publicly available grants or fellowships: Yes/No
Particular strengths in teaching and research, up to three, entered using
    a standardized list of keywords. If no program strengths appear, this
    generally indicates the program had wide interests and knowledges, and
    chose not to specify particular strengths.
Degrees conferred in the history of science
Number of graduate students in 2002. (This number may also include
    students not pursuing a degree in the history of science)
Number of masters and/or doctoral degrees awarded since June 30th, 1998
Name and e-mail of person whom prospective students should contact
    (only listed if different from general program contact)
List of regular faculty members. Names appear as they were entered by
    respondents. A broad research interest appears only after the names of
    HSS members, if they entered this information when they joined the
    Society. (As computer programming challenges are overcome, research
    interests for nonmembers should appear beginning with the 2003 online
    edition of the *Guide*.)
Brief description of the program and its special features
URL

Australia

Deakin University (Science and Technology Studies). **Description:** This program ceased granting STS degrees in 2000. The faculty members have been moved to other departments.

University of Melbourne (Department of History and Philosophy of Science). **Contact:** Helen Verran, hrv@unimelb.edu.au **Mailing Address:** Ground Floor, Old Arts Building, University of Melbourne, Parkville, Victoria, 3010, Australia. **Tel:** 61.3.8344.6556 **Fax:** 61.3.8344.7959 **Ongoing Publication Projects:** *Historical Records of Australian Science*; *Health and History*; *Critical Horizons*. **Grants:** Yes. **Program Strengths:** Europe; Australia and Oceania. **Degrees Conferred:** Bachelor's Level, Master's Level, Doctoral Level. 62 graduate students in 2002. 18 master's degrees and 13 doctorates awarded since July, 1998. **Faculty:** *Michael Arnold*; *John Cash*; *Roderick W. Home* (18th Century, Physical Sciences); *Keith R. Hutchison* (17th Century, Europe, Astronomical Sciences); *Ross Jones*; *Janet McCalman*; *Rosemary Robins*; *John Rundell*; *Charles H. Sankey* (Philosophy and Philosophy of Science); *Neil Thomason*; *Helen Verran*. **Web Site:** http://www.hps.unimelb.edu.au/

University of New South Wales (School of History and Philosophy of Science). **Contact:** John A. Schuster, j.a.schuster@unsw.edu.au **Primary E-mail:** sts@unsw.edu.au **Mailing Address:** Morven Brown LG19, University of New South Wales, Sydney, New South Wales, 2052, Australia. **Tel:** 61.2.9385.2356 **Fax:** 61.2.9385.8003 **Ongoing Publication Projects:** *Metascience* (N. Rasmussen, Editor). **Grants:** Yes. **Program Strengths:** 17th Century, Europe, Physical Sciences; 20th and 21st Century, North America, Biological Sciences, Instruments and Techniques; 18th Century, Europe, Social Relations of Science. **Degrees Conferred:** Bachelor's Level, Master's Level, Doctoral Level. 28 graduate students in 2002. 9 master's degrees and 13 doctorates awarded since July, 1998. **Faculty:** *George Bindon*; *Paul Brown*; *Anthony Corones*; *Susan Hardy*; *Stephen Healy*; *John Merson*; *David Phillip Miller*; *David R. Oldroyd* (17th Century, Earth Sciences); *Nicolas Rasmussen*; *John Schuster*; *Peter Slezak*. **Description:** The largest such unit in Australia, it offers undergraduate, honors (4th-year research based), and postgraduate courses in history of science, medicine and technology; philosophy of science; and technology studies and policy. It hosts interdisciplinary programs in environmental studies and cognitive science. **Web Site:** http://www.arts.unsw.edu.au/sts/

University of Sydney (History and Philosophy of Science). **Contact:** Rachel Ankeny, r.ankeny@scifac.usyd.edu.au **Primary E-mail:** hps@scifac.usyd.edu.au **Mailing Address:** F07 - Carslaw, University of Sydney, Sydney, New South Wales, 2006, Australia. **Tel:** 61.2.9351.4226 **Fax:** 61.2.9351.4124 **Collections/ Resources:** The rare book collection at the University of Sydney's library includes the Deane Collection which is directly related to several aspects of the history and philosophy of science. The collection's strengths include witchcraft, demonology and significant Australiana. Manuscript collections are also available, as well as the University of Sydney Archives. The Macleay Museum contains collections that document the history of science at Sydney University and elsewhere, with particular focus on Australian science and technology and on scientific instrumentation; students have participated in the curation of exhibits. **Grants:** Yes. **Program Strengths:** 20th and 21st Century, Biological Sciences; Medical Sciences; 17th Century, Europe, Physical Sciences. **Degrees Conferred:** Bachelor's Level, Master's Level, Doctoral Level. 4 graduate students in 2002. 10 master's degrees and 2 doctorates awarded since July, 1998. **Prospective Student Contact:** Rachel A. Ankeny, hps@scifac.usyd.edu.au **Faculty:** *Rachel A. Ankeny* (Philosophy and Philosophy of Science); *Alison Bashford*; *Stephen Gaukroger* (17th Century, Physical Sciences); *Claire Hooker*; *Roy M. MacLeod* (19th Century, Europe, Natural and Human History); *Katherine Neal*; *Hans Pols* (19th Century, Europe, Medical Sciences, Instruments and Techniques); *Huw Price*; *Evelleen Richards*. **Description:** This Unit is located in the Faculty of Science, with close ties to the departments of philosophy, history, and gender studies. Research strengths include history and philosophy of biology/medicine, social studies of science, early modern science, philosophy of science (particularly philosophy of physics), and history of mathematics. **Web Site:** http://www.usyd.edu.au/su/hps/

University of Wollongong (Science, Technology and Society). **Contact:** David Mercer, david_mercer@uow.edu.au **Mailing Address:** Faculty of Arts, Northfields Avenue, Wollongong, New South Wales, 2522, Australia. **Tel:** 61.2.42.214062 **Grants:** No. **Program Strengths:** 20th and 21st Century, Technology, Social Relations of Science; 20th and 21st Century, Social Sciences, Science Policy; 20th and 21st Century, Humanistic Relations of Science. **Degrees Conferred:** Bachelor's Level, Master's Level, Doctoral Level. 11 graduate students in 2002. 7 doctorates awarded since July, 1998. **Faculty:** *David William Mercer* (North America). **Description:** Recent graduate student thesis areas include: Robotics and Aged Care in Japan; History of Wind Power in Australia; Theories of Development and Technology Transfer; Critiques of Theories of Mental Illness; History of Ecology; Scientific Controversy and Public Understanding of Science; Internet Activism; Evidence Based Medicine. **Web Site:** http://www.uow.edu.au/arts/sts/

Brazil

Pontifica Universidade Catolica de São Paolo (Graduate Program on the History of Science). **Contact:** Ana Maria Alfonso-Goldfarb, cesimahc@pucsp.br **Mailing Address:** Rua Brasilia, 46 apto. 81, São Paulo, Sao Paulo, 04534-040, Brazil. **Tel:** 55.11.3256.1622 ext. 211 **Fax:** 55.11.3256.1622 ext. 211 **Ongoing Publication Projects:** The program maintains a series called "Estudos e Documentos em História da Ciência" (Coleção CESIMA), EDUC/FAPESP. Additionally, in the last ten years, the group of researchers and professors of the program have been involved with the international "Real de Intercambios para la Historia y la Epistemología de las Ciencias Químicas y Biológicas," which publishes the series called "Estudios de Historia Social de las Ciencias Quimicas y Biologicas," edited by P. Aceves Pastrana, from the Universidad Autonoma Metropolitana of Mexico. **Collections/Resources:** Our "Centro Simão Mathias de Estudos em História da Ciência" holds a "virtual" library of originals in microforms and e-books. **Grants:** Yes. **Program Strengths:** 17th Century, Europe, Physical Sciences, Humanistic Relations of Science; 18th Century, South America, Biological Sciences, Exploration, Navigation, Expeditions; 19th Century, Transcontinental, Natural and Human History, Science Policy. **Degrees Conferred:** Bachelor's Level, Master's Level. 50 graduate students in 2002. 36 master's degrees awarded since July, 1998. **Prospective Student Contact:** Vera Cecilia Machline, cesimahc@pucsp.br **Faculty:** *Ana M. Alfonso-Goldfarb* (Physical Sciences); *Ana Maria Haddad Baptista*; *Maria Helena Roxo Beltran*; *Luzia Aurelia Castaneda*; *Ubiratan D'Ambrosio* (Social Sciences); *Marcia Helena Mendes Ferraz*; *Jose Luiz Goldfarb*; *Vera Cecilia Machline*; *Lilian Al-Chueyr Pereira Martins*; *Djalma Medeiros*; *Paulo Alves Porto*. **Description:** The program holds a research center (Centro Simão Mathias de Estudos em História da Ciência) which maintains contact with many Brazilian and international centers and programs on the history of science. **Web Site:** http://www.pucsp.br/~cesima

Universidade Federal da Bahia (Mestrado em Ensino, Filosofia e Historia das Ciencias (UFBa-UEFS)). **Contact:** Olival Freire Jr., freirejr@ufba.br **Primary E-mail:** pice@fis.ufba.br **Mailing Address:** Instituto de Fisica - Universidade Federal da Bahia, Campus de Ondina, Salvador, Bahia, 40210-340, Brazil. **Tel:** 55.71.247.2033 **Fax:** 55.71.235.5592 **Ongoing Publication Projects:** *Ideacao* edited by Feira de Santana. **Grants:** Yes. **Program Strengths:** 20th and 21st Century, Physical Sciences; 20th and 21st Century, South America; 20th and 21st Century, Biological Sciences. **Degrees Conferred:** Bachelor's Level, Master's Level. 25 graduate students in 2002. 0 master's degrees awarded since July, 1998. **Faculty:** *Amilcar Baiardi*; *Elyana Barbosa*; *Andre Mattedi Dias*; *Charbel Nino El-Hani*; *Olival Freire* (20th and 21st Century, Physical Sciences); *Maria Cristina Mesquita Martins*; *Osvaldo F. Pessoa* (20th and 21st Century, Physical Sciences); *Aurino Ribeiro*; *Joao Carlos Pires Salles*; *Jose Carlos Barreto Santana*; *Waldomiro Jose Silva Filho*; *Robinson Tenorio*; *Julio C. Ribeiro Vacsoncelos*. **Description:** The program brings together historians, philosophers, scientists, and science teaching researchers, focusing their analysis on the production and diffusion of science, especially through education. **Web Site:** http://www.fis.ufba.br/dfg/pice/

Canada

McGill University (Social Studies of Medicine). **Contact:** George Weisz, george.weisz@mcgill.ca **Mailing Address:** 3647 Peel Street, Montreal, Quebec, H3A 1X1, Canada. **Tel:** 1.514.398.6033 **Fax:** 1.514.398.1498 **Collections/ Resources:** Osler Library for the History of Medicine. **Grants:** Yes. **Program**

Strengths: 20th and 21st Century, Transcontinental, Medical Sciences; 19th Century, Europe, Medical Sciences, Social Relations of Science. **Degrees Conferred:** Bachelor's Level, Master's Level, Doctoral Level. 20 graduate students in 2002. **Faculty:** *Alberto Cambrosio* (20th and 21st Century, Medical Sciences); *Myron Echenberg*; *Margaret Lock*; *Thomas Schlich*; *Faith Wallis*; *George Weisz*; *Allan Young*. **Description:** Program offers graduate studies in the history of medicine. **Web Site:** http://www.mcgill.ca

Université du Québec à Montréal (Centre Interuniversitairede Recherche sur la Science et la Technologie, CIRST). **Contact:** Yves Gingras, gingras.yves@ uqam.ca **Mailing Address:** CIRST, Université du Québec à Montréal, CP 8888, Succursale Centre-Ville, Montréal, H3C 3P8, Canada. **Tel:** 1.514.987.3000 **Fax:** 1.514.987.7726 **Grants:** Yes. **Program Strengths:** 20th and 21st Century, North America, Institutions; 19th Century, North America, Education; 20th and 21st Century, North America, Biological Sciences. **Degrees Conferred:** Bachelor's Level, Master's Level, Doctoral Level. 34 graduate students in 2002. 26 master's degrees and 6 doctorates awarded since July, 1998. **Faculty:** *Robert Gagnon*; *Yves Gingras* (20th and 21st Century, North America, Physical Sciences); *Peter Keating*; *Lyse Roy*; *Jan Sapp* (Biological Sciences). **Description:** Students at CIRST are enrolled in departmental programs as M.A. and Ph.D. students in history, philosophy or sociology of science. The center has 22 members covering these fields, as well as innovation and science policy. **Web Site:** http://www.unites.uqam.ca/cirst

University of Alberta (Department of History and Classics). **Contact:** Robert Smith, rwsmith@ualberta.ca **Primary E-mail:** histclas@ualberta.ca **Mailing Address:** Department of History and Classics, University of Alberta, Edmonton, Alberta, T6G 2H4, Canada. **Tel:** 1.780.492.4687 **Fax:** 1.780.492.9125 **Ongoing Publication Projects:** Oxford Francis Bacon Edition Project. **Collections/Resources:** The University of Alberta Libraries are generally regarded as constituting the second best collection in Canada. **Grants:** Yes. **Degrees Conferred:** Bachelor's Level, Master's Level, Doctoral Level. 7 graduate students in 2002. 1 master's degree and 1 doctorate awarded since July, 1998. **Faculty:** *Lesley B. Cormack* (14th-16th Century, Europe, Earth Sciences); *Julian J. Martin* (17th Century, Europe, Social Sciences); *Pat Prestwich*; *Robert W. Smith* (Astronomical Sciences); *Susan L. Smith*. **Description:** Five members of the Department of History and Classics at the University of Alberta have principal research interests in the history of science and medicine. These interests range from the history of geography in the sixteenth century to the study of large-scale scientific projects in the contemporary era. **Web Site:** http://www.arts.ualberta.ca/~histclas/

University of Calgary (History and Philosophy of Science; History of Science). **Contact:** Margaret J. Osler, mjosler@ucalgary.ca **Mailing Address:** Department of History, 2500 University Drive, NW, Calgary, Alberta, T2N 1N4, Canada. **Tel:** 1.403.220.6401 **Fax:** 1.403.289.8566

Collections/Resources: Large microfilm collection of early modern materials. **Grants:** Yes. **Program Strengths:** 17th Century, Europe, Humanistic Relations of Science; 18th Century, Europe, Social Sciences; Europe, Science and Religion. **Degrees Conferred:** Bachelor's Level, Master's Level, Doctoral Level. 2 graduate students in 2002. 1 master's degree and 0 doctorates awarded since July, 1998. **Faculty:** *J. J. (Jack) MacIntosh*; *Margaret J. Osler* (17th Century, Europe, Philosophy and Philosophy of Science); *Hank Stam*; *Martin S. Staum*. **Description:** In addition to M.A. and Ph.D. programs in the History Department, there is a joint program with the Philosophy Department in History and Philosophy of Science. **Web Site:** http://hist.ucalgary.ca/graduate/GradDoc.htm#indexten

University of Toronto (Institute for the History and Philosophy of Science and Technology (IHPST)). **Contact:** Janis Langins, jlangins@chass.utoronto.ca **Primary E-mail:** ihpst.info@utoronto.ca **Mailing Address:** Room 316, Victoria College, 91 Charles Street West, Toronto, Ontario, M5S 1K7, Canada. **Tel:** 1.416.978.4950 **Fax:** 1.416.978.3003 **Ongoing Publication Projects:** *Annals of Science* (Trevor H. Levere, Editor; Sungook Hong, Book Review Editor); *Historia Mathematica* (Craig G. Fraser, Editor). **Collections/Resources:** University of Toronto Museum of Scientific Instruments (UTMuSi) operated by the IHPST (*see entry under Museums*); Science and Technology Image Database Project (IMAGO); IHPST Reference Library; University of Toronto Robarts Library, Gerstein Science Information Centre Library and Thomas Fisher Rare Book Library containing S. Drake Galileo collection, Hannah collection in history of medicine, Darwin collection, K.O. May collection in history of mathematics. **Grants:** Yes. **Program Strengths:** Biological Sciences; Physical Sciences; Technology. **Degrees Conferred:** Bachelor's Level, Master's Level, Doctoral Level. 41 graduate students in 2002. 38 master's degrees and 10 doctorates awarded since July, 1998. **Prospective Student Contact:** Craig G. Fraser, cfraser@chass.utoronto.ca **Faculty:** *Brian S. Baigrie*; *Craig G. Fraser* (Mathematics); *Bert S. Hall* (5th-13th Century, Europe, Technology); *Sungook Hong* (19th Century, Europe, Physical Sciences); *Alexander Jones*; *Janis Langins*; *Trevor H. Levere*; *Pauline M. H. Mazumdar*; *Mary P. Winsor* (Biological Sciences). **Description:** Scientific developments investigated according to theoretical and experimental content, and within social and cultural context. IHPST offers graduate courses in the histories of biology, mathematics, medicine, physics, chemistry, and technology, as well as philosophy of science. **Web Site:** http://www.chass.utoronto.ca/ihpst/

York University (Science and Society). **Contact:** Bernie Lightman, lightman@yorku.ca **Mailing Address:** Room 309 Bethune College, York University, 4700 Keele St., Toronto, Ontario, M3J 1P3, Canada. **Tel:** 1.416.736.5164 ext. 22028 **Ongoing Publication Projects:** *Canadian Journal of Zoology*. **Grants:** No. **Faculty:** *Katharine M. Anderson* (19th Century, Europe, Earth Sciences); *Steve Bailey*; *Raymond E. Fancher* (Cognitive Sciences); *Martin Fichman* (19th Century, Biological Sciences); *Ernst Hamm* (Europe, Earth Sciences);

Bernard V. Lightman (19th Century, Europe, Humanistic Relations of Science); *Joan S. Steigerwald* (18th Century, Europe, Medical Sciences). **Web Site:** http://www.yorku.ca

China

Inner Mongolia Normal University (Institute for the History of Science). **Contact:** Guo Shirong, hissci@public.hh.nm.cn **Mailing Address:** Inner Mongolia Normal University, Department of History of Science and of Scientific and Technological Administration, Huhehot, Inner Mongolia Autonomous Region, 010022, China. **Tel:** 86.471.439.2029 **Ongoing Publication Projects:** *Journal of Cultural History of Mathematics*; *Studies in the History of Mathematics.* **Collections/Resources:** About 1000 old books of mathematics, physics, medicine, and astronomy written in Chinese before 1900. **Grants:** No. **Program Strengths:** Asia, Mathematics, Humanistic Relations of Science; Mathematics; Asia, Astronomical Sciences, Instruments and Techniques. **Degrees Conferred:** Bachelor's Level, Master's Level. 12 graduate students in 2002. 6 master's degrees awarded since July, 1998. **Faculty:** *Mo De*; *Li Di*; *Te Gus*; *Luo Jianjin*; *Han Jingfang*; *Deng Kehui*; *Feng Lisheng*; *Dai Qin*; *Sa Rina*; *Guo Shirong*; *Zhang Ziwen.* **Description:** The institute was established in 1983 and is one of the major research centers of history of science in China. It publishes two journals on the history of mathematics. The stronger study fields are history of mathematics, astronomical instruments, science and technology of Chinese minorities, and historical exchanges of mathematics among areas influenced by Chinese culture. **Web Site:** http://www.imnu.edu.cn/academics/kxglykxx/index.htm

Institute for the History of Natural Science (IHNS) (Program in History of Science). **Contact:** Dun Liu, dliu@95777.com **Primary E-mail:** webmaster@ihns.ac.cn **Mailing Address:** 137 Chao Nei Street, Institute for the History of Natural Science, Beijing, 100010, China. **Tel:** 86.10.6401.9661 **Fax:** 86.10.6401.7637 **Ongoing Publication Projects:** *Studies in the History of Natural Sciences*; *China Historical Materials of Science and Technology.* **Collections/Resources:** A highly specialized library that is well-known in China. In particular, a collection of hundreds of volumes of ancient Chinese mathematical works, donated by the late Professor Li Yan, which is unparalleled in the world. **Grants:** Yes. **Degrees Conferred:** Master's Level, Doctoral Level. 30 graduate students in 2002. 16 master's degrees and 12 doctorates awarded since July, 1998. **Prospective Student Contact:** Hongli Zhang, zhanghl@ihns.ac.cn **Description:** IHNS is the Chinese national institute in the history of science. Its library holds more than 140,000 volumes, of which close to 3,000 are thread-bound ancient Chinese books. Since 1978, the IHNS has trained more than 130 graduates students for both Ph.D. and Master degrees. In the 45 years since its founding, the members of IHNS have published more than 280 research works and have presented close to 6400 papers on the history of science. **Web Site:** http://www.ihns.ac.cn

Northwest University (Center for History of Mathematics and Sciences). **Contact:** Anjing Qu, qaj@sein.sxgb.com.cn **Primary E-mail:** hs@nwu.edu.cn **Mailing Address:** Department of Mathematics, Northwest University, Xi'an, Shaanxi, 710069, China. **Tel:** 86.29.830.3334 **Collections/Resources:** Historical Chinese materials in sciences. **Grants:** No. **Degrees Conferred:** Bachelor's Level, Master's Level, Doctoral Level. 10 graduate students in 2002. 3 master's degrees and 6 doctorates awarded since July, 1998. **Description:** History of mathematics; history of exact sciences in ancient and medieval times; management and policy of sciences. **Web Site:** http://hismath.go.163.com

The Czech Republic

Charles University (Department of Philosophy and History of Natural Science). **Contact:** Stanislav Komárek, filosof@natur.cuni.cz **Primary E-mail:** komarek@ natur.cuni.cz **Mailing Address:** Vinicna 7, Prague, 128 44, Czech Republic. **Tel:** 42.2.2195.3212 **Fax:** 42.2.2491.9704 **Ongoing Publication Projects:** Emanuel Rádl - scientist and philosopher. **Collections/Resources:** Archives for history of mimicry research; Archives of Emanuel Rádl. **Grants:** No. **Degrees Conferred:** Bachelor's Level, Doctoral Level. 7 graduate students in 2002. 0 master's degrees and 1 doctorate awarded since July, 1998. **Web Site:** http://www.natur.cuni.cz/SECT/PHIL

Denmark

University of Aarhus (History of Science Department). **Contact:** Louis Klostergaard, ivhlk@ifa.au.dk **Mailing Address:** Ny Munkegade, Building 521, Aarhus C, DK-8000, Denmark. **Tel:** 45.8942.3512 **Fax:** 45.8942.3510 **Collections/Resources:** The Bengt Strömgren Archives; The Ejnar Hertzsprung Correspondence; The P. O. Pedersen Archives. **Grants:** No. **Degrees Conferred:** Bachelor's Level, Master's Level, Doctoral Level. 16 graduate students in 2002. 20 master's degrees and 4 doctorates awarded since July, 1998. **Prospective Student Contact:** Ole Knudsen, ivhok@ifa.au.dk **Faculty:** *Kirsti Andersen*; *Anja Skaar Jacobsen*; *Ole Knudsen*; *Helge S. Kragh* (20th and 21st Century, Astronomical Sciences); *Anita Kildebæk Nielsen*; *Henry Nielsen*; *Kurt Møller Pedersen*. **Web Site:** http://www.ifa.au.dk/ivh/home.htm

University of Copenhagen (Department of History of Medicine). **Contact:** Thomas Söderqvist, t.soderqvist@pubhealth.ku.dk **Primary E-mail:** med.hist.museum@mhm.ku.dk **Mailing Address:** Bredgade 62, Copenhagen, 1260, Denmark. **Tel:** 45.3532.3800 **Fax:** 45.3532.3816 **Collections/ Resources:** The Medical History Museum has rich collections, including 19th and 20th century medical instruments, Danish hospital records, and a rare book collection (*see entry under Museums*). **Grants:** Yes. **Program Strengths:** 20th and 21st Century, Europe, Medical Sciences, Biography. **Degrees Conferred:** Bachelor's Level, Doctoral Level. 6 graduate students in 2002. 1 doctorate awarded since July, 1998. **Faculty:** *Frank Allan Rasmussen*

(17th Century); *Thomas Söderqvist* (20th and 21st Century, Social Sciences). **Description:** The program is under construction. Current graduate students are working on different aspects of 19th and 20th century Danish medicine. **Web Site:** http://www.pubhealth.ku.dk/amh/index-e.html

Finland

University of Oulu (History of Ideas and Science). **Contact:** Juha Manninen, juha.manninen@oulu.fi **Mailing Address:** P.O. Box 1000, University of Oulu, FIN-90014, Finland. **Tel:** 358.8.553.3303 **Fax:** 358.8.553.3315 **Ongoing Publication Projects:** Book series "Studies in the history of science and ideas," Peter Lang, Frankfurt/Main. **Collections/Resources:** Landmarks of science; Copies of the Vienna Circle collection (Haarlem, North-Holland); Numerous collections related to the history of science in Finland. **Grants:** No. **Degrees Conferred:** Bachelor's Level, Master's Level, Doctoral Level. 12 graduate students in 2002. 15 master's degrees and 3 doctorates awarded since July, 1998. **Prospective Student Contact:** Erkki Urpilainen, erkki.urpilainen@oulu.fi **Description:** The University of Oulu is the only place in Finland where the history of science can be studied at all levels.

France

Conservatoire National des Arts et Metiers. **Mailing Address:** 292, rue Saint-Martin, Paris, 75141, France. **Tel:** 33.1.40.27.20.00 **Web Site:** http://www.cnam.fr/

Universite Louis Pasteur de Strasbourg (Institut d'Histoire des Sciences). **Contact:** William Shea, william.shea@ihs-ulp.u-strasbg.fr **Mailing Address:** Universite Louis Pasteur de Strasbourg, 7, rue de l'Universite, Strasbourg, 67000, France. **Tel:** 33.3.90.24.05.80 **Fax:** 33.3.90.24.05.81 **Grants:** No. **Degrees Conferred:** Bachelor's Level, Master's Level, Doctoral Level. 2 graduate students in 2002. 3 master's degrees and 0 doctorates awarded since July, 1998.

Germany

Note: An excellent directory of history of science programs Germany is produced by the German National Committee of the International Union for the History and Philosophy of Science, Division of the History of Science. Information is listed in both English and German. See: Christoph Meinel, ed., *History of Science, Technology, and Medicine in Germany, 1997-2000* (Weinheim: Wiley-VCH, 2001).

German graduate programs are alphabetized by city, which is how they are more commonly known.

Rheinisch-Westfälische Technische Hochschule Aachen (Lehrstuhl für Geschichte der Technik). **Contact:** Walter Kaiser, kaiser@histech.rwth-aachen.de **Mailing Address:** Kopernikusstrasse 16, Aachen, D-52074, Germany. **Tel:** 49.241.80.23666 **Fax:** 49.241.80.22302 **Grants:** No. **Program Strengths:** 20th and 21st Century, Europe, Technology. **Degrees Conferred:** Bachelor's Level, Doctoral Level. **Description:** The Aachen Chair for the History of Technology deals with the development of technology in modern Europe, especially since the Industrial Revolution. Along with the interaction of science and technology, aspects of economic history and of general history are also considered. Instruction aims to give students in the field of engineering a chance to leave their restricted field for once during their studies and to learn how technology develops in a complicated field of forces. Students of general history are encouraged to include history of science and technology. **Web Site:** http://www.histech.rwth-aachen.de/

Universität Augsburg (Lehrstuhl für Philosophie und Wissenschaftstheorie). **Contact:** Klaus Mainzer, klaus.mainzer@phil.uni-augsburg.de **Mailing Address:** Universitätsstr. 10, Augsburg, D-86159, Germany. **Tel:** 49.821.5.98.55.68 **Fax:** 49.821.5.98.55.84 **Grants:** Yes. **Program Strengths:** 20th and 21st Century, Physical Sciences; 20th and 21st Century, Mathematics; 20th and 21st Century, Cognitive Sciences. **Degrees Conferred:** Bachelor's Level, Master's Level, Doctoral Level. 56 graduate students in 2002. 4 master's degrees and 6 doctorates awarded since July, 1998. **Faculty:** *Theodor Leiber*; *Cornelia Liesenfeld*; *Klaus Mainzer*; *Elena Tatievskaia*. **Description:** The Augsburg program mainly concerns philosophy and history of science in the fields of mathematics, physics, and computer science. Special features are complex systems and nonlinear dynamics in nature and society. Computational models in these fields (e.g., artificial intelligence, Internet) are of special interest. **Web Site:** http://www.philso.uni-augsburg.de/web2/Philosophie2/index.htm

Freie Universität Berlin and Humboldt-Universität Berlin (Institut für Geschichte der Medizin im Zentrum für Human- und Gesundheitswissenschaften der Berliner Hochschulmedizin). **Contact:** Johanna Bleker, johanna.bleker@medizin.fu-berlin.de **Mailing Address:** Institut für Geschichte der Medizin im ZHGB, Klingsorstr. 119, 12203 Berlin, Germany. **Tel:** 49.30.830092.20 **Fax:** 49.30.830092.37 **Ongoing Publication Projects:** Managing Editor: *Medizinhistorisches Journal*; Co-Editor: *Sudhoffs Archiv*; Editor: *Abhandlungen zur Geschichte der Medizin und der Naturwissenschaften*. **Collections/Resources:** The library contains approximately 70,000 volumes. **Program Strengths:** Europe, Medical Sciences; 19th Century, Medical Sciences, Instruments and Techniques; 20th and 21st Century, Europe, Medical Sciences, Social Relations of Science. **Degrees Conferred:** Bachelor's Level. 27 doctorates awarded since July, 1998. **Faculty:** *Johanna Bleker*; *Eva Brinkschulte*; *Guido Jüttner*; *Ilona Marz*; *Thomas Müller* (North America, Technology, Social Relations of Science); *Udo Schagen*; *Sabine Schleiermacher*;

Hess Volker; *Rolf Winau*. **Description:** The institute teaches history of medicine to medical students of both Berlin medical faculties and supervises doctoral dissertations. **Web Site:** http://www.medizin.fu-berlin.de/igm/

Technische Universität Berlin (Institut für Philosophie, Wissenschaftstheorie, Wissenschafts- und Technikgeschichte). **Contact:** Wolfgang Koenig, Hannelore.Rumi@tu-berlin.de **Primary E-mail:** martin@kgw.tu-berlin.de **Mailing Address:** Sekr.TEL 12-1, Ernst-Reuter-Platz 7, Berlin, D-10587, Germany. **Tel:** 49.30.3.142.4841 **Ongoing Publication Projects:** *Technikgeschichte*. **Grants:** No. **Degrees Conferred:** Bachelor's Level, Master's Level, Doctoral Level. 60 graduate students in 2002. 10 master's degrees and 6 doctorates awarded since July, 1998. **Faculty:** *Eberhard Knobloch*; *Hans-Werner Schütt* (Physical Sciences); *Burghard O. Weiss*. **Description:** This Institute of the Technical University treats history of science and history of technology from its antiquity to the present. It collaborates with the Philosophy Department at the TU Berlin. Specifically, philosophical questions play an important role in teaching and research. **Web Site:** http://www-philosophie.kgw.tu-berlin.de/philosophie

University of Bielefeld (Institute for Science and Technology Studies). **Contact:** Schulze Petra, office@iwt.uni-bielefeld.de **Primary E-mail:** office@uni-bielefeld.de **Mailing Address:** IWT, Postfach 10 01 31, Bielefeld, D-33501, Germany. **Tel:** 49.521.106.6898 **Fax:** 49.521.106.6418 **Grants:** Yes. **Degrees Conferred:** Master's Level, Doctoral Level. 25 graduate students in 2002. 25 doctorates awarded since July, 1998. **Prospective Student Contact:** Peter Weingart, peter.weingart@uni-bielefeld.de **Web Site:** http://www.uni-bielefeld.de/iwt/

Ruhr-Universität Bochum (Lehrstuhl für Geschichte der Medizin). **Contact:** Irmgard Müller, geschichte.medizin@ruhr-uni-bochum.de **Mailing Address:** Malakowturm, Markstr. 258A, Bochum, Nordrhein-Westfalen, D-44799, Germany. **Tel:** 49.234.3.22.33.94 **Degrees Conferred:** Doctoral Level. **Web Site:** http://www.ruhr-uni-bochum.de/malakow

Rheinische Friedrich-Wilhelms-Universität Bonn (Medizinhistorisches Institut). **Contact:** Heinz Schott, schott@mailer.meb.uni-bonn.de **Primary E-mail:** Heinz.Schott@ukb.uni-bonn.de **Mailing Address:** Medizinhistorisches Institut, Sigmund-Freud-Str. 25, Bonn, North-Rhine-Westfalia, D-53105, Germany. **Tel:** 49.228.287.5000 **Fax:** 49.228.287.5006 **Collections/Resources:** Special library for the history of medicine and allied sciences, about 38,000 volumes. **Grants:** No. **Degrees Conferred:** Bachelor's Level, Doctoral Level. **Faculty:** *Walter Bruchhausen*. **Description:** History of Medicine is not an independent discipline in Germany; it is a sub-discipline of the medical curriculum. But it is possible to graduate with a medico-historical dissertation (Dr. Med./M.D.). We offer lectures, seminars and courses, mainly for medical students, but there is no special (independent) curriculum. **Web Site:** http://www.meb.uni-bonn.de/mhi/

Technische Universität Braunschweig (Historisches Seminar). **Contact:** Herbert Mehrtens, h.mehrtens@tu-bs.de **Mailing Address:** Schleinitzstr. 13, Braunschweig, D-38106, Germany. **Tel:** 49.531.391.3091 **Fax:** 49.531.391.8162 **Collections/Resources:** Herzog August Bibliothek Wolfenbüttel. **Grants:** No. **Program Strengths:** 19th Century, Europe, Social Relations of Science; 20th and 21st Century, Europe, Mathematics. **Degrees Conferred:** Bachelor's Level, Doctoral Level. 7 graduate students in 2002. 3 doctorates awarded since July, 1998. **Faculty:** *Herbert Mehrtens*; *H. Otto Sibum*. **Description:** General history department, specialization in history of science and/or technology for Masters degree is possible. Doctoral level (Dr. Phil.) in cooperation with History of Science Department (Prof. B. Wahrig). Broad transdisciplinary orientation. **Web Site:** http://www.tu-bs.de/institute/geschichte/

Technische Universität Braunschweig (Abteilung für Geschichte der Naturwissenschaften mit Schwerpunkt Pharmaziegeschichte). **Contact:** Bettina Wahrig, B.Wahrig@tu-bs.de **Mailing Address:** Abteilung Geschichte der Naturwissenschaften Technische Universität, Pockelsstr. 14, D-38023 Braunschweig, Niedersachsen, Germany. **Tel:** 49.0531.391.5990 **Fax:** 49.0531.391.599 **Collections/Resources:** Sammlung Schneider: Collection of drugs and pharmaceutical chemicals; Documents on the Ratsapotheke Lehrte; Library on History of Pharmacy and Sciences. **Grants:** No. **Degrees Conferred:** Bachelor's Level, Doctoral Level. 8 graduate students in 2002. 4 doctorates awarded since July, 1998. **Faculty:** *Gabriele Beisswanger*; *Frank Leimkugel*; *Gabriele Wacker*; *Bettina R. Wahrig* (18th Century, Europe, Medical Sciences). **Description:** The program combines graduate and undergraduate teaching. Emphasis is on history of biology, public health and gender studies. **Web Site:** http://www.tu-bs.de/institute/pharmtech/pharmgesch/

Darmstadt University of Technology (History of Technology, Department of History). **Contact:** Mikael Hard, hard@ifs.tu-darmstadt.de **Primary E-mail:** sekrtg@ifs.tu-darmstadt.de **Mailing Address:** Schloss, Darmstadt, DE-64283, Germany. **Tel:** 49.6151.16.67.22 **Fax:** 49.6151.16.39.92 **Degrees Conferred:** Bachelor's Level, Master's Level, Doctoral Level. 5 graduate students in 2002. 4 master's degrees and 1 doctorate awarded since July, 1998. **Description:** The research foci of the section include: automobility and society; the history of sanitation and public health and its implications for technology and science; the history of food chemistry; the history of urban technology; the history of consumption. **Web Site:** http://www.ifs.tu-darmstadt.de/geschichte/index.html

Heinrich-Heine-Universität Düsseldorf (Institut für Geschichte der Medizin). **Contact:** Alfons Labisch, histmed@uni-duesseldorf.de **Mailing Address:** Universitaetsstr. 1 (Geb.23.12), Duesseldorf, D-40225, Germany. **Tel:** 49.211.8.11.39.40 **Fax:** 49.211.8.11.39.49 **Collections/Resources:** Library of the Institute for the History of Medicine (some 20,000 books); Art collection Man and Death (danse macabre). **Grants:** No. **Program Strengths:** 20th and 21st Century, Europe, Medical Sciences, Social Relations

of Science; 20th and 21st Century, Europe, Medical Sciences, Science
Policy; 19th Century, Europe, Natural and Human History, Instruments and
Techniques. **Degrees Conferred:** Bachelor's Level, Master's Level, Doctoral
Level. 15 graduate students in 2002. 6 master's degrees and 10 doctorates
awarded since July, 1998. **Prospective Student Contact:** Ulrich Koppitz,
koppitz@uni-duesseldorf.de **Faculty:** *Christoph auf der Horst; Fritz Dross;
Michael K. H. Elies; Barbara Elkeles; Silke Fehlemann; Carmen Götz; Uwe
Heyll; Norbert Kohnen; Ulrich H. Koppitz; Alfons Labisch* (20th and 21st
Century, Europe, Medical Sciences); *Norbert W. Paul; Hans Schadewaldt; Eva
Schuster; Silke Stelbrink; Joerg P. Voegele; Wolfgang W. Woelk.* **Description:**
Main areas of research are: Social History of Medicine; Illness, Health and
Body Perception; Public Health and Health Policy; Medicine and National
Socialism; Historical Demography and Epidemiology; Hospitals, Nursing and
Social Policy; Complementary Medicine (Naturheilkunde); Tropical Medicine
(Malariology); Environmental History; Man and Death in Graphical Arts. **Web
Site:** http://www.uni-duesseldorf.de/WWW/MedFak/HistMed/welcome.htm

Universität Freiburg (Institut für Geschichte der Medizin). **Contact:**
Ulrich Tröhler, medgesch@igm.uni-freiburg.de **Mailing Address:** Stefan-
Meier-Str. 26, Freiburg, D-79104, Germany. **Tel:** 49.761.2.03.50.33
Fax: 49.761.2.03.50.39 **Ongoing Publication Projects:** Members of the faculty
are coeditors of *Medizinhistorisches.* **Collections/Resources:** The Institute
Library holds about 40,000 books, periodicals, and off prints on the history
of medicine. The Medical Association of the Land of Baden-Württemberg
has incorporated its medical ethics library in the "Center for Ethics and Law
in Medicine" which to date holds about 4,500 books and over 20 periodicals
relating to medical ethics. The Institute's archives hold several thousand issues
of a complete and catalogued collection of off-prints by the leading German
pathologist of his time, Ludwig Aschoff (1866-1942), as well as the case
histories of a Swiss general practitioner between 1950 and 1985. **Grants:** No.
Program Strengths: 20th and 21st Century, Europe, Medical Sciences; 19th
Century, Europe, Medical Sciences; 400 B.C.E - 400 C.E., Europe, Medical
Sciences. **Degrees Conferred:** Bachelor's Level, Master's Level, Doctoral
Level. **Prospective Student Contact:** University of Freiburg, international-
office@verwaltung.uni-freiburg.de **Faculty:** *Hans-Georg Hofer; Karl-Heinz
Leven; Lutz D.H. Sauerteig; Ulrich Tröhler.* **Description:** The Institute for
the History of Medicine is part of the Medical Faculty; its chairman is also a
member of the History Faculty. A lecture course "History of Medicine" is held
for third-year medical students. However, this lecture course as well as a number
of other lecture courses and seminars on varying topics ranging from antiquity
to the present are open to students of all departments. The Institute is integrated
in the combined study programs of Molecular Medicine, Historical/Biological
Anthropology, and Gender Studies. Medical History can also be included
as a minor in the Master's and Doctoral programs of the History Faculty.
Web Site: http://www.uni-freiburg.de/igm/indexigm.html

Georg August Universität Göttingen (Abteilung Ethik und Geschichte der Medizin). **Contact:** Claudia Wiesemann, cwiesem@gwdg.de **Mailing Address:** Humboldtallee 36, Göttingen, D-37073, Germany. **Tel:** 49.551.39.90.06 **Fax:** 49.551.39.95.54 **Collections/Resources:** The library holds 21,783 books and journals. Services include online searching of the library catalog via the state and University Library Göttingen. The institute also has a museum, called the "Armamentarium obstetricium Göttingense." In 1751, the first European maternity hospital which was a university institution was founded in Göttingen. The directors of this hospital collected obstetric instruments, wax models and other exhibits for teaching and research purposes. A special exhibition in our institution illustrates the practice of obstetrics and its ethical conflicts throughout the last 250 years. **Grants:** No. **Program Strengths:** 14th-16th Century, Europe, Medical Sciences; 20th and 21st Century, Medical Sciences, Humanistic Relations of Science; Medical Sciences. **Degrees Conferred:** Bachelor's Level, Doctoral Level. **Prospective Student Contact:** Volker Zimmermann, vzimmer@gwdg.de **Faculty:** *Roberto Andorno*; *Nikola Biller-Andorno*; *Heiner Fangerau*; *Andreas Frewer*; *Jörg Janssen*; *Christian Lenk*; *Karl-Heinz Stubenrauch*; *Claudia Wiesemann*; *Volker Zimmermann*. **Description:** In medical history, faculty research focuses upon: medicine in the Middle Ages, renaissance physiology, medicine in national socialist time, history of medical ethics in the 20th century, history of organ transplantation and brain death, history of addiction. In medical ethics: euthanasia, gene and reproductive technologies, brain death, human experimentation, bioethics from a gender perspective, evidence-based medicine. **Web Site:** http://www.gwdg.de/paracelsus

Georg August Universität Göttingen (Institut für Wissenschaftsgeschichte). **Contact:** Nicolaas Rupke, nrupke@gwdg.de **Primary E-mail:** eeck@gwdg.de **Mailing Address:** Institut für Wissenschaftsgeschichte, Humboldtallee 11, Göttingen, D-37073, Germany. **Tel:** 49.551.39.94.66 **Fax:** 49.551.39.97.48 **Collections/Resources:** Research Library for the History of Science, with 150,000 selected volumes from the period 1600-1900 in open stacks; Manuscript Division, with papers of Göttingen scientists, including Johann Friedrich Blumenbach, Carl Friedrich Gauss, and David Hilbert. **Grants:** No. **Program Strengths:** 19th Century, Europe, Biological Sciences, Biography; Science and Religion; Social Relations of Science. **Degrees Conferred:** Bachelor's Level, Doctoral Level. 7 graduate students in 2002. **Faculty:** *Norbert Elsner*; *Jan Kornelis Oosthoek*; *Nicolaas A. Rupke* (19th Century, North America, Biological Sciences); *Bernd Weisbrod*; *Karen Elizabeth Wonders*. **Web Site:** http://www.gwdg.de/~uhwg/

Universität Greifswald (Institut für Geschichte der Medizin). **Contact:** Heinz-Peter Schmiedebach, geschmed@uni-greifswald.de **Mailing Address:** Greifswald, 17487, Germany. **Tel:** 49.3834.865.780 **Fax:** 49.3834.865.782 **Collections/Resources:** Collections of medical instruments from the 19th-20th centuries, including devices of the German Democratic Republic. **Grants:** Yes.

Program Strengths: 19th Century, Europe, Natural and Human History, Social Relations of Science; 14th-16th Century, Europe, Natural and Human History, Philosophy or Philosophy of Science; 20th and 21st Century, Europe, Medical Sciences, Science Policy. **Degrees Conferred:** Doctoral Level. 25 graduate students in 2002. 0 master's degrees and 5 doctorates awarded since July, 1998. **Faculty:** *Thomas Beddies*; *Mariacarla Bondio Gadebusch*; *Gabriele Moser*; *Heinz-Peter Schmiedebach*. **Description:** The program deals with medical history of the 19th-20th centuries and emphasizes public health, medical expertise, psychiatric care and medicine in national socialistic Germany. **Web Site:** http://www.medizin.uni-greifswald.de/geschichte

Martin-Luther Universität Halle-Wittenberg (Fachgruppe Geschichte der Naturwissenschaften und der Technik). **Contact:** Andreas Kleinert, kleinert@physik.uni-halle.de **Mailing Address:** Fachbereich Physik, Kroellwitzer Str. 44, Halle/ Saale, 06120, Germany. **Tel:** 49.34.5552.5420 **Fax:** 49.34.5552.7126 **Ongoing Publication Projects:** Correspondence of Leonhard Euler (1707-1783); Scientific correspondence of Emil Fischer (1852-1919). **Grants:** No. **Program Strengths:** 18th Century, Europe, Physical Sciences, Biography; 20th and 21st Century, Europe, Physical Sciences, Biography; 19th Century, Europe, Physical Sciences. **Degrees Conferred:** Bachelor's Level, Doctoral Level. 11 graduate students in 2002. 3 doctorates awarded since July, 1998. **Faculty:** *Horst Remane*. **Web Site:** http://www.physik.uni-halle.de/Fachgruppen/history/index.html

Universität Hamburg, Universitäts-Krankenhaus Eppendorf (Institut für Geschichte der Medzin). **Contact:** Kai Sammet, sammet@uke.uni-hamburg.de **Mailing Address:** Martinistr. 52, Hamburg, Germany, D-20246, Germany. **Tel:** 49.40.42803.2140 **Fax:** 49.40.42803.2462 **Collections/ Resources:** Historical Photo-Archive of the Universitats-Krankenhaus Eppendorf. **Grants:** No. **Program Strengths:** Medical Sciences. **Degrees Conferred:** Doctoral Level. 21 graduate students in 2002. 0 master's degrees and 3 doctorates awarded since July, 1998. **Faculty:** *Kai Sammet*. **Description:** Old medicine: Antiquity and Latin Middle Ages, Arabic-Islamic Middle Ages. History of Psychiatry in the 19th/20th Century. Medicine in Hamburg, especially History of Universitäts-Krankenhaus Eppendorf. **Web Site:** http://www.uke.uni-hamburg.de/institute/geschichte_medizin.html

Medizinischen Hochschule Hannover (Abteilung Medizingeschichte, Ethik und Theoriebildung). **Contact:** Brigitte Lohff, lohff.brigitte@mh-hannover.de **Mailing Address:** Carl-Neuberg-Straße 1, Medizinische Hochschule Hannover, Hannover, Niedersachsen, D-30625, Germany. **Tel:** 49.511.5.32.42.78 **Fax:** 49.511.532.5650 **Collections/Resources:** Complete collection of Literature of Leibniz, Ethic and the Nazi-Regime. **Grants:** No. **Degrees Conferred:** Bachelor's Level, Master's Level, Doctoral Level. 25 graduate students in 2002. 2 master's degrees and 18 doctorates awarded since July, 1998. **Faculty:** *Brigitte Lohff* (19th Century, Europe, Medical Sciences); *Gerald Neitzke*;

Sigrid Stöckel; *Angelika Voss*. **Description:** History of medicine in the 19th and 20th century in Europe. Special topics: public-health history 1900 until 1970, history of physiology and epistemological and ethical aspects of medicine. **Web Site:** http://www.mh-hannover.de/institute/medizingeschichte/

Friedrich Schiller Universität Jena (Institut für Geschichte der Medizin, Naturwissenschaften und Technik). **Contact:** Gerhard Wiesenfeldt, G.Wiesenfeldt@uni-jena.de **Mailing Address:** Berggasse 7, Jena, Thuringia, D-07745, Germany. **Tel:** 49.3641.94.95.00 **Fax:** 49.3641.949502 **Ongoing Publication Projects:** Sonderforschungsbereich "Ereignis Weimar/Jena - Kultur um 1800" (interdisciplinary research group); *Theorielabor*; Book Series "Ernst-Haeckel-Haus Studien"; *Theory in Bioscience*. **Collections/ Resources:** Ernst Haeckel archive (more than 30,000 letters, manuscripts and other documents relating to Haeckel and biology ca. 1900); Extensive library resources on the 17th and 18th centuries. **Grants:** Yes. **Program Strengths:** 19th Century, Europe, Biological Sciences; 18th Century, Europe, Humanistic Relations of Science; 20th and 21st Century, Biological Sciences, Philosophy or Philosophy of Science. **Degrees Conferred:** Bachelor's Level, Master's Level, Doctoral Level. 22 graduate students in 2002. 0 master's degrees and 0 doctorates awarded since July, 1998. **Faculty:** *Thomas Bach*; *Olaf Breidbach*; *Maurizio di Bartolo*; *Jan Frercks*; *Uwe Hossfeld*; *Joachim Schult*; *Gerhard Wiesenfeldt*; *Susanne Zimmermann*. **Description:** The institute offers a 9 semester M.Sc. course, in which history of science must be combined with one science and one other humanity. Its focus relates to the research topics of the institute (biology around 1900; romanticism, the sciences in Jena around 1800). Furthermore, there is the opportunity to enter a Ph.D. program in the history of science within the biological faculty. **Web Site:** http://www.uni-jena.de/biologie/ehh/haeckel.htm

Universität zu Lübeck (Institut für Medizin- und Wissenschaftsgeschichte). **Contact:** Dietrich von Engelhardt, v.e@imwg.mu-luebeck.de **Mailing Address:** Königstr. 42, IMWG, Lübeck, Schleswig-Holstein, D-23552, Germany. **Tel:** 49.451.707998.12 **Fax:** 49.451.707998.99 **Ongoing Publication Projects:** *Acta Biotheoretica*; *Balint*; *BioLogica*; Deutsche Biographische Enzyklopädie; *Medicine and Mind*; *NTM*; *Physis*; *Pluriverso (Ceruti)Naturforscher und Ärzte*; *Schriftenreihe zu Psychopathologie, Kunst und Literatur*; *Zeitschrift für medizinische Ethik*. **Collections/Resources:** research library of 25,000 volumes; Landmarks of Science; Bibliothek des Ärztlichen Vereins zu Lübeck. **Grants:** No. **Program Strengths:** Medical Sciences, Social Relations of Science; Biological Sciences, Social Relations of Science; Physical Sciences, Institutions. **Degrees Conferred:** Bachelor's Level, Doctoral Level. 8 graduate students in 2002. 0 master's degrees and 5 doctorates awarded since July, 1998. **Faculty:** *Dietrich von Engelhardt*; *Kai T. Kanz* (18th Century, Europe, Natural and Human History); *Volker Roelcke*; *Burghard O. Weiss*. **Description:** Students of medicine, or the natural sciences have the opportunity to choose a historical topic for their M.D. or Ph.D. thesis; however, there is no

specific graduate program in the history of science, medicine, and technology.
Web Site: http://www.imwg.mu-luebeck.de

Johannes Gutenberg Universität Mainz (Medizinhistorisches Institut).
Contact: Werner F. Kuemmel, wekuemme@mail.uni-mainz.de **Mailing
Address:** Universitätsklinikum, Am Pulverturm 13, Mainz, D-55101, Germany.
Tel: 49.6131.393.73.55/56 **Fax:** 49.6131.393.6682 **Ongoing Publication
Projects:** The Diaries of Samuel Thomas Soemmerring (1804/05-1830), Franz
Dumont, editor. **Collections/Resources:** Special literature on "Biologism"
in the late 19th and 20th centuries and on medicine and National Socialism.
Grants: No. **Program Strengths:** 5th-13th Century, Europe, Medical
Sciences, Education; 18th Century, Europe, Medical Sciences; 20th and 21st
Century, Europe, Medical Sciences. **Degrees Conferred:** Bachelor's Level,
Master's Level, Doctoral Level. 25 graduate students in 2002. 11 doctorates
awarded since July, 1998. **Prospective Student Contact:** Klaus-Dietrich
Fischer, kdfisch@mail.uni-mainz.de **Faculty:** *Klaus-Dietrich Fischer*;
Werner Friedrich Kümmel; *Michael Kutzer*; *Georg Lilienthal*; *Sabine Sander*;
Klaus-Dieter Thomann. **Description:** The Institute was founded in 1947
by Paul Diepgen. The library holdings are rich in literature on medicine in
antiquity and the Middle Ages as well as in the history of gynecology and
psychiatry. Diepgen's collection, part of the library, deserves special mention.
Web Site: http://www.uni-mainz.de/FB/Medizin/Medhist/Welcome.html

Philipps Universität Marburg (Institut für Geschichte der Pharmazie).
Contact: Christoph Friedrich, Ch.Friedrich@mailer.uni-marburg.de
Primary E-mail: igphmr@mailer.uni-marburg.de **Mailing Address:**
Roter Graben 10, Marburg, D-35032, Germany. **Tel:** 49.6421.2.82.28.29
Fax: 49.6421.2.82.28.78 **Ongoing Publication Projects:** *Quellen und Studien
zur Geschichte der Pharmazie* (Friedrich/Krafft); *Pharmaziehistorische
Forschunge* (Dilg). **Collections/Resources:** Specialized library for the history
of pharmacy and related sciences; Johann Bartholomaeus Trommsdorff
Collection. **Grants:** No. **Program Strengths:** 18th Century, Europe, Natural
and Human History, Biography; 18th Century, Europe, Natural and Human
History, Philosophy or Philosophy of Science; 20th and 21st Century, Europe,
Natural and Human History, Institutions. **Degrees Conferred:** Bachelor's
Level, Doctoral Level. 9 graduate students in 2002. 14 doctorates awarded
since July, 1998. **Prospective Student Contact:** Katja Schmiederer,
schmiede@mailer.uni-marburg.de **Faculty:** *Hartmut Bettin*; *Peter Dilg*;
Christoph Friedrich; *Fritz A. Krafft* (14th-16th Century, Europe, Humanistic
Relations of Science); *Tanja Pommerening*; *Daniela Schierhorn*; *Katja
Schmiederer*. **Description:** Three-term course of history of science and
pharmacy for graduates of natural sciences, especially pharmacists, to prepare
a dissertation in the field of the history of pharmacy or natural sciences.
Web Site: http://staff-www.uni-marburg.de/~igphmr/

Munich Center for the History of Science and Technology (Graduate Program in History of Science and Technology). **Contact:** Helmuth Trischler, h.trischler@deutsches-museum.de **Mailing Address:** Deutsches Museum, München, Bavaria, D-80306, Germany. **Degrees Conferred:** Master's Level, Doctoral Level. 8 graduate students in 2002. 15 master's degrees and 12 doctorates awarded since July, 1998. **Prospective Student Contact:** Menso Folkerts, m.folkerts@lrz.uni-muenchen.de **Faculty:** *Ralph Boch*; *Peter Dorsch*; *Michael Eckert*; *Menso Folkerts* (5th-13th Century, Europe, Mathematics); *Wilhelm Fossl*; *Alexander Gall*; *Ulf Hashagen* (19th Century, Europe, Mathematics); *Martina Hessler*; *Martin Kintzinger*; *Eva A. Mayring*; *Arne Schirrmacher*; *Ivo H. Schneider* (Europe, Mathematics); *Jürgen Teichmann*; *Helmuth Trischler*; *Ulrich Wengenroth*; *Juliane C. Wilmanns*. **Description:** Created in December, 1997, the Center brings together faculty and resources from the three Munich universities. In addition to continuing the graduate program at the Deutsches Museum, the center is about to create new master's courses for the history of science, medicine and technology. The Center also participates in a Master's Program in "Social Science of Technology" at the University of Technology of Munich. Several large research projects are under preparation. Most of the faculty of the Munich Center have their offices at the Deutsches Museum. **Web Site:** http://www.mzwtg.mwn.de/

Ludwig Maximilians Universität München (Lehrstuhl für Geschichte der Naturwissenschaften). **Contact:** Menso Folkerts, m.folkerts@lrz.uni-muenchen.de **Primary E-mail:** ign@lrz.uni-muenchen.de **Mailing Address:** Museumsinsel 1, München, D-80538, Germany. **Tel:** 49.89.2180.3252 **Fax:** 49.89.2180.3162 **Ongoing Publication Projects:** 2 series in history of science: *Boethius*; *Algorismus*. **Collections/Resources:** Microfilm collection of medieval Western manuscripts on the mathematical sciences (about 5000 items); Nachlaesse (manuscripts, correspondence etc.) of (mostly 19th century) mathematicians; also historians of mathematics (K. Vogel, J. E. Hofmann). **Grants:** No. **Program Strengths:** Mathematics; Biological Sciences; Physical Sciences. **Degrees Conferred:** Bachelor's Level, Master's Level, Doctoral Level. 7 graduate students in 2002. 10 master's degrees and 10 doctorates awarded since July, 1998. **Prospective Student Contact:** S. Kirschner, s.kirsch ner@lrz.uni-muenchen.de **Faculty:** *Menso Folkerts* (5th-13th Century, Europe, Mathematics); *Bernhard Fritscher*; *Brigitte Hoppe*; *Martin Kintzinger*; *Stefan Kirschner*; *Wolfgang Kokott*; *Andreas Kuehne*; *Paul Kunitzsch*; *Richard Paul Lorch*; *Claus Priesner*; *Felix Schmeidler*; *Michael Segre*; *Juergen Teichmann*. **Description:** The program offers courses on the history of mathematics, astronomy, physics, biology and chemistry. *This program is part of the Munich Center for the History of Science and Technology.* **Web Site:** http://www.ign.uni-muenchen.de

Technische Universität München (Social Science of Technology). **Contact:** Ulrich Wengenroth, ulrich.wengenroth@lrz.tum.de **Primary E-mail:** zigt@lrz.tum.de **Mailing Address:** c/o Deutsches Museum, München, D-80306,

Germany. **Tel:** 49.89.2.17.94.02 **Fax:** 49.89.2.17.94.08 **Grants:** No.
Program Strengths: 20th and 21st Century, Europe, Social Sciences. **Degrees Conferred:** Master's Level. **Faculty:** *Gwen Bingle* (Biological Sciences); *Martina Blum*; *Margot Fuchs* (Philosophy and Philosophy of Science); *Stephan Lindner* (17th Century, Europe, Humanistic Relations of Science); *Heike Weber* (Philosophy and Philosophy of Science); *Ulrich Wengenroth*; *Thomas Wieland* (20th and 21st Century, Europe, Philosophy and Philosophy of Science); *Juliane C. Wilmanns*. **Description:** Research focuses on 20th century history of technology. "Innovation culture" and "consumption and technology" are the main research fields. *This program is part of the Munich Center for the History of Science and Technology.* **Web Site:** http://www.zigt.ze.tu-muenchen.de/

Universität der Bundeswehr München. **Contact:** Ivo Schneider, Ivo.Schne ider@unibw-muenchen.de **Mailing Address:** Werner-Heisenberg-Weg 39, Neubiberg, D-85577, Germany. **Tel:** 49.89.60.04.3342 **Fax:** 49.89.60.04.3342 **Collections/Resources:** Database for the history of stochastics. **Grants:** No. **Program Strengths:** Europe. **Degrees Conferred:** Bachelor's Level, Master's Level, Doctoral Level. 3 graduate students in 2002. 5 master's degrees and 2 doctorates awarded since July, 1998. **Faculty:** *Brigette Hoppe*; *Ivo H. Schneider* (Europe, Mathematics); *Rudolf Seising*; *Falk Seliger*; *Carsten Trinitis*. **Description:** *This program is part of the Munich Center for the History of Science and Technology.* **Web Site:** http://www.unibw-muenchen.de/campus/ SOWI/instfak/wige/wige.html

Westfälische Wilhelms Universität Münster (Institut für Theorie und Geschichte der Medizin). **Contact:** Hans-Peter Kröner, kroener@uni-muenster.de **Mailing Address:** Waldeyerstr. 27, Münster, D-48149, Germany. **Tel:** 49.251.83.55291 **Fax:** 49.251.83.55339 **Ongoing Publication Projects:** *Artificial Intelligence,* edited by Kazem Sadegh-Zadeh. **Collections/Resources:** Collection of off-prints of the former Kaiser-Wilhelm Institute for anthropology, human genetics and eugenics; Collection of old medical books from the 17th, 18th and 19th century. **Grants:** No. **Program Strengths:** 20th and 21st Century, Europe, Medical Sciences; Medical Sciences, Philosophy or Philosophy of Science; Medical Sciences, Humanistic Relations of Science. **Degrees Conferred:** Bachelor's Level, Doctoral Level. 30 graduate students in 2002. 0 master's degrees and 7 doctorates awarded since July, 1998. **Prospective Student Contact:** Peter Hucklenbroich, hucklen@uni-muenster.de **Faculty:** *Petra Gelhaus*; *Peter Hucklenbroich*; *Hans-Peter Kroener*; *Daniela Mergenthaler*; *Heike Petermann*; *Kazem Sadegh-Zadeh*. **Description:** Research of the faculty focuses upon: medicine and National Socialism, history of eugenics, history of human genetics, history of population science, philosophy of medicine, medical ethics, and fuzzy logic. **Web Site:** http://medweb.uni-muenster.de/institute/itgm/

Universität Regensburg (Lehrstuhl für Wissenschaftsgeschichte). **Contact:** Christoph Meinel, christoph.meinel@psk.uni-regensburg.de **Mailing Address:** Universität Regensburg, Regensburg, D-93040, Germany.

Tel: 49.941.9.43.36.59 **Fax:** 49.941.9.43.19.85 **Collections/Resources:** collection of 18th and 19th century scientific instruments. **Grants:** No. **Program Strengths:** 19th Century, Europe, Physical Sciences; 17th Century, Europe, Physical Sciences; 20th and 21st Century, Physical Sciences. **Degrees Conferred:** Bachelor's Level, Master's Level, Doctoral Level. 12 graduate students in 2002. 3 master's degrees and 3 doctorates awarded since July, 1998. **Faculty:** *Lis Brack-Bernsen*; *Christoph Meinel* (Europe, Physical Sciences); *Carsten Reinhardt* (20th and 21st Century, Physical Sciences). **Description:** Research includes the history of 18th- to 20th- century chemistry, early modern science, and mathematical aspects of Babylonian astronomy. Interdisciplinary studies are particularly encouraged, and joint courses with colleagues from philosophy, physics and German literature are offered. Ph.D. dissertations in languages other than German can be accepted. **Web Site:** http://www.uni-regensburg.de/Fakultaeten/phil_Fak_I/Philosophie/Wissenschaftsgeschichte/

Universität-GH Wuppertal (Professur für Mathematikgeschichte). **Contact:** Erhard Scholz, scholz@math.uni-wuppertal.de **Mailing Address:** Gaussstr. 20, Wuppertal, D-42097, Germany. **Tel:** 49.202.4.39.25.26 **Fax:** 49.202.4.39.3778 **Ongoing Publication Projects:** Editorial participation in: *Hausdorff*-Edition Science Networks, *Birkhauser*, *Basel*, and *Revue d'Histoire des Mathématiques*. **Grants:** No. **Program Strengths:** 19th Century, Europe, Mathematics; 20th and 21st Century, Europe, Mathematics. **Degrees Conferred:** Master's Level, Doctoral Level. 1 graduate student in 2002. 1 master's degree and 1 doctorate awarded since July, 1998. **Description:** History of modern mathematics, relation between theoretical mathematics and physics, philosophical and cultural relations of mathematics, history of mathematics in teacher education. **Web Site:** http://www.math.uni-wuppertal.de/rd/didactics/index_de.html#ge

Bayerische Julius-Maximilians Universität Würzburg (Institut für Geschichte der Medizin). **Contact:** Gundolf Keil, gesch.med@mail.uni-wuerzburg.de **Mailing Address:** Oberer Neubergweg 10a, Röntgenring 10, Würzburg, Bayern, D-97074, Germany. **Tel:** 49.931.79.67.80 **Fax:** 49.931.79.67.87.8 **Ongoing Publication Projects:** *Sudhoffs Archiv: Zeitschrift für Wissenschaftsgeschichte*; *Würzburger Medizinhistorische Forschungen*; *Würzburger Medizinhistorische Mitteilungen*; *Text und Wissen*; *Der Würzburger Kreis*; *Wissensliteratur im Mittelalter*; *Die Deutsche Literatur im Mittelalter. Verfasserlexikon*. **Collections/Resources:** 135,000 volumes, perhaps the largest history of medicine book collection in continental Europe. **Grants:** Yes. **Degrees Conferred:** Bachelor's Level, Doctoral Level. 1 master's degree and 1 doctorate awarded since July, 1998. **Prospective Student Contact:** Monika Reininger, monika.reininger@mail.uni-wuerzburg.de **Faculty:** *Christian Andree*; *Josef Domes* (North America); *Michael Freyer*; *Werner E. Gerabek*; *Cornelia Gräff* (Medical Sciences); *Dominik Groß*; *Hilde-Marie Groß*; *Marianne Halbleib*; *Brigitte Hohmann*; *Johannes Gottfried Mayer*; *Reinhard Platzek*; *Waltraud Prestel*; *Anne Rappert*; *Monika Reininger* (North America, Medical Sciences);

Michael Sachs; Alexander Schütz; Doris Schwarzmann-Schafhauser; Jan Steinmetzer; Ralf Windhaber; Christine Wolf (Asia). **Description:** History of Ecclesiastical Medicine in early medieval cloisters; Fachprosaforschung, i.e. German and Dutch scientific and medical literature of the middle ages; the History of Medicine of the German speaking Silesia (now Western Poland and North Eastern Czechia; History of Phytotherapy; Lexicography of the history of medicine; Lexicology of ancient and modern dentistry and medicine. **Web Site:** http://www.uni-wuerzburg.de/medizingeschichte/

Hungary

Eotvos University, Budapest (Department of History and Philosophy of Science). **Contact:** George Kampis, gk@hps.elte.hu **Mailing Address:** Pazmany Peter s. 1., H-1117, Budapest, H-1518, PoB 32., Hungary. **Tel:** 36.1.372.2924 **Fax:** 36.1.372.2924 **Collections/Resources:** Hungarian Darwiniana; Lakatos archives. **Grants:** No. **Degrees Conferred:** 0 master's degrees and 0 doctorates awarded since July, 1998. **Faculty:** *George Kampis; Gabor Kutrovatz; Miklos Redei* (20th and 21st Century, Europe, Physical Sciences); *Laszlo Ropolyi; Peter Szegedi; Andras Szigeti; Miklos Zagoni.* **Description:** Our Department's main mission is teaching in history and philosophy of science and research related to these fields. Currently no degree program in HPS (M.Sc. pending), but undergraduate and graduate level cognitive science programs. **Web Site:** http://hps.elte.hu

Israel

Bar Ilan University (Graduate Program for the History & Philosophy of Science, Technology and Medicine). **Contact:** Noah J. Efron, efron@mail.biu.ac.il **Mailing Address:** Graduate Program for the History & Philosophy of Science, Technology & Medicine, Committee for Interdisciplinary Studies, Bar Ilan University, Ramat Gan, 52900, Israel. **Tel:** 972.3.531.7756 **Fax:** 972.3.535.4389 **Grants:** Yes. **Program Strengths:** Science and Religion; Physical Sciences, Philosophy and Philosophy of Science; Biological Sciences, Social Relations of Science. **Degrees Conferred:** Master's Level, Doctoral Level. 29 graduate students in 2002. **Faculty:** *Raz Chen-Morris; Noah J. Efron; Avshalom Elitzur* (17th Century); *Menachem Fisch; Snait Gisis; Joseph Hodara; Y. Tzvi Langermann; Eyval Ramati; David Rier.* **Description:** Program features an STS approach, emphasizing social studies of science in historical perspective. **Web Site:** http://www.biu.ac.il/hps

Hebrew University (Program for the History, Philosophy and Sociology of Science). **Contact:** Otniel E. Dror, otniel@md.huji.ac.il **Mailing Address:** Program for the History, Philosophy and Sociology of Science, Faculty of Humanities, The Hebrew University of Jerusalem, Jerusalem, 91905, Israel. **Prospective Student Contact:** Neta Zinger, netaz@savion.huji.ac.il

Faculty: *Mara Beller*; *Yemima Ben-Menahem*; *Otniel E. Dror* (Medical Sciences); *Itamar Pitowsky*.

Hebrew University-Hadassah Medical School (Manuel M. Glazier M.D. Institute of the History of Medicine). **Contact:** Otniel E. Dror, otniel@md. huji.ac.il **Mailing Address:** History of Medicine, The Hebrew University Medical School, P. O. Box 12272, Jerusalem, 91120, Israel. **Tel:** 972.2.6757162 **Fax:** 972.2.6784010 **Ongoing Publication Projects:** *Korot: Journal of Medicine and Judaism*. **Collections/Resources:** Museum for History of Medicine; History of Medicine book collection. **Grants:** Yes. **Program Strengths:** 20th and 21st Century, North America, Medical Sciences; 19th Century, Europe, Biological Sciences; Medical Sciences, Science and Religion. **Degrees Conferred:** Bachelor's Level, Master's Level, Doctoral Level. **Faculty:** *Otniel E. Dror* (Medical Sciences); *Samuel Kottek*. **Description:** The program resides in the Medical Faculty. It is not an independent program and students interested in a graduate degree must be enrolled in another department (history, philosophy, anthropology, etc.). The program offers courses in the cultural history of the life sciences, 18th-20th centuries, and courses in medicine and Judaism.

Italy

University of Bologna. **Contact:** Giuliano Pancaldi, pancaldi@alma.unibo.it **Primary E-mail:** cis@philo.unibo.it **Mailing Address:** CIS-Department of Philosophy, Via Zamboni 38, Bologna, Italy, I-40126, Italy. **Tel:** 39.051.2098331 **Fax:** 39.051.2098670 **Ongoing Publication Projects:** Bologna Studies in History of Science (9 volumes since 1989); *Universitas*: Newsletter of the International Center for the History of Universities and Science (13 issues since 1991). **Collections/Resources:** The libraries and archives of the University of Bologna, established in the 11th century. **Grants:** Yes. **Program Strengths:** 17th Century, Europe, Humanistic Relations of Science; 18th Century, Europe, Instruments and Techniques; 19th Century, Europe, Social Relations of Science. **Degrees Conferred:** Bachelor's Level, Doctoral Level. 3 graduate students in 2002. 2 doctorates awarded since July, 1998. **Faculty:** *Anna Guagnini*; *Giuliano Pancaldi* (18th Century, Europe, Biological Sciences); *Pietro Redondi*. **Description:** The program is part of a joint initiative of the Universities of Bari, Bologna, Genova and Lecce. Together with Berkeley, Paris and Uppsala, the University of Bologna also promotes, since 1988, the International Summer School in History of Science. **Web Site:** http://www.cis.unibo.it

Japan

Hokkaido University (History of Science). **Contact:** Shigeo Sugiyama, sugiyama@hps.sci.hokudai.ac.jp **Mailing Address:** Division of Physics, Graduate School of Science, Hokkaido University, Sapporo, Hokkaido, 060-0810, Japan. **Tel:** 81.11.706.4421 **Fax:** 81.11.706.4421 **Collections/**

Resources: Journals of a professor named Hori Takeo written while he was in Copenhagen in 1920s to work at the Niels Bohr Institute. **Grants:** No. **Program Strengths:** 19th Century, Asia; 20th and 21st Century, Asia; 20th and 21st Century, Asia, Education. **Degrees Conferred:** Doctoral Level. 8 graduate students in 2002. 6 master's degrees and 1 doctorate awarded since July, 1998. **Web Site:** http://hps.sci.hokudai.ac.jp/history.html

University of Tokyo (Department of History and Philosophy of Science). **Contact:** Takuji Okamoto, cotakuji@mail.ecc.u-tokyo.ac.jp **Mailing Address:** 3-8-1 Komaba, Meguro-ku, Tokyo, 153-8902, Japan. **Ongoing Publication Projects:** Archive for Philosophy and the History of Science; *The Japanese Journal for the History of Science and Technology.* **Grants:** No. **Degrees Conferred:** Bachelor's Level, Master's Level, Doctoral Level. 40 graduate students in 2002. 15 master's degrees and 3 doctorates awarded since July, 1998. **Faculty:** *Takehiko Hashimoto* (Technology, Instruments and Techniques); *Yoshiyki Hirono*; *Takuji Okamoto*; *Chikara Sasaki.*

The Netherlands

University of Twente (Science, Technology and Society). **Contact:** J. P. van Diepen, j.p.vandiepen@wmw.utwente.nl **Primary E-mail:** wwts@wmw.utwente.nl **Mailing Address:** Drienerlolaan 5, Enschede, 7522 NB, Netherlands. **Tel:** 31.53.489.4393 **Fax:** 31.53.489.2255 **Grants:** No. **Program Strengths:** 17th Century, Europe, Mathematics, Humanistic Relations of Science; 18th Century, Europe, Instruments and Techniques; 19th Century, Europe. **Degrees Conferred:** Bachelor's Level, Master's Level, Doctoral Level. 60 graduate students in 2002. 25 master's degrees and 5 doctorates awarded since July, 1998. **Faculty:** *H. Floris Cohen* (17th Century); *Fokko J. Dijksterhuis* (17th Century, Europe, Physical Sciences); *Paul Lauxtermann*; *Lissa Roberts.* **Description:** This program offers three study directions: philosophy of technology; history of science and technology; sociology of science and technology. The above information relates only to the history division of the program. **Web Site:** http://www.wmw.utwente.nl

University of Utrecht (Institute for the History and Foundations of Science). **Contact:** A. Van Helden, a.vanhelden@phys.uu.nl **Primary E-mail:** w.vanputten@phys.uu.nl **Mailing Address:** P.O. Box 80.000, Utrecht, 3508 TA, Netherlands. **Tel:** 31.30.253.8040 **Fax:** 31.30.253.6313 **Grants:** Yes. **Program Strengths:** Physical Sciences; Biological Sciences. **Degrees Conferred:** Bachelor's Level, Master's Level, Doctoral Level. 5 graduate students in 2002. **Prospective Student Contact:** Wilca van Putten, w.vanputten@phys.uu.nl **Faculty:** *Cornelis de Pater* (18th Century, Europe, Physical Sciences); *Dennis Dieks*; *Lodewijk C. Palm* (17th Century, Biological Sciences); *Bert Theunissen* (19th Century, Europe, Biological Sciences); *Jos B. M Uffink*; *Albert Van Helden* (17th Century, Europe, Astronomical Sciences); *Frans H. van Lunteren*; *R. P. W. Visser.* **Description:** The Institute is part of the Faculty of Physics and Astronomy and consists of two distinct sections: The History of Mathematics and the Natural Sciences section and The Foundations of Physics section. **Web Site:** http://www.phys.uu.nl/~wwwgrnsl/

Russia

M. Lomonosov Moscow State University (History of Mathematics). **Contact:** Konstantin Rybnikov, sgs@moids.math.msu.ru **Mailing Address:** Vorob'evy gory, Moscow State University, Mathematical Department, suite 1609, Moscow, 119899, Russia. **Tel:** 7.95.939.3860

South Korea

Korea University (Program in Science, Technology & Society). **Contact:** Mun-Cho Kim, muncho@korea.ac.kr **Primary E-mail:** science7@orgio.net **Mailing Address:** Department of Sociology, Korea University, 5 Anam-dong, Sungbuk-ku, Seoul, 136-701, South Korea. **Tel:** 82.2.3290.1604 **Fax:** 82.2.929.1957 **Ongoing Publication Projects:** *Journal of Science and Technology Studies.* **Grants:** Yes. **Program Strengths:** 20th and 21st Century, Transcontinental, Technology, Science Policy; 20th and 21st Century, Transcontinental, Institutions; Asia, Natural and Human History. **Degrees Conferred:** Bachelor's Level, Master's Level, Doctoral Level. 59 graduate students in 2002. 20 master's degrees and 7 doctorates awarded since July, 1998. **Description:** The program offers MA, MS, and Ph.D. courses in five fields of STS including: history of science, philosophy of science, sociology of science, science communication, and science policy. **Web Site:** http://web.korea.ac.kr/~science

Seoul National University (Program in History and Philosophy of Science). **Contact:** Yung Sik Kim, kysik@plaza.snu.ac.kr **Mailing Address:** San 56-1 Shilim-dong, Kwanak-ku, Seoul, 151-742, South Korea. **Tel:** 82.2.880.6637 **Fax:** 82.2.873.0418 **Grants:** No. **Degrees Conferred:** Bachelor's Level, Master's Level, Doctoral Level. 35 graduate students in 2002. 8 master's degrees and 3 doctorates awarded since July, 1998. **Faculty:** *In Rae Cho*; *Hwe-Ik Zhang.* **Web Site:** http://phps.snu.ac.kr

Spain

Universitat Autònoma de Barcelona (Programa Interuniversitari de Doctorat en Història de les Ciències). **Contact:** Xavier Roqué, Xavier.Roque@uab.es **Primary E-mail:** cehic@uab.es **Mailing Address:** Centre d'Estudis d'Història de les Ciències, Edifici Cc, Universitat Autònoma de Barcelona, Bellaterra, Barcelona, 08193, Spain. **Tel:** 34.93.581.1308 **Fax:** 34.93.581.2003 **Collections/ Resources:** Sources for history of quantum mechanics; Landmarks of Science; Fons Millàs Vallicrosa; Arxiu Ferran Sunyer i Balaguer. **Grants:** No. **Program Strengths:** Physical Sciences, Instruments and Techniques; Medical Sciences,

Social Relations of Science; Technology, Instruments and Techniques.
Faculty: *Marià Baig*; *Manuel G. Doncel* (19th Century, Physical Sciences);
Albert Dou; *Anna Estany*; *Mercè Izquierdo*; *Alvar Martinez*; *Jorge Molero*;
Annette Mulberger; *Agusti Nieto-Galan* (19th Century, Europe, Physical
Sciences); *Martí Pumarola*; *Xavier Roque* (20th and 21st Century, Physical
Sciences); *Milagros Saiz*. **Description:** The Centre was created in 1995 from the
Seminari d'Història de les Ciències (established in 1983). It gathers 12 lecturers
and professors who do research in the History of Science, Technology
and Medicine, and coordinates a PhD program in the History of Science.
Web Site: http://www.uab.es/cehic/

Sweden

Royal Institute of Technology (Teknik- och vetenskapshistoria). **Contact:**
Arne Kaijser, arnek@tekhist.kth.se **Mailing Address:** Avd. for teknik-och
vetenskapshistoria, KTH, Stockholm, S-100 44, Sweden. **Tel:** 46.8.790.62.32
Fax: 46.8.24.62.63 **Ongoing Publication Projects:** *Polhem*. **Degrees
Conferred:** 15 graduate students in 2002. **Description:** The Department of
History of Science and Technology carries on research and higher education
on technical, scientific and industrial change from a historical perspective.
The Department has two chairs, one in the history of technology and one in
industrial heritage research; and in addition it also has seven senior researchers,
fifteen doctoral students and a secretary. The staff members have different
backgrounds and research interests within history of technology, industrial
heritage research, history of science, political science and environmental history.
Web Site: http://130.237.51.248/tekhist/index_eng.html

United Kingdom

Note: More information on U.K. graduate programs can be found in the *Guide to
History of Science Courses in Britain*, produced by the British Society for the
History of Science. It is available online at http://www.chstm.man.ac.uk/bshs/
bshscour.htm.

Imperial College, University of London (Centre for the History of Science,
Technology and Medicine). **Contact:** Caroline Treacey, c.treacey@ic.ac.uk
Mailing Address: Sherfield Building 446, London, SW7 2AZ, United Kingdom.
Tel: 44.20.7594.9360 **Fax:** 44.20.7594.9353 **Ongoing Publication Projects:**
The Newton Project (web publication of all Newton's manuscripts); *History
of Science* (journal editorial office). **Collections/Resources:** Science Museum
Library (600,000 volumes); Science Museum; Natural History Museum;
Imperial College Archives; Wellcome Library; British Library. **Grants:** No.
Program Strengths: Europe. **Degrees Conferred:** Bachelor's Level, Master's
Level, Doctoral Level. 31 graduate students in 2002. 58 master's degrees and
2 doctorates awarded since July, 1998. **Faculty:** *Serafina Cuomo*; *David E. H.
Edgerton*; *Hannah Gay*; *Robert Iliffe*; *Lisbet Koerner*; *J. Andrew Mendelsohn*;

Andrew C. Warwick. **Description:** Faculty and doctoral research interests range from antiquity to the present, covering physical and life sciences, mathematics, technology, and medicine. Master's program taught by 18 faculty; run jointly with the STS Department at University College London and the Wellcome Trust Centre for the History of Medicine at UCL. **Web Site:** http://www.hstm.ic.ac.uk

Institute of Railway Studies (University of York and National Railway Museum) (Railway Studies). **Contact:** Colin Divall, cd11@york.ac.uk **Mailing Address:** Dept of History, University of York, Heslington, York, YO1O 5DD, United Kingdom. **Tel:** 44.1904.432990 **Fax:** 44.1904.432986 **Ongoing Publication Projects:** *Working Papers in Railway Studies.* **Collections/ Resources:** The Institute enjoys access to the rich archival, library and material culture collections of the National Railway Museum as well as the facilities of the University of York. **Grants:** Yes. **Program Strengths:** Technology, Institutions; Technology, Instruments and Techniques; Technology, Humanistic Relations of Science. **Degrees Conferred:** Bachelor's Level, Master's Level, Doctoral Level. 12 graduate students in 2002. 6 master's degrees and 2 doctorates awarded since July, 1998. **Prospective Student Contact:** Tim Owston, tjo3@york.ac.uk **Faculty:** *Colin Divall*; *Ralph Harrington*; *Barbara Schmucki.* **Description:** The Institute of Railway Studies offers unique opportunities for research and graduate training in the history and public history of railways and other forms of transport. **Web Site:** http://www.york.ac.uk/inst/irs/

Lancaster University (Department of History). **Contact:** Ghil O'Neill, g.o'neill@lancaster.ac.uk **Mailing Address:** Postgraduate Secretary, Lancaster, LA1 4YG, United Kingdom. **Tel:** 44.1524.592549 **Fax:** 44.1524.846102 **Ongoing Publication Projects:** The Correspondence and Manuscript Papers of James Clerk Maxwell (P. M. Harman, Ed.). **Collections/Resources:** Ruskin Library (collection of manuscripts of John Ruskin). **Grants:** Yes. **Degrees Conferred:** Bachelor's Level, Master's Level, Doctoral Level. 2 graduate students in 2002. 6 master's degrees and 2 doctorates awarded since July, 1998. **Faculty:** *Peter Michael Harman*; *Paolo S. Palladino* (20th and 21st Century, Europe, Biological Sciences); *Stephen Pumfrey.* **Description:** The main faculty's experise is in early modern science, 18th and 19th century natural philosophy, and 19th and 20th century human and environmental sciences. **Web Site:** http://www.lancs.ac.uk/users/history/histwebsite/homepage.htm

London Centre for the History of Science, Medicine and Technology. **Description:** This one-year master's degree program of the University of London is run jointly by the Centre for the History of Science, Technology and Medicine at Imperial College, the Department of Science and Technology Studies at University College London, and the Wellcome Trust Centre for the History of Medicine at UCL. It comprises a regular teaching staff of some 20 faculty in the field and attracts about 20 students per year, some of whom continue as doctoral students at one of the three constituent institutions. *See also*

Imperial College, University of London, and University College London in this section. The Wellcome Trust Centre is listed under Research Centers.

University College London (Department of Science and Technology Studies). **Contact:** Jon Turney, j.turney@ucl.ac.uk **Primary E-mail:** h.chang@ucl.ac.uk **Mailing Address:** Gower Street, London, WC1E 6BT, United Kingdom. **Tel:** 44.20.7679.1328 **Fax:** 44.20.7916.2425 **Collections/Resources:** College library houses the Graves collection (bequeathed in 1870) of 14,000 items in mathematics and physical science from 16th century onwards; Karl Pearson papers, among many others; Microfilm collection of complete papers and manuscripts of Isaac Newton; Close proximity to Wellcome Library for History of Medicine and British Library. **Grants:** Yes. **Degrees Conferred:** Bachelor's Level, Master's Level, Doctoral Level. 27 graduate students in 2002. 72 master's degrees and 5 doctorates awarded since July, 1998. **Prospective Student Contact:** Hasok Chang, h.chang@ucl.ac.uk **Faculty:** *Gregory Andrew*; *Brian Balmer*; *Joe Cain* (20th and 21st Century, North America, Biological Sciences); *Hasok Chang* (Philosophy and Philosophy of Science); *Jane Gregory*; *Arthur Miller*; *Steve Miller*; *Jon Turney*. **Description:** The department combines interests in history, philosophy and sociology of science, science communication and science policy. Research interests range from ancient science and philosophy to history of physics and twentieth-century Darwinian theory. The master's program is taught jointly with Imperial College and the Wellcome Trust Centre for History of Medicine. **Web Site:** http://www.ucl.ac.uk/sts/

University of Bath (Science Studies Centre). **Contact:** Patricia Sechi-Johnson, pssps@bath.ac.uk **Primary E-mail:** psychology-enquiries@bath.ac.uk **Mailing Address:** Science and Culture, c/o Patricia Sechi-Johnson, Postgraduate Coordinator, Science Studies Centre, Department of Psychology, Bath, BA2 7AY, United Kingdom. **Tel:** 44.1225.383041 **Fax:** 44.1225.826752 **Grants:** Yes. **Program Strengths:** 20th and 21st Century, Cognitive Sciences; 19th Century, Europe, Physical Sciences; 20th and 21st Century, Instruments and Techniques. **Degrees Conferred:** Bachelor's Level, Master's Level, Doctoral Level. 16 graduate students in 2002. 25 master's degrees and 4 doctorates awarded since July, 1998. **Faculty:** *Wendy Barnaby*; *Jeff Gavin*; *David C. Gooding* (North America); *Willem Hackmann*; *Helen Haste*; *Frank A. James* (19th Century, Europe, Physical Sciences); *Karl Allan Rogers*; *David Weltman*. **Description:** The Centre exists to further research and teaching in history, philosophy and social studies of science and technology. It offers several degree programs. The masters program in "Science, Culture and Communication" combines study of the history, philosophy and the cultural context of science with the career opportunities of communications and media courses and professional placements. An ideal preparation for the Ph.D., it includes research training based on our ESRC recognized M.Res. program and a substantial research dissertation. The Centre also offers an M.Phil. and Ph.D. by research. **Web Site:** http://www.bath.ac.uk/~hssdcg/SCandC_2.html

University of Cambridge (Department of History and Philosophy of Science). **Contact:** David Thompson, hps-admin@lists.cam.ac.uk **Mailing Address:** Free School Lane, Cambridge, CB2 3RH, United Kingdom. **Tel:** 44.1223.334500 **Fax:** 44.1223.334554 **Ongoing Publication Projects:** *Studies in History and Philosophy of Science*; *Studies in Biological and Biomedical Sciences*. **Collections/Resources:** Whipple Museum of the History of Science; Whipple Library; Cambridge University Library. **Grants:** No. **Degrees Conferred:** Bachelor's Level, Master's Level, Doctoral Level. **Faculty:** *David Corfield*; *John Forrester*; *Sarah Hodges*; *Nick D. Hopwood* (19th Century, Biological Sciences); *Nicholas Jardine*; *Lauren Kassell*; *Martin Kusch*; *Peter Lipton* (Philosophy and Philosophy of Science); *Simon Schaffer*; *James A. Secord* (19th Century, Europe, Biological Sciences); *Liba Taub* (Pre-400 B.C.E, Astronomical Sciences). **Description:** The largest of its kind in the UK, the department has an outstanding international reputation. It is built around the Whipple Museum, a world-class collection of scientific instruments. The large collections of the Whipple Library, including some rare scientific books, provide the basis for research and teaching. **Web Site:** http://www.hps.cam.ac.uk/

University of Edinburgh (Science Studies Unit). **Contact:** Carole Tansley, carole.tansley@ed.ac.uk **Mailing Address:** 21 Buccleuch Place, Edinburgh, EH 8 9LN, United Kingdom. **Tel:** 44.131.650.4256 **Fax:** 44.131.650.6886 **Collections/Resources:** Edinburgh University Library, Department of Special Collections: various archive sources on history of Scottish science, technology and medicine; Edinburgh University Library, Lothian Health Services Archive: extensive archive sources on history of medicine and health care in Scotland; National Library of Scotland: various archive sources on history of science, technology and medicine in Scotland; Edinburgh is home to many other public and institutional archive sources relating to history of science, technology and medicine. **Grants:** Yes. **Program Strengths:** Social Relations of Science; Gender and Science; Technology, Social Relations of Science. **Degrees Conferred:** Bachelor's Level, Master's Level, Doctoral Level. 16 graduate students in 2002. 23 master's degrees and 6 doctorates awarded since July, 1998. **Prospective Student Contact:** Steve Sturdy, s.sturdy@ed.ac.uk **Faculty:** *David Bloor*; *Wendy Faulkner*; *John Henry*; *Donald MacKenzie*; *Steven Sturdy*; *Robin Williams*. **Description:** The Science Studies Unit, together with the Edinburgh University Institute for the Study of Science, Technology and Innovation, offers a broad-based Masters-level training in history, sociology and philosophy of science, technology and medicine, plus Ph.D. opportunities in various specializations within these fields. **Web Site:** http://www.ssu.ssc.ed.ac.uk/

University of Leeds (Division of History and Philosophy of Science). **Contact:** Gregory Radick, G.M.Radick@leeds.ac.uk **Primary E-mail:** HPS@leeds.ac.uk **Mailing Address:** The School of Philosophy, University of Leeds, Leeds, West Yorkshire, LS2 9JT, United Kingdom. **Tel:** 44.113.343.3260 **Fax:** 44.113.343.3265 **Ongoing Publication Projects:** The SciPer (Science in the Nineteenth-Century Periodicals) Project. **Collections/Resources:** Extensive holdings in history and philosophy of science in the university library, including manuscript collections (notably the Astbury papers) and the All Souls Science Collection in the Special Collections facility of the Brotherton Library; Material and archival collections of the Thackray Medical Museum in Leeds. **Grants:** Yes. **Program Strengths:** Biological Sciences; Technology; Science and Religion. **Degrees Conferred:** Bachelor's Level, Master's Level, Doctoral Level. 23 graduate students in 2002. 15 master's degrees and 6 doctorates awarded since July, 1998. **Prospective Student Contact:** Josie Green, phljog@leeds.ac.uk **Faculty:** *Geoffrey N. Cantor* (Europe); *John Christie*; *Steven French* (Philosophy and Philosophy of Science); *Graeme J. Gooday* (19th Century, Technology, Instruments and Techniques); *Jonathan Hodge*; *Chris Kenny*; *Joseph Melia*; *Jack Morrell*; *Gregory Radick*; *Jonathan R. Topham* (19th Century, Europe, Humanistic Relations of Science); *Helen Valier*; *Adrian Wilson*. **Description:** Comprising twelve staff members and a large and lively graduate community, the Leeds HPS Division offers wide scope for stimulating graduate studies in all aspects of the discipline. The one-year M.A. emphasizes the development of analytic and interpretive skills. At the Ph.D. level, students participate with the staff in weekly research seminars. **Web Site:** http://www.philosophy.leeds.ac.uk./html/hps.htm

University of Manchester (Centre for the History of Science, Technology and Medicine). **Contact:** Michael Worboys, chstm@man.ac.uk **Mailing Address:** Mathematics Tower, University of Manchester, Manchester, M13 9PL, United Kingdom. **Tel:** 44.161.275.5850 **Fax:** 44.161.275.5699 **Ongoing Publication Projects:** Series for Palgrave publishers: *Science, Technology, and Medicine in Modern History*. **Collections/Resources:** John Rylands University Library; National Archive for the History of Computing; Manchester Medical Archive; Partington Collection for History of Chemistry; Labour History Museum and Archive; People's History Library; Museum of Science and Industry, Manchester; Manchester Central Library. **Grants:** Yes. **Degrees Conferred:** Bachelor's Level, Master's Level, Doctoral Level. 28 graduate students in 2002. 30 master's degrees and 5 doctorates awarded since July, 1998. **Description:** The interests of CHSTM staff lie mainly in the 19th and 20th centuries, in Britain, Europe, and USA. The Centre includes the Wellcome Unit for the History of Medicine and the National Archive for the History of Computing. We mainly focus on the social and comparative history of HSTM. **Web Site:** http://www.man.ac.uk/chstm

University of Manchester Institute of Science & Technology (History of Science and Technology Group). **Contact:** J. O. Marsh, joe.marsh@umist.ac.uk **Mailing Address:** D38 Main Bldg., UMIST, PO Box 88, Sackville Street, Manchester, M60 1QD, United Kingdom. **Tel:** 44.161.200.3948 **Fax:** 44.161.200.3947 **Collections/Resources:** James Prescott Joule (1818-1889) Notebooks; Selected records from Manchester Machine Tool companies; Large departmental library. **Grants:** No. **Program Strengths:** 19th Century, Europe, Technology, Education; 19th Century, Europe, Technology, Biography; 19th Century, Europe,

Physical Sciences. **Degrees Conferred:** Bachelor's Level, Master's Level, Doctoral Level. 4 graduate students in 2002. 0 master's degrees and 1 doctorate awarded since July, 1998. **Faculty:** *Richard William Cox*; *Ronald Fitzgerald*; *Richard Leslie Hills*; *Joseph Oliver Marsh*; *Thomas Swailes*; *Rajkumari Williamson-Jones*; *Gordon Woodward.* **Web Site:** http://www.umist.ac.uk

University of Oxford (History of Science, Medicine, and Technology). **Contact:** Robert Fox, robert.fox@history.ox.ac.uk **Mailing Address:** Modern History Faculty, Broad Street, Oxford, Oxfordshire, OX1 3BD, United Kingdom. **Tel:** 44.1865.277277 **Fax:** 44.1865.277277 **Collections/Resources:** Museum of the History of Science; Bodleian and Radcliffe Science Libraries; Specialist library in the Wellcome Unit for the History of Medicine; Major college libraries. **Grants:** Yes. **Degrees Conferred:** Bachelor's Level, Master's Level, Doctoral Level. 40 graduate students in 2002. 15 master's degrees and 8 doctorates awarded since July, 1998. **Prospective Student Contact:** Hubert Stadler, hubert.stadler@history.ox.ac.uk **Faculty:** *J. A. Bennett*; *Robert Fox* (19th Century, Europe, Physical Sciences); *John L. Heilbron* (17th Century, Europe, Physical Sciences); *Stephen Johnston*; *John D. North* (Astronomical Sciences); *J. Robertson*; *G. L. E. Turner* (Instruments and Techniques); *Steve Woolgar.* **Description:** Oxford's history of science, medicine, and technology program offers a range of degrees at the master's and doctoral level. These degrees allow students to exploit the exceptionally rich resources of the University of Oxford, while the program's wide range of expertise makes it possible to offer supervision in most areas of the history of science, medicine, and technology since the Renaissance, with special strengths in the history of instrumentation, tropical medicine, and the relations between science and industrial technology in modern Europe. **Web Site:** http://www.history.ox.ac.uk/hsmt

University of Oxford (Master's Program at the Museum of the History of Science). **Contact:** Jim Bennett, jim.bennett@mhs.ox.ac.uk **Primary E-mail:** museum@mhs.ox.ac.uk **Mailing Address:** Museum of the History of Science, Broad Street, Oxford, OX1 3AZ, United Kingdom. **Tel:** 44.1865.277280 **Fax:** 44.1865.277288 **Collections/Resources:** Collection and Archive of the Museum of the History of Science; Bodleian Library. **Grants:** No. **Degrees Conferred:** Master's Level. 2 graduate students in 2002. 14 master's degrees awarded since July, 1998. **Faculty:** *J. A. Bennett.* **Description:** The course uses the museum's collection to focus on the role of instruments in the history of science. It also covers collecting and the place of museums in the history of science. Students have opportunities to become involved with the Museum's programs of documentation, display and exhibitions. **Web Site:** http://www.mhs.ox.ac.uk/course/index.htm

University of the West of England (School of Interdisciplinary Sciences). **Description:** The School of Interdisciplinary Sciences and the program in history of science have been discontinued.

United States

California

Stanford University (Program in the History of Science). **Contact:** Timothy Lenoir, tlenoir@stanford.edu **Mailing Address:** Building 200-33, Stanford University, Stanford, CA, 94305-2024, United States. **Tel:** 1.650.725.0714 **Fax:** 1.650.725.0597 **Collections/Resources:** Stanford is surrounded by archives for the recent history of science and technology. Stanford University Libraries has rich holdings in its Special Collections for the Scientific Revolution, as well as the modern and contemporary study of science and technology. The university is in close proximity to some of the most interesting public science museums in the country: the California Academy of Sciences, the Exploratorium, the Computer History Museum, and the Tech Museum. Graduate students can take advantage of faculty, classes, and archives at UC Berkeley through Stanford's exchange program. **Grants:** No. **Degrees Conferred:** Bachelor's Level, Master's Level, Doctoral Level. 6 graduate students in 2002. 1 master's degree and 3 doctorates awarded since July, 1998. **Prospective Student Contact:** The Registrar's Office, ck.gaa@forsythe.stanfo rd.edu **Faculty:** *Keith Baker; Barton Bernstein; Ahmad Dallal; Paula Findlen* (17th Century, Europe, Institutions); *Michael Friedman; Sarah Jain; Timothy Lenoir; Reviel Netz; Jessica G. Riskin* (18th Century, Europe, Philosophy and Philosophy of Science); *Michael Strevens* (Philosophy and Philosophy of Science); *Richard White.* **Description:** This is an interdisciplinary, non-degree program focusing on the historical and contemporary aspects of science, medicine, and technology. Degrees are not conferred by the program itself, but rather through the departments in which core faculty teach, principally Classics, Cultural and Social Anthropology, History, and Philosophy. Undergraduate degrees are offered through the departments of History and Philosophy, and through the Program in Human Biology. Undergraduate and graduate courses span the period from antiquity to the late 20th century, with special emphasis on ancient and Islamic science; Renaissance science; the scientific revolution; history of medicine and the body; history and philosophy of biology; history and philosophy of modern physics; history of computers and information sciences; and gender, science, and technology. These courses are designed both for students looking for a humanistic perspective on the sciences and for students trying to understand the relationship of the sciences to humanistic knowledge. The core of the community is the colloquium series which brings together faculty and students several times a quarter to discuss the work of invited speakers on topics of broad concerns to science and technology studies. **Web Site:** http://www.stanford.edu/dept/HPS/

University of California, Berkeley (Department of History). **Contact:** Barbara Hayashida, histgao@socrates.berkeley.edu **Mailing Address:** History, 3229 Dwinelle Hall, University of California, Berkeley, CA, 94720-2550, United States. **Tel:** 1.510.642.2378 **Fax:** 1.510.643.5323 **Ongoing Publication**

Projects: *See the entry for the Office for History of Science and Technology, listed under Research Centers.* **Collections/Resources:** The Office for History of Science and Technology maintains research collections and arranges special events and graduate student support. **Grants:** Yes. **Program Strengths:** 18th Century, Europe; 20th and 21st Century, Biological Sciences; 20th and 21st Century, Physical Sciences. **Degrees Conferred:** Bachelor's Level, Master's Level, Doctoral Level. 12 graduate students in 2002. 5 master's degrees and 6 doctorates awarded since July, 1998. **Prospective Student Contact:** Cathryn Carson, clcarson@socrates.berkeley.edu **Faculty:** *Warwick H. Anderson* (20th and 21st Century, Asia, Medical Sciences); *Cathryn L. Carson* (20th and 21st Century, Physical Sciences); *John L. Heilbron* (17th Century, Europe, Physical Sciences); *David A. Hollinger* (20th and 21st Century, North America, Social Sciences); *Thomas Laqueur* (Medical Sciences); *John E. Lesch* (20th and 21st Century, Europe, Medical Sciences); *Barbara J. Shapiro* (17th Century, Europe, Humanistic Relations of Science). **Description:** Students have full access to the intellectual resources of the Berkeley History Department. Along with the history of science, they receive training in a conventional field of history and an outside field (e.g., philosophy, anthropology, a science). In history of medicine they may work with faculty at UC San Francisco. **Web Site:** http://history.berkeley.edu

University of California, Los Angeles (Department of History: Science, Medicine and Technology Field). **Contact:** Theodore Porter, tporter@history.ucla.edu **Mailing Address:** History Department, 6265 Bunche Hall, UCLA, Los Angeles, CA, 90095-1473, United States. **Tel:** 1.310.825.4601 **Fax:** 1.310.206.9630 **Collections/Resources:** History of Neuroscience Archive; History of Pain Collection; William Andrews Clark Memorial Library; Huntington Library. **Grants:** No. **Program Strengths:** 18th Century, Europe; Physical Sciences; Humanistic Relations of Science. **Degrees Conferred:** Bachelor's Level, Master's Level, Doctoral Level. 15 graduate students in 2002. 4 master's degrees and 7 doctorates awarded since July, 1998. **Prospective Student Contact:** Barbara Bernstein, bbernste@history.ucla.edu **Faculty:** *Joel Braslow*; *Brian P. Copenhaver* (14th-16th Century, Europe, Philosophy and Philosophy of Science); *Robert G. Frank*; *Margaret C. Jacob* (18th Century, Europe, Humanistic Relations of Science); *Theodore M. Porter*; *Peter H. Reill*; *Mary Terrall* (18th Century, Europe, Physical Sciences); *Sharon Traweek*; *Jessica Wang*; *Dora B. Weiner* (18th Century, Europe, Medical Sciences, Instruments and Techniques); *M. Norton Wise* (Europe, Physical Sciences). **Description:** The graduate program in history of science at UCLA is fully integrated into a large, excellent and wide-ranging History Department. Its particular strengths include Enlightenment science, history of physics, history of medicine, history of quantification and of social science, and scientific culture as it pertains to technologies, states, and economies. **Web Site:** http://www.sscnet.ucla.edu/history

University of California, San Diego (Science Studies Program). **Contact:** Chandra Mukerji, cmukerji@ucsd.edu **Primary E-mail:** ssadmin@ucsd.edu **Mailing Address:** 9500 Gilman Drive, Science Studies #0104, La Jolla, CA, 92093-0104, United States. **Tel:** 1.858.534.0491 **Fax:** 1.858.534.7283 **Collections/Resources:** Hill Collection of Pacific Voyages. **Grants:** Yes. **Program Strengths:** 17th Century, Europe, Social Relations of Science; 20th and 21st Century, Cognitive Sciences, Philosophy or Philosophy of Science; 20th and 21st Century, North America, Social Relations of Science. **Degrees Conferred:** Doctoral Level. 26 graduate students in 2002. 12 doctorates awarded since July, 1998. **Prospective Student Contact:** Sheri Bullock, ssadmin@ucsd.edu **Faculty:** *Michael Bernstein*; *Geoffrey Bowker*; *Charles Briggs*; *Craig Callender*; *Nancy Cartwright* (Philosophy and Philosophy of Science); *Paul M. Churchland*; *Gerald Doppelt*; *Steve Epstein*; *Marta E. Hanson* (18th Century, Asia, Medical Sciences); *Martha Lampland*; *Chandra Mukerji*; *Naomi Oreskes*; *Roddey Reid*; *Andrew Scull*; *Steven Shapin*; *Leigh Star*; *Robert S. Westman* (14th-16th Century, Europe, Astronomical Sciences). **Description:** The strengths of the program lie in early modern science and technology, the history of earth sciences in the 19th-20th centuries, gender and science, medicine and social policy, history of infrastructure, and new information technologies. **Web Site:** http://sciencestudies.ucsd.edu

University of California, San Francisco (Department of Anthropology, History, and Social Medicine). **Contact:** Warwick Anderson, wanders@itsa.ucsf.edu **Primary E-mail:** rebecca@itsa.ucsf.edu **Mailing Address:** Department of Anthropology, History and Social Medicine, UCSF, 3333 California St, Suite 485, San Francisco, CA, 94143-0850, United States. **Tel:** 1.415.476.7234 **Fax:** 1.415.476.6715 **Collections/Resources:** UCSF Special Collections and Archives; UC-Berkeley Bancroft Library. **Grants:** Yes. **Program Strengths:** 20th and 21st Century, Medical Sciences. **Degrees Conferred:** Master's Level, Doctoral Level. 3 graduate students in 2002. **Faculty:** *Adele E. Clarke* (20th and 21st Century, North America, Medical Sciences); *Brian Dolan*; *Jack Lesch*; *Dorothy Porter*; *Guenter B. Risse* (20th and 21st Century, North America, Medical Sciences); *Nancy Rockafellar*. **Description:** The history of health sciences program uniquely combines training in historical and social science methods. The main emphasis is on 19th and 20th century health sciences. The regional foci are North America, Western Europe and the Asia-Pacific. The program engages with the Office for the History of Science and Technology at UC-Berkeley, and the joint medical anthropology program of UCSF and Berkeley. **Web Site:** http://www.ucsf.edu/dahsm/index.html

University of California, Santa Barbara (Program in History of Science, Technology, and Medicine). **Contact:** Michael A. Osborne, osborne@history. ucsb.edu **Mailing Address:** Department of History, UCSB, Santa Barbara, CA, 93106-9410, United States. **Tel:** 1.805.893.2901 **Fax:** 1.805.893.8795 **Ongoing Publication Projects:** Anita Guerrini is book review co-editor of *Early Science and Medicine* (Brill). Michael Osborne is co-review editor of *Science, Technology & Society: An International Journal Devoted to the Developing World* (Sage). **Collections/Resources:** In the Special Collections of the Davidson Library: Marie Stopes collection; Darwin collection; Trade Catalog Collection; American Religions Collection. **Grants:** Yes. **Program Strengths:** 19th Century, Biological Sciences; 18th Century, Europe, Medical Sciences; 20th and 21st Century, Biological Sciences. **Degrees Conferred:** Bachelor's Level, Master's Level, Doctoral Level. 7 graduate students in 2002. 2 master's degrees and 2 doctorates awarded since July, 1998. **Faculty:** *Charles Bazerman* (Humanistic Relations of Science); *Francesca Bray* (Asia); *Anita Guerrini* (18th Century, Europe, Medical Sciences); *Michael A. Osborne* (19th Century, Europe, Medical Sciences). **Description:** This program is part of the History Department at UCSB. The core faculty (Michael Osborne and Anita Guerrini) have joint appointments with the Program in Environmental Studies. It is hoped that a third faculty member will be hired to replace Lawrence Badash, who retired in 2002. Current interests of core faculty include biological science and medicine from the Scientific Revolution to the present. **Web Site:** http://www.history.ucsb.edu/fields/hos.htm

Connecticut

Yale University (History of Medicine and History of Science). **Contact:** Daniel J. Kevles, daniel.kevles@yale.edu **Mailing Address:** Department of History, P.O. Box 208324, New Haven, CT, 06520-8324, United States. **Tel:** 1.203.432.1356 **Fax:** 1.203.436.4624 **Collections/Resources:** Relevant manuscript resources in the Yale Archives; The collections of the Sterling Library; The holdings of the Medical Historical Library. **Grants:** Yes. **Program Strengths:** Medical Sciences; Biological Sciences; Physical Sciences. **Degrees Conferred:** Bachelor's Level, Master's Level, Doctoral Level. 15 graduate students in 2002. **Prospective Student Contact:** John Warner, John.warner@yale.edu **Faculty:** *John L. Heilbron* (17th Century, Europe, Physical Sciences); *Frederic L. Holmes*; *Daniel J. Kevles* (20th and 21st Century, North America, Biological Sciences); *Susan E. Lederer* (North America, Medical Sciences); *Naomi Rogers* (20th and 21st Century, North America, Medical Sciences); *William C. Summers* (20th and 21st Century, Biological Sciences); *John Warner*. **Description:** Anchored in both the Yale History Department and the Section for the History of Medicine in the Medical School, the program offers training in history of science, of medicine, of their relationships to each other and to broader historical developments. It introduces students to the diverse recent approaches in these fields and encourages innovative research. **Web Site:** http://www.med.yale.edu/histmed/

Delaware

University of Delaware (Graduate Program in History, and University of Delaware-Hagley Program in the History of Technology and Industrialization). **Contact:** Patricia Orendorf, pato@udel.edu **Mailing Address:** 201 Munroe Hall,

University of Delaware, Newark, DE, 19716, United States. **Tel:** 1.302.451.8226 **Collections/Resources:** Hagley Museum and Library. **Grants:** Yes. **Program Strengths:** Transcontinental, Technology; 19th Century, North America; 18th Century, North America. **Degrees Conferred:** Doctoral Level. 60 graduate students in 2002. 15 master's degrees and 10 doctorates awarded since July, 1998. **Faculty:** *Arwen Palmer Mohun*; *Susan Strasser*. **Description:** The University of Delaware History Department offers M.A. and Ph.D. degrees in a variety of fields. Within the department, the UD-Hagley Program focuses on the social and cultural history of technology and industrialization. Students enrolled in any of the department's programs may also work towards a museum studies certificate. **Web Site:** http://www.udel.edu/history

Florida

University of Florida (Graduate Program in History of Science). **Primary E-mail:** hist-sci-l@lists.ufl.edu **Mailing Address:** P.O. Box 117320, Gainesville, FL, 32611-7320, United States. **Tel:** 1.352.392.0271 **Fax:** 1.352.392.6927 **Collections/Resources:** Very rich library primary sources in 17th century science and philosophy (including Early English Books Online, which contains all printed books in English to 1700), as well as 18th and 19th century German journals. **Grants:** Yes. **Program Strengths:** 19th Century, Europe, Physical Sciences, Science and Religion; 17th Century, Europe, Astronomical Sciences, Institutions; 20th and 21st Century, North America, Biological Sciences, Biography. **Degrees Conferred:** Bachelor's Level, Master's Level, Doctoral Level. 6 graduate students in 2002. 5 master's degrees and 1 doctorate awarded since July, 1998. **Prospective Student Contact:** Frederick Gregory, histsci@grove.ufl.edu **Faculty:** *Robert D'Amico*; *Donald A. Dewsbury* (20th and 21st Century, Cognitive Sciences); *Antoinette Emch-Deriaz*; *Stephen Gottesman*; *Frederick Gregory* (19th Century, Social Relations of Science); *Robert A. Hatch* (17th Century); *Chuang Liu*; *David McCally*; *Stephen A. McKnight* (14th-16th Century, Europe); *Stephen Noll*; *Harry Paul*; *Charlotte M. Porter* (Natural and Human History); *Vassiliki B. Smocovitis* (20th and 21st Century, North America, Biological Sciences). **Description:** Interests of the faculty range widely from the development of scientific ideas themselves to analysis of the mutual interaction of science and culture. Faculty from the Departments of Astronomy, Philosophy, Psychology, and Zoology are affiliated with the graduate program. **Web Site:** http://web.history.ufl.edu/1-fac-staff/1-fs-hps.html

Georgia

Georgia Institute of Technology (History of Technology). **Contact:** Andrea Tone, andrea.tone@hts.gatech.edu **Mailing Address:** School of History, Technology, and Society, 685 Cherry St., Atlanta, GA, 30332-0345, United States. **Tel:** 1.404.894.2182 **Fax:** 1.404.894.0535 **Ongoing Publication Projects:** *History and Technology*; *Journal of American Ethnic History*. **Grants:** Yes. **Program Strengths:** Technology; Medical Sciences; Social

Sciences. **Degrees Conferred:** Bachelor's Level, Master's Level, Doctoral Level. 19 graduate students in 2002. 6 master's degrees and 3 doctorates awarded since July, 1998. **Faculty:** *Mike Allen* (20th and 21st Century, North America, Technology); *Mary Frank Fox*; *Gus Giebelhaus*; *Maren Klawiter*; *Kenneth J. Knoespel*; *John Krige* (20th and 21st Century, Europe, Physical Sciences); *Willie Pearson*; *Sue Rosser*; *Helen M. Rozwadowski* (20th and 21st Century, Earth Sciences); *Andrea Tone* (20th and 21st Century, Europe, Social Sciences); *Steve Usselman.* **Description:** The School of History, Technology and Society is an interdisciplinary unit that comprises both historians and sociologists. The School offers both the M.S. and Ph.D. degrees, with an emphasis on the history and sociology of science, technology, and medicine. **Web Site:** http://www.hts.gatech.edu

Illinois

Northwestern University (Science in Human Culture Program). **Contact:** Ken Alder, k-alder@northwestern.edu **Mailing Address:** Department of History, Northwestern University, Evanston, IL, 60208, United States. **Tel:** 1.847.491.7260 **Fax:** 1.847.491.1393 **Grants:** Yes. **Program Strengths:** Technology; 18th Century, Europe, Physical Sciences, Social Relations of Science; 19th Century, North America, Cognitive Sciences, Humanistic Relations of Science. **Degrees Conferred:** Bachelor's Level, Doctoral Level. 6 graduate students in 2002. 0 master's degrees and 4 doctorates awarded since July, 1998. **Faculty:** *Ken Alder* (18th Century, Europe, Technology); *Francesca M. Bordogna* (19th Century, North America, Philosophy and Philosophy of Science). **Description:** The program admits graduate students to study the history of science and technology as a member of the doctoral program of Northwestern's History Department. It offers generous fellowships. The program's strengths lie in the history of technology, early modern history, and the history of psychology. **Web Site:** http://www.shc.northwestern.edu/

University of Chicago (Conceptual and Historical Studies of Science). **Contact:** Robert J. Richards, r-richards@uchicago.edu **Primary E-mail:** cfs-sec@ uchicago.edu **Mailing Address:** 1126 East 59th Street, Chicago, IL, 60637, United States. **Tel:** 1.773.702.8391 **Fax:** 1.773.743.8949 **Collections/ Resources:** Crerar Science Library: rich holdings in scientific and medical books published since the Renaissance; University Archives: manuscripts and documents dealing with the Manhattan Project, Encyclopedia of Unified Science, DuBois-Reymond's library, and papers of distinguished scientists (Fermi, Dewey, Chandrasekar, etc.). *See also the entry for the University of Chicago Library, Special Collections Research Center.* **Grants:** Yes. **Degrees Conferred:** Bachelor's Level, Master's Level, Doctoral Level. 25 graduate students in 2002. 2 master's degrees and 4 doctorates awarded since July, 1998. **Faculty:** *Arnold Davidson*; *Jan Goldstein*; *John Haugland*; *Adrian Johns*; *Robert Perlman*; *Robert J. Richards* (20th and 21st Century, North America, Biological Sciences); *Stephen M. Stigler*; *George W. Stocking* (20th and 21st Century, North America, Natural and Human History); *Noel Swerdlow*; *Leigh Van Valen*; *William C. Wimsatt* (20th and 21st Century, Biological Sciences); *Alison Winter*. **Description:** The Committee on Conceptual and Historical Studies of Science is a Ph.D. granting unit of the University of Chicago. It provides focal studies in the following areas: Ancient science and mathematics, history of astronomy, history of probability and statistics, early modern science, history of medicine since the Renaissance, history of biology and evolutionary theory, history of anthropology and psychology, history of the book and communications technology, history of psychiatry, history of Romantic science, 19th century British science, 18th century German Science, philosophy of biology and genetics, and philosophy of history. **Web Site:** http://humanities.uchicago.edu/chss/

University of Illinois at Urbana-Champaign (Graduate Program for Studies of Science, Technology, Information, and Medicine). **Contact:** Lillian Hoddeson, hoddeson@uiuc.edu **Mailing Address:** 309 Gregory Hall, 810S. Wright Street, Urbana, IL, 61801, United States. **Tel:** 1.217.244.8412 **Fax:** 1.217.333.2297 **Collections/Resources:** National Center for Supercomputing Applications; Beckman Institute for Advanced Science and Technology; Medical Scholars Program. **Grants:** No. **Degrees Conferred:** Bachelor's Level, Master's Level, Doctoral Level. 10 graduate students in 2002. 5 doctorates awarded since July, 1998. **Prospective Student Contact:** Clare Crowston, crowston@uiuc.edu **Faculty:** *Richard W. Burkhardt* (Biological Sciences); *C. L. Cole*; *Max Edelson*; *Barrington Edwards*; *Fernando Irving Elichirigoity*; *Michael Goldman*; *Lillian Hoddeson* (Physical Sciences); *Stephen Levinson*; *Patrick Maher* (Philosophy and Philosophy of Science); *Evan M. Melhado* (Earth Sciences); *Andrew R. Pickering* (Social Relations of Science); *Richard Powers*; *Cynthia Radding*; *Leslie Reagan*; *Daniel Schneider*; *Paula Treichler*. **Description:** The STIM Program includes a broad network of faculty members from across the university, a range of academic projects, and links to other campus units. The program includes a workshop series, informal reading groups, a core graduate seminar, and other advanced courses. **Web Site:** http://www.uiuc.edu/unit/STIM/

Indiana

Indiana University (Department of History and Philosophy of Science). **Contact:** William R. Newman, wnewman@indiana.edu **Primary E-mail:** hpscdept@indiana.edu **Mailing Address:** Goodbody Hall, Rm. 130, Bloomington, IN, 47405, United States. **Tel:** 1.812.855.3071 **Fax:** 1.812.855.3631 **Ongoing Publication Projects:** *Philosophy of Science*. **Collections/Resources:** Lilly Library of Rare Books and Manuscripts. **Grants:** Yes. **Program Strengths:** 17th Century, Europe; 20th and 21st Century, Europe, Physical Sciences, Philosophy or Philosophy of Science; 19th Century, Europe, Biological Sciences. **Degrees Conferred:** Bachelor's Level, Master's Level, Doctoral Level. 21 graduate students in 2002. 21 master's degrees and 17 doctorates awarded since July, 1998. **Prospective Student Contact:**

Becky Wood, Beckyw@indiana.edu **Faculty:** *Domenico Bertoloni Meli* (17th
Century, Europe, Physical Sciences); *James H. Capshew* (Cognitive Sciences);
Ann Carmichael (14th-16th Century, Medical Sciences); *Jordi Cat* (Physical
Sciences, Philosophy and Philosophy of Science); *Frederick B. Churchill*
(19th Century, Europe, Biological Sciences); *W. Michael Dickson* (20th and
21st Century, Physical Sciences); *Sander J. Gliboff* (20th and 21st Century,
Europe, Biological Sciences); *Edward Grant* (5th-13th Century, Europe);
Noretta Koertge (Philosophy and Philosophy of Science); *Elisabeth A. Lloyd*
(Biological Sciences); *Christopher A. Martin* (Physical Sciences, Philosophy
and Philosophy of Science); *William Newman* (5th-13th Century, Europe).
Description: Graduate study in HPS at Indiana involves a mixture of courses in
history and philosophy of science, with emphasis on one field or the other. The
Department is especially strong in history of late medieval and early modern
science, including disciplines ranging from alchemy and medicine to mechanics
and mathematics. **Web Site:** http://www.indiana.edu/~hpscdept/

University of Notre Dame (Program in History and Philosophy of Science).
   Contact: Don Howard, don.a.howard.43@nd.edu **Primary E-mail:** nd.reilly.31@
nd.edu **Mailing Address:** 346 O'Shaughnessy, University of Notre Dame, Notre
Dame, IN, 46556, United States. **Tel:** 1.800.813.2304 **Fax:** 1.574.631.3985
Grants: Yes. **Program Strengths:** 19th Century, Europe, Medical Sciences,
Social Relations of Science; 19th Century, Biological Sciences; 20th and 21st
Century, Physical Sciences. **Degrees Conferred:** Master's Level, Doctoral
Level. 25 graduate students in 2002. 4 master's degrees and 9 doctorates
awarded since July, 1998. **Faculty:** *J. Matthew Ashley*; *Michael J. Crowe* (19th
Century, Astronomical Sciences); *Christopher Fox*; *Paul Franks*; *Gary Gutting*;
Christopher S. Hamlin (19th Century, Europe, Medical Sciences); *David
Harley*; *Don Howard* (Philosophy and Philosophy of Science); *Anja Jauernig*;
Lynn Joy; *Janet A. Kourany* (Gender and Science); *Edward Manier*; *Vaughn R.
McKim* (Social Sciences); *Ernan V. McMullin* (Philosophy and Philosophy of
Science); *Philip E. Mirowski*; *Lenny Moss* (19th Century, Europe, Biological
Sciences); *Philip L. Quinn* (Philosophy and Philosophy of Science); *William
M. Ramsey*; *Esther M. Sent* (20th and 21st Century, Social Sciences); *Kristin
Shrader-Frechette*; *Phillip R. Sloan* (18th Century, Europe, Natural and Human
History); *James Turner*. **Description:** With twenty-two faculty, representing
six departments, Notre Dame's HPS Program is especially strong in the social
history of medicine, the history of biology in the nineteenth and twentieth
centuries, the history of early modern science, the history of philosophy of
science, philosophy of physics, philosophy of biology, and science and ethics.
Web Site: http://www.nd.edu/~hps

Iowa

Iowa State University (Program in History of Technology and Science).
   Contact: Alan Marcus, aimarcus@iastate.edu **Mailing Address:** 603 Ross Hall,
Ames, IA, 50011, United States. **Tel:** 1.515.294.7286 **Fax:** 1.515.294.6390

Collections/Resources: Archives of Women in Science and Engineering; Archives of American Veterinary Medicine; Association of Official Analytical Chemists; American Statistical Association Repository; Archives of Factual Film; Roswell Garst Papers on hybrid corn and dissemination of agricultural technology; Maytag Design Records; Rath Packing Records. **Grants:** No. **Program Strengths:** 20th and 21st Century, North America, Technology; 19th Century, North America, Biological Sciences, Science Policy; 19th Century, Europe, Natural and Human History, Philosophy or Philosophy of Science. **Degrees Conferred:** Bachelor's Level, Master's Level, Doctoral Level. 12 graduate students in 2002. 3 master's degrees and 5 doctorates awarded since July, 1998. **Faculty:** *James T. Andrews* (20th and 21st Century, Europe, Social Relations of Science); *Amy S. Bix* (Technology); *Hamilton Cravens* (20th and 21st Century, North America); *Alan I. Marcus* (19th Century, North America); *Bernhard Wolfgang Rieger*; *David B. Wilson* (18th Century, Europe, Physical Sciences). **Description:** The Program in History of Technology and Science (HOTS) emphasizes the history of technology, science, and medicine in Europe and America since the late 18th century. Its core faculty all teach and write in this chronological period, thus providing extraordinary opportunity for graduate study. Each, moreover, is broad gauged, expert in several areas within this general focus. Each takes ideas seriously as the basis for understanding science, technology, medicine and social action. **Web Site:** http://www.public.iastate.edu/~history_info/hots/

Kansas

Kansas State University (Department of History, History of Science Studies). **Contact:** Jack Holl, jackholl@ksu.edu **Mailing Address:** Eisenhower Hall, Manhattan, KS, 66506, United States. **Tel:** 1.785.532.6730 **Fax:** 1.785.532.7004 **Collections/Resources:** Students may work in connection with the Institute of Military History and 20th Century Studies which has cooperative agreements with the Eisenhower Presidential Library, Abilene, and the US Army Command and General Staff College, Ft. Leavenworth. **Grants:** Yes. **Program Strengths:** 20th and 21st Century, North America, Physical Sciences, Institutions; 19th Century, North America, Biological Sciences, Biography; 14th-16th Century, Europe, Social Sciences, Science and Religion. **Degrees Conferred:** Bachelor's Level, Doctoral Level. 3 graduate students in 2002. 0 master's degrees and 0 doctorates awarded since July, 1998. **Prospective Student Contact:** David Graff, dgraff@ksu.edu **Faculty:** *Louise Breen*; *Jack M. Holl*. **Description:** The history of science program focuses on 20th Century science, environmental and agricultural history, nuclear history, military history, and science and religion. **Web Site:** http://www.ksu.edu/history/science/

University of Kansas. **Contact:** Robert K. DeKosky, rdekosky@ku.edu **Mailing Address:** Department of History, Wescoe Hall, Lawrence, KS, 66045, United States. **Tel:** 1.785.864.3569 **Collections/Resources:** Spencer Research Library on the Lawrence campus has extensive holdings in the history of natural history and Great Plains history. Library at the Medical Center in Kansas City, Kansas has unique holdings in the history of medicine. **Grants:** Yes. **Program Strengths:** North America, Natural and Human History; Europe, Physical Sciences. **Degrees Conferred:** Bachelor's Level, Doctoral Level. 14 graduate students in 2002. 3 master's degrees and 4 doctorates awarded since July, 1998. **Prospective Student Contact:** Tom Lewin, tomlewin@ku.edu **Faculty:** *Karl Brooks*; *Robert K. DeKosky* (19th Century, Europe, Physical Sciences); *Paul Kelton*; *Donald Worster*. **Description:** Environmental history is the major area of the history of science with graduate offerings at KU. Courses on other aspects of the history of science are taught on the Lawrence campus. Courses on the history of medicine are taught at the KU Medical Center in Kansas City, Kansas.

Kentucky

University of Kentucky. **Contact:** Eric Christianson, ehchri01@pop.uky.edu **Mailing Address:** Office Tower 1715, Lexington, KY, 40506-0027, United States. **Tel:** 1.859.257.6861 **Fax:** 1.859.323.3885 **Collections/Resources:** Rare Books and Archives of Transylvania University and the University of Kentucky; 19th and 20th century environmental and health resources. **Grants:** Yes. **Degrees Conferred:** Bachelor's Level, Master's Level, Doctoral Level. 3 graduate students in 2002. 0 master's degrees and 0 doctorates awarded since July, 1998. **Prospective Student Contact:** Ellen Furlough, furloug@pop.uky.edu **Description:** The M.A. and Ph.D. programs are in the history department. **Web Site:** http://www.uky.edu/AS/History/

Maryland

Johns Hopkins University (Department of History of Science, Medicine and Technology). **Contact:** Edna Ford, eford@jhu.edu **Mailing Address:** Johns Hopkins University, 3400 North Charles, Baltimore, MD, 21218-2690, United States. **Tel:** 1.401.516.7501 **Fax:** 1.401.516.7502 **Ongoing Publication Projects:** *Bulletin of the History of Medicine*, editorial office at Johns Hopkins School of Medicine, Welch Library, 3rd floor. **Collections/Resources:** Alan M. Chesney Archives of the Johns Hopkins Medical Institutions, containing major manuscript collections of medicine at Hopkins; The Institute of the History of Medicine contains an extensive historical collection; The university is within easy reach of the National Library of Medicine, National Archives, Smithsonian Institution, Dibner Institute Library, and Folger Library in the Washington area; the Beckman Center for History of Chemistry, the American Philosophical Society, and College of Physicians of Philadelphia in the Philadelphia area. **Grants:** Yes. **Program Strengths:** 17th Century, Europe; 19th Century, North America; 20th and 21st Century. **Degrees Conferred:** Bachelor's Level, Master's Level, Doctoral Level. 19 graduate students in 2002. 1 master's degree and 12 doctorates awarded since July, 1998. **Faculty:** *Gert H. Brieger*; *Jerome Bylebyl*; *Mary Fissell*; *James D. Goodyear*; *Owen Hannaway* (14th-16th Century, Physical Sciences); *Robert*

H. Kargon; *Sharon E. Kingsland* (20th and 21st Century, North America, Biological Sciences); *Stuart W. Leslie* (20th and 21st Century, North America, Technology); *Harry M. Marks* (20th and 21st Century, North America, Medical Sciences); *Randall Packard*; *Lawrence M. Principe* (Physical Sciences); *Daniel P. Todes* (19th Century, Europe, Biological Sciences). **Description:** The Department was created in 1992 through the merger of the Institute of the History of Medicine and the Department of the History of Science, each with a long history of graduate training. Faculty interests range from the cultural and intellectual history of medicine in classical Greece to the social history of technology in the 20th century. The program is particularly strong in the history of medicine and science in the early modern period; science, medicine, and technology in the United States, 19th-20th centuries; history of public health and colonial medicine; and Russian and Soviet science. **Web Site:** http://www.hopkinsmedicine.org/graduateprograms/history_of_science/

University of Maryland, College Park (History Department, Specialization in History of Science). **Contact:** Stephen G. Brush, brush@ipst.umd.edu **Mailing Address:** Department of History, University of Maryland, College Park, MD, 20742, United States. **Tel:** 1.301.405.4846 **Fax:** 1.301.314.9363 **Collections/Resources:** Niels Bohr Library & Center for History of Physics, American Center for Physics, College Park; National Archives (Archives II center in College Park); National Agricultural Library, Beltsville, MD (including extensive holdings in chemistry and biology as well as agriculture); Other national libraries and archives in Washington, DC area, easily accessible by rapid transit from College Park: Library of Congress, National Library of Medicine, Smithsonian Institution. **Grants:** Yes. **Program Strengths:** 20th and 21st Century, Physical Sciences; 19th Century, Europe, Physical Sciences; 20th and 21st Century, Biological Sciences. **Degrees Conferred:** Bachelor's Level, Master's Level, Doctoral Level. 4 graduate students in 2002. 1 master's degree and 0 doctorates awarded since July, 1998. **Faculty:** *Stephen G. Brush* (20th and 21st Century, Physical Sciences); *Lindley Darden* (20th and 21st Century, Biological Sciences); *Robert Friedel*. **Description:** At present, new students are accepted only at the M.A. level in history of science (with occasional exceptions). A new program in history of technology (M.A. and Ph.D.) is being developed; contact Robert Friedel (rf27@umail.umd.edu) for information. **Web Site:** http://inform.umd.edu/HIST/graduate.html

Massachusetts

Harvard University (Department of the History of Science). **Contact:** Judith Lajoie, jlajoie@fas.harvard.edu **Primary E-mail:** biscoe@fas.harvard.edu **Mailing Address:** Science Center 235, 1 Oxford Street, Cambridge, MA, 02138, United States. **Tel:** 1.617.495.3741 **Fax:** 1.617.495.3344 **Collections/Resources:** The Collection of Historical Scientific Instruments. **Grants:** No. **Program Strengths:** 20th and 21st Century, Europe, Biological Sciences, Gender and Science; 19th Century, North America, Cognitive Sciences; 18th

Century, Medical Sciences, Institutions; 17th Century, Natural and Human History, Instruments and Techniques; 14th-16th Century, Physical Sciences, Social Relations of Science. **Degrees Conferred:** Bachelor's Level, Master's Level, Doctoral Level. 44 graduate students in 2002. 15 master's degrees and 22 doctorates awarded since July, 1998. **Prospective Student Contact:** Michele Biscoe, biscoe@fas.harvard.edu **Faculty:** *Bridie J. Andrews* (Asia, Medical Sciences); *Mario Biagioli*; *Robert Brain*; *Allan M. Brandt* (20th and 21st Century, North America, Medical Sciences); *Peter L. Galison* (20th and 21st Century, Europe, Social Sciences); *Anne Harrington*; *Sarah Jansen*; *Everett I. Mendelsohn* (19th Century, Biological Sciences); *John E. Murdoch*; *Katharine Park* (14th-16th Century, Europe, Medical Sciences); *Charles Rosenberg*; *Charis Thompson*. **Description:** Historical study of the sciences from antiquity through the present. Graduate students are deeply involved in teaching in the undergraduate program, including independent seminars. Faculty, visiting scholars and graduate students meet regularly in content focused research groups to share their own work and that of visitors. The specialized research libraries at Harvard are rich in printed works, documents and archives. **Web Site:** http://www.fas.harvard.edu/~hsdept/

Massachusetts Institute of Technology (Program in Science, Technology, and Society). **Contact:** Rosalind Williams, rhwill@mit.edu **Primary E-mail:** stsweb@mit.edu **Mailing Address:** Room E51-185, 77 Massachusetts Avenue, Cambridge, MA, 02139, United States. **Tel:** 1.617.253.3452 **Fax:** 1.617.258.8118 **Collections/Resources:** Both the MIT Museum and the Special Collection contain apparatus and archival materials relating to the history of science and technology at MIT, including material relevant to the history of computing, the history of the physical and biological sciences, and the relations among the government, the military, academia and industry throughout the past 150 years. The program has a close relationship with the Dibner Institute of which the Burndy Library has a superb collection of rare texts in the History of Science. **Grants:** Yes. **Degrees Conferred:** Bachelor's Level, Doctoral Level. 23 graduate students in 2002. 0 master's degrees and 19 doctorates awarded since July, 1998. **Prospective Student Contact:** Hugh Gusterson, guster@mit.edu **Faculty:** *Deborah K. Fitzgerald*; *Loren R. Graham* (20th and 21st Century, Europe, Physical Sciences); *Evelynn Hammonds*; *David Kaiser* (20th and 21st Century, North America, Physical Sciences); *Evelyn F. Keller* (20th and 21st Century, Biological Sciences); *Harriet Ritvo* (18th Century, Europe, Natural and Human History); *Merritt Roe Smith* (19th Century, North America, Technology). **Description:** The program is broadly and deeply interdisciplinary. It is sponsored by three faculties: History, Anthropology, and STS. Teachers and students work together to understand technology and science as human experiences, in their political, social, historical, psychological, and cultural dimensions. One strength of the doctoral program is the wide range of related programs and activities at MIT and in the Boston area. **Web Site:** http://web.mit.edu/sts

Michigan Technological University (Program in Industrial History and Archaeology). **Contact:** Patrick Martin, PEM-194@mtu.edu **Mailing Address:** 1400 Townsend Drive, Houghton, MI, 49931-1295, United States. **Tel:** 1.986.487.2113 **Fax:** 1.906.487.2468 **Ongoing Publication Projects:** *IA: Journal of the Society for Industrial Archeology.* **Collections/Resources:** The Michigan Tech library contains significant and detailed records of the region's copper mining industry, which was nationally significant from approximately 1840 to 1940. **Grants:** Yes. **Program Strengths:** 19th Century, North America, Technology; 20th and 21st Century, North America, Technology. **Degrees Conferred:** Bachelor's Level, Master's Level. 8 graduate students in 2002. 13 master's degrees awarded since July, 1998. **Faculty:** *Hugh Gorman; Alison Kim Hoagland; Larry D. Lankton; Patrick E. Martin; Susan R. Martin; Terry Scott Reynolds; Bruce E. Seely; Scarlett Timothy.* **Description:** Michigan Tech's interdisciplinary program in industrial archaeology is one of the few in the world and the only one actively preparing graduate students in that field in the United States. The national headquarters of the Society for Industrial Archeology are also located at Michigan Tech. **Web Site:** http://www.ss.mtu.edu/IA/iahm.html

Michigan

University of Michigan (Department of History). **Contact:** Sonya Rose, sorose@umich.edu **Mailing Address:** 1029 Tisch Hall, University of Michigan, Ann Arbor, MI, 48109-1003, United States. **Tel:** 1.734.764.6305 **Fax:** 1.734.647.4881 **Grants:** Yes. **Program Strengths:** 20th and 21st Century, Transcontinental, Technology; 20th and 21st Century, North America, Medical Sciences; Africa. **Degrees Conferred:** Bachelor's Level, Doctoral Level. **Prospective Student Contact:** Sheila Williams, shewms@umich.edu **Faculty:** *John Carson* (North America); *Paul Edwards; Dario Gaggio; Gabrielle Hecht; Joel D. Howell* (19th Century, Medical Sciences); *Martin S. Pernick* (20th and 21st Century, North America, Medical Sciences); *Michael Wintroub.* **Description:** We engage graduate students in individual and collective interdisciplinary investigations of historical thought, and in research on the historical experiences of humanity in different times and places. We expect our graduate students to become outstanding teachers at universities throughout the world, and to make innovative and enlightening contributions to the study of history. **Web Site:** http://www.lsa.umich.edu/history/

Minnesota

University of Minnesota (History of Medicine and the Biological Sciences). **Contact:** John M. Eyler, eyler001@tc.umn.edu **Primary E-mail:** soria001@ umn.edu **Mailing Address:** 505 Essex Street SE, 506 UMHC, Minneapolis, MN, 55455, United States. **Tel:** 1.612.624.4416 **Fax:** 1.612.625.7938 **Collections/Resources:** Owen H. Wangensteen Historical Library of Biology and Medicine (60,000 pre-1920 volumes); Social Welfare History Archives; Immigration History Archives; Minnesota History Center. **Grants:** Yes.

Program Strengths: 19th Century, North America, Medical Sciences; 19th Century, Europe, Medical Sciences; 20th and 21st Century, North America, Medical Sciences. **Degrees Conferred:** Bachelor's Level, Master's Level, Doctoral Level. 8 graduate students in 2002. 1 master's degree and 0 doctorates awarded since July, 1998. **Faculty:** *John Beatty* (20th and 21st Century, Biological Sciences); *C. Carlyle Clawson*; *John M. Eyler*; *Jennifer L. Gunn*; *Jon M. Harkness*; *Sally Gregory Kohlstedt* (North America, Gender and Science); *Elaine Tyler May*; *David J. Rhees*; *Jole R. Shackelford* (14th-16th Century, Europe, Medical Sciences). **Description:** Our students enter with diverse backgrounds in medicine, the sciences, or history. The curriculum is flexible, and every effort is made to tailor the student's training to his/her background and academic interests. The program's particular strengths are in modern European and American subjects. **Web Site:** http://www.med.umn.edu/history/home.htm

University of Minnesota (Program in History of Science and Technology). **Contact:** Sally Gregory Kohlstedt, sgk@tc.umn.edu **Primary E-mail:** eastwold@ physics.spa.umn.edu **Mailing Address:** Tate Laboratory of Physics, 116 Church Street SE, Minneapolis, MN, 55455, United States. **Tel:** 1.612.624.7069 **Fax:** 1.612.624.4578 **Ongoing Publication Projects:** Charles Babbage Center for the History of Computing has several publications and projects. **Collections/ Resources:** Philip Wangensteen Library for History of Science and Technology; Bell Library and Special Collections; Charles Babbage Library and Archives. **Grants:** Yes. **Program Strengths:** Physical Sciences; Biological Sciences; Technology. **Degrees Conferred:** Bachelor's Level, Master's Level, Doctoral Level. 31 graduate students in 2002. 1 master's degree and 6 doctorates awarded since July, 1998. **Faculty:** *Jennifer K. Alexander* (Europe, Technology); *John Beatty* (20th and 21st Century, Biological Sciences); *Michel H. Janssen* (20th and 21st Century, Physical Sciences); *Sally Gregory Kohlstedt* (North America, Gender and Science); *Robert W. Seidel* (20th and 21st Century, North America, Physical Sciences); *Alan E. Shapiro* (17th Century, Europe, Physical Sciences); *Roger H. Stuewer* (20th and 21st Century, Europe, Physical Sciences). **Description:** The Program has seven full-time faculty members, plus affiliates, whose particular strengths are in the history of physics, the history of biological and natural science, the history of technology, and the history of science in America. The program has a regular colloquium lecture series in conjunction with the Minnesota Center for the Philosophy of Science, works closely with the historians of medicine, and participates in a minor in Studies of Science and Technology. **Web Site:** http://www.physics.umn.edu/~hsci/

Montana

Montana State University (Program in the History and Philosophy of Science). **Contact:** Robert Rydell, rwrydell@montana.edu **Mailing Address:** Wilson Hall, Bozeman, MT, 59717, United States. **Tel:** 1.406.994.4395 **Fax:** 1.406.994.6879 **Collections/Resources:** 2000 books and pamphlets regarding world's fairs; Significant collections regarding Yellowstone National Park. **Grants:** Yes.

Program Strengths: 19th Century, Europe, Natural and Human History, Social Relations of Science; 20th and 21st Century, North America, Natural and Human History, Social Relations of Science; Europe, Physical Sciences. **Degrees Conferred:** Bachelor's Level, Master's Level. 5 graduate students in 2002. 5 master's degrees and 0 doctorates awarded since July, 1998. **Prospective Student Contact:** David Cherry, zpi7001@montana.edu **Faculty:** *Prasanta Bandyopadhyay*; *Gordon Brittan*; *Robert B. Campbell*; *Pierce Mullen*; *Sara B. Pritchard*; *Michael S. Reidy*; *Robert Rydell*; *Lynda Sexson*; *Brett Walker*. **Description:** The department of history and philosophy is currently developing a graduate option dedicated to science, technology, society, and the environment. **Web Site:** http://www.montana.edu

New Hampshire

University of New Hampshire (History of Psychology Program). **Contact:** Benjamin F. Harris, bh5@unh.edu **Primary E-mail:** bh5@cisunix. unh.edu; woodward@cisunix.unh.edu **Mailing Address:** Conant Hall, Durham, NH, 03824, United States. **Tel:** 1.603.862.4107; 1.603.862.3199 **Fax:** 1.603.862.4986 **Ongoing Publication Projects:** Individual faculty are on the editorial boards of various journals (e.g., *History of Psychology, Journal of the History of the Behavioral Sciences, Psychologie und Geschichte, Revista de Historia de la Psicologia,* and *Zeitschrift für Humanontogenetik*) and facilitate graduate student involvement with these journals. William Woodward is co-editor of the *Cambridge Studies in the History of Psychology* (Cambridge University Press). **Collections/Resources:** Our location (approximately one hour north of Boston) gives researchers easy access to archival collections in the Boston area, plus university libraries from Providence R.I. to Hanover, N.H. Our university library houses the papers of Fred Keller, pioneering behaviorist and educational innovator. **Grants:** No. **Program Strengths:** 20th and 21st Century, Cognitive Sciences. **Degrees Conferred:** Bachelor's Level, Master's Level, Doctoral Level. 4 graduate students in 2002. 2 master's degrees and 1 doctorate awarded since July, 1998. **Prospective Student Contact:** Janice Chadwick, janicec@cisunix.unh.edu **Faculty:** *Benjamin F. Harris*; *William R. Woodward*. **Description:** We host the only Ph.D. program in the history of psychology in the United States. Our faculty and graduates helped establish the history of psychology as a research specialty within the history of science. Our institutional offspring include the Cheiron Society and the *Journal of the History of the Behavioral Sciences.* **Web Site:** http://www.unh.edu/psychology/Historyd.html

New Jersey

Princeton University (Program in History of Science). **Contact:** Angela Creager, creager@princeton.edu **Primary E-mail:** vtgt@princeton.edu **Mailing Address:** 129 Dickinson Hall, Princeton, NJ, 08544, United States. **Tel:** 1.609.258.6705 **Fax:** 1.609.258.5326 **Grants:** No. **Program Strengths:** Biological Sciences; Physical Sciences; Mathematics. **Degrees Conferred:** Bachelor's Level,

Master's Level, Doctoral Level. 18 graduate students in 2002. 12 master's degrees and 5 doctorates awarded since July, 1998. **Prospective Student Contact:** Vicky Glosson, vtgt@princeton.edu **Faculty:** *D. Graham Burnett* (Earth Sciences); *Angela N. H. Creager* (20th and 21st Century, Biological Sciences); *Benjamin A. Elman*; *Daniel E. Garber* (17th Century, Europe, Physical Sciences); *Charles C. Gillispie*; *Michael D. Gordin* (19th Century, Europe, Physical Sciences); *Anthony Grafton*; *Elizabeth A. Lunbeck* (20th and 21st Century, North America, Gender and Science); *Michael S. Mahoney* (Mathematics); *Gyan Prakash*; *Helen Tilley*. **Description:** Under the aegis of the Department of History, the Program in History of Science trains graduate students for the professional responsibilities of teaching and research. Drawing on the methods of intellectual, cultural, and social history, we seek to understand the development of the sciences, technology, and medicine in a global context. **Web Site:** http://www.princeton.edu/~hos

Rutgers University, New Brunswick (Program in History of Technology, Environment and Health). **Contact:** Paul Israel, pisrael@rci.rutgers.edu **Mailing Address:** Thomas A. Edison Papers, 44 Road 3, Piscataway, NJ, 08854-8049, United States. **Tel:** 1.732.445.8511 **Fax:** 1.732.445.8512 **Collections/Resources:** Thomas A. Edison Papers; Center for the History of Electrical Engineering (IEEE); Institute for Health, Health Care Policy, and Aging Research; Industrial Environments program of the Rutgers Center for Historical Analysis (2001-03). **Grants:** Yes. **Program Strengths:** 20th and 21st Century, Medical Sciences, Science Policy; Technology; Gender and Science. **Degrees Conferred:** Bachelor's Level, Doctoral Level. 2 graduate students in 2002. 0 doctorates awarded since July, 1998. **Faculty:** *Michael Adas*; *Lauren Benton*; *Janet Golden*; *Paul B. Israel* (19th Century, North America, Technology); *Julie Livingston*; *Margaret Marsh*; *Philip J. Pauly* (19th Century, North America, Biological Sciences); *James Reed*; *Beryl Satter*; *Susan Schrepfer*; *Phil Scranton*; *Richard Sher*; *Gail Triner*; *Keith Wailoo*. **Description:** This Ph.D. program emphasizes environmental politics and environmental transformations, technology in culture and society, and the evolution of health in national and global contexts, with a special interest in the study of history in relation to policy. Faculty strengths include: American, African-American, African, colonial history, and gender studies. **Web Site:** http://hteh.rutgers.edu

Rutgers University, Newark and New Jersey Institute of Technology (History of Technology, Environment, and Medicine/Health). **Contact:** Neil M. Maher, maher@njit.edu **Mailing Address:** NJIT Deptartment of History, University Heights, Newark, NJ, 07102-9895, United States. **Tel:** 1.973.596.6348 **Fax:** 1.973.642.4689 **Grants:** Yes. **Program Strengths:** Medical Sciences; Technology; Natural and Human History. **Degrees Conferred:** Bachelor's Level, Master's Level. 5 graduate students in 2002. 5 master's degrees and 0 doctorates awarded since July, 1998. **Faculty:** *Michael Adas*; *Paul B. Israel* (19th Century, North America, Technology); *Jan Lewis*; *Julie Livingston*; *Neil Maher*; *John O'Connor*; *Susan Schrepfer*; *Phil Scranton*;

Doris Sher; *Richard Sher*. **Description:** The History of Technology, Environment, and Medicine/Health (HisTEM) is a field of concentration in the M.A. degree in history offered jointly by Rutgers University, Newark and NJIT. Students matriculate at Rutgers University, Newark. Faculty in the program are also affiliated with the Ph.D. major field in the History of Technology, Environment, and Health (HTEH) offered by Rutgers University, New Brunswick. The program has ongoing relationships with the Edison Papers, the IEEE History Center, and the Institute for Health, Health Care Policy, and Aging Research at Rutgers University, New Brunswick. **Web Site:** http://www.njit.edu/Directory/Academic/History/grad.html

New York

City University of New York (Specialization in History of Science within Ph.D. Program in History). **Contact:** Joseph W. Dauben, jdauben@worldnet.att.net **Mailing Address:** Ph.D. Program in History, The Graduate Center, CUNY, 365 Fifth Avenue at 34th Street, New York, NY, 10016-4309, United States. **Tel:** 1.212.817.8430 **Fax:** 1.212.817.1523 **Collections/Resources:** New York Public Library; New York Academy of Sciences; New York Academy of Medicine; New York Botanical Garden; American Museum of Natural History. **Grants:** No. **Program Strengths:** Medical Sciences; Mathematics; Physical Sciences. **Degrees Conferred:** Doctoral Level. 5 graduate students in 2002. 1 doctorate awarded since July, 1998. **Faculty:** *Evelyn Ackerman*; *Timothy Alborn*; *Bruce Chandler*; *Alberto Cordero-Lecca* (Philosophy and Philosophy of Science); *Joseph W. Dauben*; *Daniel Gasman*; *Dolores Greenberg*; *James R. Jacob*; *Arnold Koslow*; *Suzanne C. Quellette*; *Rosamond Rhodes*; *Barbara Katz Rothman*; *Nancy Siraisi*; *Martin Tamny* (17th Century, Astronomical Sciences, Philosophy and Philosophy of Science). **Description:** Offers a small number of select graduate students the opportunity to work with a diverse faculty, with the resources of major institutions at their disposal for research purposes, including the various museums, libraries, botanical gardens, and scientific academies located in New York City.

Columbia University (Center for the History and Ethics of Public Health). **Contact:** David Rosner, dr289@columbia.edu **Primary E-mail:** hphm@ columbia.edu **Mailing Address:** Columbia University School of Public Health, 722 West 168th Street, New York, NY, 10032, United States. **Tel:** 1.212.305.0092 **Fax:** 1.212.342.1986 **Ongoing Publication Projects:** Series in the History & Ethics of Public Health (in planning). **Collections/ Resources:** Columbia has an excellent collection in both the history of science and the history of medicine and public health. Special collections are housed on both the Health Sciences and liberal arts campuses and include 200,000 rare books and manuscripts. Of particular interest are materials on the history of psychiatry, social welfare and children's services, plastic surgery, and a host of other manuscripts that have been gathered by an excellent archival staff or donated by various faculty and friends. The breadth of the collections available to researchers cannot be overstated. **Grants:** Yes. **Program Strengths:** 20th and 21st Century, North America, Medical Sciences, Social Relations of Science; 20th and 21st Century, South America, Medical Sciences, Social Relations of Science; 20th and 21st Century, North America, Social Sciences, Science Policy. **Degrees Conferred:** Bachelor's Level, Master's Level, Doctoral Level. 37 graduate students in 2002. 17 master's degrees and 0 doctorates awarded since July, 1998. **Prospective Student Contact:** Martina Lynch, RML17@columbia.edu **Faculty:** *Lerner Barron*; *Ronald Bayer*; *Amy Fairchild*; *Ruth Fischbach*; *Gerald Markowitz*; *Gerald Oppenheimer*; *Samuel K. Roberts*; *David Rosner*; *David Rothman*; *Sheila Rothman*; *Nancy Leys Stepan*; *Marcia Wright*. **Description:** This new Center is home to a unique program combining resources of Columbia's faculties in history and Schools of Medicine and Public Health. Students receive an MPH degree and/or Ph.D., while preparing for both academic and professional careers. It is unique in offering the MPH specialization in history and ethics. **Web Site:** http://cpmcnet.columbia.edu/dept/hphm

Cornell University (Department of Science and Technology Studies). **Contact:** Trevor Pinch, tjp2@cornell.edu **Primary E-mail:** stsgradfield@ cornell.edu **Mailing Address:** 632 Clark Hall, Cornell University, Ithaca, NY, 14853, United States. **Tel:** 1.607.255.3810 **Fax:** 1.607.255.6044 **Ongoing Publication Projects:** *Isis*; *Social Studies of Science*. **Grants:** No. **Program Strengths:** Social Relations of Science. **Degrees Conferred:** Bachelor's Level, Doctoral Level. 30 graduate students in 2002. 0 master's degrees and 9 doctorates awarded since July, 1998. **Prospective Student Contact:** Judy Yonkin, jly5@cornell.edu **Faculty:** *Richard Boyd*; *Joan Jacobs Brumberg*; *Peter R. Dear* (17th Century, Europe, Physical Sciences); *Stephen Hilgartner*; *Ronald Kline* (20th and 21st Century, North America, Technology); *Bruce V. Lewenstein* (Social Relations of Science); *Michael E. Lynch*; *Helene Mialet*; *Trevor J. Pinch*; *Alison Power*; *William B. Provine* (20th and 21st Century, North America, Biological Sciences); *Judith Reppy*; *Margaret W. Rossiter* (20th and 21st Century, North America, Gender and Science); *Phoebe Sengers*. **Description:** S&TS is concerned with understanding science and technology as historical and cultural productions. Inquiry in this field therefore requires the ability to lay bare the ways scientific knowledge, authority, and expertise are established in different social contexts, and their changing historical meanings. **Web Site:** http://www.sts.cornell.edu/CU-STS.html

Rensselaer Polytechnic Institute (Department of Science and Technology Studies). **Contact:** David Hess, hessd@rpi.edu **Primary E-mail:** vumbak@ rpi.edu **Mailing Address:** STS Dept., Sage Building 5th Floor, 110 8th St., Rensselaer Polytechnic Institute, Troy, NY, 12180-3590, United States. **Tel:** 1.518.276.6574 **Fax:** 1.518.276.4871 **Ongoing Publication Projects:** Linda Layne edits *Technoscience*, the newsletter of the Society for Social Studies of Science; Sal Restivo is editor-in-chief of the *Oxford Encyclopedia of Science, Technology, and Society*. **Collections/Resources:** Library includes several major collections: Eban Horsford papers (chemist who invented baking powder), John Roebling and sons collection (builder of Brooklyn Bridge), Paul Harteck papers (German atomic physicist). See http://www.lib.rpi.edu for additional collections. **Grants:** Yes. **Program Strengths:** 20th and 21st Century, Transcontinental, Technology, Science Policy; North America, Medical Sciences, Institutions; Europe, Natural and Human History, Gender and Science. **Degrees Conferred:** Bachelor's Level, Master's Level, Doctoral Level. 40 graduate students in 2002. 15 master's degrees and 10 doctorates awarded since July, 1998. **Prospective Student Contact:** Kathie Vumbacco, vumbak@rpi.edu **Faculty:** *Atsushi Akera; Sharon Anderson-Gold; Steve Breyman; Linnda Caporael; Ron Eglash; Kim Fortun; Mike Fortun; Ray Fouche; Jeff Hannigan; David Hess; Linda Layne; Tom Phelan; Sal Restivo; Sharra Vostral; Langdon Winner; Edward Woodhouse.* **Description:** The graduate program offers an M.S. and Ph.D. in STS, which embraces history, social studies, and policy studies. **Web Site:** http://www.rpi.edu/dept/sts

State University of New York at Stony Brook (History of Science, Technology, and Medicine). **Contact:** Nancy Tomes, nancy.tomes@sunysb.edu **Mailing Address:** History Department, SUNY at Stony Brook, Stony Brook, NY, 11794-4348, United States. **Tel:** 1.631.632.7500 **Fax:** 1.631.632.7367 **Collections/Resources:** Stony Brook is affiliated with Brookhaven National Laboratory, which has an extensive archive. Cold Spring Harbor lab is also nearby. **Grants:** Yes. **Program Strengths:** 20th and 21st Century, North America, Medical Sciences; 20th and 21st Century, Physical Sciences; Earth Sciences. **Degrees Conferred:** Master's Level, Doctoral Level. **Prospective Student Contact:** Christopher Sellers, csellers@notes.cc.sunysb. edu **Faculty:** *Alix Cooper* (17th Century, Europe, Natural and Human History); *Ruth Schwartz Cowan* (19th Century, North America, Technology); *Elizabeth Garber* (18th Century, Europe, Physical Sciences); *Helen Rodnite Lemay; Wolf Schafer; Christopher Sellers* (20th and 21st Century, North America, Biological Sciences); *Nancy Jane Tomes.* **Description:** The Stony Brook history department has seven faculty who specialize in different areas of the history of medicine, technology, and the environment. HSTM is integrated into an innovative graduate program that emphasizes comparative and global history. **Web Site:** http://www.sunysb.edu/history

North Carolina

Duke University (Duke-UNC Program in the History of Science, Medicine and Technology). **Contact:** Seymour Mauskopf, shmaus@duke.edu **Mailing Address:** Department of History, Duke University, Box 90719, Durham, NC, 27708-0719, United States. **Tel:** 1.919.684.2581 **Fax:** 1.919.681.7670 **Ongoing Publication Projects:** *Journal of the History of Medicine* (editor, Margaret Humphreys). **Grants:** Yes. **Program Strengths:** 19th Century, Europe, Physical Sciences; 20th and 21st Century, North America, Medical Sciences; 20th and 21st Century, North America, Technology. **Degrees Conferred:** Bachelor's

Level, Master's Level, Doctoral Level. 9 graduate students in 2002. 2 master's degrees and 1 doctorate awarded since July, 1998. **Faculty:** *Peter English*; *Margaret E. Humphreys* (19th Century, North America, Medical Sciences); *John Kasson*; *Seymour H. Mauskopf* (18th Century, Europe, Physical Sciences); *Michael R. McVaugh* (5th-13th Century, Europe, Medical Sciences); *Alex Roland* (Technology). **Web Site:** http://www-history.aas.duke.edu/history.htm

University of North Carolina (Duke-UNC Program in the History of Science, Medicine and Technology). **Contact:** Michael McVaugh, mcvaugh@email. unc.edu **Mailing Address:** Department of History, CB 3195, University of North Carolina, Chapel Hill, NC, 27599-3195, United States. **Tel:** 1.919.962.2373 **Fax:** 1.919.962.1403 **Ongoing Publication Projects:** *Journal of the History of Medicine* (at Duke). **Grants:** Yes. **Program Strengths:** 5th-13th Century, Europe, Medical Sciences; 17th Century, Europe, Medical Sciences; 19th Century, North America. **Degrees Conferred:** Bachelor's Level, Master's Level, Doctoral Level. 9 graduate students in 2002. 2 master's degrees and 1 doctorate awarded since July, 1998. **Faculty:** *Peter English*; *Margaret E. Humphreys* (19th Century, North America, Medical Sciences); *John Kasson*; *Seymour H. Mauskopf* (18th Century, Europe, Physical Sciences); *Michael R. McVaugh* (5th-13th Century, Europe, Medical Sciences); *Alex Roland* (Technology). **Description:** The history departments of UNC and Duke have formed a collaborative program for graduate study in history of science, medicine, and technology. Graduate students in this program pursue the general study of history at one or the other of these universities while also pursuing the more specialized program with faculty from both universities. **Web Site:** http://www.unc.edu/depts/history/science/index.html

Ohio

Case Western Reserve University (Program in History of Science, Technology, Environment, and Medicine). **Contact:** Carroll Pursell, cxp7@po.cwru.edu **Mailing Address:** Case Western Reserve University, Cleveland, OH, 44106, United States. **Tel:** 1.216.368.2380 **Fax:** 1.216.368.4681 **Grants:** Yes. **Program Strengths:** 20th and 21st Century, North America, Technology; 19th Century, Europe; 20th and 21st Century, Medical Sciences. **Degrees Conferred:** Bachelor's Level, Master's Level, Doctoral Level. 7 graduate students in 2002. 3 master's degrees and 6 doctorates awarded since July, 1998. **Faculty:** *Miriam R. Levin*; *Carroll Pursell*; *Alan J. Rocke* (19th Century, Europe, Physical Sciences); *Jonathan Sadowsky*; *Ted Steinberg*. **Description:** Our program embraces—and emphasizes connections between—the history of science, technology, environment, and medicine, concentrating on the nineteenth and twentieth centuries. Specialties include the cultural history of technology and medicine, and relations among these fields and modern American culture and values. Expertise extends also to European fields, especially France and Germany. **Web Site:** http://www.cwru.edu/artsci/hsty/

Oklahoma

University of Oklahoma (Department of the History of Science). **Contact:** Steven Livesey, slivesey@ou.edu **Mailing Address:** 622 Physical Sciences Building, 601 Elm, Room 622, Norman, OK, 73019-3106, United States. **Tel:** 1.405.325.2213 **Fax:** 1.405.325.2363 **Ongoing Publication Projects:** *Isis Current Bibliography*; *Newsletter of Commission on History of Science and Technology in Islamic Civilization*; EarlyScience Listserv. **Collections/ Resources:** History of Science Collections, containing nearly 90,000 volumes, including 50 incunabula and over 900 16th-century titles. The collections' holdings emphasize both the intellectual and the social contexts of scientific inquiry, ranging from the works of individual scientists to such supporting materials as textbooks, popular works on science, and journals of scientific societies and academies, as well as current publications in the history of science. See http://libraries.ou.edu/depts/HistScience/. The History of Science Program provides short-term travel assistance to pre- and postdoctoral scholars who wish to use the collections. See the Web site for the Andrew W. Mellon Travel Fellowship Program, http://libraries.ou.edu/depts/HistScience/ mellon/index.html. **Grants:** Yes. **Program Strengths:** 14th-16th Century, Transcontinental; 20th and 21st Century, North America, Natural and Human History; 18th Century, Europe, Earth Sciences. **Degrees Conferred:** Bachelor's Level, Master's Level, Doctoral Level. 16 graduate students in 2002. 11 master's degrees and 3 doctorates awarded since July, 1998. **Prospective Student Contact:** Kenneth L. Taylor, ktaylor@ou.edu **Faculty:** *Peter Barker* (14th-16th Century, Astronomical Sciences); *Hunter A. Crowther-Heyck* (20th and 21st Century, North America); *Kathleen Crowther-Heyck* (14th-16th Century, Europe, Gender and Science); *Ralph R. Hamerla* (19th Century, North America, Physical Sciences); *Steven J. Livesey* (5th-13th Century, Europe, Institutions); *Kerry V. Magruder* (17th Century, Astronomical Sciences); *Katherine A. Pandora* (North America); *Jamil Ragep* (5th-13th Century, Asia, Astronomical Sciences); *Kenneth L. Taylor* (Earth Sciences); *Sarah Tracy* (North America, Medical Sciences); *Stephen P. Weldon* (Science and Religion). **Description:** In 2000, the History of Science Program celebrated its Golden Jubilee, making it one of the oldest programs in the United States. History of Science at the University of Oklahoma offers graduate students the opportunity to acquire a comprehensive knowledge of scientific inquiry from antiquity to the modern era, along with the skills and background necessary to conduct specialized research. The research expertise of the department's faculty is especially strong in the following areas: medieval and early modern science in Islam and Western Europe, including issues of the transmission of science within and between these cultures (Professors Barker, K. Crowther-Heyck, Livesey, Magruder and Ragep); natural and social sciences in the modern world, especially biology and ecology, natural history, psychology, and geology (Professors Barker, H. Crowther-Heyck, Magruder, Ogilvie, Pandora, Taylor and Weldon); history of technology, and pedagogical applications for technology in history of science (H. Crowther-Heyck, Magruder, Pandora). The department also recognizes the diverse opportunities for students with training in the history of science. Since 1993, students at OU have had the opportunity to enroll in a dual degree program for master's degrees in the History of Science and Library and Information Studies, one of the few such dual degree programs in the country. **Web Site:** http://www.ou.edu/cas/hsci

Oregon

Oregon State University (History of Science). **Contact:** Mary Jo Nye, nyem@ucs.orst.edu **Primary E-mail:** mbethman@orst.edu **Mailing Address:** Program in History of Science, Department of History, 306 Milam Hall, Oregon State University, Corvalis, OR, 97311-5104, United States. **Tel:** 1.541.737.3421 **Fax:** 1.541.737.1257 **Collections/Resources:** Horning Endowment and annual Horning lecture series (several events each quarter); Horning conferences in the history of science; Ava and Linus Pauling papers and History of Atomic Energy collection, Special Collections, Valley Library, OSU; Oregon State University archives; OSU Hatfield Marine Sciences Center Library. **Grants:** Yes. **Program Strengths:** Europe; North America. **Degrees Conferred:** Bachelor's Level, Master's Level, Doctoral Level. 7 graduate students in 2002. 2 master's degrees and 1 doctorate awarded since July, 1998. **Faculty:** *Mina Carson*; *Ronald E. Doel* (20th and 21st Century, Earth Sciences); *Paul L. Farber* (18th Century, Biological Sciences); *Gary B. Ferngren* (400 B.C.E-400 C.E., Europe, Medical Sciences); *Paul Kopperman*; *Ben Mutschler*; *Mary Jo Nye* (Physical Sciences); *Robert A. Nye*; *William G. Robbins*; *Lisa T. Sarasohn* (17th Century, Europe, Social Relations of Science); *Jeffrey Sklansky*. **Description:** This program connects the humanities, social sciences, and natural sciences by interpreting scientific developments within particular historical settings and analyzing their changing roles within modern cultures. Emphasis is on scientific traditions in Europe and North America after 1500, the physical, biological, medical, environmental, and social sciences, and environmental history. **Web Site:** http://osu.orst.edu/dept/history

Pennsylvania

Carnegie Mellon University (Science, Technology, and Business). **Contact:** David Hounshell, hounshel@andrew.cmu.edu **Mailing Address:** Department of History, Baker Hall 240, Pittsburgh, PA, 15213, United States. **Tel:** 1.412.268.2880 **Fax:** 1.412.268.1019 **Ongoing Publication Projects:** *Social Science History*, edited by Katherine Lynch. **Grants:** Yes. **Program Strengths:** 20th and 21st Century, North America, Technology, Institutions; 19th Century, North America, Technology, Institutions; 17th Century, Europe, Medical Sciences, Humanistic Relations of Science. **Degrees Conferred:** Bachelor's Level, Doctoral Level. 5 graduate students in 2002. 0 master's degrees and 4 doctorates awarded since July, 1998. **Prospective Student Contact:** David W. Miller, dwmiller@andrew.cmu.edu **Faculty:** *Caroline J. Acker*; *Edward W. Constant*; *David Allen Hounshell*;

Robert W. Kiger (Biological Sciences); *Mary Lindemann*; *John Soluri*; *Joel A. Tarr*. **Description:** The program in science, technology, business, and environmental history is a subspecialty of the department's Ph.D. Program in History and Policy. **Web Site:** http://www.history.cmu.edu

Drexel University (Science, Technology, and Society). **Contact:** Kathryn Steen, steenk@drexel.edu **Mailing Address:** Department of History and Politics, 3141 Chestnut Street, Philadelphia, PA, 19104-2875, United States. **Tel:** 1.215.895.2463 **Fax:** 1.215.895.6614 **Grants:** No. **Program Strengths:** Social Relations of Science; Technology. **Degrees Conferred:** Master's Level. 8 graduate students in 2002. 0 master's degrees awarded since July, 1998. **Faculty:** *Eric Dorn Brose*; *Richardson Dilworth*; *Christian Hunold*; *Scott Gabriel Knowles*; *Erik P. Rau* (20th and 21st Century, North America, Technology); *Richard Rosen*; *Amy Slaton*; *Kathryn Steen*; *Donald F. Stevens*. **Description:** The M.S. program in Science, Technology, and Society is new and growing, drawing on the research of the several historians of technology in the department, as well as on political scientists engaged in the study of science and technology. In keeping with Drexel's tradition, the program requires a practicum and is geared towards professionals. University fellowships may be available for outstanding masters applicants. **Web Site:** http://www.drexel.edu/academics/coas/depts/histpol/grad.htm

Lehigh University (Program in the History of Technology and Science). **Contact:** John Smith, jks0@lehigh.edu **Mailing Address:** 9 W. Packer Ave., Bethleham, PA, 18015, United States. **Tel:** 1.610.758.3365 **Fax:** 1.610.758.6554 **Ongoing Publication Projects:** *Science, Technology, and Society Newsletter*. **Collections/Resources:** Library has excellent collection of books on science and engineering. **Grants:** Yes. **Degrees Conferred:** Bachelor's Level, Master's Level, Doctoral Level. 40 graduate students in 2002. 15 master's degrees and 10 doctorates awarded since July, 1998. **Faculty:** *Stephen H. Cucliffe*; *Steven L. Goldman* (20th and 21st Century, North America, Technology); *Tom F. Peters*; *John K. Smith*. **Description:** Lehigh's graduate program includes the history of science within its focus on Industrial America. The Ph.D. program at Lehigh requires students to develop expertise in four broadly defined fields of history. **Web Site:** http://www.lehigh.edu/~inhis/inhis.html

Pennsylvania State University (Science, Medicine, and Technology in Culture). **Contact:** Robert N. Proctor, rnp5@psu.edu **Primary E-mail:** LLS10@ psu.edu **Mailing Address:** History Dept., Weaver Bldg, Pennsylvania State University, University Park, PA, 16802, United States. **Tel:** 1.814.863.8943 **Fax:** 1.814.863.7840 **Ongoing Publication Projects:** Tobacco and Health History Project; The Historian as Expert Witness Workshop; Nanotechnology and Culture Project; Human Origins Research Project (& Darwin Group); Rhetoric of Science Project; Disability Rhetoric Project; *Environmental History*; NASA Space Grant Link (w/ Astrobiology Center); *Philosophy and Rhetoric*; VRG—Virtual Research Group; SETI Oracle Studies Project.

Collections/Resources: Astrobiology Center with archives; Materials Research Laboratory archives; Nazi Health and Medicine Archives. **Grants:** Yes. **Program Strengths:** 20th and 21st Century, Transcontinental, Natural and Human History, Exploration, Navigation, Expeditions; 19th Century, Europe, Medical Sciences, Gender and Science; 18th Century, Africa, Technology. **Degrees Conferred:** Bachelor's Level, Master's Level, Doctoral Level. 6 graduate students in 2002. 2 master's degrees and 2 doctorates awarded since July, 1998. **Prospective Student Contact:** Carol Reardon, car9@psu.edu **Faculty:** *Gary Cross*; *Richard Doyle*; *Robert N. Proctor*; *Adam Rome*; *Paul L. Rose*; *Londa Schiebinger* (Gender and Science); *Susan Squier*; *Judy Wakhungu*; *Kenneth Weiss*. **Description:** Faculty interests range from astrobiology and the origin of life, to virtual reality, genome diversity, Nazi medicine, public health history, expert witnessing, gender in colonial botany, science rhetoric, Darwin studies, science fiction, neocatastrophism and "the social construction of ignorance" (agnatology). Students can enter SMTC via History, English, or one of the sciences (Anthropology, Geosciences, etc.). Students should note that PSU offers a number of competitive fellowships for graduate training, including several National Science Foundation graduate fellowships for the academic years 2002-2004. **Web Site:** http://faculty.la.psu.edu/ssps/smtc.html

University of Pennsylvania (Department of History and Sociology of Science). **Contact:** Henrika Kuklick, hkuklick@sas.upenn.edu **Mailing Address:** Logan Hall 303, 249 S. 36th Street, Philadelphia, PA, 19104-6304, United States. **Tel:** 1.215.898.8400 **Fax:** 1.215.573.2231 **Grants:** Yes. **Degrees Conferred:** Bachelor's Level, Master's Level, Doctoral Level. 25 graduate students in 2002. 7 master's degrees and 12 doctorates awarded since July, 1998. **Prospective Student Contact:** Mark Adams, madams@sas.upenn.edu **Faculty:** *Mark B. Adams* (Biological Sciences); *Robert Aronowitz*; *David S. Barnes* (19th Century, Europe, Medical Sciences); *Ruth Schwartz Cowan* (19th Century, North America, Technology); *Nathan Ensmenger*; *Steven Feierman*; *Robert E. Kohler* (Biological Sciences); *Henrika Kuklick*; *Susan Lindee* (20th and 21st Century, North America, Biological Sciences); *Nathan Sivin*; *Janet A. Tighe* (19th Century, North America, Medical Sciences). **Description:** Although faculty within the department and elsewhere in the university have expertise in pre-modern science, the department is especially strong in the study of developments in the natural sciences, the human sciences, technology, and medicine from the nineteenth century to the present. The department is truly interdisciplinary. The subjects in which the faculty earned advanced degrees include American Studies, Anthropology, Chemistry, Medicine, and Sociology, as well as History of Science. The department also possesses unusually rich faculty resources for students wishing to undertake research on non-European and non-North American topics. **Web Site:** http://ccat.sas.upenn.edu/hss

University of Pittsburgh (Department of History and Philosophy of Science). **Contact:** John Norton, jdnorton@pitt.edu **Primary E-mail:** hpsdept@pitt.edu **Mailing Address:** 1017 Cathedral of Learning, University of Pittsburgh,

Pittsburgh, PA, 15260, United States. **Tel:** 1.412.624.5896 **Fax:** 1.412.624.6825 **Ongoing Publication Projects:** *Studies in History & Philosophy of Modern Physics*; philsci-archive.pitt.edu. **Collections/Resources:** Archive for Scientific Philosophy which includes papers of Rudolf Carnap, Carl Hempel, Hans Reichenbach, Frank Ramsey, Heinrich Hertz, Bruno deFinetti and Wilfred Sellars; Microfilm copy of the Archive for History of Quantum Theory. **Grants:** No. **Degrees Conferred:** Bachelor's Level, Doctoral Level. 30 graduate students in 2002. 7 master's degrees and 7 doctorates awarded since July, 1998. **Prospective Student Contact:** Joann McIntyre, vanna@pitt.edu **Faculty:** *John S. Earman*; *Paul E. Griffiths* (Biological Sciences); *James G. Lennox* (19th Century, Europe, Biological Sciences); *Peter Machamer* (17th Century, Europe); *James E. McGuire*; *Sandra D. Mitchell* (20th and 21st Century, Philosophy and Philosophy of Science); *John Norton* (20th and 21st Century, Physical Sciences). **Description:** The Department has broad strengths in most areas of HPS. Special strengths are modern physical science, including relativity theory, statistical physics and quantum theory; biological sciences, including evolutionary theory and the emotions; early modern science, including Galileo, Descartes and Newton; and ancient science, including biology and cosmology. **Web Site:** http://www.pitt.edu/~hpsdept

Rhode Island

Brown University (Department of the History of Mathematics). **Contact:** David Pingree **Mailing Address:** Box 1900, Dept. of the History of Mathematics, Brown University, Providence, RI, 02912, United States. **Tel:** 1.401.863.2101 **Ongoing Publication Projects:** Census of the Exact Sciences in Sanskrit; Catalogues of South and West Asian manuscripts (American Committee for South Asian Manuscripts). **Collections/Resources:** Primary and secondary sources on history of the exact sciences in Akkadian, Egyptian, Greek, Latin, Arabic, Persian, and Sanskrit. **Grants:** No. **Degrees Conferred:** Bachelor's Level, Master's Level, Doctoral Level. 3 graduate students in 2002. 0 master's degrees and 1 doctorate awarded since July, 1998. **Faculty:** *David E. Pingree* (Pre-400 B.C.E, Astronomical Sciences); *Kim L. Plofker* (Astronomical Sciences); *Alice Slotsky*. **Description:** Focuses on history of exact sciences (mathematics, astronomy, other quantitative disciplines such as astrology and divination) from the ancient Near East up through Renaissance Europe, particularly intercultural transmission of scientific ideas. Students must learn at least two relevant classical languages (Akkadian, Greek, Latin, Sanskrit, Arabic, Persian) plus French and German. Joan Richards (Victorian mathematics) is a member of the History Department. **Web Site:** http://www.brown.edu/ Departments/History_Mathematics/gradstudy.html

Tennessee

Vanderbilt University (Department of History). **Contact:** Arleen Tuchman, arleen.m.tuchman@vanderbilt.edu **Mailing Address:** Vanderbilt University,

History Department, Box 1802-B, Nashville, TN, 37235, United States. **Tel:** 1.615.322.2575 **Fax:** 1.615.343.6002 **Grants:** Yes. **Program Strengths:** 19th Century, Europe, Medical Sciences; 18th Century, Europe, Medical Sciences; 20th and 21st Century, North America, Medical Sciences. **Degrees Conferred:** Bachelor's Level, Master's Level, Doctoral Level. 1 graduate student in 2002. 0 master's degrees and 0 doctorates awarded since July, 1998. **Faculty:** *Dennis Dickerson*; *Marshall C. Eakin* (20th and 21st Century, South America, Social Relations of Science); *Matthew Ramsey*; *Arleen Tuchman.* **Description:** Vanderbilt currently has three faculty members who teach and conduct research on the history of modern American and European medicine and public health. Other members of the department have interests in the history of technology, robotics, and artificial intelligence. **Web Site:** http://www.vanderbilt.edu/AnS/history/contents.htm

Texas

University of Houston (Department of History). **Contact:** Susan Kellogg, skellogg@uh.edu **Mailing Address:** Department of History, 524 Agnes Arnold Hall, University of Houston, Houston, TX, 77204-3003, United States. **Tel:** 1.713.743.3095 **Fax:** 1.713.743.3216 **Grants:** Yes. **Program Strengths:** Medical Sciences, Social Relations of Science. **Degrees Conferred:** Bachelor's Level, Master's Level, Doctoral Level. 2 graduate students in 2002. 0 master's degrees and 0 doctorates awarded since July, 1998. **Prospective Student Contact:** Daphyne Pitre, dpitre@mail.uh.edu **Faculty:** *Roberta E. Bivins.* **Description:** Within the history of science, the University of Houston specializes in the study of the history of medicine. We have several faculty members interested in this area and Houston's Medical Center offers an opportunity for in-depth historical research. **Web Site:** http://vi.uh.edu

University of Texas, Austin (Program in the History and Philosophy of Science). **Contact:** Sahotra Sarkar, sarkar@mail.utexas.edu **Primary E-mail:** philsci@ uts.cc.utexas.edu **Mailing Address:** Waggener 316, University of Texas at Austin, Austin, TX, 78712, United States. **Tel:** 1.512.232.7101 **Fax:** 1.512.471.4806 **Ongoing Publication Projects:** *Encyclopedia of the Philosophy of Science*; Annotated Bibliographies for the History and Philosophy of Contemporary Biology. **Collections/Resources:** Harry Ransom Collection. **Grants:** Yes. **Program Strengths:** 20th and 21st Century, Biological Sciences; 20th and 21st Century, Biological Sciences; 20th and 21st Century, Biological Sciences. **Degrees Conferred:** Bachelor's Level, Master's Level, Doctoral Level. 9 graduate students in 2002. 4 master's degrees and 1 doctorate awarded since July, 1998. **Prospective Student Contact:** Jill Glenn, jilljg@mail.utexas.edu **Faculty:** *Roger Hart*; *Bruce J. Hunt* (19th Century, Europe, Physical Sciences), *Sahotra Sarkar.* **Description:** Graduate students receive degrees from the Department of Philosophy with a specialization in the History and Philosophy of Science. The Program strongly encourages interdisciplinary work. Faculty members from the Architecture, History, Integrative Biology,

Philosophy, Psychology and Physics Departments are involved in the program. **Web Site:** http://uts.cc.utexas.edu/~philsci/index1.html

Virginia

Virginia Polytechnic Institute and State University (Graduate Program in Science and Technology Studies). **Contact:** Valerie Hardcastle, valerie@ vt.edu **Mailing Address:** Center for Interdisciplinary Studies, Virginia Tech, Blacksburg, VA, 24061-0027, United States. **Tel:** 1.540.231.7615 **Fax:** 1.540.231.7013 **Degrees Conferred:** Master's Level, Doctoral Level. **Faculty:** *Barbara Allen*; *Henry Bauer*; *Daniel Breslau*; *Daryl Chubin*; *Eileen Crist*; *Alexandra Cuffel*; *Mordechai Feingold* (17th Century, Europe, Education); *Anne Fitzpatrick*; *Ellsworth R. Fuhrman*; *Marjorie Grene*; *Valerie Hardcastle*; *Richard F. Hirsh* (20th and 21st Century, North America, Physical Sciences); *Kathleen Jones*; *Ann LaBerge*; *Muriel Lederman*; *Harlan Miller*; *Barbara J. Reeves*; *Doris T. Zallen* (20th and 21st Century, Biological Sciences). **Description:** STS at Virginia Tech is a cooperative venture of the Science and Technology Studies faculty in the Center for Interdisciplinary Studies, and the Departments of History, Philosophy, Political Science, and Sociology. STS is also a member unit of the School of Public and International Affairs. Courses leading to an M.S. or a Ph.D. in STS are available at two sites: Virginia Tech's main campus in Blacksburg and the Northern Virginia Center (NVC) in the greater D.C. metro area. **Web Site:** http://www.cis.vt.edu/stshome/

Washington

University of Washington (Program in the History of Science, Technology, and Medicine). **Contact:** Bruce Hevly, bhevly@u.washington.edu **Mailing Address:** Dept. of History, Box 353560, University of Washington, Seattle, WA, 98195, United States. **Tel:** 1.206.543.5790 **Fax:** 1.206.543.9451 **Grants:** Yes. **Degrees Conferred:** Bachelor's Level, Master's Level, Doctoral Level. 6 graduate students in 2002. **Faculty:** *Thomas L. Hankins* (Biography); *Bruce Hevly*; *Simon Werrett*.

West Virginia

West Virginia University (Department of History). **Contact:** Robert Maxon, rmaxon@wvu.edu **Mailing Address:** P.O. Box 6303, Morgantown, WV, 26506-6303, United States. **Tel:** 1.304.293.2421 **Fax:** 1.304.293.3616 **Ongoing Publication Projects:** *Earth Sciences History*; *Archaeoastronomy: The Journal of Astronomy*; Stephen McCluskey serves as listowner of HASTRO-L, an electronic discussion group on research and teaching in the History of Astronomy. **Grants:** Yes. **Degrees Conferred:** Bachelor's Level, Master's Level, Doctoral Level. **Prospective Student Contact:** Steven Zdatny, szdatny@wvu.edu **Faculty:** *Gregory A. Good* (20th and 21st Century, Earth

Sciences); *Stephen C. McCluskey* (5th-13th Century, Europe, Astronomical
Sciences); *Michael McMahon.* **Web Site:** http://www.as.wvu.edu/history

Wisconsin

University of Wisconsin (History of Science). **Contact:** Lynn Nyhart,
lknyhart@facstaff.wisc.edu **Primary E-mail:** mail@histsci.wisc.edu **Mailing
Address:** 7143 Social Science Building, 1180 Observatory Drive, Madison,
WI, 53706-1393, United States. **Tel:** 1.608.262.1406 **Fax:** 1.608.262.3984
Ongoing Publication Projects: *Cambridge History of Science* (eds. Lindberg
and Numbers). **Collections/Resources:** History of Science rare book collection;
History of Science bibliographer; History of Medicine collection (incl. rare
books); History of Medicine bibliographer; History of Pharmacy; State
Historical Library (American History). **Grants:** No. **Degrees Conferred:**
Bachelor's Level, Master's Level, Doctoral Level. 32 graduate students in
2002. 12 doctorates awarded since July, 1998. **Prospective Student Contact:**
Eileen Ward, mail@histsci.wisc.edu **Faculty:** *Thomas H. Broman* (18th
Century, Europe, Medical Sciences); *Jane R. Camerini* (20th and 21st Century,
Exploration, Navigation and Expeditions); *Victor L. Hilts*; *Judith A. Houck*;
Florence Hsia; *Richard Keller*; *Judith W. Leavitt* (19th Century, North America,
Medical Sciences); *David C. Lindberg* (5th-13th Century, Europe, Physical
Sciences); *Gregg Mitman* (20th and 21st Century, North America, Biological
Sciences); *Ronald L. Numbers* (North America); *Lynn K. Nyhart* (19th Century,
Europe, Biological Sciences); *Robin E. Rider* (Social Relations of Science);
Eric Schatzberg; *Michael H. Shank* (5th-13th Century, Europe, Astronomical
Sciences); *Richard Staley* (Physical Sciences). **Description:** This is the oldest
department of its kind in North America and a leader in graduate education in
the history of science, technology and medicine from medieval times to the
present. **Web Site:** http://polyglot.lss.wisc.edu/histsci/histsci.html

University of Wisconsin (Medical History and Bioethics). **Contact:** Ronald
Numbers, rnumbers@med.wisc.edu **Primary E-mail:** medhisteth@
med.wisc.edu **Mailing Address:** 1300 University Avenue, Room 1432
MSC, Madison, WI, 53706-1532, United States. **Tel:** 1.608.262.1460
Fax: 1.608.262.2327 **Ongoing Publication Projects:** *Cambridge History
of Science*- 8 volumes. **Collections/Resources:** Extensive slide collection;
Extensive video collection. **Grants:** Yes. **Degrees Conferred:** Bachelor's
Level, Master's Level, Doctoral Level. 12 graduate students in 2002. 7 master's
degrees and 2 doctorates awarded since July, 1998. **Prospective Student
Contact:** History of Science Dept., mail@histsci.wisc.edu **Faculty:** *Thomas H.
Broman* (18th Century, Europe, Medical Sciences); *Judith A. Houck*; *Richard
Keller*; *Judith W. Leavitt* (19th Century, North America, Medical Sciences);
Gregg Mitman (20th and 21st Century, North America, Biological Sciences);
Ronald L. Numbers (North America). **Description:** History of Medicine
Department was the second such department to be established in the U.S.
(1950). **Web Site:** http://www.medsch.wisc.edu/medhist

Research Centers

The following definition for research centers was chosen by the HSS executive committee: "Research centers actively support research in the history of science. They may do this by hosting fellows, organizing symposia and conferences, or publishing journals. They do not commonly grant academic degrees."
43 respondents described their institutions as history of science research centers. Entries are organized by country, and then by state for programs within the United States. Not all institutions responded with complete data. All entries contain some portion of the following information:

Institution name
Name of program in history of science, technology, or medicine
Primary contact person's name, title, and e-mail address
Primary e-mail address, if different from the contact person's email
Mailing address
Telephone number
Facsimile number
Journal editorial offices and ongoing serial publication projects
Special collections and research resources
Does the institution offer publicly available grants or fellowships (yes/no)
Particular research focus, entered using a standardized list of keywords
List of staff members, including research interests. Names appear as they were entered by respondents. A broad research interest appears only after the names of HSS members, if they entered this information when they joined the Society. (As computer programming challenges are overcome, research interests for non-members should appear beginning with the 2003 online edition of the *Guide*.)
Brief description of the center and its special features
URL

Brazil

State University of Campinas (Grupo de História e Teoria da Ciência).
Contact: Roberto de Andrade Martins, Rmartins@ifi.unicamp.br **Primary**
E-mail: ghtc@ifi.unicamp.br **Mailing Address:** IFGW - Unicamp, Caixa
Postal 6165, Campinas, SP, 13084-971, Brazil. **Tel:** 55.19.3788.5516
Fax: 55.19.3788.5512 **Grants:** Yes. **Staff:** *Raphael de Souza Brolasse*; *Fabiana*
Guariglia; *Rosangela de Jesus Silva*. **Description:** Main research subjects of
the Brazilian Group of History and Theory of Science: general epistemology;
history and philosophy of physics and biology; history of science, medicine
and technology in Portugal and Brazil (15th to 19th century). The Group has a
large database (80,000 entries) on Portuguese and Brazilian primary sources.
Web Site: http://www.ifi.unicamp.br/~ghtc/

Universidade Federal do Rio Grande do Sul (Grupo Interdisciplinar em Filosofia
e História das Ciências). **Contact:** Aldo Mellender de Araujo, aldomel@
portoweb.com.br **Primary E-mail:** gifhc@ilea.ufrgs.br **Mailing Address:**
UFRGS - Campus do Vale, ILEA, Av. Bento Gonçalves, 9500 - Prédio
43322, sala 104, Porto Alegre, RS, 91509-900, Brazil. **Tel:** 55.51.3316.6945
Fax: 55.51.3316.7155 **Ongoing Publication Projects:** *Episteme: Filosofia*
e História das Ciências. **Grants:** No. **Program Strengths:** 20th and 21st
Century, South America, Biological Sciences; 20th and 21st Century,
South America, Earth Sciences; 19th Century, Natural and Human
History. **Staff:** *Aldo Mellender Araujo*; *Russel Teresinha Dutra da Rosa*
(Natural and Human History); *Fernando Lang da Silveira* (20th and 21st
Century, Biological Sciences); *Daniel Sander Hoffmann* (19th Century,
Europe, Physical Sciences); *Rualdo Menegat* (Education); *Lauro Nardi*;
Anna Carolina K. P. Regner; *Maria Lúcia Castagna Wortmann* (Physical
Sciences). **Description:** GIFHC congregates researchers from different
fields working on the history and philosophy of science, publishes a journal
(http://www.ilea.ufrgs.br/episteme), and promotes national and international
meetings. **Web Site:** http://www.ilea.ufrgs.br/gifhc/

France

Centre National de la Recherche Scientifique Archives. **Contact:** Louis
Cosnier, louis.cosnier@cnrs-dir.fr. **Mailing Address:** 3 Rue Michel-Ange
75794 Paris Cedex 16, Paris, 75794, France. **Tel:** 33.1.44.96.44.35
Fax: 33.1.44.96.49.32 **Grants:** Yes. **Program Strengths:** 20th and 21st
Century, Europe. **Web Site:** http://www.cnrs.fr/Archives/

Museum National d'Histoire Naturelle (Centre Alexandre Koyre).
Contact: Dominique Pestre, pestre@ehess.fr
Mailing Address: Pavillon Chevreul, 57, rue Cuvier - 75231, Paris,
cedex 05, France. **Tel:** 33.1.43.36.70.69 **Fax:** 33.1.43.31.34.49
Web Site: http://www.ehess.fr/centres/koyre/Centre_A_KOYRE.html

Germany

Eberhard-Karls-Universität Tübingen (Institut für Geschichte der Medizin).
Contact: Urban Wiesing, urban.wiesing@uni-tuebingen.de **Primary**
E-mail: igm@uni-tuebingen.de **Mailing Address:** Goethestr. 6, Tübingen,
D-72076, Germany. **Tel:** 49.7071.2.97.29.50 **Fax:** 49.7071.551.784 **Ongoing**
Publication Projects: Sigmund Freud: Letters to Minna and Martha Bernays;
Binswanger-Clinic Bellevue: Case Histories. **Collections/Resources:** Research
library: history of psychiatry and psychoanalysis; Binswanger Library;
PHS and IWD Databases; Special resources in history of psychoanalysis.
Grants: No. **Program Strengths:** Medical Sciences, Humanistic Relations
of Science; Natural and Human History, Institutions; Medical Sciences,
Biography. **Staff:** *Gerhard Fichtner; Albrecht Hirschmüller; Urban Wiesing.*
Web Site: http://www.uni-tuebingen.de/igm/

Friedrich Wilhelms Universität, Bonn (Medizinhistorisches Institut).
Contact: H Schott, heinz.schott@ukb.uni-bonn.de **Mailing Address:**
Sigmund-Freud-Str. 25, D-53105, Bonn, Nordrhein Westfalen, 53105, Germany.
Tel: 49.228.287.5000 **Fax:** 49.228.287.5006 **Collections/Resources:** A
special library for medical history. **Grants:** No. **Program Strengths:** Natural
and Human History. **Description:** The Institute is concerned with medical
humanities, especially cultural history of medicine, medical anthropology
and medical ethics. There are a series of research projects, e.g. in regard
to the work of Paracelsus, the history of psychosomatic medicine, medical
anthropology in East Africa, Peru/Ecuador and the medical faculty of Bonn
during the Third Reich. The staff members belong to various disciplines
(medicine, history, philosophy, German studies, psychology). The institute
offers a changing teaching program (lectures, seminars, colloquia) besides the
obligatory courses in medical terminology and lectures in history of medicine.
This includes seminars in medical ethics for all stages of undergraduate
medical education; seminars are open for students of theoretical medicine
and cultural anthropology from the faculty of arts; lectures in the history
and anthropology of complementary medicine for the general public. It is
possible for external scholars to work with the excellent institute library.
Web Site: http://www.meb.uni-bonn.de/mhi

Friedrich-Alexander-Universität Erlangen-Nürnberg (Institut für Geschichte
und Ethik der Medizin). **Contact:** Renate Wittern-Sterzel, mfgm01@
gesch.med.uni-erlangen.de **Mailing Address:** Glückstr. 10, Erlangen, Bavaria,
D-91054, Germany. **Tel:** 49.9131.8.52.23.08 **Fax:** 49.9131.8.52.28.52
Grants: No. **Program Strengths:** 400 B.C.E - 400 C.E., Europe, Medical
Sciences; 14th-16th Century, Europe, Medical Sciences; 19th Century,
Europe, Medical Sciences. **Staff:** *Marion Maria Ruisinger; Frank Stahnisch.*
Web Site: http://www.gesch.med.uni-erlangen.de

Max-Planck-Institut für Wissenschaftsgeschichte. **Contact:** Jochen Schneider, jsr@mpiwg-berlin.mpg.de **Primary E-mail:** zentrale@mpiwg-berlin.mpg.de **Mailing Address:** Wilhelmstr. 44, Berlin, 10117, Germany. **Tel:** 49.30.226.670 **Fax:** 49.30.226.67.299 **Grants:** Yes. **Program Strengths:** Philosophy or Philosophy of Science; Instruments and Techniques; Social Relations of Science. **Staff:** *Peter Damerow*; *Peter Beurton*; *Giuseppe Castagnetti*; *Lorraine J. Daston* (18th Century, Europe, Social Sciences); *Sven Dierig*; *Mechthild Fend*; *Peter Geimer*; *Michael Hagner* (19th Century, Europe, Medical Sciences); *Dieter Hoffmann*; *Horst Kant*; *Ursula Klein* (18th Century, Europe, Physical Sciences); *Wolfgang Lefèvre*; *Peter C. McLaughlin*; *Staffan Müller-Wille* (20th and 21st Century, Europe, Biological Sciences); *Annik Pietsch*; *J. Renn*; *Hans-Jorg Rheinberger* (20th and 21st Century, Biological Sciences); *Simone Rieger*; *Henning Schmidgen*; *Markus Schnoepf*; *Volkmar Schüller*; *H. Otto Sibum*; *Anke te Heesen*; *Fernando Vidal*; *Annette Vogt*; *Renate Wahsner*. **Description:** The institute was established in March 1994. Its research is primarily devoted to a theoretically oriented history of science, principally of the natural sciences, but with methodological perspectives drawn from the cognitive sciences and from cultural history. All three departments of the institute aim at the construction of a "historical epistemology" of the sciences. **Web Site:** http://www.mpiwg-berlin.mpg.de

Robert Bosch Stiftung (Institut für Geschichte der Medizin). **Contact:** Robert Jütte, robert.juette@igm-bosch.de **Primary E-mail:** igm.bosch@ t-online.de **Mailing Address:** Straussweg 17, Stuttgart, D-70184, Germany. **Tel:** 49.711.46.08.4171 **Fax:** 49.711.46.08.4181 **Ongoing Publication Projects:** Hahnemann-Krankenjournal-Edition; Quellen und Studien zur Homöopathiegeschichte; *Medizin, Gesellschaft und Geschichte*. **Collections/Resources:** Homeopathic Archives; Hahnemann papers; archives of the central association of German homeopathic physicians. **Grants:** Yes. **Description:** Research in the institute concentrates on the history of homeopathy and the social history of medicine. **Web Site:** http://www.igm-bosch.de

Universität Hannover (Zentrum für Zeitgeschichte von Bildung und Wissenschaft). **Contact:** Manfred Heinemann, m.heinemann@zzbw. uni-hannover.de **Primary E-mail:** postmaster@zzbw.uni-hannover.de **Mailing Address:** Wunstorfer Straße 14, Hannover, D-30453, Germany. **Tel:** 49.511.7.62.94.12 **Fax:** 49.511.762.9418 **Ongoing Publication Projects:** *Edition Bildung und Wissenschaft*; *Bildung und Erziehung*. **Collections/ Resources:** collections of files of central German academic agencies; collections of personal files. **Grants:** No. **Program Strengths:** 20th and 21st Century, Europe, Institutions; 20th and 21st Century, Europe, Natural and Human History, Education. **Description:** ZZBW offers unique collections about contemporary education history since 1945. The research center may host scholars and assist them with research in Germany. **Web Site:** http://www.zzbw.uni-hannover.de/

Universität Leipzig (Karl-Sudhoff-Institut für Geschichte der Medizin und der Naturwissenschaften). **Contact:** Ortrun Riha, riha@medizin.uni-leipzig.de **Mailing Address:** Augustusplatz 10/11, Leipzig, D-04109, Germany. **Tel:** 49.341.9.72.56.00 **Fax:** 49.341.9612458 **Ongoing Publication Projects:** *NTM-Zeitschrift für Geschichte und Ethik der Naturwissenschaften, Technik und Medizin*; *Quellen zur Geschichte Sibiriens und Alaskas aus russischen Archiven*; *Deutsch-russische Beziehungen in Medizin und Naturwissenschaften.* **Collections/Resources:** DFG (Deutsche Forschungsgemeinschaft); Special editions in the general history of medicine and science. **Grants:** No. **Program Strengths:** 18th Century, Europe, Mathematics, Humanistic Relations of Science; 5th-13th Century, Europe, Medical Sciences, Gender and Science; 5th-13th Century, Europe, Medical Sciences, Philosophy or Philosophy of Science. **Web Site:** http://www.uni-leipzig.de/~ksi/

Universität zu Köln (Institut für Geschichte und Ethik der Medizin). **Contact:** Klaus Bergdolt, bergdolt@uni-koeln.de **Mailing Address:** Joseph-Stelzmann-Str. 9, Gebäude 29, Köln, D-50931, Germany. **Tel:** 49.221.478.52.66 **Fax:** 49.221.478.6794 **Grants:** No. **Staff:** *Klaus Bergdolt.* **Web Site:** http://www.uni-koeln.de/med-fak/igem

Greece

National Hellenic Research Foundation (History and Philosophy of Science Programme). **Contact:** Efthymios Nicolaides, efnicol@eie.gr **Primary E-mail:** gvlahakis@eie.gr **Mailing Address:** 48, Vas. Constantinou Av., Athens, 116 35, Greece. **Tel:** 30.1.7273.557.9 **Fax:** 30.1.7246.212 **Ongoing Publication Projects:** *Newsletter for the History of Science in Southeastern Europe.* **Collections/Resources:** Hellenic Archives of Scientific Instruments; Manuscripts and old/rare science books and microfilms; 18th-19th century scientific books database. **Grants:** No. **Program Strengths:** 18th Century, Europe, Physical Sciences; 19th Century, Europe, Physical Sciences. **Staff:** *Yannis Karas*; *Efthymios Nicolaides*; *George X. Vlahakis* (18th Century, Europe, Physical Sciences). **Description:** The program focuses its activities on the study of science in Southeastern Europe during the 16th-20th century. Leader of the program is Yannis Karas. The program collaborates with other relevant institutions in Southeastern Europe to promote history of science in this region. It also carries out joint research projects with other European research centers in Serbia, Italy, Russia, and France. **Web Site:** http://www.eie.gr/institutes/kne/ife/

Israel

Hebrew University of Jerusalem (The Sidney M. Edelstein Center for the History and Philosophy of Science, Technology and Medicine). **Contact:** Anthony Travis, travis@shum.huji.ac.il **Mailing Address:** Edmond Safra Campus, Givat RamJerusalem, 91 904, Israel. **Tel:** 972.2.658.5652 **Fax:** 972.2.658.6709 **Ongoing Publication Projects:** *Aleph: Historical Studies* *in Science and Judaism*. **Collections/Resources:** Edelstein Collections: history of chemistry, alchemy, and dyeing. Einstein Archive. Theological and alchemical papers: Isaac Newton. Waldemar M. Haffkine papers. Freidenwald Collection: History of medicine. Archive materials on toxic chemicals/industrial hygiene. **Grants:** Yes. **Program Strengths:** Physical Sciences. **Description:** The Edelstein Center encourages and supports all areas of history and philosophy of science and technology. It is associated with the Sidney M. Edelstein Library, and offers postdoctoral and senior research fellowships, as well as the Edelstein International studentship and fellowship program operated jointly with Chemical Heritage Foundation, Philadelphia. **Web Site:** http://sites.huji.ac.il/edelstein/

Tel Aviv University (The Cohn Institute for the History and Philosophy of Science and Ideas). **Contact:** Naomi Diamant, naomid@post.tau.ac.il **Mailing Address:** Tel Aviv University, Gilman Building Rooms 383/384, Tel Aviv, 69789, Israel. **Tel:** 972.3.6409198 **Fax:** 972.3.6409463 **Ongoing Publication Projects:** *Science in Context*. **Web Site:** http://www.tau.ac.il/humanities/cohn/

Italy

Physics Laboratory and Museum of Scientific Instruments, Italy. **Contact:** Flavio Vetrano, vetrano@fis.uniurb.it **Primary E-mail:** gabfis@ uniurb.it **Mailing Address:** Piazza Della Repubblica 3, Urbino, PU, I-61029, Italy. **Tel:** 39.722.4146 **Fax:** 39.722.327857 **Grants:** No. **Program Strengths:** 19th Century, Europe, Physical Sciences, Instruments and Techniques. **Staff:** *Roberto Mantovani*; *Flavio Vetrano*. **Description:** Collection of more than six hundred instruments related to research and educational activity in universities from 18th to 20th centuries. The instruments are by the most illustrious instrument makers in Europe in that period. All instruments have been restored; an historical catalog has been recently published. **Web Site:** http://www.uniurb.it/PhysLab/Museum.html

The Netherlands

Vrije Universiteit Amsterdam (School of Medicine). **Contact:** Eddy Houwaart, e.houwaart.medhistory@med.vu.nl **Mailing Address:** Department Metamedica VUmc, Van der Boechorstraat 7, Amsterdam, Noord-Holland, 1081 BT, Netherlands. **Tel:** 31.20.444.8218 **Fax:** 31.20.444.8258 **Collections/Resources:** One of the few open access libraries in the Netherlands for the history of medicine. **Grants:** No. **Program Strengths:** 20th and 21st Century, Europe, Medical Sciences, Instruments and Techniques; 20th and 21st Century, Europe, Medical Sciences, Social Relations of Science; 19th Century, Europe, Medical Sciences. **Staff:** *E.S. Houwaart*; *Toine Pieters*. **Description:** Also offers undergraduate courses for medical students. **Web Site:** http://www.metamedica.nl

Norway

Norges Teknisk-Naturvitenskapelige Universitet (Institutt for Tverfaglige Kulturstudier). **Contact:** Knut H. Sorensen, knut.sorensen@hf.ntnu.no **Mailing Address:** Institutt for tverrfaglige kulturstudier, NTNU, Trondheim, 7493, Norway. **Fax:** 73.59.13.27 **Grants:** No. **Staff:** *Per Østby*; *Knut H. Sorensen* (20th and 21st Century, Europe, Technology). **Description:** The department hosts the Centre for Technology and Society, which hosts around 25 researchers in history, sociology, anthropology and political science, working in science and technology studies. The department runs a doctoral program, and a masters program is planned for the fall of 2003. **Web Site:** http://www.hf.ntnu.no/itk/

Russia

Russian Academy of Sciences (Institute for the History of Science and Technology). **Contact:** Olga Sokolova, postmaster@ihst.ru **Mailing Address:** Staropanskii per., 1/5, Institute for the History of Science and Technology, Moscow, 103012, Russian Federation. **Tel:** 7.95.928.1307 **Fax:** 7.95.925.9911 **Ongoing Publication Projects:** *Voprosy Istorii Estestvoznania i Tekhniki (Problems of History of Natural Sciences and Technology).* **Grants:** No. **Web Site:** http://www.ihst.ru/

Sweden

Lund University (Research Policy Institute). **Contact:** Rikard Stankiewicz, Rikard.Stankiewicz@fpi.lu.se **Primary E-mail:** fpi@fpi.lu.se **Mailing Address:** Scheelvägen 15, Ideon Alfa 1, Lund, SE-223 63, Sweden. **Fax:** 46.46.146986 **Ongoing Publication Projects:** *VEST*-Nordic Journal of Science Studies. **Collections/Resources:** Science policy library and policy documentation. **Grants:** No. **Program Strengths:** 20th and 21st Century, Technology, Science Policy. **Staff:** *Wilhelm Agrell*; *Mats Benner*; *Bo Göransson*; *Anders Granberg*; *Gustav Holmberg*; *Rikard Stankiewicz.* **Web Site:** http://www1.ldc.lu.se/fpi/

Switzerland

University of Geneva (Institute for the History of Medicine and Health). **Contact:** Bernardino Fantini, Bernardino.Fantini@medecine.unige.ch **Mailing Address:** Institute for the History of Medicine and Health, University of Geneva, CMU, Geneva 4, 1211, Switzerland. **Tel:** 41.22.702.57.90 **Fax:** 41.22.702.57.92 **Ongoing Publication Projects:** Editorial office of *History and Philosophy of the Life Sciences.* **Grants:** Yes. **Program Strengths:** Medical Sciences; Biological Sciences; Social Relations of Science. **Staff:** *Marino Buscaglia*; *Andrea Carlino*; *Joîlle Droux*; *Bernardino Fantini*; *Marc Geiser*; *Concetta Pennuto*; *Mark Ratcliff*; *Philip Rieder*; *Bruno Strasser*; *Alexandre Wenger.* **Description:** Strengths and research areas of the faculty include: medical science, STS, cultural and social history of medicine, history of life

sciences, and intellectual history of infectious diseases, 15th-21th century.
Web Site: http://www.medecine.unige.ch/newfacmed/jeantet.html

United Kingdom

Keele University (Centre for Social Theory and Technology). **Contact:** Rolland
Munro, mna13@keele.ac.uk **Primary E-mail:** t.wood@mngt.keele.ac.uk
Mailing Address: Darwin Building r1.15, Keele University, Keele,
Staffordshire, ST5 5BG, United Kingdom. **Tel:** 44.1782.58.4273
Fax: 44.1782.58.4272 **Program Strengths:** Technology, Social Relations of
Science. **Staff:** *Fred Botting*; *Robert Cooper*; *Miriam David*; *Valérie Fournier*;
Mihaela Kelemen; *Matthias Klaes*; *Nick Lee*; *Simon Lilley*; *Rolland Munro*;
Martin Parker; *Richard Sparks*; *Rob Walker*. **Description:** The aim of CSTT
is to foster interdisciplinary research and teaching in the social theory of
technology and organizations. It is especially concerned with themes that
foreground the special nature of contemporary technology-organization systems,
using a range of theoretical approaches including cybernetic models, postmodern
theory, and complexity theory. **Web Site:** http://www.keele.ac.uk/depts/stt/

University of York (Science and Technology Studies Unit). **Contact:** Andrew
Webster, ajw25@york.ac.uk **Mailing Address:** SATSU, University of
York, York, Yorkshire, YO10 5DD, United Kingdom. **Tel:** 44.1904.434.740
Fax: 44.1904.433.043 **Collections/Resources:** National repository for
research on innovative health technologies related to the UK ESRC's national
program. **Grants:** Yes. **Program Strengths:** 20th and 21st Century, Europe,
Social Sciences, Social Relations of Science. **Staff:** *Nik Brown* (North
America); *Anne Kerr*; *Graham Lewis*; *Paul Rosen* (Technology); *Richard
Tutton*; *Brian Woods*. **Description:** Conducts work in three main areas:
innovation dynamics, technology and culture, and risk and futures analysis.
Web Site: http://www.york.ac.uk/org/satsu/

**The Wellcome Trust Centre for the History of Medicine at University College
London.** **Contact:** Harold Cook, h.cook@wellcome.ac.uk **Primary E-mail:**
bettina.plettenberg@ucl.ac.uk **Mailing Address:** 24 Eversholt Street, London,
NW1 1AD, United Kingdom. **Tel:** 44.679.8167 **Fax:** 44.679.8194 **Ongoing
Publication Projects:** *Medical History*, edited by W. F. Bynum and Anne
Hardy. Assistant Editor, Caroline Tonson-Rye; *Medical History Supplements*
(an annual supplement to the journal); Occasional Publications (five high
quality monographs a year). **Collections/Resources:** *See entry for: The
Wellcome Library for the History and Understanding of Medicine.* **Grants:** Yes.
Staff: *Sanjoy Bhattacharya*; *Janet Browne*; *William F. Bynum* (19th Century,
Europe, Medical Sciences); *Harold J. Cook* (17th Century, Europe, Medical
Sciences); *Anne Hardy*; *Stephen Jacyna*; *Christopher Lawrence*; *Vivienne Lo*;
Michael Neve; *Tilli Tansey*; *Andrew Wear*; *Dominik Wujastyk*. **Description:** The
Wellcome Trust Centre for the History of Medicine at UCL exists primarily as a
center for research into the history of medicine. It is supported by the Wellcome

Trust but forms part of the Department of Anatomy and Developmental Biology at UCL. The Centre's chief aim is the promotion of high quality scholarly research in the history of medicine, but it is also involved in undergraduate and postgraduate teaching, the organization of a varied program of lectures and symposia, and a range of other activities designed to raise the profile of the discipline amongst the academic community and beyond. The Centre's objective of establishing itself as the foremost international institute for research in the history of medicine is underpinned not only by the work of its permanent staff, who are all leading scholars in the field, but also by its policy of promoting visits from scholars from across the world. **Web Site:** http://www.ucl.ac.uk/histmed

United States

California

The Huntington Library, Art Collections, and Botanical Gardens. **Contact:** Dan Lewis, dlewis@huntington.org **Mailing Address:** 1151 Oxford Road, San Marino, CA, 91108, United States. **Tel:** 1.626.405.2206 **Fax:** 1.626.449.5720 **Ongoing Publication Projects:** Planning and research underway for a printed guide to the history of civil engineering holdings at the Huntington. **Collections/Resources:** 200 separate manuscript collections related to the history of science, including: Edwin Hubble Papers; Mt. Wilson Observatory Director's Papers, and many other historical documents by scientists from a variety of fields, including geology and civil engineering. **Grants:** Yes. **Program Strengths:** 20th and 21st Century, North America, Astronomical Sciences, Humanistic Relations of Science; 19th Century, North America, Earth Sciences, Exploration, Navigation, Expeditions; 20th and 21st Century, North America, Technology. **Description:** We issue the largest number of fellowships for full-time study of any research institution in the U.S. (136 for 2002-03). We strongly encourage applications for fellowships in the history of science, particularly in the fields of the history of engineering, the history of medicine, the history of astronomy, and the history of chemistry. **Web Site:** http://www.huntington.org

University of California, Berkeley (Office for History of Science and Technology). **Contact:** Cathryn Carson, clcarson@socrates.berkeley.edu **Primary E-mail:** office@ohst7.berkeley.edu **Mailing Address:** 543 Stephens Hall, #2350, University of California, Berkeley, CA, 94720-2350, United States. **Tel:** 1.510.642.4581 **Fax:** 1.510.643.5321 **Ongoing Publication Projects:** *Historical Studies in the Physical and Biological Sciences*; Berkeley Papers in the History of Science. **Collections/Resources:** Archive for History of Quantum Physics; Reading room with journals and space for visiting researchers. **Grants:** No. **Description:** Along with its publication and research activities, OHST sponsors conferences and colloquia and belongs to the network of institutions that host the International Summer School in History of Science. Graduate affiliates come from the Department of History as well as other

Berkeley departments. OHST welcomes visiting scholars upon application.
Web Site: http://ohst.berkeley.edu/

University of California, Berkeley (Regional Oral History Office). **Contact:**
Sally Hughes, shughes@library.berkeley.edu **Primary E-mail:** roho@library.
berkeley.edu **Mailing Address:** The Bancroft Library, University of California
at Berkeley, Berkeley, CA, 94720-6000, United States. **Tel:** 1.510.642.7395
Fax: 1.510.643.2548 **Collections/Resources:** Oral history series on medical
physics, AIDS, biotechnology, and physical sciences are available for research in
the Bancroft Library or for purchase at cost. **Grants:** Yes. **Program Strengths:**
20th and 21st Century, North America, Physical Sciences, Biography; 20th
and 21st Century, North America, Biological Sciences, Biography; 20th and
21st Century, North America, Medical Sciences, Biography. **Staff:** *Richard
Candida Smith*; *Roger Hahn* (18th Century, Europe, Physical Sciences);
Sally Smith Hughes. **Description:** The office specializes in the production of
researched, edited, and indexed oral histories in many fields. Under the new
director, the oral history office is taking on a heavier teaching and research
orientation. Collaborative research projects using oral history methodology
are sought with other institutions. Please contact the office for discussion.
Web Site: http://library.berkeley.edu/BANC/ROHO

District of Columbia

George Washington University (Center for History of Recent Science).
Contact: Horace Freeland Judson, hfjudson@gwu.edu **Mailing Address:**
801 22d. St., NW, Washington, DC, 20052, United States. **Grants:** Yes.
Program Strengths: Biological Sciences, Science Policy. **Staff:** *Nathaniel
Comfort* (20th and 21st Century, North America, Biological Sciences); *Horace
Freeland Judson*. **Description:** The Center's defined mission is research in
the immediate histories of sciences of the present day. A fairly new institution,
the Center's current research focuses upon modern biological and biomedical
sciences and their social and policy ramifications. We are broadening our
activities into the physical sciences as historians, coming to the Center as senior
visiting fellows or as post-doctoral fellows, bring robust projects. Throughout,
the interaction of science with public policy is inseparable from the work
of the Center, for the history of recent science is, of course, shaped by the
many-faceted politics of science. **Web Site:** http://recentscience.gwu.edu/

Smithsonian Institution (Joseph Henry Papers Project). **Contact:** Marc
Rothenberg, rothenbergm@osia.si.edu **Mailing Address:** A&I Room
2188, 900 Jefferson Dr., SW, Washington, DC, 20560-0429, United States.
Tel: 1.202.357.1421 ext. 18 **Fax:** 1.202.786.2878 **Ongoing Publication
Projects:** 11 volume edition of the papers of Joseph Henry. **Collections/
Resources:** Database of over 125,000 Henry documents. Bell-Henry Library,
containing over 3,000 volumes from the personal libraries of Alexander Graham
Bell and Joseph Henry. **Grants:** Yes. **Program Strengths:** 19th Century, North

America, Physical Sciences, Institutions; 19th Century, Transcontinental, Biological Sciences, Exploration, Navigation, Expeditions; 19th Century, Science Policy. **Description:** Graduate, pre-doc, and post-doc fellowships are available through the Smithsonian Institution Office of Fellowships and Grants. **Web Site:** http://www.si.edu/archives/ihd/jhp/index.htm

Illinois

Illinois Institute of Technology, Chicago-Kent College of Law (Institute for Science, Law and Technology). **Contact:** Lori Andrews, landrews@kentlaw.edu **Mailing Address:** 565 W. Adams St, Chicago, IL, 60661, United States. **Tel:** 1.312.906.5359 **Fax:** 1.312.906.5388 **Grants:** No. **Staff:** *Lori B. Andrews*; *Ellen Mitchell*; *Laurie Rosenow*. **Description:** New possibilities in biotechnology, such as cloning and genetic engineering, raise difficult questions about ethics and challenge traditional legal concepts. The Institute for Science, Law & Technology, as part of a major technological university with additional strengths in law, business, architecture, psychology and design, provides cross-disciplinary approaches to the challenges of new technologies. **Web Site:** http://www.kentlaw.edu/islt

University of Chicago (Morris Fishbein Center for the History of Science and Medicine). **Contact:** Robert Richards, r-richards@uchicago.edu **Mailing Address:** 1126 East 59th Street, Chicago, IL, 60637, United States. **Tel:** 1.773.702.8391 **Fax:** 1.773.743.8949 **Collections/Resources:** Special Collections in History of Science, Regenstein Library; Papers connected with Encyclopedia of Unified Science. **Grants:** Yes. **Program Strengths:** 14th-16th Century, Europe, Astronomical Sciences; 19th Century, Europe, Biological Sciences, Philosophy or Philosophy of Science; 19th Century, Europe, Medical Sciences. **Faculty:** *Arnold Davidson*; *Jan Goldstein*; *Adrian Johns*; *Howard Margolis*; *Robert J. Richards*; *George W. Stocking* (20th and 21st Century, North America, Natural and Human History); *Noel Swerdlow*; *Alison Winter*. **Description:** The Fishbein Center for History of Science and Medicine provides graduate fellowships and postdoctoral fellowships for students formally enrolled in the Committee on Conceptual and Historical Studies of Science and in the History Department. **Web Site:** http://www.uchicago.edu/ssd/fishbein/

Maryland

American Institute of Physics (Center for History of Physics). **Contact:** Spencer Weart, sweart@aip.org **Primary E-mail:** chp@aip.org **Mailing Address:** One Physics Ellipse, College Park, MD, 20740-3843, United States. **Tel:** 1.301.209.3165 **Fax:** 1.301.209.0882 **Ongoing Publication Projects:** *AIP History of Physics* Newsletter. **Collections/Resources:** Niels Bohr Library; Emilio Segre Visual Archives; International Catalog of Sources for History of Physics and Allied Fields; oral history interviews. **Grants:** Yes. **Program Strengths:** 20th and 21st Century, Physical Sciences; 19th Century, Physical

Sciences; 20th and 21st Century, Astronomical Sciences. **Staff:** *R. Joseph Anderson*; *Spencer R. Weart* (20th and 21st Century, Physical Sciences). **Description:** The AIP Center for History of Physics works internationally with scientists, archivists, historians and others to preserve personal papers, institutional records, and other source materials for the history of modern physics, astronomy and geophysics, and to provide access to these materials for scholarly research, education, and public information. Specific inquiries may be addressed to photos@aip.org. **Web Site:** http://www.aip.org/history/

Massachusetts

Boston University (Center for Philosophy and History of Science). **Contact:** Alfred Tauber, ait@bu.edu **Mailing Address:** 745 Commonwealth Avenue, Boston, MA, 02215, United States. **Tel:** 1.617.353.2604 **Fax:** 1.617.353.6085 **Staff:** *Alisa N. Bokulich* (Philosophy and Philosophy of Science); *Tian Cao*; *Robert S. Cohen* (20th and 21st Century, North America, Philosophy and Philosophy of Science); *Juliet Floyd*; *Thomas F. Glick* (19th Century, South America, Biological Sciences); *Thomas Hawkins*; *Jaakko J. Hintikka*; *Jon H. Roberts* (19th Century, North America, Humanistic Relations of Science); *Peter H. Schwartz* (Biological Sciences); *John Stachel*; *Alfred I. Tauber* (20th and 21st Century, Europe, Philosophy and Philosophy of Science); *Judson Webb*; *Wesley Wildman*. **Description:** The Center seeks to examine, in the broadest humanistic and social context, the factors that govern the theory and practice of science. The Center is devoted to research, academic programs, special conferences of scholars, public education, and publication in the philosophy and history of sciences. Also sponsors the Boston Colloquium for Philosophy of Science. **Web Site:** http://www.bu.edu/philo/centers/cphs/CPHS.html

Dibner Institute for History of Science and Technology. **Contact:** Trudy Kontoff, dibner@mit.edu **Mailing Address:** MIT E56-100, 38 Memorial Drive, Cambridge, MA, 02139, United States. **Tel:** 1.617.253.8721 **Fax:** 1.617.253.9858 **Grants:** Yes. **Program Strengths:** Astronomical Sciences; Physical Sciences; Biological Sciences; Technology; Mathematics. **Description:** The Dibner Institute for the History of Science and Technology is an international center for advanced research in the history of science and technology. The Institute's consortium includes MIT, the host institution; Boston University; and Harvard University. The Dibner Institute provides senior and postdoctoral fellowships to scholars and hosts conferences in these disciplines. **Web Site:** http://dibinst.mit.edu

Michigan

University of Michigan (The Historical Center for the Health Sciences). **Contact:** Howard Markel, howard@umich.edu **Primary E-mail:** hchs@umich.edu **Mailing Address:** 100 Simpson Memorial Institute, 102 Observatory, Ann Arbor, MI, 48109-0725, United States. **Tel:** 1.734.647.6914

Fax: 1.734.647.6915 **Collections/Resources:** Numerous books and publications relating to the history of medicine at the University of Michigan. **Grants:** No. **Program Strengths:** Medical Sciences, Humanistic Relations of Science; Medical Sciences, Social Relations of Science. **Staff:** *Howard Markel* (19th Century, North America, Medical Sciences); *Alexandra Minna Stern*. **Description:** Specializes in the history of the health sciences from 1850 to the present at the University of Michigan, in the state of Michigan, and more broadly, in American society. **Web Site:** http://www.med.umich.edu/HCHS/

Minnesota

Charles Babbage Center for the History of Information Technology. **Contact:** Arthur L. Norberg, norberg@cs.umn.edu **Primary E-mail:** cbi@tc.umn.edu **Mailing Address:** 211 Andersen Library, University of Minnesota, Minneapolis, MN, 55455, United States. **Tel:** 1.612.624.5050 **Fax:** 1.612.625.8054 **Ongoing Publication Projects:** *Iterations: An Interdisciplinary Journal of Software History.* **Collections/Resources:** Archival collection of more than 6,000 cubic feet of corporate records, manuscript materials, oral history interviews, product literature, periodicals, photographs, film, and video on the history of computing, software, and networking. **Grants:** Yes. **Program Strengths:** 20th and 21st Century, North America, Technology. **Staff:** *Arthur L. Norberg* (20th and 21st Century, North America, Technology). **Description:** The Charles Babbage Institute is a research center and archives dedicated to the history of information technology. For more than two decades it has been the international leader in conducting and promoting research and preserving materials on computers, software, and networking. **Web Site:** http://www.cbi.umn.edu

New Jersey

IEEE Center for the History of Electrical Engineering. **Contact:** Robert Colburn, r.colburn@ieee.org **Primary E-mail:** history@ieee.org **Mailing Address:** Rutgers, The State University of New Jersey, 39 Union Street, New Brunswick, NJ, 08901-8538, United States. **Tel:** 1.732.932.1066 **Fax:** 1.732.932.1193 **Ongoing Publication Projects:** Tri-annual newsletter; Monographs on history of electrical and computer history. **Collections/Resources:** IEEE (AIEE and IRE) corporate archives; Historical photo archive with over 4000 images; Sydney Darlington Papers. **Grants:** Yes. **Staff:** *Michael Geselowitz*. **Description:** Preserves and promotes the study of the history of electrical and computing technologies. **Web Site:** http://www.ieee.org/organizations/history_center/

New York

Columbia University (Center for the Study of Society and Medicine). **Contact:** Jill Conte, jc1466@columbia.edu **Mailing Address:** 630 West 168th Street, P & S Box 11, New York, NY, 10032, United States. **Tel:** 1.212.305.4184

Fax: 1.212.305.6416 **Grants:** No. **Program Strengths:** Social Sciences, Social Relations of Science. **Staff:** *Lerner Barron*; *Sherry Brandt-Rauf*; *Ruth Fischbach*; *David Rothman*; *Sheila Rothman*. **Description:** An interdisciplinary institution examining the complex interactions between medicine and society. Center faculty use historical, sociological, literary, and philosophical methods to analyze clinical practices and biomedical research to broaden the training for health care professionals. Through an understanding of medicine in its full social and cultural context, the Center works to inform patient care and public policy. **Web Site:** http://www.societyandmedicine.org

Pennsylvania

Carnegie Mellon University (Hunt Institute for Botanical Documentation). **Contact:** Robert W. Kiger, rkiger@andrew.cmu.edu **Primary E-mail:** huntinst@ andrew.cmu.edu **Mailing Address:** Hunt Institute, Carnegie Mellon University, 5000 Forbes Avenue, Pittsburgh, PA, 15213-3890, United States. **Tel:** 1.412.268.2434 **Fax:** 1.412.268.5677 **Grants:** No. **Program Strengths:** Biological Sciences. **Staff:** *Gavin D. R. Bridson*; *Terry D. Jacobsen*; *Robert W. Kiger* (Biological Sciences); *Charlotte A. Tancin*; *Angela L. Todd*; *Frederick H. Utech*; *James J. White*. **Web Site:** http://huntbot.andrew.cmu.edu/

Chemical Heritage Foundation. **Contact:** Arnold Thackray, athackray@ chemheritage.org **Mailing Address:** 315 Chestnut Street, Philadelphia, PA, 19106, United States. **Tel:** 1.215.925.2222 **Fax:** 1.215.925.1954 **Ongoing Publication Projects:** *Chemical Heritage*. **Grants:** Yes. **Program Strengths:** Physical Sciences. **Description:** The Chemical Heritage Foundation (CHF) seeks to advance the heritage and public understanding of the chemical and molecular sciences by operating a historical research library; discovering and disseminating information about historical resources; encouraging research, scholarship, and popular writing; publishing historical materials; conducting oral histories; creating exhibits; and taking other appropriate steps to make known the achievements of chemical and molecular scientists and engineers and of related sciences, technologies, and industries. **Web Site:** http://www.chemheritage.org/

The College of Physicians of Philadelphia (Francis Clark Wood Institute for the History of Medicine). **Contact:** Edward Morman, emorman@collphyphil.org **Mailing Address:** 19 South 22nd Street, Philadelphia, PA, 19103, USA. **Tel:** 1.215.563.3737 ext. 305 **Fax:** 1.215.569.0356 **Grants:** Yes. **Staff:** *Edward T. Morman* (19th Century, North America, Medical Sciences); *Gabriela Zoller*. **Description:** The Wood Institute is the historical programming arm of the Colege of Physicians of Philadelphia and serves to promote use of the collections of the College Library and the Mutter Museum. The Institute provides grants for use of the College's collections and sponsors seminars, lectures, and conferences. **Web Site:** http://www.collphyphil.org/

University of Pennsylvania (Center for the Study of the History of Nursing). **Contact:** Karen Buhler-Wilkerson, nhistory@nursing.upenn.edu **Mailing Address:** School of Nursing, 420 Guardian Drive, 307 Nursing Education Building, Philadelphia, PA, 19104-6096, United States. **Tel:** 1.215.898.4502 **Fax:** 1.215.573.2168 **Ongoing Publication Projects:** *The Chronicle*, a bi-annual newsletter. **Collections/Resources:** Manuscript collections (1,400 ln. ft.) include records of hospitals, schools of nursing, nursing alumni associations, voluntary non-profit associations, professional associations and military associations; personal papers of individual practitioners, collectors, and researchers; Philadelphia General Hospital, Mercy-Douglass Hospital, and Visiting Nurse Society of Philadelphia photographic images (3,000); textbooks, manuals, and histories (1,000 volumes). **Grants:** Yes. **Staff:** *Karen Buhler-Wilkerson; Gail Farr; Joan Lynaugh*. **Description:** The Center is a repository for primary source materials pertinent to the development of nursing and health care in the Mid-Atlantic region and as a national center for visiting nurse association materials. It also offers two $2,500 fellowships which support residential study using the Center's collections. **Web Site:** http://www.nursing.upenn.edu/history/

West Virginia

West Virginia University (Institute for the History of Technology and Industrial Archaeology). **Contact:** Dan Bonenberger, bberger@wvu.edu **Primary E-mail:** ihtia@wvu.edu **Mailing Address:** Bicentennial House, 1535 Mileground, Morgantown, WV, 26505, United States. **Tel:** 1.304.293.2513 **Fax:** 1.304.293.2449 **Ongoing Publication Projects:** IHTIA Monograph Series; IHTIA Technical Report Series. **Collections/Resources:** Frank Duff McEnteer Collection; Thomas Hahn Collection; Domestic Coke Corporation Collection; Roland Parker Davis Collection. **Grants:** No. **Program Strengths:** 19th Century, North America, Technology, Instruments and Techniques; 20th and 21st Century, North America, Technology, Instruments and Techniques; 18th Century, North America, Technology, Instruments and Techniques. **Staff:** *Dan Bonenberger; Emory L. Kemp; Michael E. Workman*. **Description:** IHTIA develops techniques for recording and preserving industrial sites and engineering works. It provides guidance to the National Park Service, the Federal Highways Administration, and other clients, public and private. Its mission includes research, teaching, service, and technology. **Web Site:** http://www.as.wvu.edu/ihtia/

Venezuela

Central University of Venezuela (Grupo Venezolano de Historia de la Ciencia [Venezuelan Group for History and Sociology of Science]). **Contact:** Yolanda Texera Arnal, ytexera@telcel.net.ve **Mailing Address:** c/o Members of the Venezuelan Group for Social History and Sociology of Science, Apartado postal 47607, Caraca 1041A, Caracas, Distrito Federal, 1010, Venezuela.

Tel: 58.212.753.6854 **Fax:** 58.212.578.3570 **Grants:** No. **Staff:** *Alfonso Arellano-Cardenas*; *Pedro R. Chalbaud Cardona* (20th and 21st Century, South America, Astronomical Sciences); *Alfredo Cilento-Sarli*; *Dora Dávila*; *Yajaira Freites*; *Ana Teresa Gutiérrez*; *José Antonio León*; *Juan José Martín-Frechilla*; *Pascual Mora-Garcia*; *Humberto Ruiz-Calderon*; *Susana Strozzi*; *Yolanda Texera-Arnal*; *Hebe Vessuri*. **Description:** This group focuses upon historical studies of the individuals, disciplines, and institutions that have contributed to the development of science and technology in Venezuela. **Web Site:** http://www.arq.ucv.ve/~infodoc/

Libraries, Archives, and Special Collections

The HSS Executive Committee endorsed the following definition: "For inclusion in the *Guide*, a Library, Archive or Special Collection must aim to identify, collect and preserve the published and unpublished records of the history of science. These institutions typically provide access for research or teaching purposes." 85 respondents described their institutions as a library, archive or special collection. Entries are organized by country, and then by state for programs within the United States. Not all institutions responded with complete data. All entries contain some portion of the following information:

Institution name
Name of program in history of science, technology, or medicine
Primary contact person's name, title, and e-mail address
Primary e-mail address, if different from the contact person's email
Mailing address
Telephone number
Facsimile number
Journal editorial offices and ongoing serial publication projects
Special collections and research resources
Does the institution offer publicly available grants or fellowships (yes/no)
Particular research focus, entered using a standardized list of keywords
List of staff members, including research interests. Names appear as they were entered by respondents. A broad research interest appears only after the names of HSS members, if they entered this information when they joined the Society. (As computer programming challenges are overcome, research interests for non-members should appear beginning with the 2003 online edition of the *Guide*.)
Brief description of the library, archive, or collection and its special features
URL

Australia

Basser Library. **Contact:** Rosanne Walker, lb@science.org.au
   Mailing Address: The Shine Dome, Gordon Street, Acton, ACT, 2600,
   Australia. **Tel:** 61.2.6247.9024 **Fax:** 61.2.6257.4620 **Collections/
   Resources:** 210 manuscript collections. An alphabetical listing of these
   collections, together with listings for most of them, can be found on
   the Web. **Grants:** No. **Program Strengths:** Australia and Oceania.
   Web Site: http://www.science.org.au/academy/basser/bass_inf.htm

The Czech Republic

Academy of Sciences of the Czech Republic (Archives). **Contact:** Ludmila
   Sulitkova, ludmila.sulitkova@seznam.cz **Mailing Address:** V Zamcich 56/76,
   Prague, 181 00, Czech Republic. **Tel:** 420.2.3354.1765 **Fax:** 420.2.3354.1560
   Collections/Resources: Documents of the predecessors of the Czechoslovak
   Academy of Sciences including the Royal Learned Bohemian Society, the Czech
   Academy of Sciences and Arts, die Gesellschaft zur Foerderung deutscher
   Wissenschaft, Kunst und Literatur, the Masaryk Academy of Work, the State
   Observatory in Prague. Documents of the Czechoslovak Academy of Sciences
   and its boards and institutes, and learned specialized societies. Personal papers
   of significant Czech and German scientists and scholars. Collections of photos,
   medals, films, personal files of CSAS members, CSAS honors and awards.
   Grants: No. **Web Site:** http://www.archiv.cas.cz

France

Institut Pasteur (Archives). **Contact:** Stephane Kraxner or Daniel
   Demellier, skraxner@pasteur.fr or ddemelli@pateur.fr **Mailing Address:**
   28 Rue Du Dr Roux, 75724 Cedex 15, Paris, France. **Tel:** 33.01.456881.12
   Fax: 33.01.40.61.34.05 **Program Strengths:** 19th Century, Europe, Biological
   Sciences, Institutions; 20th and 21st Century, Africa, Medical Sciences,
   Exploration, Navigation, Expeditions; 20th and 21st Century, Transcontinental,
   Biological Sciences. **Web Site:** http://www.pasteur.fr/infosci/archives/

Germany

Universität Passau (Institut für Geschichte der Psychologie). **Contact:**
   Horst Gundlach, gundlach@uni-passau.de **Mailing Address:** Leopoldstr. 4,
   Passsau, D-94030, Germany. **Tel:** 49.851.5609.860 **Fax:** 49.851.5609.860
   Collections/Resources: The Institute has the following collections relating
   to the history of psychology: instruments (around 4500 items), psychological
   tests, and archival documents. **Grants:** No. **Program Strengths:** 19th Century,
   Cognitive Sciences, Instruments and Techniques; 20th and 21st Century,
   Cognitive Sciences, Instruments and Techniques; 20th and 21st Century,

Cognitive Sciences, Institutions. **Staff:** *Horst Gundlach* (Cognitive Sciences). **Web Site:** http://www.phil.uni-passau.de/igp/

Greece

The Hellenic Archives of Scientific Instruments (History and Philosophy of Science Programme, National Hellenic Research Foundation). **Contact:** Efthymios Nicolaidis, efnicol@eie.gr **Primary E-mail:** kne@eie.gr **Mailing Address:** National Hellenic Research Foundation, Vassileos Constantinou av. 48, Athens, 11638, Greece. **Tel:** 30.1.0727.3557 **Fax:** 30.1.0724.6212 **Ongoing Publication Projects:** Virtual museum of scientific instruments; Virtual historical experiments. **Collections/Resources:** 19th and early 20th century Greek instruments. **Grants:** No. **Description:** The Hellenic Archives of Scientific Instrument (HASI) belong to the History and Philosophy of Science Programme of the National Hellenic Research Foundation. The collections of HASI are presented on the Web. Alternate Web address: http://www.weblab.gr/hasi. **Web Site:** http://www.eie.gr/hasi

Israel

Hebrew University of Jerusalem (Albert Einstein Archives, Jewish National & University Library). **Contact:** Barbara Wolff, barbaraw@savion.huji.ac.il **Mailing Address:** The Jewish National & University Library, The Hebrew University of Jerusalem, PO B 34165, Jerusalem, 91341, Israel. **Tel:** 972.2.658.5781 **Fax:** 972.2.658.6910 **Ongoing Publication Projects:** The Einstein Papers Project, Cal Tech, USA. **Collections/Resources:** Albert Einstein's personal papers; Albert Einstein photograph collection. **Grants:** No. **Program Strengths:** 20th and 21st Century, Europe, Physical Sciences; 20th and 21st Century, North America, Physical Sciences. **Description:** The Albert Einstein Archives contain the largest collection of original manuscripts by Einstein and include his vast correspondence. It comprises the most exhaustive compilation of material about Albert Einstein. **Web Site:** http://www.albert-einstein.org/

Italy

University of Bologna, Department of Astronomy (Archives). **Contact:** Marina Zuccoli, zuccoli@bo.astro.it **Mailing Address:** Dipartimento di Astronomia, via Ranzani 1, Bologna, 40127, Italy. **Tel:** 39.51.209.5792 **Fax:** 39.51.209.5700 **Grants:** No. **Program Strengths:** Astronomical Sciences, Instruments and Techniques; Mathematics. **Web Site:** http://www.bo.astro.it/~biblio/Archives/copertina.html

South Korea

Yonsei Medical College (Library and Museum of Medical History).
Contact: Hyoung-Woo Park, hwoopark@yumc.yonsei.ac.kr **Primary**
E-mail: medhistory@yumc.yonsei.ac.kr **Mailing Address:** Sinchon-dong
134, Seodaemoon-gu, Seoul, 120-752, South Korea. **Tel:** 82.2.361.5705
Fax: 82.2.393.1885 **Ongoing Publication Projects:** *Yonsei Journal of Medical
History.* **Program Strengths:** Asia, Medical Sciences. **Staff:** *In Sok Yeo.*
Description: The library is affiliated with the Medical Museum, which is unique
in Korea. It is the only museum which depicts the introduction and development
of westernized medicine in Korea.

Sweden

Royal Swedish Academy of Sciences (Center for History of Science).
Contact: Karl Grandin, karlg@kva.se **Primary E-mail:** center@kva.se **Mailing
Address:** Box 50005, Stockholm, SE-10405, Sweden. **Tel:** 46.8.673.95.00
Fax: 46.8.673.95.98 **Collections/Resources:** The archives of the Royal Swedish
Academy of Sciences; the Nobel Archives in Physics and Chemistry; several
Swedish scientist's archives; scientific Instrument Collection; old scientific
books. **Grants:** Yes. **Program Strengths:** 19th Century, Europe; 18th Century,
Europe; 20th and 21st Century, Europe. **Staff:** *Tore Frangsmyr* (18th Century,
Europe, Earth Sciences); *Karl O. Grandin* (20th and 21st Century, Europe,
Physical Sciences); *Urban Wrakberg.* **Description:** The Royal Swedish
Academy of Sciences was founded in 1739 by Linnaeus, among others, and has
played an important role in promoting science in Sweden. It is mostly known
today for being responsible for awarding of the Nobel prizes in physics and
chemistry as well as the prize in economics. **Web Site:** http://www.center.kva.se

Switzerland

CERN Historical and Scientific Archives. **Contact:** Anita Hollier,
Anita.Hollier@cern.ch **Mailing Address:** Organisation Europeenne Pour La
Recherche Nucleaire, Geneve 23, CH-1211, Switzerland. **Tel:** 41.22.76.74953
Fax: 41.22.76.72860 **Ongoing Publication Projects:** Scientific Correspondence
with Bohr, Einstein, Heisenberg, and others, by Wolfgang Pauli. Editor: K.
von Meÿenn. **Collections/Resources:** Administrative, scientific and technical
records that document the history of CERN. Documents representing the
scientific legacy of Wolfgang Pauli. **Grants:** No. **Description:** The CERN
Archive includes documents produced by the CERN Council and its
subordinate committees; files of former Directors-General and other senior
staff; and material documenting the work of CERN Divisions, and selected
experiments and committees. It also contains the Pauli Archive, a collection
of correspondence and manuscripts, scientific books, reprints, and awards
representing the scientific legacy of Wolfgang Pauli, Nobel Laureate, 1945.
Web Site: http://library.cern.ch/archives/index.html

Swiss Federal Institute of Technology (Archives and History of Science Collections). **Contact:** Stefan Gemperli, gemperli@library.ethz.ch **Primary E-mail:** archiv@library.ethz.ch **Mailing Address:** Raemistrasse 101, Zurich, 8092, Switzerland. **Tel:** 41.1.632.21.82 **Fax:** 41.1.632.10.87 **Collections/Resources:** Historical School Board Archive; Manuscript Division with personal papers of numerous noted scientists; Carl Gustav Jung Archive. **Grants:** No. **Program Strengths:** 20th and 21st Century, Europe, Mathematics, Biography; 14th-16th Century, Europe, Astronomical Sciences; 20th and 21st Century, Europe, Technology. **Description:** Printed inventories of personal papers given to the Archives of the ETH Zurich are available. **Web Site:** http://www.ethbib.ethz.ch/eth-archiv/index_e.html

Universität Bern (Medizin-historisches Institut). **Contact:** Anne Ott, anne.ott@mhi.unibe.ch **Mailing Address:** Buhlstrasse 26, Postfach CH-3000, Bern 9, Switzerland. **Tel:** 41.31.631.84.86 **Fax:** 41.31.631.84.91 **Description:** Contains the Haller Project and miscellaneous archives. **Web Site:** http://www.mhi.unibe.ch

United Kingdom

Armagh Observatory (Historical Collections and Archives). **Contact:** John McFarland, jmf@star.arm.ac.uk **Primary E-mail:** admin@star.arm.ac.uk **Mailing Address:** College Hill, Armagh BT61 9D6, Armagh, N. Ireland, BT61 9DG, United Kingdom. **Tel:** 44.28.3752.2928 **Fax:** 44.28.3752.7174 **Collections/Resources:** Armagh Observatory Meteorological Observations 1795-present; Armagh Observatory Archives up to 1920; Armagh Observatory Scientific Instrument Collection. **Grants:** No. **Program Strengths:** Astronomical Sciences; Earth Sciences; Mathematics. **Staff:** *Mark Edward Bailey*; *Christopher John Butler*; *Apostolos Alexandros Christou*; *John Gerard Doyle*; *Christopher Simon Jeffery*; *John McFarland*; *Michael David Smith*. **Description:** Armagh Observatory is an independent public body, founded by Archbishop Richard Robinson, Church of Ireland Primate of All-Ireland, and incorporated by an Act of the Irish Parliament in 1791. Its principal role is to undertake research into astronomy and related sciences. It currently receives its core funding from the Department of Culture, Arts and Leisure for N. Ireland. **Web Site:** http://www.arm.ac.uk/home.html

Cambridge University (Whipple Library). **Contact:** Jill Whitelock, jw330@ cam.ac.uk **Primary E-mail:** wsm1@ula.cam.ac.uk **Mailing Address:** Free School Lane, Cambridge, CB2 3RH, United Kingdom. **Tel:** 44.1223.334547 **Fax:** 44.1223.334554 **Staff:** *Sonia Hollins*; *Dawn Moutrey*; *Jill Whitelock*. **Description:** The Whipple Library was founded on the gift of Robert Stewart Whipple of his rare scientific books to Cambridge University in 1944. It is the largest specialist library in the history and philosophy of science and medicine in the UK. It functions as the departmental library for the Cambridge History and Philsophy of Science department and provides the basis for research and teaching at both undergraduate and graduate level. **Web Site:** http://www.hps.cam.ac.uk/library/

National Archive for the History of Computing. **Mailing Address:** Mathematics Tower, The University, Manchester M13 9PL, United Kingdom. **Tel:** 44.161.275.5845 **Grants:** No. **Program Strengths:** 20th and 21st Century, Europe, Technology. **Description:** This archive specializes in documents (correspondence, printed and unprinted material, photographs) invaluable in studying the history of British computing. **Web Site:** http://www.chstm.man.ac.uk/nahc

Needham Research Institute (The East Asian History of Science Library). **Contact:** John Moffett, jm10019@cus.cam.ac.uk **Mailing Address:** Needham Research Institute, 8 Sylvester Road, Cambridge, CB 3 9AF, United Kingdom. **Tel:** 44.1223.311545 **Fax:** 44.1223.362703 **Ongoing Publication Projects:** Joseph Needham's multi-volume series "Science and Civilisation in China," published by Cambridge University Press; The "Needham Research Institute Studies" series, published by Curzon Press. **Grants:** No. **Description:** The library and its associated collections of offprints, photographs and archives is the product of the decades of work of Dr. Joseph Needham and his collaborators on the history of traditional East Asian science, technology and medicine. The Library continues to collect research materials in this field. **Web Site:** http://www.nri.org.uk/library

Royal Society of Chemistry (Library and Information Centre). **Contact:** Nicola Best, library@rsc.org **Mailing Address:** Burlington House, Piccadilly, London, WIJ OBA, United Kingdom. **Tel:** 44.20.7437.8656 **Fax:** 44.20.7287.9798 **Collections/Resources:** The LIC holds some 3000 historical chemistry books from 16th to 19th centuries, including 41 books by Boyle, 14 by Faraday and 7 by Dalton. Within this collection are two smaller groupings: the Nathan Collection with nearly 1000 items on explosives & firearms and the Roscoe Collection of 96 items. **Web Site:** http://www.rsc.org/lic/history

University of Bath (National Cataloguing Unit for the Archives of Contemporary Scientists). **Contact:** Peter Harper, lispbh@bath.ac.uk **Primary E-mail:** ncuacs@bath.ac.uk **Mailing Address:** The Library, University of Bath, Claverton Down, Bath, BA2 7AY, United Kingdom. **Tel:** 44.1225.383.522 **Fax:** 44.1225.386.229 **Grants:** No. **Program Strengths:** 20th and 21st Century. **Description:** The focus of the library is in locating and processing scientific archives for deposit in academic repositories throughout the UK. **Web Site:** http://www.bath.ac.uk/ncuacs/

The Wellcome Library for the History and Understanding of Medicine. **Contact:** Sue Gold, contact@wellcome.ac.uk **Primary E-mail:** library@ wellcome.ac.uk **Mailing Address:** The Wellcome Building, 183 Euston Road, London, NW1 2BE, United Kingdom. **Tel:** 44.20.7611.8582

Fax: 44.20.7611.8369 **Collections/Resources:** c. 66,000 pre-1851 publications; 600 incunabula and c. 5000 16th century works; 11,000 oriental medical manuscripts in 43 languages; Archives of major medical organizations; Manuscripts of eminent figures including Jenner and Lister; 100,000 prints, drawings, paintings; 600,000 medical textbooks from 1850 to 1980; Substantial ephemera collections. **Grants:** No. **Description:** The Wellcome Library seeks to preserve the record of medicine past and present in order to foster understanding of medicine, its history, and its impact in society. *See also: The Wellcome Trust Centre for the History of Medicine, in the section on research centers.* **Web Site:** http://www.wellcome.ac.uk/library

United States

Alabama

University of Alabama (Reynolds Historical Library). **Contact:** Katie Oomens, koomens@uab.edu **Mailing Address:** UAB Lister Hill Library, LHL 301, 1530 3rd Ave South, Birmingham, AL, 35294-0013, United States. **Tel:** 1.205.934.4475 **Fax:** 1.205.975.8476 **Collections/Resources:** Civil War medicine; 19th century Western diagnosis & therapeutics; Electrotherapeutics; Dentistry; Botanical medicine; Development of Western surgery & anatomy; Southern medicine; Classics of Western medicine. **Grants:** Yes. **Program Strengths:** Medical Sciences. **Description:** The Reynolds Historical Library, a collection of rare books and manuscripts pertaining to the history of medicine, is part of the Historical Collections Unit at the University of Alabama at Birmingham's Lister Hill Library. The Unit also comprises the Alabama Museum of the Health Sciences and UAB Archives. **Web Site:** http://www.uab.edu/reynolds

California

California Institute of Technology (Archives). **Contact:** Judith Goodstein, jrg@caltech.edu **Primary E-mail:** archives@caltech.edu **Mailing Address:** Mail Code 015A-74, California Institute of Technology, Pasadena, CA, 91125, United States. **Tel:** 1.626.395.2700 **Fax:** 1.626.7938756 **Ongoing Publication Projects:** On-line guides to research collections; active oral history program and pilot program for putting select interviews online. **Collections/Resources:** Substantial numbers of rare books in the history of science, mathematics, and engineering dating from the 16th century. More than 5000 photographs, engravings, and prints of scientists, scientific apparatus and illustrations are searchable on-line. Extensive rare book collection in the history of astronomy, physics and seismology; many 19th and early 20th century civil engineering instruments, physics laboratory apparatus. **Grants:** Yes. **Program Strengths:** 20th and 21st Century, Physical Sciences; 20th and 21st Century, Biological Sciences; 20th and 21st Century, North America, Earth Sciences. **Staff:** *Charlotte Erwin; Kevin Knox.* **Description:** Personal and scientific papers of more than 150 distinguished scientists including George Beadle, Max Delbruck, Lee DuBridge, George Hale, Robert Millikan, Richard Feynman, William Fowler, Olga Taussky Todd, Harry Bateman, C. C. Lauritsen, Gerry Neugebauer, Marshall Cohen, Jesse Greenstein, Robert Sharp, Clair Patterson, and H. P. Robertson. **Web Site:** http://archives.caltech.edu

Stanford University (Lane Medical Library, Special Collections). **Contact:** Heidi Heilemann, heilemann@stanford.edu **Primary E-mail:** arch@lanelib. stanford.edu **Mailing Address:** Lane Medical Library, L-109, Stanford University Medical Center, Palo Alto, CA, 94305-5123, United States. **Tel:** 1.650.725.4582 **Fax:** 1.650.725.7471 **Grants:** No. **Description:** The Special Collections and Archives at Lane Medical Library include monographs, rare journals, portraits, photographs, personal papers, and medical instruments. Monographs and periodicals published 1850 or earlier and later materials, singled out because of their historical value, are the core of Special Collections. There are over 6,000 titles including 21 incunabula. This collection is non-circulating. Materials, depending on their condition, may be viewed in the reading room by appointment. Use the Lane Catalog to locate specific items in the collections including finding aids. **Web Site:** http://lane.stanford.edu

Stanford University Libraries (History of Science & Technology Collections). **Contact:** Henry Lowood, lowood@stanford.edu **Mailing Address:** Green Library, 557 Escondido Mall, Palo Alto, CA, 94305-6004, United States. **Tel:** 1.650.723.4602 **Fax:** 1.650.725.1068 **Collections/Resources:** Silicon Valley Archives; Barchas and Brasch-Newton rare book collections; Athanasius Kircher collection; Cabrinety Collection of the History of Microcomputing; Stanford University Archives; Papers of numerous individuals, including Buckminster Fuller, Douglas Engelbart, Frederick Terman, and organizations, such as Apple Computer, Ampex Corporation, System Development Foundation, *AI Magazine*. **Grants:** No. **Program Strengths:** 20th and 21st Century, North America, Technology; 20th and 21st Century, North America, Physical Sciences; 17th Century, Europe, Physical Sciences. **Staff:** *Henry Lowood* (Education). **Web Site:** http://www-sul.stanford.edu/depts/hasrg/histsci/scihome.html

University of California, Davis (Carlson Health Sciences Library, Special Collections). **Contact:** Jo Anne Boorkman, jaboorkman@ucdavis.edu **Primary E-mail:** hslref@ucdavis.edu **Mailing Address:** One Shields Avenue, University of California, Davis, CA, 95616-5291, United States. **Tel:** 1.530.752.6383 **Fax:** 1.530.752.4718 **Collections/Resources:** Historical veterinary materials. **Grants:** No. **Program Strengths:** Medical Sciences. **Description:** All materials in our Special Collections are accessible by searching WorldCat or the California Digital Library's (CDL) MELVYL union catalog. **Web Site:** http://www.lib.ucdavis.edu/healthsci/

University of California, Los Angeles (Louise M. Darling Biomedical Library, History and Special Collections Division). **Contact:** Alison Bunting,

abunting@library.ucla.edu **Mailing Address:** UCLA, 12-007 Center for the
Health Sciences, Box 951798, Los Angeles, CA, 90095-1798, United States.
Tel: 1.310.825.5781 **Collections/Resources:** Arabic and Persian medical
manuscripts; John C. Liebeskind History of Pain Collection; Watercolor
paintings of plants of the Southwest; Botanical Drawings; Photographic
Collections; Japanese and Chinese books on the art of medicine and the
science of natural history; Neuroscience History Archives. As collections are
processed, the finding aids (detailed descriptions) are added to the Online
Archive of California (OAC) and cataloging records are added to ORION2,
UCLA's online catalog. **Grants:** Yes. **Program Strengths:** 14th-16th Century,
Europe, Medical Sciences, Humanistic Relations of Science; 14th-16th Century,
Europe, Medical Sciences, Institutions; 14th-16th Century, Europe, Medical
Sciences, Instruments and Techniques. **Staff:** *Robert G. Frank*; *Marcia L.
Meldrum*; *Ynez V. O'Neill*; *Dora B. Weiner* (18th Century, Europe, Medical
Sciences, Instruments and Techniques). **Description:** The History & Special
Collections Division of the Louise M. Darling Biomedical Library, formally
established in 1964, now contains close to 30,000 rare books. In addition to the
rare books, the Division collects and curates manuscripts, prints and portraits,
and museum objects. We collect materials in two broad subject areas: medicine
and life sciences. In medicine we attempt to document the history of medical
thought and practice from the earliest times to the recent past. In the life sciences
we emphasize natural history, zoology, botany, evolution and evolutionary
theory. The collecting period is 15th century to the early 20th century.
Web Site: http://www.library.ucla.edu/libraries/biomed/his/hisdiv.html

University of California, San Francisco (Archives and Special Collections).
Contact: Valerie Wheat, archives-info@library.ucsf.edu **Mailing Address:**
The Library and Center for Knowledge Management, 530 Parnassus
Avenue, San Francisco, CA, 94143-0840, United States. **Tel:** 1.415.476.8112
Fax: 1.415.476.4653 **Collections/Resources:** AIDS History Project; Tobacco
Control Archives; East Asian Collection. **Grants:** No. **Program Strengths:**
20th and 21st Century, North America, Medical Sciences; Asia, Medical
Sciences. **Description:** Archives and Special Collections at UCSF preserves
and maintains unique materials to support research and teaching in the history
of the health sciences. These materials include Contemporary Collecting
Projects/Digital Archives in the health sciences; UCSF Archives; and Special
Collections including manuscripts, rare books and the East Asian collection.
Web Site: http://www.library.ucsf.edu/sc/

University of California, Santa Barbara (Special Collections, History of
Science). **Contact:** David C. Tambo, tambo@library.ucsb.edu **Primary E-mail:**
special@library.ucsb.edu **Mailing Address:** Special Collections, Davidson
Library, UCSB, Santa Barbara, CA, 93106, United States. **Tel:** 1.805.893.3420
Fax: 1.805.893.5749 **Collections/Resources:** Numerous collections of
personal papers; Special collections on Darwin and evolution, Get Out Oil, and
Community Development and Conservation. **Grants:** Yes. **Program Strengths:**

20th and 21st Century, North America. **Description:** Friends of the UCSB Library Travel Grants are available. **Web Site:** http://www.library.ucsb.edu

Connecticut

Yale University (Harvey Cushing/John Hay Whitney Medical Library, Historical Library). **Contact:** Toby A. Appel, toby.appel@yale.edu **Mailing Address:** 333 Cedar Street, P.O. Box 208014, New Haven, CT, 06520-8014, United States. **Tel:** 1.203.785.4354 **Fax:** 1.203.785.5636 **Collections/Resources:** Special strengths are the works of Hippocrates, Galen, Vesalius, Boyle, Harvey, and S. Weir Mitchell; Works on anesthesia, and smallpox inoculation and vaccination; Over 300 medical incunabula; Fine prints and drawings from the 16th century to the 20th century on medical subjects; Manuscripts of the 19th century medical missionary Peter Parker and paintings by the artist Lam-Qua of patients at Canton Hospital; Edward Clark Streeter Collection of Weights and Measures is one of the most comprehensive and extensive collections of its kind in the world. **Grants:** No. **Program Strengths:** Europe, Medical Sciences; North America, Medical Sciences; Biological Sciences. **Staff:** *Toby A. Appel* (20th and 21st Century, North America, Biological Sciences). **Description:** The Historical Library contains a large and unique collection of rare medical books, medical journals to 1920, pamphlets, prints, and photographs, as well as current works on the history of medicine. **Web Site:** http://www.med.yale.edu/library/historical/

Delaware

The Hagley Museum and Library. **Contact:** Michael Nash, mikenash@ udel.edu **Mailing Address:** P.O. Box 3630, Wilmington, DE, 19807-0630, United States. **Tel:** 1.302.658.2400 ext. 301 **Fax:** 1.302.658.0568.Admini stration. **Collections/Resources:** Extenive holdings of corporate records; Archives documenting history of the computer industry including records of Sperry-UNIVAC, Engineering Research Associates, IBM Anti-Trust Suit; Records of trade associations, including the Society for the Plastics Industry, National Association of Manufacturers, and American Iron and Steel Institute. **Grants:** Yes. **Description:** The Hagley Museum and Library is an independent research library specializing in the history of business and technology. Hagley is the repository for the records of more than 1,000 firms and trade associations. Hagley's Center for the History of Business, Technology, and Society offers short term research grants. It sponsors two conferences a year and a monthly research seminar. Housing is available on the Hagley grounds. **Web Site:** http://www.hagley.lib.de.us/

District of Columbia

Carnegie Institution of Washington (DTM-Geophysical Laboratory Library and Archives). **Contact:** Shaun J. Hardy, hardy@dtm.ciw.edu **Primary E-mail:** library@dtm.ciw.edu **Mailing Address:** 5241 Broad Branch

Road N.W., Washington, DC, 20015, United States. **Tel:** 1.202.478.7960
Fax: 1.202.478.8821 **Collections/Resources:** History of geomagnetism (to
1940); History of volcanology and petrology (ca. 1880-1940); Early 20th
century exploration and travel; Archives contain the records of the Department
of Terrestrial Magnetism (founded 1904) and Geophysical Laboratory
(founded 1905); records of the research vessel "Carnegie" (1909-1929);
and 35,000 photographs documenting DTM's World Magnetic Survey
(1904-1946) as well as field and laboratory studies of both departments.
Grants: No. **Program Strengths:** 20th and 21st Century, Earth Sciences,
Exploration, Navigation, Expeditions; 20th and 21st Century, Physical Sciences,
Instruments and Techniques. **Staff:** *Shaun Hardy*. **Description:** The 40,000
volume library supports advanced research in the physical and earth sciences.
Web Site: http://www.ciw.edu/library/

Library of Congress (Science, Technology and Business Division). **Contact:**
William Sittig, wsit@loc.gov **Primary E-mail:** www@loc.gov **Mailing
Address:** 101 Independence Avenue, S.E., Science, Technology and Business
Division, Washington, DC, 20540-4750, United States. **Tel:** 1.202.707.5664
Fax: 1.202.707.1925 **Ongoing Publication Projects:** "Tracer Bullets," informal
literature guides on topics of current interest in the sciences. **Grants:** Yes.
Program Strengths: North America. **Description:** Founded in 1800, the
Library of Congress is the nation's oldest federal cultural institution. With
more than 124 million items in various languages, disciplines, and formats,
the Library of Congress is the world's largest repository of knowledge and
creativity. The Library of Congress serves the U.S. Congress and the nation both
on-site in its 21 reading rooms on Capitol Hill and through its award-winning
Web site. The Library is one of the world's most significant repositories of
scientific literature. Its general and rare book collections hold most of the
world literature relevant to the history of science and technology. The project
to acquire titles listed in Horblit's "One hundred books famous in science"
(1964) is within several volumes of completion. The manuscript collections
are the most extensive and significant extant regarding American science
and technology, ranging in time from the papers of such figures as Benjamin
Franklin and Thomas Jefferson to Samuel F. B. Morse, Alexander Graham Bell
and the Wright Brothers, and including many contemporary figures such as
Glenn Seaborg and E. O. Wilson. **Web Site:** http://lcweb.loc.gov

Smithsonian Institution (Anthropology Collections & Archives Program).
Contact: Jake Homiak, homiak.jake@nmnh.si.edu **Primary E-mail:** sklarj@
nmnh.si.edu **Mailing Address:** Museum Support Center, 4210 Silver Hill Road,
Suitland, MD, 20746, United States. **Tel:** 1.301.2382873 **Fax:** 1.301.238.2883
Collections/Resources: Voluminous collections in anthropology and ethnology,
including millions of artifacts, hundreds of thousands of photographs, and
tens of thousands of linear feet of manuscripts; The records of the American
Anthropological Association, the Society of American Archaeology, and over
20 other anthropological organizations. **Grants:** Yes. **Program Strengths:**

North America, Social Sciences. **Description:** The Anthropology Collections & Archives Program (including the National Anthropological Archives, Human Studies Film Archives, and the Anthropology Collections Unit), supports research across the four fields of anthropology (ethnology, linguistics, archaeology, and physical anthropology). The Program hosts four or five pre- and post-doctoral fellows each year and serves approximately 850 researchers from around the world. Research topics of recurrent interest include the history of anthropology, various topics in North American ethnology, the study of material culture and the relationship between cultural production and identity, and the study of endangered languages, history of ethnographic film and photography. **Web Site:** http://www.nmnh.si.edu/naa/

Smithsonian Institution (Archives). **Contact:** Pamela Henson, hensonp@ osia.si.edu **Primary E-mail:** osiaref@osia.si.edu **Mailing Address:** Arts and Industries Building, Room 2135, MRC 414, P. O. Box 37012, Washington, DC, 20013-7012, United States. **Tel:** 1.202.357.1420 **Fax:** 1.202.357.2395 **Ongoing Publication Projects:** The Papers of Joseph Henry, edited by Marc Rothenberg. **Collections/Resources:** Collections which document the full range of Smithsonian activities, including American history, art history, science and art related exhibitions, astrophysics, botany, ecology, tropical biology and zoology, particularly strong in 19th century American science; Collections which complement the official records of the Smithsonian concerning expeditions, international expositions, scientists, collectors, professional societies, projects and institutions; Papers of Joseph Henry and Alexander Graham Bell. **Grants:** Yes. **Program Strengths:** 19th Century, North America, Astronomical Sciences; 19th Century, North America, Biological Sciences; 19th Century, North America, Earth Sciences. **Staff:** *Alan Bain*; *William E. Cox*; *Kathleen W. Dorman*; *Edie W. Hedlin*; *Pamela M. Henson* (19th Century, North America, Natural and Human History); *Deborah Y. Jeffries*; *Frank Millikan*; *Marc Rothenberg* (North America). **Description:** The Smithsonian Institution Archives is a repository for records and papers of historic value about the Smithsonian and the fields of science, art, history, and the humanities. The Smithsonian Archives is open to all researchers. *A Guide to the Smithsonian Archives* is accessible at http://www.siris.si.edu. Staff supervise the research of interns, graduate students, and predoctoral, postdoctoral, and visiting fellows. For fellowship information, visit http://siofg.si.edu. **Web Site:** http://www.si.edu/archives

Smithsonian Institution (Libraries). **Contact:** Ronald Brashear, brashearr@ si.edu **Primary E-mail:** libmail@sil.si.edu **Mailing Address:** Dibner Library of History of Science & Technology, Smithsonian Institution NMAH 1041, Washington, DC, 20560-0672, United States. **Tel:** 1.202.357.1568 **Fax:** 1.202.633.9102 **Ongoing Publication Projects:** Publication of the Dibner Library Lecture, an annual series of lectures on topics relating to the collections in the Dibner Library of History of Science and Technology. Dibner Library News, a semiannual newsletter about the collections and research in the Dibner

Library. **Collections/Resources:** The Dibner Library of History of Science and Technology (25,000 rare books and 2,000 manuscripts dating from the 13th to the 19th centuries); Joseph F. Cullman 3rd Library of Natural History (10,000 rare books dating from the 16th to the 19th centuries); Manufacturers' commercial trade catalogs (285,000 pieces representing 30,000 companies); Extensive collections of early air and space history with strengths in ballooning, rocketry, and aviation. **Grants:** Yes. **Description:** With its 22 branch libraries, the Smithsonian Institution Libraries (SIL) plays an essential role in the research conducted at the Institution and in its exhibitions and programs. The collection of 1.5 million volumes with 40,000 rare books includes especially strong holdings in most of the Institution's historical disciplines. The library catalog is on the Internet, accessible at www.siris.si.edu. In addition to providing library and information services to Smithsonian staff, researchers, and the public, SIL presents public programs, including exhibitions, lectures, and symposia. SIL also provides a number of grants of $2500 per month to scholars to work in residence with the special collections. **Web Site:** http://www.sil.si.edu/research

Florida

National Aeronautics and Space Administration (Historical Archives, Kennedy Space Center). **Contact:** James Dumoulin, dumoulin@titan.ksc.nasa.gov **Mailing Address:** YA-D6, Kennedy Space Center, FL, 32899, United States. **Fax:** 1.321.867.7578 **Grants:** No. **Program Strengths:** 20th and 21st Century, North America, Technology, Exploration, Navigation, Expeditions; 20th and 21st Century, North America, Earth Sciences, Exploration, Navigation, Expeditions. **Web Site:** http://www.ksc.nasa.gov/

University of Florida (Archives). **Contact:** Carl Van Ness, carvann@mail. uflib.ufl.edu **Mailing Address:** George A. Smathers Libraries, University of Florida, Gainesville, FL, 32611-7007, United States. **Tel:** 1.352.392.6547 **Fax:** 1.352.846.2746 **Collections/Resources:** The University of Florida Archives contains records related to its agricultural and engineering experiment stations as well as the papers of prominent scientists on the faculty. The latter includes the papers of marine biologist Archie Carr and nuclear scientist George K. Davis. **Grants:** No. **Description:** The University of Florida Archives is devoted to the preservation of the university's historic role as a leader in research, education and service to the people of Florida. **Web Site:** http://www.uflib.ufl.edu/spec/archome/

Georgia

Medical College of Georgia (Special Collections). **Contact:** Susanna Joyner, sjoyner@mail.mcg.edu **Mailing Address:** Medical College of Georgia, Greenblatt Library, AB-225, 1120 15th Street, Augusta, GA, 30912-4400, United States. **Tel:** 1.706.721.3444 **Fax:** 1.706.721.2018 **Collections/Resources:** British, American, and French medical literature of the late 18th and early

19th century; Historic photographs; Landmarks in Modern Medicine book collection; Papers of Robert B.Greenblatt. **Grants:** No. **Staff:** *Susanna Joyner.* **Description:** The Robert B. Greenblatt, M.D. Library Special Collections contain the original Medical College of Georgia 19th century library, a Landmarks in Modern Medicine collection, several medical museum items and artifacts, the Greenblatt Archive, and a variety of MCG institutional publications. **Web Site:** http://www.mcg.edu/library/Services/SpeCol.html

Hawaii

Hawaii Medical Library (Mamiya Medical Heritage Center). **Contact:** Judith A. Kearney, kearney@hml.org **Mailing Address:** Hawaii Medical Library, 1221 Punchbowl Street, Honolulu, HI, 96813, United States. **Tel:** 1.808.536.9302 ext. 112 **Fax:** 1.808.524.6956 **Collections/Resources:** Archives of the Hawaii Medical Association and the Honolulu County Medical Society; Over 3000 files on individual physicians who have practiced in Hawaii. **Grants:** No. **Program Strengths:** 20th and 21st Century, Medical Sciences, Institutions; 19th Century, Medical Sciences, Institutions; 19th Century, Medical Sciences, Instruments and Techniques. **Description:** The Mamiya Medical Heritage Center is dedicated to collecting, organizing, preserving, and providing access to materials documenting the history of medicine in Hawaii. The Center is the repository of the Charles S. Judd, Jr. History of Medicine Collection comprised of primary and secondary sources relevant to the study of the history of medicine, including one of the few extant copies of Gerrit Parmele Judd's "Anatomia," the first medical book in Hawaiian, published in 1838. **Web Site:** http://hml.org/mmhc/

Illinois

Southern Illinois University School of Medicine (Special Collections and Archives). **Contact:** Fran Kovach, fkovach@siumed.edu **Mailing Address:** Southern Illinois University, 801 N. Rutledge, Springfield, IL, 62794-9625, United States. **Tel:** 1.217.545.2658 **Fax:** 1.217.545.0988 **Grants:** No. **Web Site:** http://www.siumed.edu/lib

University of Chicago (Special Collections Research Center). **Contact:** Alice Schreyer, schreyer@uchicago.edu **Primary E-mail:** specialcollections@lib. uchicago.edu **Mailing Address:** 1100 East 57th, Chicago, IL, 60637, United States. **Tel:** 1.773.702.0095 **Fax:** 1.773.702.3728 **Collections/Resources:** Extensive collections of rare books in the history of science and medicine, from the 15th century to the 20th, including first and early editions of Galileo, Kepler, Descartes, Francis Bacon, Newton, Darwin and Einstein; Manuscript and archival collections relating to the Manhattan Project and atomic scientists; Manuscripts and personal papers of many noted scientists. **Grants:** No. **Description:** The Center has voluminous holdings that may be of interest to historians of science, from the 15th to the 20th century. More information can be found on the Center's Web site. **Web Site:** http://www.lib.uchicago.edu/e/spcl/

The Wood Library and Museum of Anesthesiology. Contact: Patrick P. Sim, wlm@asahq.org **Mailing Address:** 520 N. Northwest Highway, Park Ridge, IL, 60068-2573, United States. **Tel:** 1.847.825.5586 **Fax:** 1.847.825.1692 **Collections/Resources:** 8,500 books, 100 foreign and domestic journal titles, related newsletters, comprehensive biography and photography vertical files; Archives of the American Society of Anesthesiologists; Audiovisual collection of historical interviews recording the contemporary history of anesthesiology; Manuscripts, letters, artifacts, portraits, pictures and works of special contributions to anesthesia. **Program Strengths:** 20th and 21st Century, Medical Sciences. **Staff:** *Patrick P. Sim.* **Description:** The Wood Library-Museum of Anesthesiology collects and preserves literature and equipment pertaining to anesthesiology and makes available to the anesthesiology community, other medical professionals, and the public the most comprehensive educational, scientific and archival resources in anesthesiology. **Web Site:** http://www.asahq.org/wlm/

Indiana

Indiana University School of Medicine (Special Collections). **Contact:** Nancy Eckerman, billings@iupui.edu **Primary E-mail:** rlmladmin@iupui.edu **Mailing Address:** Indiana University School of Medicine, 975 W. Walnut Street IB100, Indianapolis, IN, 46202-5121, United States. **Tel:** 1.317.274.2076 **Fax:** 1.317.278.2349 **Collections/Resources:** 19th Century Indiana Medical publications; Database of 19th Century Indiana Physicians. **Grants:** No. **Program Strengths:** 19th Century, North America, Medical Sciences. **Staff:** *Nancy Pippen Eckerman.* **Description:** Due to limited trained staff, please contact us before planning to research here. **Web Site:** http://www.medlib.iupui.edu/hom/

Iowa

Iowa State University (Special Collections). **Contact:** Tanya Zanish-Belcher, tzanish@iastate.edu **Mailing Address:** 403 Parks Library, Iowa State University, Ames, IA, 50011-2140, United States. **Tel:** 1.515.294.6672 **Fax:** 1.515.294.5525 **Ongoing Publication Projects:** Women in Chemistry oral history project. **Collections/Resources:** Archives of Women in Science and Engineering (collected personal papers of women scientists and engineers and records of national and regional women's organizations in these fields); Papers and artifacts related to John Vincent Atanasoff, early computer pioneer; Major scientific collections include the American Phytopathological Society, American Association of Cereal Chemistry, Iowa Academy of Science, as well as the personal papers of Iowa State scientists, such as Paul Errington, John V. Atanansoff, Darleane C. Hoffman and others. The Archives of Women in Science and Engineering is a collecting area, focusing on the papers of women scientists in a variety of areas, and the organizational records of scientific groups, such as the Association for Women Geoscientists and Iota Sigma Pi.

Grants: No. **Program Strengths:** 20th and 21st Century, North America; 20th and 21st Century, North America, Gender and Science. **Description:** Two collections of particular interest are the Archives of Women in Science and Engineering and the John V. Atanasoff Collection. The Archives of Women in Science and Engineering seeks to preserve the historical heritage of American women in science and engineering. To do this, the Archives solicits, collects, arranges, and describes the personal papers of women scientists and engineers as well as the records of national and regional women's organizations in these fields. The Atanasoff Collection is complemented by a Web site sponsored by the Department of Computer Science at Iowa State: http://www.cs.iastate.edu/jva/jva-archive.shtml. **Web Site:** http://www.lib.iastate.edu/spcl/index.html

Kansas

University of Kansas (Kenneth Spencer Research Library). **Contact:** Richard W. Clement, rclement@ku.edu **Mailing Address:** Kenneth Spencer Research Library, University of Kansas, 1450 Poplar Lane, Lawrence, KS, 66045-7616, United States. **Tel:** 1.785.864.4334 **Fax:** 1.785.864.5803 **Collections/Resources:** The Ellis collection of literature pertaining to natural history consists of some 15,000 bound volumes, as well as a very large quantity of pamphlets, letters, original drawings, manuscripts, and other miscellanea (ca. one-third of the collection is focused on ornithology). The extensive botanical collections are based on some 8000 volumes from the library of the late Thomas Jefferson Fitzpatrick. The Linnaeus collection includes well over 2000 volumes of works by Linnaeus and items of Linnaeana. The Herrick Collection on the History of Neurology. **Grants:** No. **Program Strengths:** Natural and Human History, Exploration, Navigation, Expeditions; 19th Century, Australia and Oceania, Natural and Human History; 18th Century, Europe, Natural and Human History. **Description:** The Kenneth Spencer Research Library is the regional history, archive, rare books and manuscripts library of the University of Kansas. It is open to all researchers. **Web Site:** http://spencer.lib.ku.edu

Louisiana

Tulane University School of Medicine (Rudolph Matas Medical Library). **Contact:** Patricia Copeland, copeland@.tulane.edu **Primary E-mail:** medlib@tulane.edu **Mailing Address:** Tulane University School of Medicine, 1430 Tulane Avenue Box SL-86, New Orleans, LA, 70112-2699, United States. **Tel:** 1.504.588.5155 **Fax:** 1.504.587.7417 **Collections/Resources:** Elizabeth Bass Collection on Women in Medicine; Louisiana Physicians File; Art and Portraits Collections; Weinstein Medallion Collection; Light Collection (Scientific Letters); Registre du Comite Medicale de la Nouvelle Orleans, 1816-1854 (Minutes of the licensing board of Eastern Louisiana). **Grants:** No. **Description:** Strengths of the collection are medical Americana, military medicine, practice of medicine in Louisiana and women in medicine. **Web Site:** http://www.tulane.edu/~matas/historical.html

Maryland

American Institute of Physics (Emilio Segre Visual Archives). **Mailing Address:** American Institute of Physics, One Physics Ellipse, College Park, MD, 20740-3843, United States. **Tel:** 1.301.209.3184 **Fax:** 1.301.209.3144 **Collections/Resources:** Over 25,000 photographs and illustrations, mostly portraits, relating to physics, astronomy and geophysics. **Grants:** No. **Program Strengths:** 20th and 21st Century, Physical Sciences; 19th Century, Physical Sciences; 20th and 21st Century, Astronomical Sciences. **Description:** A few thousand of the images are available online on the Web site. **Web Site:** http://www.aip.org/history/esva/

American Institute of Physics (Niels Bohr Library). **Contact:** R. Joseph Anderson, chp@aip.org **Primary E-mail:** nbl@aip.org **Mailing Address:** One Physics Ellipse, College Park, MD, 20740-3843, United States. **Tel:** 1.301.209.3177 **Fax:** 1.301.209.3144 **Ongoing Publication Projects:** International Catalog of Sources for History of Physics and Allied Fields; *AIP History of Physics* Newsletter. **Collections/Resources:** Books, pamphlets, catalogs and other publications; Journals published by the American Institute of Physics; Records of physics and astronomy societies; Collections of personal papers of physical scientists; Oral history interviews; Biographical files; Emilio Segre Visual Archives. **Grants:** Yes. **Description:** The Niels Bohr Library holds outstanding collections documenting the history of physics, astronomy and geophysics since the 19th century, internationally. Help is offered to scholars, educators and the media to create and locate printed, archival, and audiovisual materials. **Web Site:** http://www.aip.org/history

American Society for Microbiology (Albin O. Kuhn Library and Gallery). **Contact:** Jeff Karr, jkarr@asmusa.org **Mailing Address:** Albin O. Kuhn Library, UMBC, 1000 Hilltop Circle, Baltimore, MD, 21250, United States. **Tel:** 1.410.455.3601 **Fax:** 1.410.455.1088 **Collections/Resources:** Anne Sayre Collection of Rosalind Franklin Materials. **Grants:** No. **Program Strengths:** 19th Century, Biological Sciences; 20th and 21st Century, Biological Sciences. **Description:** The American Society for Microbiology Archives maintains the records of the Society (founded in 1899 as the Society of American Bacteriologists), as well as materials relating to the history of the microbiological sciences. Of particular note is the 10,000 volume book collection, including complete runs of major textbooks, lab manuals, and monographs. **Web Site:** http://www.asmusa.org/mbrsrc/archive/intro

Johns Hopkins University (Institute of the History of Medicine-Historical Collections). **Contact:** Christine Ruggere, ruggere@jhmi.edu **Mailing Address:** Johns Hopkins University, 1900 East Monument Street, Baltimore, MD, 21202, United States. **Tel:** 1.410.955.3159 **Grants:** No. **Program Strengths:** Medical Sciences. **Description:** Research library for the history of medicine and allied sciences. **Web Site:** http://www.welch.jhu.edu/ihm/iohmlibrary.html

National Library of Medicine (History of Medicine Division). **Contact:** Paul Theerman, paul_theerman@nlm.nih.gov **Primary E-mail:** hmdref@ nlm.nih.gov **Mailing Address:** 8600 Rockville Pike, Room 1-E21, Bethesda, MD, 20894, United States. **Tel:** 1.301.402.8878 **Fax:** 1.301.402.0872 **Ongoing Publication Projects:** Profiles in Science (web publication of manuscripts of biomedical researchers and, in the future, figures in public health and health policy); HISTMED (on-line bibliography in the history of medicine); Islamic Medical Manuscripts at the National Library of Medicine (on-line catalogue of approximately 300 manuscripts); Reports of the Surgeon General (on-line access to full text of reports from 1964 to the present). **Collections/Resources:** Extensive holdings of pre-1914 books, journals, pamphlets and dissertations; East Asian manuscript collection; Audiovisual materials from 1917-1970; Modern manuscripts and oral histories. **Grants:** Yes. **Program Strengths:** Medical Sciences; Biological Sciences. **Staff:** *Simon Baatz*; *James H. Cassedy* (North America, Physical Sciences); *Elizabeth Fee*; *Walter Hickel*; *Michael Sappol*; *Susan Speaker*; *Philip Teigen*; *Paul Theerman* (Medical Sciences). **Description:** The History of Medicine Division collects, preserves, and makes available historically significant materials in the history of medicine. It promotes and conducts scholarly research and publication, and sponsors exhibitions, lectures, film programs, and symposia. It holds one of the best collections in the United States of materials in the history of medicine and allied sciences. **Web Site:** http://www.nlm.nih.gov/hmd/hmd.html

National Oceanic and Atmospheric Administration (Central Library). **Contact:** Albert (Skip) E. Theberge, albert.e.theberge.Jr@noaa.gov **Primary E-mail:** reference@nodc.noaa.gov **Mailing Address:** NOAA Central Library, 1315 East West Highway, 2nd Floor, Silver Spring, MD, 20910, United States. **Tel:** 1.301.713.2600.X.115 **Fax:** 1.301.713.4599 **Ongoing Publication Projects:** Three major Web-based projects: NOAA Photo Library, NOAA History, History of Ocean Exploration. **Collections/Resources:** 1.5 million documents spanning five centuries and all continents; Collections covering the history of geodesy, geophysics, oceanography, meteorology, climatology, fisheries science, scientific exploration, and related subject matter; Over 25,000 digitized public domain photos capturing elements of the history of the Coast and Geodetic Survey, Weather Bureau, and Bureau of Commercial Fisheries. **Grants:** No. **Description:** The library has 1.5 million documents. At least 40% of the collection is rare and/or unique. The collection covers most elements of the oceanic, atmospheric, and geophysical sciences. **Web Site:** http://www.history.noaa.gov/

University of Maryland (Historical Collections, Health Sciences and Human Services Library). **Contact:** Richard Behles, rbehles@hshsl.umaryland.edu **Mailing Address:** University of Maryland, 601 W. Lombard Street, Baltimore, MD, 21201, United States. **Tel:** 1.410.706.5048 **Fax:** 1.410.706.3101 **Ongoing Publication Projects:** We host and moderate CADUCEUS-L, the discussion list for the history of medicine and allied health sciences. **Collections/Resources:**

Resources pertinent to the history of the University of Maryland and its place in Maryland medical history. Historical book collections in allied health sciences disciplines, such as medicine, pharmacy, dentistry, and the Florence Kendall Collection of books in physical therapy. **Grants:** No. **Program Strengths:** Medical Sciences. **Web Site:** http://www.hshsl.umaryland.edu

Massachusetts

Burndy Library. **Contact:** Anne Battis, abattis@mit.edu **Primary E-mail:** dibner@mit.edu **Mailing Address:** MIT E56-100, 38 Memorial Library, Cambridge, MA, 02139, United States. **Tel:** 1.617.253.7381 **Fax:** 1.617.253.9858 **Ongoing Publication Projects:** Publications of the Burndy Library (series). **Collections/Resources:** Collections of papers and works by noted scientists, including Faraday, Smith, Tyndall and Volta; Special holdings in heralds of science, color theory, incunabula, instruments, obelisks, piranesi, and portraits. **Grants:** No. **Description:** The Burndy Library's collections of rare books and secondary materials provide Dibner Institute Fellows and visitors with research facilities in specialized and general areas of science and technology history. The library catalog (http://burndydibner.mit.edu) is fully searchable online. **Web Site:** http://www.dibinst.mit.edu/

Harvard University (Countway Library of Medicine, Rare Books and Special Collections). **Contact:** Thomas Horrocks, thomas_horrocks@hms.harvard.edu **Mailing Address:** 10 Shattuck Street, Boston, MA, 02115, United States. **Tel:** 1.617.432.2170 **Fax:** 1.617.432.4737 **Grants:** Yes. **Program Strengths:** 14th-16th Century, Europe, Medical Sciences; 19th Century, Medical Sciences; 20th and 21st Century, North America, Medical Sciences. **Description:** The Library offers one or two fellowships annually through the New England Regional Fellowship Consortium. For information on the consortium's fellowship program, see http://www.masshist.org/fellowships.html. **Web Site:** http://www.countway.med.harvard.edu/rarebooks/

Smithsonian Astrophysical Observatory and Harvard College Observatory. **Contact:** Owen Gingerich, ginger@cfa.harvard.edu **Primary E-mail:** library@cfa.harvard.edu **Mailing Address:** Wolbach Library, Harvard-Smithsonian Center for Astrophysics, Cambridge, MA, 02138, United States. **Tel:** 1.617.496.5769 **Fax:** 1.617.495.7199 **Collections/Resources:** Outstanding collection of astronomical periodicals and monographs including many older titles. Rare astronomy books, such as the world's second largest collection of Kepler titles, are found in Harvard's Houghton Library. Harvard Observatory's historical archives are found in the Harvard University Archives. Smithsonian Observatory's historical archives are found in the Smithsonian Institution Archives in Washington, DC. **Grants:** No. **Program Strengths:** Astronomical Sciences. **Staff:** *Owen Gingerich* (Astronomical Sciences). **Web Site:** http://www.harvard.edu/sao-home

Minnesota

Mayo Foundation History of Medicine Library. **Contact:** Alexander E. Lucas, lucas.alexander2@mayo.edu **Mailing Address:** Plummer Bldg, Room 1507, 200 1st Street S.W., Rochester, MN, 55905, United States. **Tel:** 1.507.284.3676 **Fax:** 1.507.266.4910 **Collections/Resources:** Several thousand volumes of rare medical classics (from 1479) and early journal literature (from 1665); Early medical imprints (pre-1875) and more recently published histories, biographies, facsimiles, and other support material; *Vanity Fair* caricatures of physicians and scientists, photographs of selected Nobel laureates and Mayo Clinic Staff, caricatures of medical specialists by Bavarian wood-carvers, Mayo Clinic related-cartoons, medical philately, and heraldry. **Grants:** No. **Program Strengths:** 20th and 21st Century, North America, Medical Sciences. **Description:** The Mayo History of Medicine Library is a specialized library within the Mayo Medical Library which houses important collections in the history of medicine and allied sciences. Resources in the Mayo History of Medicine Library may be used by special arrangement by calling for an appointment. Contemporary support materials (recently published historical accounts, biographies, etc.) are available for circulation to Mayo employees. Help in identifying historical works and reference verification and other literature research assistance can be arranged. **Web Site:** http://www.mayo.edu/medlib/lib_exhibit/histmed

University of Minnesota (Wangensteen Historical Library of Biology and Medicine). **Contact:** Elaine Challacombe, e-chal@umn.edu **Mailing Address:** University of Minnesota, 568 Diehl Hall, 505 Essex Street S.E., Minneapolis, MN, 55455, United States. **Tel:** 1.612.626.6881 **Fax:** 1.612.626.6500 **Grants:** No. **Program Strengths:** Medical Sciences. **Staff:** *John M. Eyler; Jennifer L. Gunn*. **Description:** The collections number approximately 60,000 dating from the early 1400s to 1930. Holdings are particularly strong in surgery, anatomy, pharmacy, botany, natural history, ob/gyn, pediatrics, and journal holdings. Limited collections of medical artifacts. **Web Site:** http://www.biomed.lib.umn.edu/wang/wangmain.html

Missouri

Linda Hall Library of Science, Engineering & Technology. **Contact:** Bruce Bradley, bradleyb@lindahall.org **Mailing Address:** 5109 Cherry Street, Kansas City, MO, 64110-2498, United States. **Tel:** 1.816.926.8737 **Fax:** 1.816.926.8790 **Collections/Resources:** The History of Science Collection is the library's special collection of rare books on science, engineering, and technology. It includes printed books from the 15th century to the present. Additional materials to support historical research are available in the library's general collections of over one million volumes. Two rare book rooms adjacent to the library's main reading room provide space for readers, visiting scholars, and group visits. **Grants:** Yes. **Program Strengths:** Astronomical Sciences; Earth Sciences;

Physical Sciences. **Staff:** *William B. Ashworth* (North America, Technology); *Bruce Bradley*. **Description:** The Linda Hall Library invites applications for research in the library's collections on the history and philosophy of science, engineering, and technology. Short-term fellowships for up to eight weeks offer weekly stipends to assist researchers with travel and living expenses. These fellowships support advanced and independent studies, dissertation research, and postdoctoral research. **Web Site:** http://www.lindahall.org/

New Jersey

Institute for Advanced Study (Archives). **Contact:** Lisa Coats, lcoats@ias.edu **Primary E-mail:** webmaster@ias.edu **Mailing Address:** Einstein Drive, Princeton, NJ, 08540, United States. **Tel:** 1.609.734.8375 **Fax:** 1.609.951.4515 **Collections/Resources:** Rare Book Room of the Rosenwald Collection of the History of Science. **Grants:** Yes. **Description:** The Institute for Advanced Study is an independent, private institution dedicated entirely to the encouragement, support and patronage of learning through fundamental research and definitive scholarship. It consists of the Schools of Historical Studies, Mathematics, Natural Sciences, Social Science, and a newly created program in Theoretical Biology. **Web Site:** http://www.admin.ias.edu/hs/lib/archives

Rutgers, The State University of New Jersey (Thomas A. Edison Papers). **Contact:** Paul Israel, taep@rci.rutgers.edu **Mailing Address:** 44 Road 3, Piscataway, NJ, 08854-8049, United States. **Tel:** 1.732.445.8511 **Fax:** 1.732.445.8512 **Ongoing Publication Projects:** Selective editing and publishing of primary documents and secondary studies in image and book editions. **Collections/Resources:** Microfilm collections; Databases for Edison documents. **Staff:** *Louie Carlat* (North America, Technology, Social Relations of Science); *Theresa M. Collins* (North America); *Linda Eikmeier-Endersby* (Technology); *David Hochfelder* (North America, Technology); *Paul B. Israel* (19th Century, North America, Technology); *Thomas E. Jeffrey* (North America); *Brian C. Shipley* (North America). **Description:** Co-sponsored by Rutgers University, the National Park Service's Edison National Historic Site, the New Jersey Historical Commission, and the Smithsonian Institution. The project researches 19th and early 20th century technology, science, and business, as well as Edison's correspondence and activities. **Web Site:** http://edison.rutgers.edu/

New York

Columbia University (Health Sciences Library, Archives and Special Collections). **Contact:** Stephen E. Novak, sen13@columbia.edu **Primary E-mail:** hs-library@ columbia.edu **Mailing Address:** Lower Level 1, Augustus C. Long Health Sciences Library, Columbia University, 701 W. 168th Street, New York, NY, 10032, United States. **Tel:** 1.212.305.7931 **Fax:** 1.212.305.6097 **Collections/ Resources:** Archives of the Columbia University Health Science Schools (medicine, dentistry, nursing and public health); Jerome P. Webster Library of Plastic Surgery; Auchincloss Florence Nightingale Collection; Freud Library; Hyman Collection in the History of Anesthesiology. **Grants:** No. **Program Strengths:** Medical Sciences; Biological Sciences. **Description:** Archives & Special Collections is the archival repository for Columbia University's four health science schools: medicine (1767), nursing (1892), dentistry (1916) and public health (1922). It actively collects personal papers of biomedical scientists, physicians, nurses and others in health care or research. Important collections include the papers of Nobel Laureate Andre Cournand, psychiatrist Viola Bernard, dyslexia researchers Samuel T. and June Lyday Orton and physiologist Frederic Schiller Lee, among others. The department also houses a 15,000 volume library in the history of the health sciences dating from the 15th to the 20th century. Strengths are anatomy, anesthesiology, physiology, plastic and general surgery. **Web Site:** http://cpmcnet.columbia.edu/library/archives/index.html

Cornell University Library (Division of Rare Books and Manuscript Collections). **Contact:** David W. Corson, dwc4@cornell.edu **Primary E-mail:** rareref@cornell.edu **Mailing Address:** 2B Kroch Library, Cornell University, Ithaca, NY, 14853-5302, United States. **Tel:** 1.607.255.3530 **Fax:** 1.607.255.9524 **Collections/Resources:** The collection is strong in 17th and 18th century physics and chemistry, especially in Lavoisier, Boyle, and Newton. In the biological sciences, the library has extensive holdings in anatomy, embryology, and physiology, especially in 17th and 18th century Italian anatomists such as Malpighi and Fabricius. In the history of technology, the collection's strengths include the Hollister Collection (300 volumes), which focuses on civil engineering, and the Cooper Collection on American railroad bridges, which includes original blueprints. **Grants:** No. **Program Strengths:** 18th Century, Europe, Physical Sciences; 18th Century, Europe, Biological Sciences; North America, Natural and Human History, Exploration, Navigation, Expeditions. **Description:** RMC holdings include 300,000 printed books, over 40,000 linnear feet of documents, and another million photographs, prints, and other visual media. The collection chronicles American history, medieval and Renaissance studies, the Reformation, 18th century France and England, Anglo-American literature, Icelandic culture, history of science, and human sexuality, as well as Cornell University's Archives. The division plays an active teaching role at Cornell and promotes access to its collections through a variety of programs, including instruction, tours, and exhibitions. **Web Site:** http://rmc.library.cornell.edu

New York Academy of Medicine Library (Historical Collections). **Contact:** Christian Warren, cwarren@nyam.org **Mailing Address:** 1216 Fifth Avenue, New York, NY, 10029, United States. **Tel:** 1.212.822.7314 **Fax:** 1.212.423.0273 **Collections/Resources:** Approximately 49,000 volumes on the history of medicine, science and other health-related disciplines. Materials dating from 1700 B.C. to A.D. 1800 number approximately 32,000 volumes; approximately 100,000 medical pamphlets; Strong holdings for epidemic diseases, public health, and medical institutions in New York City. **Grants:** Yes. **Description:** The Historical Collections Department makes the Academy Library's collection of published and archival materials in the history of medicine and public health available to researchers and the general public. It sponsors two short-term research fellowships in the history of medicine and medical humanities, and hosts an annual series of lectures, exhibitions and symposia. **Web Site:** http://www.nyam.org

The Rockefeller Archive Center. **Contact:** Darwin Stapleton, stapled@ mail.rockefeller.edu **Primary E-mail:** archive@mail.rockefeller.edu **Mailing Address:** 15 Dayton Avenue, Pocantico Hills, Sleepy Hollow, NY, 10591, United States. **Tel:** 1.914.631.4505 **Fax:** 1.914.631.6017 **Ongoing Publication Projects:** *Rockefeller Archive Center Newsletter*; *Research Reports from the Rockefeller Archive Center*. **Collections/Resources:** Rockefeller family archives; Rockefeller Foundation Archives; Rockefeller University Archives; Commonwealth Fund Archives; Markle Foundation Archives; Russell Sage Foundation Archives; Social Science Research Council Archives. **Grants:** Yes. **Program Strengths:** 20th and 21st Century; 19th Century, North America; 20th and 21st Century, Institutions. **Staff:** *Lee R. Hiltzik*; *Kenneth W. Rose*; *Darwin H. Stapleton* (19th Century, North America, Technology). **Description:** The Rockefeller Archive Center has important collections in the history of science in the 20th century, particularly in the archives of the Rockefeller Foundation and Rockefeller University (formerly the Rockefeller Institute for Medical Research). The Center assists researchers who visit the Center, located 25 miles north of mid-Manhattan. **Web Site:** http://www.rockefeller.edu/archive.ctr/

North Carolina

Duke University Medical Center Library (History of Medicine Collections). **Contact:** Suzanne Porter, porte004@mc.duke.edu **Mailing Address:** Box 3702, Durham, NC, 27710, United States. **Tel:** 1.919.660.1143 **Fax:** 1.919.681.7599 **Grants:** No. **Program Strengths:** Medical Sciences. **Description:** The History of Medicine Collections, which include the Trent Collection, cover the history of the health sciences and number 20,000 volumes, consisting of monographs, serials, manuscripts, medical instruments, artifacts, prints, photographs, and ephemera. Everything from incunabula to current secondary sources is represented. **Web Site:** http://www.mclibrary.duke.edu/

Ohio

Case Western Reserve University (Special Collections). **Contact:** Norma Sue Hanson, nsh2@po.cwru.edu **Mailing Address:** Kelvin Smith Library/ University Library, 11055 Euclid Avenue, Cleveland, OH, United States. **Tel:** 1.216.368.2993 **Fax:** 1.216.368.6950 **Collections/Resources:** Charles F. Brush Collection; Warner & Swasey Collection; Editions of many significant works from 16th century to the present; Journals and encyclopedias from the

17th century to the present. **Grants:** No. **Description:** Houses rare books, manuscripts and special collections of the University Library: History of the Book, German, French, English and American literature and history, Science and Technology materials, notable works in the Fine Arts, and collections of local interest. **Web Site:** http://www.cwru.edu/UL/SpecColl/histtech

Lloyd Library and Museum. Contact: Kathleen D. Connick, kdconnick@ lloydlibrary.org **Mailing Address:** 917 Plum Street, Cincinnati, OH, 45202, United States. **Tel:** 1.513.721.3707 **Fax:** 1.513.721.6575 **Collections/ Resources:** Extensive personal papers of 19th and 20th century biologists and doctors; Drug Price Lists and Circulars, 1870-1970; Colleges of Pharmacy Announcements & Catalogues, 1865-1965; Seed and Plant Catalogs, 1890-1980; Records of Mergers & Acquisitions in the Pharmaceutical Industry, 1944-1990. French Pharmacy & Pharmacists, 19th Century. **Grants:** Yes. **Program Strengths:** 19th Century, Biological Sciences; Medical Sciences. **Description:** Cincinnati's Lloyd Library & Museum was founded in 1898 by John Uri Lloyd. It is a research library focused on botany, horticulture, pharmacy, natural products, herbal medicine, alternative and eclectic medicine. **Web Site:** http://www.lloydlibrary.org/

The Ohio State University (Goldthwait Polar Library). **Contact:** Lynn Lay, lay.1@osu.edu **Mailing Address:** 176 Scott Hall, 1090 Carmack Road, Columbus, OH, 43210-1002, United States. **Tel:** 1.614.292.6715 **Fax:** 1.614.292.4697 **Grants:** No. **Description:** Byrd Polar Research Center is a multidisciplinary center which specializes in almost all facets of polar research. **Web Site:** http://www.lib.ohio-state.edu/Lib_InfoPLR.html

University of Cincinnati Medical Center (Cincinnati Medical Heritage Center). **Contact:** Billie Broaddus, Billie.Broaddus@uc.edu **Mailing Address:** University of Cincinnati Medical Center, AIT&L, Cincinnati Medical Heritage Center, Cincinnati, OH, 45267-0574, United States. **Tel:** 1.513.558.4570 **Fax:** 1.513.558.0472 **Collections/Resources:** History of the University of Cincinnati College of Medicine; History of Pharmacy; Civil War Collection (including many Rare Books). **Grants:** No. **Program Strengths:** 19th Century, North America, Medical Sciences, Education; 20th and 21st Century, North America, Medical Sciences. **Web Site:** http://www.aitl.uc.edu

Wright State University (Fordham Health Sciences Library, Special Collections and Archives). **Contact:** Dawne Dewey, ddewey@library.wright.edu **Primary E-mail:** archive@wsuol2.wright.edu **Mailing Address:** 125 Medical Sciences Bldg., Fordham Health Sciences Library, Wright State University, Dayton, OH, 45435, United States. **Tel:** 1.937.775.2092 **Fax:** 1.937.775.4109 **Collections/Resources:** Collection emphasis is on aerospace medicine and human factors engineering. Collection highlights include the papers of Ross A. McFarland, the Howard Hasbrook Aviation Crash Injury Research Collection, and the papers of NASA physician/astronaut William Thornton. **Grants:** No.

Program Strengths: 20th and 21st Century, North America, Medical Sciences, Instruments and Techniques. **Web Site:** http://www.libraries.wright.edu/special

Oklahoma

University of Oklahoma (History of Science Collections). **Contact:** Marilyn Ogilvie, mogilvie@ou.edu **Mailing Address:** Bizzell Library, 401 W. Brooks, Room 521, Norman, OK, 73019-6030, United States. **Tel:** 1.405.325.2741 **Fax:** 1.405.325.7618 **Ongoing Publication Projects:** *ISIS Current Bibliography*; Landmarks of Science (microform); Prometheus Project Digital Libraries; *The Lynx* (newsletter). **Collections/Resources:** Comprehensive general collections in every area of the History of Science including the DeGolyer Collection, the Roller Collection, the Harlow Collection, the Lacy Collection, and many others. Specialized collections include the Klopsteg Collection in the History of Archery. **Grants:** Yes. **Program Strengths:** 14th-16th Century; 17th Century; 18th Century. **Staff:** *Kerry V. Magruder* (17th Century, Astronomical Sciences); *Marilyn B. Ogilvie* (Education). **Description:** The University of Oklahoma is a comprehensive research institution. The collections work with the Department of the History of Science to support research, teaching, and special events in the history of science by graduate students, faculty, and visiting scholars. **Web Site:** http://www-lib.ou.edu/depts/histsc/

Oregon

Oregon State University (Special Collections: History of Science). **Contact:** Cliff Mead, special.collections@orst.edu **Mailing Address:** 121 The Valley Library, Oregon State University, Corvallis, OR, 97331-4501, United States. **Tel:** 1.541.737.2075 **Fax:** 1.541.737.8674 **Collections/Resources:** Extensive collections of personal papers of noted scientists, including Linus Pauling and Ava Helen Pauling; 3,000 items on Atomic Energy Collection, from 1896 to present. **Grants:** No. **Staff:** *Clifford S. Mead.* **Description:** OSU's History of Science collection reflects the subject strengths of the library's main collection. Some of the recent acquisitions include Amadeo Avagadro's "Fisca de' Corpi Ponderabili ossia."; Neils Bohr's rare doctoral thesis; and a first appearance of Pierre Curie's papers, announcing the discovery of radium and polonium. **Web Site:** http://osu.orst.edu/dept/Special_Collections/subpages/ahp/index.htm

Pennsylvania

American Philosophical Society Library. **Contact:** Martin L. Levitt, mlevitt@amphilsoc.org **Mailing Address:** 105 South Fifth Street, Philadelphia, PA, 19106-3386, United States. **Tel:** 1.215.440.3400 **Fax:** 1.215.440.3423 **Collections/Resources:** The Society's Library, located near Independence Hall in Philadelphia, is a leading international center for research in the history of American science and technology and its European roots, as well as early American history and culture. The Library houses over 7 million manuscripts,

300,000 volumes and bound periodicals, and thousands of maps and prints. Outstanding historical collections and subject areas include the papers of Benjamin Franklin; the American Revolution; 18th and 19th-century natural history; western scientific expeditions and travel including the journals of Lewis and Clark; polar exploration; the papers of Charles Willson Peale, including family and descendants; American Indian languages; anthropology including the papers of Franz Boas; the papers of Charles Darwin and his forerunners, colleagues, critics, and successors; history of genetics, eugenics, and evolution; history of biochemistry, physiology, and biophysics; 20th-century medical research; and history of physics. **Grants:** Yes. **Staff:** *Edward C. Carter; Robert S. Cox; Roy Goodman; Martin Levitt.* **Description:** Research fellowships are awarded for use of the library's collections. **Web Site:** http://www.amphilsoc.org

The College of Physicians of Philadelphia (Historical Library). **Contact:** Ed Morman, emorman@collphyphil.org **Primary E-mail:** histref@collphyphil.org **Mailing Address:** 19 South 22nd Street, Philadelphia, PA, 19103-3097, United States. **Tel:** 1.215.563.3737 **Fax:** 1.215.569.0356 **Collections/Resources:** Major collection of early printed books (including 415 incunabula) in medicine and related subjects. Excellent holdings in clinical medicine, biomedical science, and related subjects, 1801-1990. Important archival and manuscript collections, largely (but not exclusively) related to medicine in the Philadelphia region. Important collection of the secondary literature on history of medicine and related fields. **Grants:** Yes. **Staff:** *Charles Greifenstein; Laura Guelle; Edward T. Morman* (19th Century, North America, Medical Sciences); *Christopher Stanwood.* **Description:** The College of Physicians was founded in 1787 and its library was established one year later. At the time of the Civil War, its library was the most significant medical library in the country, but was soon overtaken by the Army Medical Library (now the NLM). The College Library remained a key resource for physicians until the 1980s. It is now formally a historical medical library. **Web Site:** http://www.collphyphil.org/histpg1.shtml

Pennsylvania State University (Joseph Priestley Collection). **Contact:** Sandra Stelts, sks@psulias.psu.edu **Primary E-mail:** libinfo@psulias.psu.edu **Mailing Address:** Penn State University, 104 Paterno Library, University Park, PA, 16802-1808, United States. **Tel:** 1.814.865.1793 **Fax:** 1.814.863.5318 **Collections/Resources:** 439 books by and about Priestley; Priestley's Last Will and Testament, two holograph copies of Priestley's memoirs; The 1875 Priestley Memorial Scrapbook; letters. **Grants:** No. **Program Strengths:** 18th Century, North America, Physical Sciences; 19th Century, North America; 20th and 21st Century, North America. **Description:** The Special Collections Department at the Pennsylvania State University Libraries comprises three units: Rare Books and Manuscripts, the Penn State University Archives, and Historical Collections and Labor Archives. The department is open to all researchers Monday through Friday, 8:00 to 5:00. Contact with staff prior to visiting is recommended. **Web Site:** http://www.libraries.psu.edu/crsweb/speccol/priestley

University of Pennsylvania (History of Science Collection). **Contact:** Stephen Lehmann, lehmann@pobox.upenn.edu **Mailing Address:** Van Pelt-Dietrich Library, 3420 Walnut Street, Philadelphia, PA, 19104-6206, United States. **Tel:** 1.215.898.5999 **Fax:** 1.215.573.9079 **Collections/ Resources:** E. F. Smith Collection in the History of Chemistry. **Web Site:** http://www.library.upenn.edu/services/collections/

University of Pittsburgh (Archives of Scientific Philosophy in the Twentieth Century). **Contact:** Brigitta Arden, arden@pitt.edu **Primary E-mail:** asp@ library.pitt.edu **Mailing Address:** University of Pittsburgh, 363 Hillman Library, Pittsburgh, PA, 15260, United States. **Tel:** 1.412.648.8197 **Fax:** 1.412.648.8192 **Collections/Resources:** Papers of Rudolf Carnap, Hans Reichenbach, Frank Ramsey, Paul Hertz, Rose Rand, Herbert Feigl (on microfilm), Ludwig Wittgenstein (on microfilm); Archives for the History of Quantum Physics (AHQP) (on microfilm). **Grants:** No. **Program Strengths:** 20th and 21st Century, Philosophy or Philosophy of Science. **Description:** Most materials are housed in the Special Collections Department of the University's Hillman Library. Researchers using the Archives by mail are asked to submit questions that are as specific as possible. Written inquiries are most effectively answered when proper names and dates are included. Scholars who anticipate using material extensively without traveling to Pittsburgh or Konstanz may purchase inventories to the various collections to aid their research. Inventories to the most recently opened collections are also available online at the ASP Web site. Those planning to visit the Archives will make the most of their time on site by becoming familiar with these inventories before arriving. **Web Site:** http://www.library.pitt.edu/libraries/special/asp/archive.html

University of Pittsburgh (Falk Library of the Health Sciences, History of Medicine Division). **Contact:** Jonathon Erlen, erlen@pitt.edu **Mailing Address:** University of Pittsburgh, 200 Scaife Hall, Pittsburgh, PA, 15261, United States. **Tel:** 1.412.648.8927 **Fax:** 1.412.489.9020 **Ongoing Publication Projects:** Bibliography of dissertations in the medical humanities history of medicine; medical ethics; history of nursing; history of pharmacy; medicine and religion; medicine and art; women's health issues; philosophy and medicine. **Collections/ Resources:** Major strengths: Rare books in surgery, public health, psychiatry. **Grants:** No. **Program Strengths:** Medical Sciences. **Staff:** *Joe Alter; Cary Balaban; Charles Bender; Thomas Benedek; John Delaney; Georgia Duker; Jonathon Erlen* (20th and 21st Century, North America, Medical Sciences); *Ann Jannetta; Jason Rosenstock.* **Description:** The University of Pittsburgh currently offers 14 separate courses on various aspects of the history of medicine and health care. **Web Site:** http://www.hsls.pitt.edu/guide/histmed/history.html

Texas

Texas Medical Center Library (John P. McGovern Historical Collections and Research Center). **Contact:** Elizabeth White, bwhite@library.tmc.edu

Primary E-mail: mcgovern@library.tmc.edu **Mailing Address:** Texas Medical
Center Library, 1133 John Freeman Blvd., Houston, TX, 77030, United
States. **Tel:** 1.713.799.7139 **Fax:** 1.713.790.7052 **Collections/Resources:**
Atomic Bomb Casualty Commission Collections; Burbank/Fraser Collection
on Arthritis, Rheumatism and Gout; McGovern Collection on the History
of Medicine; Cora and Webb Mading Collection on Public Health; Hilde
Bruch Papers; Philip Showalter Hench Papers; Johnson Space Center-Space
Life Sciences Archive; R. Lee Clark Papers; Texas State Board of Medical
Examiners Collection. **Grants:** No. **Program Strengths:** 20th and 21st Century,
North America, Medical Sciences; 19th Century, North America, Medical
Sciences. **Description:** The McGovern Research Center holds approximately
200 archival collections related to Texas physicians and Houston, Texas
healthcare institutions. There are approximately 5,000 photographs related to
Houston medicine and 40,000 images from the journal, *Medical World News*.
Web Site: http://mcgovern.library.tmc.edu

University of Texas Medical Branch (Moody Medical Library, History of
Medicine Collections). **Contact:** Sarita Bullard Oertling, soertlin@utmb.edu
Mailing Address: The University of Texas Medical Branch, 301 University
Blvd., Galveston, TX, 77555-1035, United States. **Tel:** 1.409.772.2397
Fax: 1.409.765.9852 **Collections/Resources:** Immunology: Centered around
the impressive Pasteur imprints, this collection of about 800 items traces the
development of the germ theory of disease and includes publications of Pasteur's
collaborators as well as such individuals as Robert Koch and Paul Ehrlich.
Also included are about 400 titles relating to the history of smallpox, works
by and about Edward Jenner, and anti-vaccination pamphlets. Anesthesiology:
this extensive collection of more than 900 books and pamphlets is rich in 18th
century publications on the chemistry of respiration. Among the 19th century
figures represented are John Snow, John C. Warren, W.T.G. Morton, and James
Y. Simpson. Psychiatry and the Behavioral Sciences: the private library of
Haskell F. Norman, M.D., forms the core of this collection of 4,600 books and
pamphlets. Works of Philippe Pinel, J. E. D. Esquirol, Jean-Martin Charcot, and
Sigmund Freud are well represented. Also noteworthy are holdings in witchcraft,
mesmerism, and phrenology. Anatomy and Surgery: holdings in this area include
many first editions and anatomical atlases, famous for their striking illustrations.
Works of Galen, Vesalius and Harvey are well represented. A major part of the
collection consists of the private libraries of Drs. William M. Crawford, Robert
J. Moes, and Truman G. Blocker, Jr. Osleriana: The Samuel X. Radbill collection
of 400 books and offprints by and about William Osler. Forensic Medicine: more
than 500 titles in the history of toxicology, criminology, and legal medicine.
Grants: No. **Program Strengths:** Biological Sciences. **Description:** The
Blocker Collections, the largest such collection in the history of medicine
and allied sciences in the southern United States, consist of rare books, visual
materials, archives and manuscripts, microscopes, non-prescription drugs, and
medical and surgical instruments. **Web Site:** http://library.utmb.edu/blocker.htm

Virginia

University of Virginia (Health Sciences Library). **Contact:** Joan E. Klein, jre@virginia.edu **Mailing Address:** Box 800722, University of Virginia Health System, Charlottesville, VA, 22908-0722, United States. **Tel:** 1.434.924.0052 **Fax:** 1.434.243.5873 **Collections/Resources:** Subjects of special collections include: yellow fever, the American Lung Association of Virginia, the Eighth Evacuation Hospital, surgical instruments from antiquity, and interdisciplinary resources for the study of health systems, epidemiology, public health, and health care research. **Grants:** No. **Program Strengths:** 20th and 21st Century, North America, Medical Sciences; 19th Century, North America, Medical Sciences. **Description:** Thomas Jefferson himself selected the original medical books to be included in the library of his fledgling University of Virginia. Today the philosophical foundation of the Claude Moore Health Sciences Library continues to be rooted in Jefferson's belief that the arts and sciences are inextricably linked. Their marriage has produced exquisite works represented in the medical library's historical collections, along with other materials that illustrate the connections between past, present, and future–another Jeffersonian idea. Works that instruct and inspire and that demonstrate the importance of the roots of health sciences disciplines are preserved and made available for research uses. Historical Collections includes rare books, journals containing classic and landmark articles, manuscript materials and archival records, photographs, and artifacts such as surgical instruments, bloodletting devices, and World War II uniforms. **Web Site:** http://www.med.virginia.edu/hs-library/historical/

Virginia Polytechnic Institute and State University (Special Collections, Archives of American Aerospace Exploration). **Contact:** Jennifer Gunter, specref@vt.edu **Mailing Address:** Virginia Polytechnical Institute, PO Box 90001, Blacksburg, VA, 24062-9001, United States. **Tel:** 1.540.231.6308 **Fax:** 1.540.231.9263 **Collections/Resources:** International Archive of Women in Architecture; Collections on the Civil War, Norfolk Western and Southern Railroad, culinary history, and heraldry. **Grants:** No. **Program Strengths:** North America, Astronomical Sciences, Exploration, Navigation, Expeditions; North America, Natural and Human History; 20th and 21st Century, Social Sciences, Instruments and Techniques. **Web Site:** http://www.spec.lib.vt.edu/

Wisconsin

University of Wisconsin, Madison (Health Sciences Library, History Collections). **Contact:** Micaela Sullivan-Fowler, micaela@library.wisc.edu **Mailing Address:** University of Wisconsin-Madison, 1305 Linden Drive, Madison, WI, 53706-1593, United States. **Tel:** 1.608.262.2402 **Fax:** 1.608.262.4732 **Collections/Resources:** Exemplary electronically cataloged collection of books and journals from 16th through 21st century on anatomy, vaccination, women's health, unorthodox practice, surgery, public health, diseases, anesthesia, etc. Uncataloged collection of reprints of 18th and 19th century

European medical journals. Uncataloged collection of American pamphlets, promotional material, and other ephemera from 1850 to the present on health sciences topics. **Grants:** No. **Program Strengths:** 19th Century, North America, Medical Sciences, Instruments and Techniques; 19th Century, Europe, Medical Sciences, Institutions; 18th Century, Europe, Medical Sciences. **Staff:** *Rima D. Apple* (20th and 21st Century, Gender and Science). **Description:** The majority of books (1492-present) are electronically cataloged at http://madcat.library.wisc.edu. Journals (though not the journal content) are cataloged there as well. We do not have an archive, though we have an unprocessed collection of 1980's-present AIDS material. **Web Site:** http://www.hsl.wisc.edu/

University of Wisconsin, Madison (History of Science Special Collections). **Contact:** Jill Rosenshield, rosen@doit.wisc.edu **Mailing Address:** Memorial Library Room 976, 728 State Street, Madison, WI, 53706-1494, United States. **Tel:** 1.608.262.3243 **Fax:** 1.608.265.2754 **Grants:** Yes. **Web Site:** http://www.library.wisc.edu/libraries.SpecialCollections/

Wyoming

University of Wyoming (American Heritage Center). **Contact:** Ginny Kilander, papyrus@uwyo.edu **Primary E-mail:** ahc@uwyo.edu **Mailing Address:** American Heritage Center, P.O. Box 3924, Laramie, WY, 82071, United States. **Tel:** 1.307.766.4114 **Fax:** 1.307.766.5511 **Ongoing Publication Projects:** *American Heritage Center Heritage Highlights*; *American Heritage Center Annual Report*; *Annals of Wyoming*. **Collections/Resources:** Anaconda Geological Documents Collection (records of the Anaconda Copper Mining Company); plus collections in the fields of life sciences, social sciences, physical sciences, medicine, and geology. Please see the on-line catalog or contact the Reference Department for additional details about these holdings. **Grants:** Yes. **Description:** The University of Wyoming American Heritage Center is a research facility and manuscript repository which houses 7,000 collections of papers, maps, audiovisual materials, rare books, and artifacts related to Wyoming and the West, petroleum and mining industries, transportation, American culture, conservation, politics and world affairs, University Archives and Rare Books. **Web Site:** http://www.uwyo.edu/ahc

Museums

As defined by the HSS executive committee, "museums primarily preserve and collect the material relics of science. These institutions typically provide access for research and educational purposes." 43 respondents described their institutions as history of science museums. Entries are organized by country, and then by state for museums within the United States. Not all institutions responded with complete data. All entries contain some portion of the following information:

> Institution name
> Name of program in history of science, technology, or medicine
> Primary contact person's name, title, and e-mail address
> Primary e-mail address, if different from the contact person's email
> Mailing address
> Telephone number
> Facsimile number
> Journal editorial offices and ongoing serial publication projects
> Special collections and research resources
> Does the institution offer publicly available grants or fellowships (yes/no)
> Particular research focus, entered using a standardized list of keywords
> List of staff members, including research interests. Names appear as they were entered by respondents. A broad research interest appears only after the names of HSS members, if they entered this information when they joined the Society. (As computer programming challenges are overcome, research interests for non-members should appear beginning with the 2003 online edition of the *Guide*.)
> Brief description of the museum and its special features
> URL

Australia

Macleay Museum, University of Sydney. Contact: Julian Holland, julian@
macleay.usyd.edu.au **Primary E-mail:** macleay@macleay.usyd.edu.au
Mailing Address: Macleay Museum, University of Sydney, Sydney, New South
Wales, 2006, Australia. **Tel:** 61.2.9351.2274 **Fax:** 61.2.9351.5646 **Ongoing
Publication Projects:** The Museum issues a newsletter twice a year (April
and October) outlining its activities. Other more substantial publications are
sometimes issued in conjunction with exhibitions. **Collections/Resources:**
The Invertebrate Collection of the Museum, primarily collected by Alexander
Macleay in London around 1800-1825 and augmented by later members of his
family; Historic photographs associated with scientific teaching and research,
including extensive series of lantern slides. **Grants:** Yes. **Program Strengths:**
19th Century, Australia and Oceania, Natural and Human History; 19th Century,
Australia and Oceania, Physical Sciences, Instruments and Techniques.
Description: The Museum's main collection areas are natural history
(invertebrate and vertebrate), ethnography, scientific instruments, and historic
photographs. **Web Site:** http://www.usyd.edu.au/su/macleay/

Brazil

Museu de Astronomia e Ciencias Afins. Contact: Alfredo Tolmasquim, alfredo@
mast.br **Primary E-mail:** mast@mast.br **Mailing Address:** Rua General
Bruce 586, Rio de Janeiro, RJ, 20.921-030, Brazil. **Tel:** 55.21.2580.9432
Fax: 55.21.2580.4531 **Collections/Resources:** Collection of scientific
instruments used from the 18th Century to the beginning of the 20th Century.
Personal archives of scientists and of scientific institutions in Brazil. The library
specializes in history of science and science education. Brazilian Bibliography
of History of Science is available for consulting on-line. **Grants:** No.
Staff: *Ana Maria Ribeiro de Andrade*; *Christina da Motta Barboza*; *Heloisa M.
Bertol Domingues* (19th Century, South America, Biological Sciences); *Carlos
Ziller Camenietzki*; *Alda Heizer*; *Henrique Lins de Barros*; *Pedro Eduardo
Marinho*; *Alfredo Tiomno Tolmasquim*. **Description:** The Museum of Astronomy
and Related Sciences (MAST) was created in 1985, and is one of the research
institutes sponsored by the Ministry of Science and Technology in Brazil. It
houses a research group on history of science, with special focus on Brazilian
science, and another on science education. **Web Site:** http://www.mast.br

Canada

Canada Science and Technology Museum. Contact: Randall Brooks, rbrooks@
nmstc.ca **Mailing Address:** PO Box 9724, Station T, Ottawa, ON, K1G 5A3,
Canada. **Tel:** 1.613.990.2804 **Fax:** 1.613.990.3636 **Ongoing Publication
Projects:** *Material History Review* (twice annually); *Transformation* (published
1-2 times per year), Geoff Rider, Managing Editor, grider@nmstc.ca.
Collections/Resources: 5000+ scientific instruments including collections from

the National Research Council of Canada, Nobel laureates, Meteorological Service of Canada, Measurement Canada, and geodetic and topographical surveys of Canada. Most of the collection dates from the middle of the 19th century. Computing technology is also well represented. The Canadian National Railway photo collection comprises almost a million images dating from the 1850s to ca. 1980, and covers all aspects of Canadian society, life and industry. **Grants:** No. **Program Strengths:** 20th and 21st Century, North America, Physical Sciences, Instruments and Techniques; 20th and 21st Century, North America, Astronomical Sciences, Instruments and Techniques; 19th Century, Europe. **Staff:** *Randall C. Brooks; Helen Graves Smith; Garth Wilson.* **Description:** Canada's only comprehensive museum dealing with and collecting in the history of science and technology and its social impact. The museum's mandate is to study the "Transformation of Canada," and the sub-themes: Canadian Context, Finding New Ways, How "Things" Work, and People, Science and Technology. **Web Site:** http://www.science-tech.nmstc.ca

University of Toronto Museum of Scientific Instruments. **Contact:** Janis Langins, utmusi@chass.utoronto.ca **Primary E-mail:** ihpst.info@utoronto.ca **Mailing Address:** Room 316, Victoria College, 91 Charles Street West, University of Toronto, Toronto, Ontario, M5S 1K7, Canada. **Fax:** 1.416.978.3003 **Description:** This web-based virtual museum gives details of the historic scientific instruments located at the University of Toronto. **Web Site:** http://www.chass.utoronto.ca/utmusi/

The Czech Republic

National Technical Museum (History of Science and Technology). **Contact:** Jaroslav Folta, info@ntm.cz **Mailing Address:** Kostelni 42, Prague, CZ-170 78, Czech Republic. **Tel:** 42.2.33377290 **Fax:** 420.2.333.71.801 **Ongoing Publication Projects:** *Acta Historiae Rereum Naturalium Necnon Technicarum; Rozpravy Narodniho Technickeho Muzea.* **Collections/Resources:** Collections focus on the history of technology (machinery, transport, electrotechnic, communications, chemistry, time measurement, textile machinery, polygraph, geodesy and exact science instruments, etc.) since the industrial revolution, mostly connected with Czech development. **Staff:** *Josef Cerny; Jaroslav Folta; Vaclav Heisler; Jan Hozak; Ivo Janousek; Milo Josefovic; Jiri Korinek; Petr Kozisek; Petr Krajci; Jiri Krohn; Toma Kucera; Radek Kyncl; Irena Laboutkova; Jana Masinova; Jana Nekvasilova; Arnost Nezmeskal; Miroslav Novak; Jana Pauly; Josef Petrik; Pavel Pitrak; Josef Priplata; Zdenek Rasl; Milada Sekyrkova; Antonin Svejda; Vladimira Valcharova; Karel Zeithammer.* **Description:** Some members of the institution teach the history of technology or individual sciences at Charles University, Technical University Prague or the Higher Special School of Information Services. NTM also hosts the Society for the History of Science and Technology of the Czech Republic. **Web Site:** http://www.ntm.cz

Denmark

Danish National Museum for the History of Science and Medicine (The Steno Museum). **Contact:** Kurt M. Pedersen, stenokmp@au.dk **Primary E-mail:** stenomus@au.dk **Mailing Address:** C.F. Mollers Alle, Bldg. 100, The University Park, Aarhus, DK-8000, Denmark. **Tel:** 45.89.42.39.86 **Fax:** 45.89.42.39.95 **Collections/Resources:** The collection includes instruments that have been used in astronomy, surveying, optics, electromagnetism, atomic and nuclear physics, radio, computing, and chemistry. Full-size reconstructions of Tycho Brahe's Armillary, a country doctor's surgery, a dental surgery, a pharmacy, an old X-ray machine, equipment for resuscitation and anaesthetizing. The Museum also has a planetarium and a medical herb garden. **Grants:** No. **Program Strengths:** Physical Sciences, Instruments and Techniques; Medical Sciences, Instruments and Techniques; Astronomical Sciences, Education. **Staff:** *Hans Buhl*; *Aase Roland Jacobsen*; *Ole J. Knudsen*; *Kurt Moeller Pedersen*; *Hanne Teglhus*. **Description:** Behind the scenes, the museum staff work in collecting, registering, restoring, reconstructing, researching, and teaching in the fields of the history of science and medicine. **Web Site:** http://www.stenomuseet.dk/index.html

Medical History Museum at the University of Copenhagen (Department of History of Medicine). **Contact:** Frank Allen Rasmussen, far@mhm.ku.dk **Primary E-mail:** med.hist.museum@mhm.ku.dk **Mailing Address:** Bredgade 62, Copenhagen, 1260, Denmark. **Tel:** 45.3532.3800 **Fax:** 45.3532.3816 **Collections/Resources:** The museum has rich collections of 18th - 20th century medical instruments, pathology specimens, an archive, and a specialized library, including a rare book collection. **Grants:** No. **Program Strengths:** 20th and 21st Century, Europe, Medical Sciences, Instruments and Techniques. **Staff:** *Frank Allan Rasmussen* (17th Century). **Web Site:** http://www.mhm.ku.dk

Germany

Adam-Ries-Museum. **Contact:** Rainer Gebhardt, dr.rainer.gebhart@t-online.de **Primary E-mail:** info@adam-ries-bund.de **Mailing Address:** PSF 100 102, Johannisgasse 23, Annaberg-Buchholz, Saxony, 09456, Germany. **Tel:** 49.37.33.42.90.86 **Fax:** 49.37.33.42.90.87 **Collections/Resources:** Papers of the mathematician Adam Ries, material on mathematics in the Middle Ages, history of the abacus, and genealogies of Adam Ries and his descendents. **Grants:** No. **Program Strengths:** Europe, Mathematics, Humanistic Relations of Science. **Web Site:** http://www.adam-ries-bund.de

Archenhold-Sternwarte. **Contact:** Dieter B. Herrmann, dbherrmann@astw.de **Mailing Address:** Alt-Treptow 1, Berlin, D-12435, Germany. **Tel:** 49.30.534.80.80 **Ongoing Publication Projects:** Complete edition of the correspondence of Gottfried Kirch (1639-1710) with comments; *Acta Historica Astronomiae*; editors Wolfgang R. Dick and Juergen Hamel (e-mail:

jhamel@t-online.de). **Collections/Resources:** Special library for astronomy with 30,000 bibliographical units from 16th century. **Program Strengths:** 18th Century, Europe, Astronomical Sciences, Biography. **Staff:** *Juergen Hamel.* **Web Site:** http://www.astw.de

The Atom Keller Museum at Haigerloch. **Contact:** Michael Thorwart, m.thorwart@tn.tudelft.nl **Primary E-mail:** kulturamt@haigerloch.de **Mailing Address:** Atomkeller Museum, Postfach 54, 72394, Haigerloch, Germany. **Tel:** 49.7474.697.27 **Collections/Resources:** Original host cavern for German atomic research project during 1944-1945, original destroyed reactor vessel, and rebuilt model of the reactor. **Grants:** No. **Program Strengths:** 20th and 21st Century, Europe, Physical Sciences, Instruments and Techniques. **Web Site:** http://www.haigerloch.de/KELLER/ekeller.htm

Christian-Albrechts-Universität Kiel (Institut für Geschichte der Medizin und Pharmazie). **Mailing Address:** Brunswiker Str. 2, Kiel, D-24105, Germany. **Tel:** 49.431.5.97.37.20 **Collections/Resources:** Museum for the history of medicine and pharmacy with several showrooms, containing many original and working artifacts, including an iron lung and a pathology cabinet. **Program Strengths:** Medical Sciences. **Staff:** *Jörn Hennig Wolf.* **Description:** Contains many medical instruments from the late 19th and early 20th centuries, and a section room with more than 500 pathology preparations from 1912 to 1961. **Web Site:** http://www.med-hist.uni-kiel.de

Deutsches Museum. **Contact:** Wolf Peter Fehlhammer, wpf@deutsches-museum.de **Mailing Address:** Deutsches Museum, München, Bavaria, D-80306, Germany. **Tel:** 49.89.2179.209 **Fax:** 49.89.2179.239 **Ongoing Publication Projects:** Journal: *Kultur & Technik* (quarterly); Series: *Abhandlungen und Berichte*; Series: *Artefacts* (in co-operation with the Smithsonian and the Science Museum, London). **Collections/Resources:** Very extensive archives, including private papers of eminent scientists, archives of major German companies and institutions, trade literature, pictures and special collections. The library of the Deutsches Museum, with about 880,000 books and about 3600 journals, is the largest library on the history of science and technology in Europe. The collections also include about 85,000 objects. **Grants:** Yes. **Staff:** *Ralph Boch*; *Peter Dorsch*; *Michael Eckert*; *Wilhelm Fossl*; *Alexander Gall*; *Ulf Hashagen* (19th Century, Europe, Mathematics); *Martina Hessler*; *Eva A. Mayring*; *Arne Schirrmacher*; *Jürgen Teichmann*; *Helmuth Trischler*; *Ulrich Wengenroth*. **Description:** The Deutsches Museum hosts the Munich Center for the History of Science and Technology. The Munich Center pools the resources of the Deutsches Museum and the three Munich universities in the history of science and technology (including the Research Institute of the Deutsches Museum). *Also see the Munich Center for the History of Science and Technology under Graduate Programs.* **Web Site:** http://www.deutsches-museum.de/

Liebig-Museum. Contact: Magnus Mueller, magnus.mueller@phys.chemie. uni-giessen.de **Mailing Address:** Liebig-Museum, Liebigstrasse 12, D-35390 Giessen, Giessen, D-35390, Germany. **Tel:** 49.641.76392 **Collections/ Resources:** Library and many letters to and from Liebig between the years 1820 and 1873. **Grants:** No. **Program Strengths:** Physical Sciences, Gender and Science. **Web Site:** http://www.uni-giessen.de/~gi04/home1.html

Italy

Astronomical and Geophysical Museum. Contact: Luca Lombroso, museoastrogeo@unimo.it **Mailing Address:** Piazza Roma 22, Modena, 41100, Italy. **Tel:** 39.059.205.6204 **Fax:** 39.059.205.6243 **Grants:** Yes. **Web Site:** http://www.museoastrogeo.unimo.it

Institute and Museum of History of Science, Florence. Contact: Paolo Galluzi, galluzzi@imss.fi.it **Primary E-mail:** imss@imss.fi.it **Mailing Address:** Piazza dei Giudici 1, Firenze, 50122, Italy. **Tel:** 39.055.265.311 **Fax:** 39.055.265.3130 **Ongoing Publication Projects:** *Nuncius. Annali di Storia della Scienza*; Biblioteca della Scienza Italiana (Series Editorial Office). **Collections/Resources:** Research library of history of science. **Grants:** Yes. **Program Strengths:** Physical Sciences. **Description:** Situated in the heart of Florence. The nucleus of the Museum is made up of the collections of Medici and Lorenese scientific instruments. The library totals 100,000 volumes. The multimedia lab has specialized personnel and advanced technologies. **Web Site:** http://www.imss.fi.it

Museo Di Fisica, Universita La Sapienza. Contact: Fabio Sebastiani, fabio_sebastiani@roma1.infn.it **Mailing Address:** P. le Aldo Moro 2, 00185, Rome, Italy. **Tel:** 39.06.49913482 **Fax:** 39.06.4463158 **Ongoing Publication Projects:** Book-catalog: La storia dell'istituto di Fisica di Roma attraverso le sue collezioni. **Collections/Resources:** Collections: Fermi collection, collection on electromagnetism, collection on spectroscopy; Resources: Archivio Amaldi, Archivio Persico, Archivio Touschek. **Grants:** No. **Program Strengths:** 19th Century, Europe, Physical Sciences, Instruments and Techniques; 20th and 21st Century, Europe, Physical Sciences, Instruments and Techniques. **Staff:** *Giovanni Battimelli* (19th Century, Europe, Physical Sciences); *Michelangelo De Maria*; *Maria Grazia Ianniello*; *Giulio Maltese*. **Description:** The Museum is a research laboratory for specialists in history of scientific instrumentation and in history of physics. Due to the limited personnel of the museum, it is normally not open to the public. **Web Site:** http://www.phys.uniroma1.it/Docs/museo/home.htm

Museo Di Informatica E Storia Del Calcolo. Contact: Baldoni Renzo, info@museoinformatica.it **Mailing Address:** via Pianacci, Pennabilli (PS), 61016, Italy. **Tel:** 39.054.192.8563 **Fax:** 39.054.192.8563 **Collections/ Resources:** sezione calcolo, sezione informatica, biblioteca, emeroteca, stazioni multimediali. **Grants:** No. **Program Strengths:** Mathematics, Instruments and Techniques; Technology, Instruments and Techniques; Physical Sciences, Education. **Description:** The museum is a research laboratory in the history of mathematics and in the history of information science. **Web Site:** http://www.museoinformatica.it/

Museo per la Storia dell'Università di Pavia. **Contact:** Alberto Calligaro, alberto.calligaro@unipv.it **Mailing Address:** Strada Nuova 65, Pavia, 27100, Italy. **Tel:** 39.03.822.9724 **Fax:** 39.03.822.9724 **Collections/Resources:** 18th and 19th century surgical instruments. Camillo Golgi's archive (scientific manuscripts, letters, documents). Alessandro Volta's school instruments. **Grants:** No. **Program Strengths:** 18th Century, Europe, Medical Sciences, Instruments and Techniques; 19th Century, Europe, Medical Sciences, Humanistic Relations of Science; 19th Century, Europe, Physical Sciences, Instruments and Techniques. **Staff:** *Fabio Bevilacqua*; *Paolo Mazzarello*. **Description:** Located in the ancient rooms of the School of Medicine, the Museum collects many documents concerning different disciplines. Two main sections are in the History of Medicine and the History of Physics. **Web Site:** http://ppp.unipv.it/museo/museo.htm

The Netherlands

Museum Boerhaave. **Contact:** M. Fournier, wetenschap@museumboerhaave.nl **Mailing Address:** P.O. Box 11280, 2301 EG Leiden, Leiden, Netherlands. **Tel:** 31.71.5214224 **Fax:** 31.71.5120344 **Ongoing Publication Projects:** Catalogs of the Collection. **Collections/Resources:** Leiden Physical Cabinet, mainly from the 18th century. Microscopes from the late 17th century through 1950, with emphasis on Dutch makers (Leeuwenhoek). 17th century mathematical instruments, many from Dutch makers. Dutch Nobel prize winners (van 't Hoff, Einthoven, Lorenz, Kamerlingh Onnes and others). Library containing some 35,000 titles on the history of science and medicine, including source materials from the 17th-20th centuries. **Grants:** No. **Program Strengths:** 17th Century, Europe, Physical Sciences, Instruments and Techniques; 18th Century, Europe, Physical Sciences, Instruments and Techniques; 20th and 21st Century, Europe, Medical Sciences, Instruments and Techniques. **Staff:** *Marian Fournier*; *Bart Grob*; *Kees Storm Grooss*; *Hans Hooijmaijers*; *Tim Huisman*; *Trienke van der Spek*. **Description:** The building dates back to the 16th century and was used as a hospital for many years. Herman Boerhaave gave his famed lectures at patients' bedsides in this building. Now the microscopes of Antoni van Leeuwenhoek stand next to ingenious devices such as the pendulum clocks of Christiaan Huygens, the helium liquefactor of Kamerlingh Onnes and the electrocardiograph of Einthoven. A unique feature is the anatomical theater with an amazing collection of skeletons. **Web Site:** http://www.museumboerhaave.nl

Poland

Nicholaus Copernicus Museum in Frombork. **Contact:** Henryk Szkop, frombork@softel.elblag.pl **Mailing Address:** Frombork, 14-530, Poland. **Tel:** 48.55.243.72.18 **Fax:** 48.55.243.72.18 **Web Site:** http://www.frombork.art.pl/Ang01

Spain

Museo Nacional de Ciencia y Tecnología. **Contact:** Amparo Sebastian Caudet, museo.mnct@mnct.mcu.es **Mailing Address:** Paseo de las Delicias, 61, Madrid, 28045, Spain. **Tel:** 34.1.530.30.01 **Fax:** 34.467.51.19 **Collections/Resources:** Scientific instruments for astronomy, topography, geodesy, physics, medicine, among others from the 16th century to the present day. Objects and industrial equipment of the 19th and 20th centuries including phonographs, telephones, and computers. Small (10,000 volumes), but specialized library in topics related to scientific instruments, museology and restoration. The collections and resources relate mainly to Spain or Spanish teaching institutions. **Grants:** Yes. **Web Site:** http://www.mnct.mcu.es/

Switzerland

Universität Zürich (Medizinhistorisches Institut und Museum). **Contact:** Christoph Moergeli, cmoergel@mhiz.unizh.ch **Primary E-mail:** mhizli@mhiz. unizh.ch **Mailing Address:** Raemistrasse 71, Zurich, CH 8006, Switzerland. **Tel:** 41.1.634.20.73 **Fax:** 41.1.634.23.49 **Collections/Resources:** Folk medicine. Ophthalmological collection. **Grants:** No. **Description:** The founder of the museum was Dr. Gustav Adolf Wehrli (1888-1949). Wehrli established a medical history collection which commands international recognition in quality and quantity. **Web Site:** http://www.medizin-museum.unizh.ch

United Kingdom

Museum of the History of Science. **Contact:** Jim Bennett, jim.bennett@ mhs.ox.ac.uk **Primary E-mail:** museum@mhs.ox.ac.uk **Mailing Address:** Broad Street, Oxford, OX1 3AZ, United Kingdom. **Tel:** 44.1865.277.280 **Fax:** 44.1865.277.288 **Collections/Resources:** The museum's collection of instruments and apparatus ranges from antiquity to the 20th century. **Grants:** No. **Staff:** *Jim Bennett*; *Stephen Johnston*. **Description:** The museum houses one of the world's finest collections of historic scientific instruments in the original home of the Ashmolean Museum, opened in 1683. It has particular strengths in early mathematical instruments and in microscopes, but ranges widely from astronomy and natural philosophy to chemistry and photography. **Web Site:** http://www.mhs.ox.ac.uk/

National Museum of Science and Industry. **Contact:** Robert Bud,
r.bud@nmsi.ac.uk **Mailing Address:** Exhibition Road, South
Kensington, London SW7 2DD, United Kingdom. **Tel:** 44.870.970.4771
Fax: 44.20.7942.4302 **Web Site:** http://www.sciencemuseum.org.uk/

The Royal Institution Michael Faraday Museum. **Contact:** Frank James,
fjames@ri.ac.uk **Primary E-mail:** ri@ri.ac.uk **Mailing Address:** Royal
Institution, 21 Albemarle Street, London, W1S 4BS, United Kingdom.
Tel: 44.020.7409.2992 **Fax:** 44.020.7629.3569 **Ongoing Publication Projects:**
The Correspondence of Michael Faraday. **Grants:** No. **Staff:** *Gertrude Mae
Prescott*. **Description:** The Faraday Museum contains much of the apparatus
used by Michael Faraday (1791-1867) in making his discoveries in natural
philosophy and chemistry. These include the first electric transformer and
generator, as well as the first sample of benzene. The Royal Institution also has
apparatus associated with Humphry Davy (electro-chemistry and the miners'
safety lamp), John Tyndall (radiant heat and why the sky is blue), James Dewar
(Dewar flasks), William Bragg and Lawrence Bragg (x-ray crystallographical
apparatus and molecular models), among others. The Royal Institution also
has large collections of the papers of each of these individuals and many
others (details can be found at http://www.aim25.ac.uk/cgi-bin/frames/
browse1?inst_id=17), as well as a large collection of portraits in different media.
Web Site: http://www.rigb.org/heritage/faradaypage.html

Science Museum Wroughton (NMSI). **Contact:** Steph Gillett, s.gillett@
nmsi.ac.uk **Primary E-mail:** m.atkinson@nmsi.ac.uk **Mailing Address:**
Science Museum, Wroughton, Wroughton Airfield, Swindon, Wiltshire,
SN4 9NS, United Kingdom. **Tel:** 44.1793.814.466 **Fax:** 44.1793.813.569
Collections/Resources: Civil aircraft, airplane engines. Motor cars, commercial
vehicles, bicycles, motorbikes. Tractors, implements. Fire fighting equipment.
Rocketry. Wood Press: former Fleet Street printing press. **Grants:** No. **Program
Strengths:** 20th and 21st Century, Technology. **Description:** The Science
Museum, Wroughton is the large object storage facility for the National Museum
of Science & Industry. We store and conserve over 18,000 large objects such
as aircraft, cars and tractors, as well as industrial plant, civil and mechanical
engineering, medical equipment, computers, fire engines and bicycles.
Web Site: http://www.sciencemuseum.org.uk/wroughton/

Whipple Museum of the History of Science, Cambridge University.
Contact: Liba Taub, lct1001@hermes.cam.ac.uk **Primary E-mail:**
hps-whipple-museum@lists.cam.ac.uk **Mailing Address:** Department of History
and Philosophy of Science, Free School Lane, Cambridge, CB2 3RH, United
Kingdom. **Tel:** 44.1223.330906 **Fax:** 44.1223.334554 **Program Strengths:**
Instruments and Techniques. **Staff:** *Liba Taub* (Pre-400 B.C.E, Astronomical
Sciences). **Description:** Founded in 1944, the museum provides a home for
many instruments which were used in the colleges and faculties of Cambridge
University from the 16th century to the present. The collection covers all

branches of science and its applications from physics to phrenology and
from magnetism to microscopy. Teaching and demonstration models used in
a wide range of fields, including mathematics and anatomy, are on display.
Web Site: http://www.hps.cam.ac.uk/whipple/

United States of America

California

Robert A. Paselk Scientific Instruments Museum, Humboldt State University.
Contact: Richard Paselk, rap1@axe.humboldt.edu **Mailing Address:**
Humboldt State University, 1 Harpst Street, Arcata, CA, 95521, United States.
Tel: 1.707.826.5719 **Fax:** 1.707.826.3279 **Grants:** No. **Program Strengths:**
20th and 21st Century, North America, Physical Sciences, Instruments and
Techniques; 20th and 21st Century, North America, Education; 20th and 21st
Century, North America, Biological Sciences, Instruments and Techniques.
Staff: *Richard Alan Paselk.* **Description:** The Museum primarily collects
scientific instruments and apparatus used at Humboldt State University. The
extensive museum Web site/catalog provides for each object: photographs;
usage/history description condition; scanned vendor's catalog description.
Information for select instruments is provided via links to early and/or
contemporary instructions on usage, histories of manufacturers, as well as other
useful information. **Web Site:** http://www.humboldt.edu/~scimus

Connecticut

The Peabody Museum of Natural History, Yale University (Collection of
Historical Scientific Instruments). **Contact:** Ellen Faller, eleanor.faller@
yale.edu **Mailing Address:** PO Box 208118, 170 Whitney Avenue,
New Haven, CT, 06520-8118, United States. **Tel:** 1.203.432.3141
Fax: 1.203.432.9816 **Grants:** No. **Staff:** *Ellen Faller; David F. Musto.*
Description: Due to renovation, the collection is in closed storage for the
near future. Work is progressing on an online database with digital images.
Web Site: http://www.peabody.yale.edu/collections/hsi/

District of Columbia

National Museum of Health and Medicine. **Contact:** J. T. H. Connor,
connorj@afip.osd.mil **Primary E-mail:** nmhminfo@afip.org **Mailing Address:**
Building 54, Walter Reed Army Medical Center, Washington, DC, 20306-6000,
United States. **Tel:** 1.202.782.2200 **Fax:** 1.202.782.3573 **Collections/**
Resources: Records of the Army Medical Museum and the Armed Forces
Institute of Pathology; Manuscripts, archives, films, prints and photographs, and
institutional records. Material is not necessarily institutionally related; Medical
instrumentation; Human and veterinary specimens of historical and pathological
interest. Neuroanatomical collections; Material on the history of embryology.

Grants: No. **Program Strengths:** 19th Century, North America, Medical Sciences, Instruments and Techniques; 20th and 21st Century, North America, Biological Sciences; 19th Century, North America, Natural and Human History, Instruments and Techniques. **Staff:** *Lenore Barbian*; *J. T. H. Connor*; *Archie Fobbs*; *Alan Hawk*; *Elizabeth Lockett*; *Adrianne Noe*; *Michael Rhode*; *Paul Sledzik*. **Description:** The National Museum of Health and Medicine was established during the Civil War as the Army Medical Museum, a center for the collection of specimens for research in military medicine and surgery. During the late 19th and early 20th centuries, museum staff engaged in various types of medical research. By World War II, research at the museum focused increasingly on pathology. In 1946 the museum became a division of the new Army Institute of Pathology, which became the Armed Forces Institute of Pathology in 1949. Today, the National Museum of Health and Medicine inspires interest in personal and public health and promotes understanding of medicine, past, present, and future, with a special emphasis on American military medicine. In addition to innovative exhibitions, the museum offers programs throughout the year aimed at adults and children, addressing medical, scientific, and historical subjects. The museum's Web site has a *Guide to the Collections* which can be used as a beginning point to search the collections. The museum is open for research to qualified researchers. We provide reference services by telephone and mail. **Web Site:** http://www.natmedmuse.afip.org

Smithsonian Institution. **Primary E-mail:** info@si.edu **Mailing Address:** PO Box 37012, SI Building, Room 153, MRC 010, Washington, DC, 20013-7012, United States. **Tel:** 1.202.357.2700 **Description:** The Smithsonian consists of several museums and thousands of collections that document many aspects of science, technology, art, and American life. *Please see each museum's entry for more detailed information.* **Web Site:** http://www.si.edu/

The Jerome and Dorothy Lemelson Center for the Study of Invention and Innovation, Smithsonian Institution. **Contact:** Arthur Molella, molellaa@ si.edu **Primary E-mail:** lemcen@si.edu **Mailing Address:** National Museum of American History, Smithsonian Institution, PO Box 37012, Rm. 1016, MRC 604, Washington, DC, 20013-7012, United States. **Tel:** 1.202.357.1593 **Fax:** 1.202.357.4517 **Ongoing Publication Projects:** Lemelson Center Series on Invention, MIT Press. **Collections/Resources:** Modern Inventor's Documentation Program (MIND). **Grants:** Yes. **Program Strengths:** 20th and 21st Century, North America, Technology, Biography; Instruments and Techniques. **Description:** Our mission is to document, interpret, and disseminate information about invention and innovation, to encourage inventive creativity in young people, and to foster an appreciation for the central role that invention and innovation play in the history of the United States. **Web Site:** http://www.si.edu/lemelson

National Air and Space Museum, Smithsonian Institution. **Contact:** Peter Golkin, golkinp@nasm.si.edu **Primary E-mail:** info@info.si.edu **Mailing**

Address: Independence Avenue at Sixth Street, SW, MRC 321, Washington, DC, 20560-0321, United States. **Tel:** 1.202.357.1552 **Fax:** 1.202.633.8174 **Ongoing Publication Projects:** The National Air and Space Museum publications program includes a range of books, from general to scholarly. NASM General Publications include exhibition books, gallery guides, catalogs, and books describing the collection, as well as books by Museum curators in their area of research. Also of interest is the Smithsonian History of Aviation and Spaceflight Series edited by Dominick A. Pisano and Allan A. Needell. **Collections/Resources:** Archives. Center for Earth and Planetary Studies. The Collections Division of the National Air and Space Museum is located primarily at the Paul E. Garber Facility in Suitland, Maryland. **Grants:** No. **Program Strengths:** 20th and 21st Century, Technology, Exploration, Navigation, Expeditions; Biological Sciences; Earth Sciences. **Description:** The Smithsonian Institution's National Air and Space Museum (NASM) maintains the largest collection of historic air and spacecraft in the world. It is also a vital center for research into the history, science, and technology of aviation and space flight. **Web Site:** http://www.nasm.si.edu/

National Museum of American History, Behring Center, Smithsonian Institution. **Contact:** Patricia Gossel **Mailing Address:** National Museum of American History, Smithsonian Information, Washington, DC, 20560, United States. **Tel:** 1.202.786.2669 **Fax:** 1.202.357.1631 **Collections/Resources:** Extensive collections of scientific instruments and objects associated with science in American life. **Grants:** Yes. **Program Strengths:** North America. **Staff:** *David Allison* (20th and 21st Century, North America, Technology); *Bernard S. Finn* (19th Century, North America, Physical Sciences); *Paul Forman* (20th and 21st Century, Physical Sciences); *Patricia L. Gossel* (19th Century, North America, Biological Sciences, Instruments and Techniques); *Barton C. Hacker* (Technology, Instruments and Techniques); *Paul F. Johnston* (North America); *Peggy A. Kidwell* (20th and 21st Century, Astronomical Sciences); *Ramunas A. Kondratas* (North America, Medical Sciences); *Arthur P. Molella* (North America); *Katherine Ott*; *G. Terry Sharrer* (Medical Sciences); *Jeffrey K. Stine* (Technology); *Steven Turner*; *Deborah J. Warner*. **Description:** The museum preserves and interprets the rich material legacy of American science and technology. Its collections place America's scientific and technical heritage into its historical context in order to better understand and explain American history, society, and culture. **Web Site:** http://americanhistory.si.edu/

Illinois

The Adler Planetarium & Astronomy Museum. **Contact:** Bruce Stephenson, astrohistory@adlernet.org **Mailing Address:** 1300 S. Lake Shore Drive, Chicago, IL, 60605-2403, United States. **Tel:** 1.312.322.0594 **Fax:** 1.312.341.9935 **Ongoing Publication Projects:** Catalog of the collection. **Collections/Resources:** Scientific instruments (approximately 1450-1850). Rare books on astronomy & scientific instruments. Works on paper on

astronomy & scientific instruments. **Grants:** No. **Program Strengths:**
14th-16th Century, Astronomical Sciences, Instruments and Techniques; 17th
Century, Astronomical Sciences, Instruments and Techniques; 18th Century,
Astronomical Sciences, Instruments and Techniques. **Staff:** *Marvin P. Bolt*
(Astronomical Sciences); *Anna F. Friedman*; *Bruce Stephenson* (17th Century,
Europe, Astronomical Sciences). **Description:** The Adler Planetarium &
Astronomy Museum was the first planetarium in the western hemisphere. Its
collections are the finest of their kind in the United States, and are tended by a
history of astronomy department. **Web Site:** http://www.adlerplanetarium.org/

Iowa

University of Iowa Hospitals and Clinics Medical Museum. **Contact:**
Adrienne Drapkin, adrienne-drapkin@uiowa.edu **Mailing Address:** 200
Hawkins Drive, Iowa City, IA, 52242, United States. **Tel:** 1.319.356.7106
Fax: 1.319.384.8141 **Grants:** No. **Program Strengths:** 20th and 21st Century,
North America, Medical Sciences, Humanistic Relations of Science; 19th
Century, North America, Medical Sciences, Instruments and Techniques; 20th
and 21st Century, North America, Social Sciences, Education. **Staff:** *Susan C.
Lawrence*. **Description:** Two galleries and six satellite exhibits throughout the
institution display various aspects of the history of medicine. The Museum's
Web site contains most of the exhibitions and exhibits the museum has created
or borrowed. **Web Site:** http://www.uihealth.com/medmuseum/

Massachusetts

Harvard Museum of Natural History. **Contact:** Emily Estock, eestock@
oeb.harvard.edu **Primary E-mail:** hmnh@oeb.harvard.edu **Mailing Address:**
26 Oxford Street, Cambridge, MA, 02138, United States. **Tel:** 1.617.384.8309
Fax: 1.617.496.8782 **Collections/Resources:** The HMNH presents to the
public the vast collections and research of Harvard University's Botanical
Museum, Museum of Comparative Zoology, and Mineralogical and Geological
Museum. Please contact the individual museums for all research-related
questions. **Grants:** No. **Program Strengths:** Natural and Human History,
Education. **Staff:** *Carl A. Francis*; *James Hanken*; *Edward O. Wilson*.
Description: The museum's seventeen galleries feature the Ware Collection
of Blaschka Glass Models of Plants, fondly called the "Glass Flowers." The
zoological galleries feature examples of animals ranging from the earliest
prehistoric creatures to today's mammals, birds, and fish from around the
world. In the mineralogical galleries, sparkling displays of gemstones
and an exceptional selection of meteorites are on display, along with an
extraordinarily comprehensive collection of minerals from around the world.
Web Site: http://www.hmnh.harvard.edu/

Harvard University Collection of Historical Scientific Instruments
(Department of the History of Science). **Contact:** Sara Schechner, schechn@

fas.harvard.edu **Primary E-mail:** chsi@fas.harvard.edu **Mailing Address:**
Science Center B-6, 1 Oxford Street, Cambridge, MA, 02138, United
States. **Tel:** 1.617.495.2779 **Fax:** 1.617.496.5932 **Collections/Resources:**
Close to 20,000 scientific instruments dating from AD 1400 to the present.
Many used for teaching and research at Harvard in the fields of astronomy,
navigation, surveying, horology, physics, chemistry, biology, geology,
meteorology, mathematics, psychology, and communications. A research
library of 6,500 titles including rare books, pamphlets, trade manuals, and
publications related to the historical apparatus. Archival materials including
the William Bond and Son papers, photographs, films, and drawings. Other
supporting documents dating back to the 17th century are held by the Harvard
University Archives. **Grants:** No. **Program Strengths:** North America,
Physical Sciences, Instruments and Techniques; North America, Cognitive
Sciences, Instruments and Techniques; North America, Astronomical
Sciences, Instruments and Techniques. **Degrees Conferred:** Bachelor's
Level, Master's Level, Doctoral Level. **Description:** We are a research and
teaching collection affiliated with the Department of the History of Science at
Harvard University, which offers academic programs. The Harvard University
Archives hold other supporting documents dating back to the 17th century.
Web Site: http://www.fas.harvard.edu/~hsdept/chsi/

Minnesota

The Bakken Library and Museum. Contact: Elizabeth Ihrig, ihrig@
thebakken.org **Mailing Address:** 3537 Zenith Avenue South, Minneapolis, MN,
55416-4623, United States. **Tel:** 1.612.926.3878 ext. 227 **Fax:** 1.612.927.7265
Collections/Resources: Books, journals, ephemera, trade catalogues,
manuscripts, and artifacts documenting the historical role of electricity and
magnetism in the life sciences and medicine; Mesmerism and animal magnetism.
Grants: Yes. **Program Strengths:** Medical Sciences, Instruments and
Techniques; Physical Sciences; Technology. **Description:** The library collection
includes approximately 11,000 items encompassing the history of electricity,
electrophysiology, electrotherapeutics and their accompanying instrumentation.
Primary sources date from the 13th century with emphasis on the 18th, 19th and
early 20th centuries. The artifact collection includes about 2500 instruments,
appliances, machines, and accessories. **Web Site:** http://www.thebakken.org

Pennsylvania

The Franklin Institute Science Museum. Mailing Address: 222 North
20th Street, Philadelphia, PA, 19103, United States. **Tel:** 1.215.448.1200
Collections/Resources: Wright Brothers Aeronautical Engineering Collection
(http://www.fi.edu/wright). **Grants:** No. **Program Strengths:** 19th Century,
North America, Technology, Instruments and Techniques; 20th and 21st Century,
North America, Technology, Instruments and Techniques; 20th and 21st Century,
North America, Physical Sciences, Institutions. **Web Site:** http://www.fi.edu

The Mütter Museum, College of Physicians of Philadelphia. **Contact:**
Gretchen Worden, gworden@collphyphil.org **Mailing Address:** 19 South
22nd Street, Philadelphia, PA, 19103, United States. **Tel:** 1.215.563.3737
Fax: 1.215.561.6477 **Staff:** *Margaret Lyman*; *Gretchen Worden*. **Description:**
The Museum's collections include over 20,000 objects, including fluid-preserved
anatomical and pathological specimens, medical instruments, anatomical and
pathological models, items of memorabilia of famous scientists and physicians,
and medical illustrations. **Web Site:** http://www.collphyphil.org/muttpg1.shtml

South Carolina

McKissick Museum, University of South Carolina. **Contact:** Lynn Robertson,
lynn.robertson@sc.edu **Mailing Address:** University of South Carolina,
Columbia, SC, 29208, United States. **Tel:** 1.803.777.7251 **Fax:** 1.803.777.2829
Collections/Resources: L. L. Smith Geology Collection; Dominick Moth
Collection. **Grants:** No. **Program Strengths:** North America, Natural and
Human History, Humanistic Relations of Science. **Description:** McKissick
is a general university museum that specializes in the study of the culture
and environment of the Southeastern United States. Material culture and
folklife studies are combined with natural history to look at how people
construct identity and establish traditions. Studies of baskets, ceramics,
and other artifacts reveal important information on natural materials.
Web Site: http://www.cla.sc.edu/mcks/

Yugoslavia

Nikola Tesla Museum. **Contact:** Maria Sesic, ntmuseum@eunet.yu **Mailing
Address:** Krunska 51, 11000, Beograd, Yugoslavia. **Tel:** 381.11.433.886
Fax: 381.11.436.408 **Collections/Resources:** Extensive collection of
materials on Nikola Tesla. Electrical instruments. Late 19th century alternating
current devices. Mechanical inventions. Electrotherapeutical instruments.
Press clippings from the 19th and 20th centuries relating to all aspects
of science (American press only), 100,000 items. **Grants:** No. **Program
Strengths:** 19th Century, North America, Physical Sciences, Instruments
and Techniques. **Description:** The Tesla museum holds Tesla's personal
archive, library, instruments, and worldly belongings, including his ashes.
Web Site: http://www.tesla-museum.org

Societies and Organizations

Preliminary research identified 255 societies and organizations potentially involved with the history of science. Several declined to be included in the *Guide*, saying they were too small to be of international interest. Of the remainder, 106 submitted data. Entries are organized alphabetically by organization name. Not all societies responded with complete data. All entries contain some portion of the following information:

Name of organization (and English equivalent, if applicable)
Acronym, if applicable
Openness of membership and whether a fee is charged
Number of members
Frequency and date of primary meeting
Contact person's name and e-mail address
Organization's primary e-mail, if different from contact's e-mail
Mailing address
Telephone number
Facsimile number
Year founded
Organization's historical focus, entered using standardized keywords
Purpose of organization
Titles of publications sponsored or produced by the organization
Does the organization offer grants/fellowships or award prizes (yes/no)
List of paid staff members involved with the history of science, including research interests. Names appear as they were entered by respondents. A broad research interest appears only after the names of HSS members, if they entered this information when they joined the Society. (As computer programming challenges are overcome, research interests for non-members should appear beginning with the 2003 online edition of the *Guide*.)
Brief description of the organization and its special features
URL

307

Societies and Organizations

Agricultural History Society. Membership is open. A membership or subscription fee is charged. 1200 members. Meets annually with the primary meeting in March-May. **Contact:** C. Fred Williams, **E-mail:** cfwilliams@ualr.edu **Address:** Department of History - UALR, 2801 S. University Ave., Little Rock, AR, 72204-1099, United States. **Tel:** 1.501.569.8782 **Fax:** 1.501.569.3059 **Founded:** 1919 **Purpose:** Stimulate interest in, promote the study of, and facilitate research and publications in the history of agriculture. **Publications:** *Agricultural History.* **Offers grants/fellowships:** No. **Awards prizes:** Yes. **Description:** The Agricultural History Society is dedicated to preserving the nation's and world's rural heritage. The Society has an international mission and is interdisciplinary, drawing its research from agricultural economics, cultural anthropology, historical geography, rural sociology, and the general field of history. **Web site:** http://www.iastate.edu/~history_info/aghissoc.html

American Association for the Advancement of Science (AAAS), Section L. Membership is open. A membership or subscription fee is charged. 135,000 members. Meets annually with the primary meeting in December-February. **Contact:** Paul Lawrence Farber, pfarber@orst.edu **Address:** Dept. of History, Oregon State University, Corvallis, OR, 97330, United States. **Tel:** 1.541.7371273 **Fax:** 1.541.7371257 **Founded:** 1848 **Purpose:** To further science, scientific education, and public understanding of science. **Publications:** *Science*; *Science Online*; *Science's Next Wave*; Selected Symposia; Newsletters;. **Offers grants/fellowships:** Yes. **Awards prizes:** Yes. **Description:** From its modest beginning in 1848, the AAAS has worked continually to advance science. Today it has extensive operations that include policy, education, and the publication of one of the major journals of scientific record. It has also become the world's largest federation of scientific and engineering societies with 275 affiliate organizations. Section L is the branch of AAAS dedicated to promoting and preserving the history of science. **Web site:** http://www.aaas.org

American Association for the Advancement of Science (AAAS), Sextion X. Membership is open. A membership or subscription fee is charged. 135,000 members. Meets annually with the primary meeting in December-February. **Contact:** Stephanie J. Bird, sjbird@mit.edu **Address:** Rm. E25-310B, Massachusetts Institute of Technology, Cambridge, MA, 02139, United States. **Tel:** 1.617.253.8024 **Fax:** 1.617.253.6692 **Founded:** 1848 **Purpose:** To further science, scientific education, and public understanding of science. **Publications:** *Science*; *Science Online*; *Science's Next Wave*; Selected Symposia; Newsletters. **Offers grants/fellowships:** Yes. **Awards prizes:** Yes. **Description:** Section X is devoted to studying the societal impacts of science and engineering. **Web site:** http://www.aaas.org

American Association for the History of Medicine (AAHM). Membership is open. A membership or subscription fee is charged. 1300 members. Meets annually with the primary meeting in March-May. **Contact:** Todd L. Savitt, savittT@mail.ecu.edu **Address:** Department of Medical Humanities, East Carolina University School of Medicine, Greenville, NC, 27858-4354, United States. **Tel:** 1.252.816.2797 **Fax:** 1.252.816.2319 **Founded:** 1925 **Focus:** Medical Sciences. **Purpose:** Promote and stimulate study, research, and teaching of the history of medicine and allied health sciences. **Publications:** *Bulletin of the History of Medicine*; *AAHM Newsletter*; *AAHM Membership Directory*. **Offers grants/fellowships:** No. **Awards prizes:** Yes. **Description:** Membership is broad-based, including healthcare practitioners, historians, laboratory scientists, librarians, and scholars from non-history disciplines. Student members are welcome. Non-members may attend and present papers at the annual meeting. **Web site:** http://www.histmed.org

American Association for the History of Nursing (AAHN). Membership is open. A membership or subscription fee is charged. 600 members. Meets annually with the primary meeting in September-November. **Contact:** Janet L. Fickeissen, aahn@aahn.org **Address:** PO Box 175, Lanoka Harbor, NJ, 08734, United States. **Tel:** 1.609.693.7250 **Fax:** 1.609.693.1037 **Founded:** 1979 **Focus:** North America, Natural and Human History. **Purpose:** Fosters the importance of history in understanding the present and guiding the future of nursing. **Publications:** *Nursing History Review*; *Bulletin of the AAHN*. **Offers grants/fellowships:** Yes. **Awards prizes:** Yes. **Description:** AAHN is a professional organization open to everyone interested in the history of nursing. Aims to: stimulate national and international interest and collaboration in the history of nursing; educate nurses and the public regarding the history and heritage of the nursing profession; encourage and support research in the history of nursing and recognize outstanding scholarly achievement in nursing history; encourage the collection, preservation, and use of materials of historical importance to nursing; serve as a resource for information about nursing history; produce and distribute educational materials related to the history and heritage of the nursing profession; promote the inclusion of nursing history in nursing curricula; foster interdisciplinary collaboration in history. **Web site:** http://www.aahn.org

American Association of Physical Anthropologists (AAPA), History Committee. Membership is open. A membership or subscription fee is charged. 1800 members. Meets annually with the primary meeting in March-May. **Contact:** Michael A. Little, Mlittle@bingvmb.cc.binghamton.edu **Address:** Department of Anthropology, State University of New York, Binghamton, NY, 13902-6000, United States. **Founded:** 1930 **Focus:** Social Sciences. **Purpose:** Fostering communication among biological anthropologists, ensuring that our interests are fairly presented before government agencies, and making the public aware of the interesting, valuable research we do. **Publications:** *American Journal of Physical Anthropology*; *Yearbook of Physical Anthropology*; *Physical Anthropology Newsletter*. **Awards prizes:** Yes. **Description:** The AAPA is the world's leading professional organization for physical anthropologists. Formed by 83 charter members in 1930, the AAPA now has an international membership of over 1,800. The Association's annual meetings draw more than a thousand scientists and students from all over the world. **Web site:** http://physanth.org

American Astronautical Society (AAS). Membership is open. A membership or subscription fee is charged. 1,500 members. Meets annually with the primary meeting in September-November. **Contact:** Michael Ciancone **E-mail:** aas@astronautical.org **Address:** 6352 Rolling Mill Place, Ste. 102, Springfield, VA, 22152-2354, United States. **Tel:** 1.703.866.0020 **Fax:** 1.703.866.3526 **Founded:** 1954 **Focus:** 20th and 21st Century, Physical Sciences. **Purpose:** Dedicated to the advancement of space science and exploration and their history. **Publications:** *The Journal of the Astronautical Sciences*; *Space Times* magazine; AAS History Series. **Offers grants/ fellowships:** No. **Awards prizes:** Yes. **Description:** The American Astronautical Society is the premier independent scientific and technical group in the United States exclusively dedicated to promoting astronautics and strengthening the global space program. **Web site:** http://www.astronautical.org

American Astronomical Society, Historical Astronomy Division. Membership is closed. A membership or subscription fee is charged. 278 members. Meets annually with the primary meeting in December-February. **Contact:** Ronald S. Brashear, brashearr@si.edu **Address:** Dibner Library NMAH 1041, Smithsonian Institution Libraries, Washington, DC, 20560-0672, United States. **Tel:** 1.202.357.1568 **Fax:** 1.202.633.9102 **Founded:** 1980 **Focus:** Astronomical Sciences. **Purpose:** To advance interest in topics relating to the historical nature of astronomy. **Publications:** *H.A.D. News*. **Offers grants/fellowships:** No. **Awards prizes:** Yes. **Description:** HAD is a division of the American Astronomical Society, but it does accept affiliate members who are members of another professional organization that has some interest in historical astronomy. HAD's interests include traditional history of astronomy, archeoastronomy, and the application of historical records to modern astrophysical problems. Members of HAD include both historians of astronomy and astronomers with an interest in the history of their profession. **Web site:** http://www.aas.org/had/had.html

American Chemical Society, Division of the History of Chemistry. Membership is open. A membership or subscription fee is charged. 800 members. Meets semiannually. **Contact:** Vera V. Mainz, mainzv@aries.scs. uiuc.edu **Address:** University of Illinois at Urbana-Champaign, 142B RAL, Box 34 Noyes Lab, 600 S. Mathews Ave, Urbana, IL, 61801, United States. **Tel:** 1.217.244.0564 **Fax:** 1.217.244.8068 **Founded:** 1922 **Focus:** Physical Sciences. **Purpose:** A forum for discussion and research in the history of chemistry and allied sciences. **Publications:** *HIST Newsletter*; *Bulletin for the History of Chemistry*. **Offers grants/fellowships:** No. **Awards prizes:** Yes. **Description:** The Division acts as a forum for the history of chemistry and archaeological chemistry. Participated in establishing Center for the History of

Chemistry (now Chemical Heritage Foundation) and ACS National Historic Chemical Landmarks Program. Annual awards: Outstanding Paper Award and Sidney M. Edelstein Award for Outstanding Achievement in the History of Chemistry. **Web site:** http://scs.uiuc.edu/~mainzv/HIST/index.htm

American Fisheries Society, Fisheries History Section. Membership is open. A membership or subscription fee is charged. Meets annually with the primary meeting in June-August. **Contact:** Gus Rassam, main@fisheries.org **Address:** 5410 Grosvenor Lane, Suite 110, Bethesda, MD, 20814, United States. **Tel:** 1.301.897.8616 **Focus:** Biological Sciences, Science and Religion. **Offers grants/fellowships:** No. **Awards prizes:** Yes. **Web site:** http://www.fisheries.org

American Geophysical Union, History of Geophysics Committee. Membership is closed. A membership or subscription fee is charged. 22 members. Meets biennially. **Contact:** Ronald E. Doel, rdoel@orst.edu **Address:** Department of Geosciences, 104 Wilkinson Hall, Oregon State University, Corvallis, OR, 97331, United States. **Tel:** 1.541.737.3469 **Fax:** 1.541.737.1257 **Founded:** 1981 **Focus:** Earth Sciences. **Purpose:** Foster interest in the history of geophysics and unite AGU Sections. **Offers grants/fellowships:** No. **Awards prizes:** No. **Description:** This Committee fosters interest in the history of geophysics within the American Geophysical Union. It also encourages interdisciplinary interaction and educates AGU members about the importance of the problems and issues in the history of geophysics. The Committee includes AGU members as well as historians, archivists and librarians interested in the Earth and space sciences. **Web site:** http://history.agu.org/

American Historical Association (AHA). Membership is open. A membership or subscription fee is charged. 15,000 members. Meets annually with the primary meeting in December-February. **Contact:** Miriam E. Hauss, mhauss@theaha.org **E-mail:** aha@theaha.org **Address:** 400 A Street SE, Washington, DC, 20003, United States. **Tel:** 1.202.544.2422 **Fax:** 1.202.544.8307 **Founded:** 1884 **Focus:** Social Sciences. **Purpose:** Promotes historical studies, collects and preserves historical documents and artifacts, and disseminates historical research. **Publications:** *American Historical Review*; *Perspectives*; *Directory of History Departments and Organizations in the United States and Canada*; several pamphlet series. **Offers grants/fellowships:** Yes. **Awards prizes:** Yes. **Description:** The American Historical Association (AHA) is a nonprofit organization founded in 1884 and incorporated by Congress in 1889. It promotes historical studies, collects and preserves historical documents and artifacts, and disseminates historical research. As the largest U.S. historical society, the AHA is an umbrella organization for historians of every period and geographical area. **Web site:** http://www.theaha.org

American Institute for the History of Pharmacy. Membership is open. A membership or subscription fee is charged. 900 members. Meets annually with the primary meeting in March-May. **Contact:** Gregory James Higby,

higby@aihp.org **E-mail:** staff@aihp.org **Address:** 777 Highland Avenue, University of Wisconsin, Madison, WI, 53705-2222, United States. **Tel:** 1.608.262.5378 **Founded:** 1941 **Focus:** Medical Sciences. **Purpose:** The AIHP is devoted to exploring the place of pharmacy in the history of civilization. **Publications:** *Pharmacy in History*; *Apothecary's Cabinet*; *AIHP Notes*. **Offers grants/fellowships:** Yes. **Awards prizes:** Yes. **Paid Staff:** *Gregory Higby*; *Elaine C. Stroud.* **Description:** In addition to our quarterly journal, *Pharmacy in History*, and our popular newsletter, *Apothecary's Cabinet*, the Institute publishes an annual pharmaco-historical calendar and small books. (Our publication catalog is available on-line). In conjunction with the University of Wisconsin, Madison, we maintain the largest manuscript and reference collection in North America on the history of pharmacy. Grants are available for visiting scholars. **Web site:** http://www.aihp.org

American Lunar Society. Membership is open. A membership or subscription fee is charged. 80 members. Meets irregularly with the primary meeting in June-August. **Contact:** William M. Dembowski, dembow@twd.net **Address:** Elton Moonshine Observatory, 219 Old Bedford Pike, Windber, PA, 15963-8905, United States. **Tel:** 1.814.262.0450 **Founded:** 1982 **Focus:** Astronomical Sciences, Education. **Purpose:** Dedicated to the study of Earth's only natural satellite through direct observation and research. **Publications:** *Selenology*. **Offers grants/fellowships:** No. **Awards prizes:** Yes. **Description:** Areas of interest include telescopic observation of the Moon, lunar exploration, and historical subjects such as lunar mythology and the history of lunar astronomy. In addition, members have the opportunity to participate in the Society's many volunteer research projects. **Web site:** http://otterdad.dynip.com/als/

American Meteorological Society, Committee for the History of Atmospheric Sciences. Membership is closed. A membership or subscription fee is charged. 12,000 members. Meets annually with the primary meeting in December-February. **Contact:** Corinne Kazarosian, ckazarosian@ametsoc.org **E-mail:** jannese@ametsoc.org **Address:** 45 Beacon Street, Boston, MA, 02108-3693, United States. **Tel:** 1.617.227.2426 **Fax:** 1.617.742.8718 **Founded:** 1919 **Focus:** Earth Sciences. **Purpose:** To advance the atmospheric sciences and their applications. **Publications:** Historical Monograph Series; *Bulletin of the American Meteorological Society*; *Meteorological and Geoastrophysical* Abstracts; nine specialized scientific journals. **Offers grants/fellowships:** Yes. **Awards prizes:** Yes. **Description:** The AMS Committee for the History of the Atmospheric Sciences administers an annual graduate fellowship in the history of science, sponsors historical sessions and projects, conducts an oral history program, and supports preservation of and access to archives in the atmospheric and related sciences. Statistics provided here are for the American Meteorological Society as a whole. **Web site:** http://www.ametsoc.org/AMS/

American Philosophical Society (APS). Membership is closed. Free. 853 members. Meets semiannually. **Contact:** Martin L. Levitt, mlevitt@amphilsoc.org **Address:** 105 S. 5th Street, Philadelphia, PA, 19106-3387, United States. **Tel:** 1.215.440.3400 **Fax:** 1.215.440.3423 **Founded:** 1743 **Focus:** 17th Century, North America, Physical Sciences. **Purpose:** Promoting useful knowledge by supporting scholarship through meetings, research grants, publications, and its research library. **Offers grants/fellowships:** Yes. **Awards prizes:** Yes. **Description:** The American Philosophical Society Library accepts applications for short-term residential fellowships for conducting research in its collections. The Society's Library, located near Independence Hall in Philadelphia, is a leading international center for research in the history of American science and technology and its European roots, as well as early American history and culture. The Library houses over 7 million manuscripts, 300,000 volumes and bound periodicals, and thousands of maps and prints. Outstanding historical collections and subject areas include the papers of Benjamin Franklin; the American Revolution; 18th and 19th Century natural history; western scientific expeditions and travel including the journals of Lewis and Clark; polar exploration; the papers of Charles Willson Peale, including family and descendants; American Indian languages; anthropology including the papers of Franz Boas; the papers of Charles Darwin and his forerunners, colleagues, critics, and successors; history of genetics, eugenics, and evolution; history of biochemistry, physiology, and biophysics; 20th Century medical research; and history of physics. The Library does not hold materials on philosophy in the modern sense. **Web site:** http://www.amphilsoc.org

American Physical Society, Forum on the History of Physics. Membership is open. A membership or subscription fee is charged. 3018 members. Meets annually with the primary meeting in March-May. **Contact:** Kenneth W. Ford, kwford@verizon.net **Address:** Forum on History of Physics, APS, One Physics Ellipse, College Park, MD, 20740, United States. **Tel:** 1.215.844.8054 **Fax:** 1.215.844.1399 **Founded:** 1980 **Focus:** Physical Sciences. **Purpose:** Encourage scholarly research in history of physics and diffusion of knowledge of this history. **Publications:** *History of Physics Newsletter*. **Offers grants/fellowships:** No. **Awards prizes:** Yes. **Description:** The Forum on History of Physics (FHP) is a sub-unit of the American Physical Society (APS). Membership is free to members of APS. In addition to publishing the *History of Physics Newsletter* twice a year, FHP sponsors history of physics symposia at the national APS meetings in March and April. Through the *Newsletter* and the programs at meetings, FHP brings physics historians and physicists together to address important issues in history of physics, including firsthand reports of important developments. **Web site:** http://www.aps.org/FHP

American Physiological Society, History of Physiology Group. Membership is closed. A membership or subscription fee is charged. 11,000 members. Meets annually with the primary meeting in March-May. **Contact:** Martin Frank, mfrank@the-aps.org **E-mail:** info@the-aps.org **Address:** American

Physiological Society, 9650 Rockville Pike, Bethesda, MD, 20814-3991, United States. **Tel:** 1.301.634.7164 **Fax:** 1.301.634.7242 **Founded:** 1887 **Focus:** Medical Sciences. **Purpose:** To promote the increase of physiological knowledge and its utilization. **Publications:** *American Journal of Physiology*; *The Physiologists*. **Offers grants/fellowships:** Yes. **Awards prizes:** Yes. **Description:** The American Physiological Society is devoted to fostering scientific research, education, and the dissemination of scientific information. The Society strives to play a role in the progress of science, and the advancement of knowledge by providing a spectrum of physiological information. The Society's primary focus is to provide current, usable information to the physiological community. The Society is focused on integrating the life sciences from molecule to organism. **Web site:** http://www.the-aps.org

American Psychological Association, Division of the History of Psychology (Division 26). Membership is open. A membership or subscription fee is charged. Approx. 700 members. Meets annually with the primary meeting in June-August. **Contact:** Laura Koppes, laura.koppes@eku.edu **E-mail:** christo@yorku.ca **Address:** Department of Psychology, Eastern Kentucky University, 521 Lancaster Ave, Richmond, KY, 40475, United States. **Tel:** 1.859.622.1564. **Founded:** 1966 **Focus:** Cognitive Sciences. **Purpose:** To extend the awareness and appreciation of the history of psychology. **Publications:** *History of Psychology*. **Offers grants/fellowships:** No. **Awards prizes:** Yes. **Description:** Through teaching and through original research and scholarship, members of Division 26 seek to extend the awareness and appreciation of the history of psychology as an aid to the understanding of contemporary psychology, of the relation of the discipline to other scientific fields, and of its role in society. **Web site:** http://www.psych.yorku.ca/orgs/apa26/

American Society for Eighteenth-Century Studies. Membership is open. A membership or subscription fee is charged. 2700 members. Meets annually with the primary meeting in April. **Contact:** David Brewer, dbrewer@umn.edu **E-mail:** asecs@wfu.edu **Address:** PO Box 7867, Wake Forest University, Winston-Salem, NC, 27109, United States. **Founded:** 1969 **Purpose:** Advance study and research in the history of the eighteenth century. **Publications:** *Eighteenth-Century Studies*; *Studies in Eighteenth-Century Culture*; *News Circular* (Newsletter);. **Offers grants/fellowships:** Yes. **Awards prizes:** Yes. **Description:** Established in 1969, the American Society for Eighteenth-Century Studies is a group dedicated to the promotion of scholarship in all aspects of the period from the later seventeenth through the early nineteenth century. ASECS is a pioneer in interdisciplinary investigation, and it therefore welcomes as members those working in all areas of scholarly inquiry pertinent to eighteenth-century studies.

American Society for Environmental History (ASEH). Membership is open. A membership or subscription fee is charged. 1,400 members. Meets annually with the primary meeting in March-May. **Contact:** Lisa Mighetto, mighetto@hrassoc.com **Address:** 119 Pine Street, Suite 207, Seattle, WA, 98101, United States. **Tel:** 1.206.343.0226 **Fax:** 1.206.343.0249 **Founded:** 1977 **Focus:** 20th and 21st Century, North America, Natural and Human History, Humanistic Relations of Science. **Purpose:** Understanding the human experience of the environment from the perspectives of history, liberal arts, and sciences. **Publications:** *Environmental History*; *ASEH News.* **Offers grants/fellowships:** Yes. **Awards prizes:** Yes. **Description:** The Society encourages cross-disciplinary dialogue on every aspect of the present and past relationship of humankind to the natural environment. Its membership comprises university and college faculty and students, environmental managers, scientists, lawyers, museum professionals, secondary school teachers, writers, and others. Members are interested in contributing to the development and dissemination of knowledge about humankind's relationship to the natural world and society's efforts to deal with the environmental consequences of urbanization, industrial development, and other aspects of social and economic change over time. **Web site:** http://www2.h-net.msu.edu/~environ/ index.html

American Studies Association. Membership is open. A membership or subscription fee is charged. 5,500-6,000 members. Meets annually with the primary meeting in September-November. **Contact:** John F. Stephens, asastaff@erols.com **Address:** 1120 19th Street NW, Washington, DC, 20036, United States. **Tel:** 1.202.467.4783 **Fax:** 1.202.467.4786 **Founded:** 1951 **Focus:** North America. **Purpose:** To support the study of American cultures in their diverse and changing meanings, to promote interdisciplinary research, and to encourage innovative methodological experimentation. **Publications:** *American Quarterly*; *American Studies Association Newsletter*; *Guide to American/American Ethnic Studies ResourcesDirectory of Graduate Programs in American/American Ethnic Studies*; *Guide for Reviewers of American Studies Programs*; *National Resource Guide for American Studies in the Secondary Schools.* **Offers grants/fellowships:** No. **Awards prizes:** Yes. **Description:** Chartered in 1951, the American Studies Association now has more than 5,000 members. They come from many fields: history, literature, religion, art, philosophy, music, science, folklore, ethnic studies, anthropology, material culture, museum studies, sociology, government, communications, education, library science, gender studies, popular culture, and others. They include persons concerned with American culture: teachers and other professionals whose interests extend beyond their specialty, faculty and students associated with American Studies programs in colleges and secondary schools, museum directors and librarians interested in all segments of American life, public officials and administrators concerned with the broadest aspects of education. They approach American culture from many directions but have in common the desire to view America as a whole rather than from the perspective of a single discipline. **Web site:** http://www.theasa.net

Anesthesia History Association. Membership is open. A membership or subscription fee is charged. 650 members. Meets semiannually with the primary meeting in September-November. **Contact:** Doris K. Cope, bloombergdj@ anes.upmc.edu **Address:** 200 Delafield Avenue, Suite 2070, Pittsburgh, PA, 15215, United States. **Tel:** 1.412.784.5343 **Fax:** 1.412.784.5350 **Founded:** 1983 **Focus:** 19th Century, North America, Medical Sciences, Instruments and Techniques. **Purpose:** To encourage the study and preservation of the history of anesthesiology, critical care and pain medicine. **Publications:** *Bulletin of Anesthesia History.* **Offers grants/fellowships:** No. **Awards prizes:** Yes. **Description:** We are interested in biographies of famous people, medical and scientific inventions, and the relationship between medicine and the arts. We meet annually at the ASA Annual Meeting and collaborate with other international history of anesthesiology societies for other meetings.

Associação de Filosofia e História da Ciência do Cone Sul (AFHIC). Membership is open. A membership or subscription fee is charged. 170 members. Meets biennially with the primary meeting in March-May. **Contact:** Roberto de Andrade Martins, Rmartins@ifi.unicamp.br **Address:** PO Box 6059, Campinas, SP, 13084-971, Brazil. **Tel:** 55.19.3788.5516 **Fax:** 55.19.3788.5512 **Founded:** 2000 **Purpose:** History and philosophy of science. Members from Argentina, Brazil, Chile and Uruguay. **Offers grants/fellowships:** Yes. **Awards prizes:** Yes. **Description:** The primary aim of the South Cone Association for the Philosophy and History of Science is to establish a strong mechanism for communication and collaboration among researchers of those two fields, especially those who live in the countries of the so-called South Cone: Argentina, Brazil, Chile, Uruguay. **Web site:** http://www.afhic.org/

Association Villard de Honnecourt for the Interdisciplinary Study of Medieval Technology (AVISTA). Contact: Lynn Courtenay, ltcourte@facstaff.wisc.edu **Address:** 3100 Lake Mendota Drive, # 504, Madison, WI, 53705. **Tel:** 1.608.238.3236 **Founded:** 1984 **Purpose:** To study and make known work in art, architecture, and technology of the Middle Ages in honor of the medieval craftsmen who worked in these fields. **Publications:** AVISTA. **Description:** AVISTA is a scholarly organization organized through the urging of the late Jean Gimpel (author of *The Medieval Machine, The Cathedral Builders*, and other works on medieval art and technology) and with the support of Lynn White, Jr. (author of *Medieval Technology and Social Change*). The society serves as an organization to promote any and all aspects of medieval topics which relate to the practical sciences or technology. Our "home" is largely the International Congress on Medieval Studies held at Western Michigan University in Kalamazoo, where each May we sponsor up to a half dozen sessions on various themes. **Web site:** http://www.avista.org/

Australasian Association for the History, Philosophy and Social Studies of Science (AAHPSSS). Membership is open. A membership or subscription fee is charged. 150 members. Meets annually with the primary meeting

in June-August. **E-mail:** aahpsss@scifac.usyd.edu.au **Founded:** 1967
Purpose: To foster scholarship in Australasia among historians, philosophers
and social scientists studying science, technology and medicine. **Publications:**
Metascience. **Offers grants/fellowships:** No. **Awards prizes:** Yes. **Description:**
AAHPSSS is the professional association of historians, philosophers and social
scientists in Australasia studying science, technology and medicine, with all
these terms being interpreted broadly. The organization's main activity is an
annual conference, which aims to foster HPS/STS scholarship and teaching,
mentoring of graduate students, and professional contacts among its members.
Web site: http://www.usyd.edu.au/hps/aahpsss/

**Australian Academy of Science, National Committee for History and
Philosophy of Science.** Membership is closed. Free. 10 members. Meets
annually with the primary meeting in June-August. **Contact:** Roderick Home,
home@unimelb.edu.au **Address:** Dept. of HPS, University of Melbourne,
Parkville, Victoria, 3010, Australia. **Tel:** 61.3.8344.6556 **Fax:** 61.3.8344.7959
Founded: 1969 **Focus:** Australia and Oceania. **Purpose:** Promoting
research and teaching in HPS in Australia; maintaining Australia's scientific
links in this field. **Offers grants/fellowships:** No. **Awards prizes:** No.
Web site: http://www.science.org.au/internat/natcomm/

Australian Science History Club. Membership is open. A membership or
subscription fee is charged. 50 members. Meets monthly. **Contact:** Julian
Holland, julian@macleay.usyd.edu.au **Address:** C/- Macleay Museum,
University of Sydney, Sydney, NSW, 2006, Australia. **Tel:** 61.2.9351.3739
Fax: 61.2.9351.5646 **Founded:** 1987 **Focus:** Australia and Oceania.
Purpose: Present talks and seminars on historical aspects of Australian science,
technology, engineering, and medicine. **Offers grants/fellowships:** No.
Awards prizes: No. **Description:** The Australian Science History Club is an
informal organization which presents evening talks, one-day seminars and
occasional excursions on themes related to the history of Australian science,
technology, engineering, medicine and related disciplines. Its evening meetings
are held (more or less) monthly between March and November. Seminars are
held once or twice a year. (Previously known as the Colonial Science Club,
1987-2000.) **Web site:** http://www.usyd.edu.au/su/macleay/ASHclub.htm

The Bartlett Society. Membership is open. Free. 200 members. Meets annually
with the primary meeting in September-November. **Contact:** Vernon
Kisling, vkisling@mail.uflib.ufl.edu **Address:** PO Box 1511, High Springs,
FL, 32655-1511, United States. **Tel:** 1.352.392.2838 **Fax:** 1.352.392.4787
Founded: 1984 **Focus:** Natural and Human History, Social Relations of
Science. **Purpose:** For the study of zoo and wild animal husbandry history.
Publications: Newsletter (North American); Newsletter (British); *The Bartlett
Society Journal* (British). **Offers grants/fellowships:** No. **Awards prizes:** No.
Paid Staff: *Vernon N. Kisling* (Natural and Human History). **Description:** An
international society of individuals with interests in the zoological and social

aspects of zoo history. Founded in London, and named in honor of Abraham Dee Bartlett, Superintendent of the London Zoo (1859-1897), there is a North American branch, a British branch, and a related organization in India called the Society for Promotion of History of Zoos and Natural History in India and Asia. **Web site:** http://www.milwaukeezoo.org/rcenter/history/bartlett.html

Botanical Society of America, Historical Section. Membership is open. A membership or subscription fee is charged. 2800 members. Meets annually with the primary meeting in July/August. **Contact:** Laurence J. Dorr, dorrl@nmnh.si.edu **Address:** Department of Botany, MRC-166, National Museum of Natural History, Smithsonian Institution, Washington, DC, 20560-0166, United States. **Tel:** 1.202.633.9106.or.202.357.2534 **Fax:** 1.202.786.2563 **Founded:** 1906 **Purpose:** Promote the study of plant science and its history. **Publications:** *American Journal of Botany*; *Careers in Botany*; *Plant Science Bulletin*. **Offers grants/fellowships:** No. **Awards prizes:** Yes. **Description:** The Botanical Society of America aims to promote research, education, and communication in the field of botany. The History Section, a division of the Society, specializes in the exploration of the history of botanical science. **Web site:** http://www.botany.org/bsa

British Agricultural History Society. Membership is open. A membership or subscription fee is charged. 950 members. Meets semiannually with the primary meeting in March-May. **Contact:** Mark Overton, bahs@exeter.ac.uk **Address:** Department of History, University of Exeter, Amory Building, Rennes Drive, Exeter, EX4 4RJ, United Kingdom. **Tel:** 44.01.392.263.284 **Fax:** 44.01.392.263.305 **Founded:** 1953 **Purpose:** To promote the study of agricultural history. **Publications:** *Agricultural History Review*. **Offers grants/ fellowships:** No. **Awards prizes:** No. **Web site:** http://www.bahs.org.uk

British Society for the History of Mathematics. Membership is open. A membership or subscription fee is charged. 390 members. Meets irregularly. **Contact:** Mary Croarken, mgc@dcs.warwick.ac.uk **Address:** 20 Dunvegan Close, Exeter, Devon, EX4 4AF, United Kingdom. **Tel:** 44.1392.420219 **Fax:** 44.1392.662921 **Founded:** 1971 **Focus:** Mathematics. **Purpose:** To promote research into the history of mathematics and its use in educational settings. **Publications:** *Newsletter of the British Society for the History of Mathematics*. **Offers grants/fellowships:** No. **Awards prizes:** Yes. **Description:** The BSHM aims to promote research into the history of mathematics at all levels, amateur and professional, and to further the use of the history of mathematics in education at all levels. Membership includes people from many backgrounds. About one third of the membership is from outside the UK. **Web site:** http://www.dcs.warwick.ac.uk/bshm

British Society for the History of Science (BSHS). Membership is open. A membership or subscription fee is charged. 750 members. Meets irregularly. **Contact:** Geoff Bennett, bshs@hidex.demon.co.uk **Address:** 31 High Street,

Stanford in the Vale, Faringdon, Oxfordshire, SN7 8LH, United Kingdom.
Tel: 44.13.6771.8963 **Fax:** 44.13.6771.8963 **Founded:** 1947 **Purpose:** Promote
and further the study of history and philosophy of science. **Publications:** *British
Journal for the History of Science*; Newsletter; Education Forum; Series
of Monographs. **Offers grants/fellowships:** No. **Awards prizes:** Yes.
Description: The Society welcomes members with interests in all branches of
the history of science. It has a broad membership - amateurs and professionals,
scientists and historians, UK and overseas; all having their interests served
through regular meetings and publications. **Web site:** http://www.bshs.org.uk

British Society for the Philosophy of Science. Membership is open. A
membership or subscription fee is charged. 350 members. Meets annually
with the primary meeting in June-August. **Contact:** Robin Findlay Hendry,
r.f.hendry@durham.ac.uk **Address:** Department of Philosophy, University of
Durham, 50 Old Elvet, Durham, DH1 3HN, United Kingdom. **Founded:** 1951
Focus: Philosophy or Philosophy of Science. **Purpose:** To study the logic, the
methods, and the philosophy of science. **Publications:** *British Journal for the
Philosophy of Science.* **Offers grants/fellowships:** Yes. **Awards prizes:** Yes.
Description: The Society holds regular meetings in London, and an annual
conference with invited and contributed papers on topics in the philosophy of
science. It supports research and teaching in the philosophy of science through
grants to UK conferences, travel grants for postgraduate students, and a textbook
prize. **Web site:** http://www.dur.ac.uk/philsophy.department/bsps/BSPSHome.html

The C.F. Reynolds Medical History Society. Membership is open.
A membership or subscription fee is charged. 226 members. Meets
bi-monthly. **Contact:** Jonathon Erlen, erlen@pitt.edu **Address:** University
of Pittsburgh, 200 Scaife Hall, Pittsburgh, PA, 15261, United States.
Tel: 1.412.6488927 **Fax:** 1.412.6489020 **Founded:** 1972 **Focus:** Medical
Sciences. **Purpose:** Promote scholarship in the history of medicine.
Offers grants/fellowships: No. **Awards prizes:** Yes. **Description:** The Society
sponsors at least 5 evening lectures on a wide variety of history of medicine
topics. All but one of these lectures is presented by a speaker from outside of
the Pittsburgh region. The Society serves the history of medicine community
in the greater Western Pennsylvania region. Society members constitute the
teaching faculty for history of medicine courses at the University of Pittsburgh,
Carneige Mellon University, Duquesne University, and Seaton Hill College.
Web site: http://www.hsls.pitt.edu/giude/histmed/cfrey.html

**Canadian Science and Technology Historical Association (Assocation pour
l'Histoire de la Science et de la Technologie au Canada).** Membership
is open. A membership or subscription fee is charged. 60 members. Meets
biennially with the primary meeting in September-November. **Contact:** Alain
Canuel, alain.canuel@nce.gc.ca **Address:** 30, rue de la Comete, Hull, Quebec,
J9A 2Y5, Canada. **Founded:** 1980 **Focus:** North America. **Purpose:** To
study the history of Canadian science and technology. **Publications:** *Scientia*

Canadensis; *Dialogues*. **Offers grants/fellowships:** No. **Awards prizes:** No.
Description: CSTHA aims to sustain a research community in the history
of Canadian science and technology; encourage respect for this heritage;
diffuse new knowledge through scholarly conferences and its refereed journal
Scientia Canadensis; and inform members through its newsletter, *Dialogues*.
Web site: http://www.er.uqam.ca/nobel/r20430/ahstc-cstha

Canadian Society for the History and Philosophy of Mathematics (CSHPM).
Membership is open. A membership or subscription fee is charged.
220 members. Meets annually with the primary meeting in March-May.
Contact: Glen Van Brummelen, gvanbrum@bennington.edu **Address:**
Bennington College, Bennington, VT, 05201, United States. **Tel:** 1.802.440.4467
Fax: 1.802.440.4461 **Founded:** 1974 **Purpose:** Encourage and promote the
study of history and philosophy of mathematics. **Publications:** Proceedings of
meetings; *CSHPM Bulletin*; *Historia Mathematica*; *Philosophia Mathematica*.
Offers grants/fellowships: No. **Awards prizes:** No. **Description:** The CSHPM
promotes research and teaching in the history and philosophy of mathematics.
Web site: http://www.cshpm.org

**Canadian Society for the History and Philosophy of Science (Société
Canadienne d'Histoire et Philosophie des Sciences).** Membership is open.
A membership or subscription fee is charged. 150 members. Meets annually
with the primary meeting in March-May. **Contact:** Bernie Lightman,
lightman@yorku.ca **Address:** Rm 309 Bethune College, York University,
4700 Keele Street, Toronto, Ontario, M3J IP3, Canada. **Tel:** 1.416.736.2100.x
.22028 **Founded:** 1959 **Purpose:** To promote throughout Canada, discussion,
research, teaching and publication in history and philosophy of science.
Publications: *Communiqué* (newsletter). **Offers grants/fellowships:** No.
Awards prizes: Yes. **Web site:** http://www.ukings.ns.ca/cshps

**Canadian Society for the History of Medicine (La Société Canadienne
d'Histoire de la Médecine).** Membership is open. A membership or
subscription fee is charged. 278 members. Meets annually with the primary
meeting in March-May. **Contact:** David Wright, dwright@mcmaster.ca
Address: Secretary/Treasurer, Canadian Society for the History of Medicine,
Health Sciences Centre-Room 3N10, McMaster University, Hamilton, Ontario,
L8N 3Z5, Canada. **Tel:** 1.905.525.9140 ext. 22752 **Fax:** 1.905.522.9509
Founded: 1950 **Focus:** Natural and Human History. **Purpose:** To promote
the study of the history of Canadian health and healthcare professions
and institutions. **Publications:** *Canadian Bulletin of Medical History*.
Offers grants/fellowships: No. **Awards prizes:** No. **Description:** The
organization and its publication are bilingual (English/French).
Web site: http://meds.queensu.ca/medicine/histm/cshmweb/cshmhome.html

**CHEIRON: International Society for the History of the Behavioral and Social
Sciences.** Membership is open. A membership or subscription fee is charged.

Meets annually with the primary meeting in June-August. **Contact:** Benjamin Harris, bh5@cisunix.unh.edu **Founded:** 1968 **Focus:** Social Sciences. **Purpose:** To promote international cooperation and multidisciplinary studies in the history of the social and behavioral sciences. **Publications:** Newsletter; Affiliated publication: *Journal for the History of the Behavioral Sciences.* **Offers grants/fellowships:** No. **Awards prizes:** Yes. **Description:** Cheiron, named after the wise centaur of Greek myth, welcomes members for whom history is a side-interest, as well as scholars with a primary commitment to historical study. The Society pursues its goals through a biannual newsletter and, especially, its annual meetings. Meetings are held in June of each year, and typically include the presentation of papers, symposia, posters, and works-in-progress. Most important of all is the ample opportunity provided for friendly interaction with scholars from different disciplines. **Web site:** http://www.psych.yorku.ca/orgs/cheiron/

Chinese Society for the History of Science and Technology (CSHST). Membership is closed. A membership or subscription fee is charged. 1300 members. Meets quadrennially with the primary meeting in June-August. **Contact:** Rongyu Su, su@ihns.ac.cn **E-mail:** webmaster@ihns.ac.cn **Address:** Chinese Society for the History of Science and Technology, 137 Chao Nei Street, Beijing, 100010, China. **Tel:** 86.10.6401.7635 **Fax:** 86.10.6401.7637 **Founded:** 1980 **Publications:** *Studies in the History of Natural Sciences*; China Historical Materials of Science and Technology. **Offers grants/fellowships:** No. **Awards prizes:** Yes. **Description:** A nationwide non-governmental organization for research in the field of the history of science. Under the society, there are 11 scientific commissions and 2 inter-society commissions. Since its founding 22 years ago, CSHST has strengthened its contacts with the international community of the history of science. Since the 16th ICHS held in Bucharest in 1981, the Society has sent a delegation to each Congress. Working with the Institute for the History of Natural Science (IHNS), the CSHST will organize the 22nd ICHS, to be held in Beijing in 2005. **Web site:** http://www.ihns.ac.cn

Club d'Histoire de la Chimie. Membership is open. A membership or subscription fee is charged. 600 members. Meets quarterly. **Contact:** Marika Blondel-Megrelis, marika.blondel-megrelis@libertysurf.fr **Address:** 250, rue St. Jacques, Paris, 75005, France. **Tel:** 33.05.61.22.4876 **Founded:** 1990 **Purpose:** To know and diffuse the history of chemistry. **Publications:** *Actualite Chemisque.* **Offers grants/fellowships:** No. **Awards prizes:** No. **Web site:** http://www.sfc.fr/GRHIST/JeuCadres.html

Columbia History of Science Group. Membership is open. A membership or subscription fee is charged. 100 members. Meets annually with the primary meeting in March-May. **Contact:** Keith R. Benson, krbenson@centurytel.net **Address:** 13423 Burma Rd. SW, Vashon Island, WA, 98070, United States. **Tel:** 1.206.567.5839 **Founded:** 1983 **Purpose:** Support history and philosophy of science in the greater Columbia River basin. **Publications:** *CHSG Newsletter.*

Offers grants/fellowships: No. **Awards prizes:** Yes. **Description:** The Columbia History of Science Group is dedicated to supporting the history and philosophy of science through an informal and enjoyable annual meeting. The group emphasizes its informal character in large part to encourage graduate students, independent scholars, and non-specialists to pursue their interests in science studies. Furthermore, it also encourages the inclusion of families in its activities, which are traditionally held at the University of Washington's Friday Harbor Laboratories.

Deutsche Gesellschaft für Geschichte der Pharmazie. Membership is open. A membership or subscription fee is charged. Meets biennially with the primary meeting in March-May. **Contact:** Klaus Meyer, meyer-kl@t-online.de **Address:** Sertürner Str. 9 B, Münster, D-48149, Germany. **Tel:** 49.251.8570.800 **Fax:.** 49.251.8570.802 **Founded:** 1926 **Offers grants/fellowships:** No. **Awards prizes:** Yes. **Web site:** http://www.dggp.de

European Association for the History of Medicine and Health (EAHMH). A membership or subscription fee is charged. **Contact:** Claudia Wiesemann, info@eahmh.org **Address:** Abteilung Ethik und Geschichte der Medizin, Universität Göttingen, Humboldtallee 36, Göttingen, 37073, Germany. **Tel:** 49.551.399.006 **Fax:** 49.551.399.554 **Founded:** 1991 **Publications:** Newsletter; Conference proceedings; Occasional papers. **Description:** EAHMH aims to look outwards in terms of fields of interest and of researchers. Indeed, EAHMH attracts members world-wide, from North America to Australasia, who are demographers, social historians, social anthropologists and medical historians. EAHMH offers a high-level interdisciplinary and international forum for studies in the history of medicine, health and disease. It also promotes and fosters research, teaching, and international scientific cooperation among individuals as well as with related national and internaional societies, and it advances the education of the public in the historical aspects of medicine, health, and disease. **Web site:** http://www.eahmh.org/

The European Association for the Study of Science and Technology (EASST). A membership or subscription fee is charged. **Contact:** Sally Wyatt, wyatt@ pscw.uva.nl **E-mail:** easst@pscw.uva.nl **Founded:** 1981 **Purpose:** To stimulate communcation, exchange and collaboration in the field of studies of science and technology. **Publications:** *EASST Review*. **Description:** EASST is an interdisciplinary scholarly society, which reflects the closeness of history, philosophy, psychology and sociology of science in recent years. It also welcomes a policy perspective on science and technology. Cross-disciplinary interaction and cross-fertilization between humanistic and policy-oriented studies are important aims. These aims are furthered through the EASST general conference, held every other year, workshops and the *EASST Review*. **Web site:** http://www.chem.uva.nl/easst/easst.html

Finnish Society for the History of Medicine (Suomen Laaketieteen Historian Seura). Membership is open. A membership or subscription fee is charged. 420 members. Meets quarterly. **Contact:** Hindrik Strandberg, hindrik.stra ndberg@helsinki.fi **Address:** PO BOX 2, Helsinki, FIN-00561, Finland. **Tel:** 358.09.191.4823 **Fax:** 358.09.191.4824 **Founded:** 1961 **Focus:** Europe, Medical Sciences, Humanistic Relations of Science. **Purpose:** To promote the history of medicine, health care, odontology, pharmacy and veterinary medicine in Finland. **Publications:** *Hippokrates Annales Societatis Historiae Medicinae Fennicae.* **Offers grants/fellowships:** No. **Awards prizes:** No. **Description:** Each year, the society holds four meetings and takes an excursion to some place of interest. The yearbook contains published articles from the different fields represented by the society. **Web site:** http://www.helsinki.fi/jarj/slhs

The Finnish Society for the History of Science and Learning. Membership is open. 165 members. Meets monthly. **Contact:** Ora Patoharju, ora.patoharju@ helsinki.fi **E-mail:** klaus.karttunen@helsinki.fi **Address:** Tehtaankatu 8 A 1, Helsinki, FIN-00140, Finland. **Tel:** 358.0.9.635.759 **Founded:** 1966 **Purpose:** To promote interest in the history, development, and growth of learning in various sciences and disciplines. **Offers grants/fellowships:** No. **Awards prizes:** No.

Finnish Society for the History of Technology. Membership is open. A membership or subscription fee is charged. 380 members. Meets semiannually. **Contact:** Lars J. Hukkinen, larsj.hukkinen@kolumbus.fi **Address:** c/o Tekniikan museo, Viikintie 1, Helsinki, FIN-00560, Finland. **Founded:** 1928 **Purpose:** Preservation and Research of the History of Technology, particularly in Finland. **Publications:** *Tekniikan Waiheita - Teknik i Tiden.* **Awards prizes:** Yes. **Paid Staff:** *Kimmo Antila; Terhi Ketolainen.* **Description:** Founded in conjunction with the Central Museum of Technology in Helsinki, the society offers technical assistance and consultation to the museum and publishes the *Finnish Journal for the History of Technology* (four issues annually). **Web site:** http://www.ths.fi

Forest History Society. Membership is open. A membership or subscription fee is charged. 2000 members. Meets irregularly. **Contact:** Steven Anderson, stevena@duke.edu **E-mail:** recluce2@duke.edu **Address:** 701 Wm. Vickers Avenue, Durham, NC, 27701-3162, United States. **Tel:** 1.919.682.9319 **Fax:** 1.919.682.2349 **Founded:** 1946 **Focus:** Natural and Human History. **Purpose:** To improve natural resource management and human welfare by bringing a historical context to environmental decision-making. **Publications:** *Environmental History; Forest History Today; Forest Timeline;* FHS Issues Series. **Offers grants/fellowships:** Yes. **Awards prizes:** Yes. **Paid Staff:** *Steven Anderson; Michele Justice; Cheryl Oakes.* **Description:** The Forest History Society believes that "by understanding our past we shape our future." To achieve its mission the Society will: (1) Preserve forest and conservation history for present and future generations; (2) Encourage scholarship in forest and conservation history; (3) Conduct a comprehensive applied history program that brings the lessons of forest history to bear on the most pressing issues in natural resource management and contributes to identifying viable solutions to them. **Web site:** http://www.lib.duke.edu/forest

Forum for the History of Science in America. Membership is open. A membership or subscription fee is charged. 130 members. Meets annually with the primary meeting in September-November. **Contact:** Julie R. Newell, jnewell@spsu.edu **Address:** Southern Polytechnic State University SIS Dept, 1100 S. Marietta Parkway SE, Marietta, GA, 30060-2896, United States. **Founded:** 1984 **Purpose:** Promote interest in and study of the history of American science. **Publications:** *History of Science in America: News and Views.* **Offers grants/fellowships:** No. **Awards prizes:** Yes. **Description:** The Forum meets annually with the History of Science Society. It sponsors a lecture by a "Distinguished Scientist and by a "Distinguished Historian" in alternate years. It awards a prize for the best book on the history of American science one year and for the best article the next. **Web site:** http://academic.udayton.edu/AmSciForum

Gauss-Gesellschaft Göttingen.
  Contact: Gudrun Wolfschmidt, wolfschmidt@math.uni-hamburg.de
  Web site: http://www.math.uni-hamburg.de/math/ign/gauss/gaussges.html

Geological Society of America, History of Geology Division (GSA). Membership is closed. A membership or subscription fee is charged. 350 members. Meets annually with the primary meeting in September-November. **Contact:** William R. Brice, wbrice@pitt.edu **Address:** University of Pittsburgh, Johnstown, Dept. Geology and Planetary Science, 250 Krebs Hall, Johnstown, PA, 15904, United States. **Tel:** 1.814.269.2942 **Founded:** 1976 **Focus:** Earth Sciences. **Purpose:** Promote the history of earth sciences within the GSA. **Publications:** Division newsletter (e-mail only). **Offers grants/fellowships:** No. **Awards prizes:** Yes. **Description:** The Division sponsors sessions with contributed and invited papers on historical topics at the GSA meeting, awards an annual prize for achievement in the history of geology, and publishes two issues a year of an electronic newsletter. Any member of GSA may join. A Division committee directs the "Rock Stars" project to publish short biographies of historically important geologists in *GSA Today.* **Web site:** http://gsahist.org

German Chemical Society, History of Chemistry Division (Gesellschaft Deutscher Chemiker). Membership is open. A membership or subscription fee is charged. 350 members. Meets biennially with the primary meeting in March-May. **Contact:** Hans-Werner Schuett, hw.schuett@tu-berlin.de **E-mail:** r.kiessling@gdch.de **Address:** Gesellschaft Deutscher Chemiker, Postfach (POB) 90 04 40, D-60444 Frankfurt/M., Germany. **Tel:** 49.6.97.917.580 **Fax:** 49.6.97.917.656 **Founded:** 1946 **Focus:** Europe,

Physical Sciences. **Purpose:** To further historical interest, especially among chemists. **Publications:** *Mitteilungen der Fachgruppe Geschichte der Chemie.* **Offers grants/fellowships:** No. **Awards prizes:** Yes. **Paid Staff:** *Hans-Werner Schütt* (Physical Sciences). **Description:** The "Fachgruppe" enhances interest in the history of chemistry among chemists. Chemistry as a science and chemical technology are treated with equal emphasis. **Web site:** http://www.gdch.de/fachgrup/geschich.htm

Gesellschaft für Wissenschafts- und Technikforschung (GWTF). Membership is open. A membership or subscription fee is charged. 67 members. Meets annually with the primary meeting in September-November. **Contact:** Jörg Strübing, joerg.struebing@tu-berlin.de **E-mail:** struebing@gwtf.de **Address:** c/o TU Berlin Institute for Sociology, Franklinstr. 28/29 (SEKR 2-5), Berlin, D-10587, Germany. **Tel:** 49.0.30.3147.9467 **Fax:** 49.0.30.3147.9494 **Founded:** 1988 **Focus:** Social Sciences. **Purpose:** Foster cross disciplinary collaboration among researchers and people working in the field of STS. **Offers grants/fellowships:** No. **Awards prizes:** No. **Description:** The GWTF supports the discourse among the specialties and disciplines involved in STS. It also serves as a forum for research on interdisciplinarity–one of the key issues in current STS. The GWTF also hopes to establish a permanent dialogue between sciences and humanities in order to overcome the unproductive schism of "the two cultures." These goals are served by organizing conferences and publishing books related to conference themes. Additionally the GWTF organizes the only German language STS-mailing list and maintains an independent Web site for STS topics. Future activities will include additional institutional support structures for trans-disciplinary research and grants and prizes for related work of high quality. **Web site:** http://www.gwtf.de

Greek Society for the History of Science and Technology. Membership is open. A membership or subscription fee is charged. 150 members. Meets annually with the primary meeting in March-May. **Contact:** George N. Vlahakis, gvlahakis@eie.gr **Address:** Andromahis 17, 151 25 Marousi, Athens, Greece. **Founded:** 1989 **Purpose:** The promotion and development of the history of science and technology in Greece. **Publications:** Newsletter (in Greek). **Offers grants/fellowships:** No. **Awards prizes:** No. **Description:** The Greek Society for the History of Science and Technology aims to promote and develop these fields in Greece as well as to strengthen the relationships between the Greek and the international communities of historians of science. It has organized one international and three national congresses and several other meetings and seminars on specific subjects.

Historical Metallurgy Society. Membership is open. A membership or subscription fee is charged. 570 members. Meets semiannually with the primary meeting in September-November. **Contact:** Peter Hutchison, hon-sec@hist-met.org **Address:** 22 Easterfield Drive, Southgate, Swansea, SA3 2DB, United Kingdom. **Tel:** 44.1792.233.223 **Founded:** 1962

Focus: Technology, Instruments and Techniques. **Purpose:** Recording the history of metals and preserving historic metallurgical sites. **Publications:** *HMS Newsletter*; *Historical Metallurgy*. **Offers grants/fellowships:** Yes. **Awards prizes:** No. **Description:** We are an international organization whose object is to educate the public by the study, preservation and recording of metallurgical sites and processes. We aim to educate the public in the uses and importance of the metal industry from its beginnings to the recent past. **Web site:** http://hist-met.org

History of Earth Sciences Society. Membership is open. A membership or subscription fee is charged. 250 members. Meets irregularly. **Contact:** Ronald Rainger, j3ron@ttacs.ttu.edu **Address:** c/o Ronald Rainger, Texas Tech University, Box 41013, Lubbock, TX, 79409, United States. **Tel:** 1.806.742.3744 **Fax:** 1.806.742.1060 **Founded:** 1980 **Focus:** Physical Sciences. **Purpose:** To promote the study and understanding of the history of the earth sciences broadly conceived. **Publications:** *Earth Sciences History*. **Offers grants/fellowships:** No.

The History of Medicine and Health Colloquium, University of Michigan Program in Society and Medicine. Membership is closed. Free. 100 members. Meets monthly. **Contact:** Martin S. Pernick, mpernick@umich.edu **Address:** Department of History, University of Michigan, Ann Arbor, MI, 48109-1003, United States. **Tel:** 1.734.647.4876 **Fax:** 1.734.647.4881 **Founded:** 1992 **Focus:** Medical Sciences. **Purpose:** Sponsor presentations and discussions on medical history. **Offers grants/fellowships:** No. **Awards prizes:** No. **Description:** The History of Medicine and Health Colloquium brings together faculty, graduate and professional school students interested in the social, cultural, technical, and intellectual histories of health, disease, and the healing professions. Members come from a broad range of fields across the University of Michigan and other area institutions. Since 1999 we have met jointly with the Science and Technology Studies colloquium (contact Paul Edwards, School of Information, University of Michigan pne@umich.edu). Meetings are held twice a month September through April, alternating between medical and non-medical topics. We also jointly sponsor several large public lectures a year. **Web site:** http://umich.edu/psm/ history.html

History of Science Society (HSS). Membership is open. A membership or subscription fee is charged. 2800 members. Meets annually with the primary meeting in September-November. **Contact:** Robert Jay Malone, info@hssonline. org **Address:** Box 351330, University of Washington, Seattle, WA, 98195, United States. **Tel:** 1.206.543.9366 **Fax:** 1.206.685.9544 **Founded:** 1924 **Purpose:** To foster worldwide interest in the history of science and its cultural influences. **Publications:** *Isis*; *Osiris*; *HSS Newsletter*; *Isis* Readers; *Isis* Current Bibliography; Syllabi Samplers. **Offers grants/fellowships:** Yes. **Awards prizes:** Yes. **Paid Staff:** *Robert J. Malone* (18th Century, North America, Earth Sciences); *Roger Turner* (20th and 21st Century, North America);

Joan Vandegrift; Alan Weber; Stephen P. Weldon. **Description:** HSS is the oldest society dedicated to the history of science and its social and cultural influences. As an international organization, it encourages communication among scholars and interested lay people worldwide. A large meeting, held in the last quarter of each year, open to non-members, consistently showcases top scholars and outstanding research. **Web site:** http://www.hssonline.org

History of Science Society, Metropolitan New York Section (MNYSHSS)
[Note: Section is no longer active.]. Membership is closed. Free.
Contact: Mary Louise Gleason, marylougleason@cs.com **Tel:** 212.595.6084
Founded: 1953 **Purpose:** To encourage and maintain within the Metropolitan New York area an active interest in the history of science. **Offers grants/ fellowships:** No. **Awards prizes:** No. **Description:** The Metropolitan New York Section of the History of Science Society (MNYSHSS) was founded in 1953 and was a vibrant and active organization of historians of science until the early 1990s. It has since then merged with the New York Academy of Sciences. Its meetings were seamlessly incorporated into the meetings of the History and Philosophy of Science Section of the New York Academy of Sciences, which shares the mandate and organizational focus of the former MNYSHSS.

The History of Science Society of Japan (HSSJ). Membership is open. A membership or subscription fee is charged. 1000 members. Meets annually with the primary meeting in March-May. **Contact:** Shuntaro Ito **Address:** West Pine Bldg. 201, Hirakawa-cho 2-15-19, Chiyoda-ku, Tokyo, 102-0093, Japan. **Tel:** 81.3.3239.0545 **Fax:** 81.3.3239.0545 **Founded:** 1941 **Publications:** *Historia Scientiarum*; *Kagakushi Kenkyu* (Journal of History of Science, Japan); *Kagakushi Tsushin* (Bulletin for members);. **Offers grants/ fellowships:** No. **Awards prizes:** Yes. **Description:** The society has eight local branches in Japan and two specialized chapters, history of biology and history of technology. Anyone interested in the history of science or technology may join the society. *Historia Scientiarum* is published in European languages; other publications are in Japanese. **Web site:** http://wwwsoc.nii.ac.jp/jshs/

Indian Society for History of Mathematics (ISHM). Membership is open. A membership or subscription fee is charged. 200 members. Meets annually with the primary meeting in December-February. **Contact:** Dr. Man Mohan, indianshm@yahoo.com **Address:** Department of Mathematics and Statistics, Ramjas College, University of Delhi, Delhi, 110007, India. **Tel:** 91.011.718.7604 **Founded:** 1978 **Focus:** Mathematics. **Purpose:** To promote study, teaching, research and education in the history of mathematical sciences. **Publications:** *Ganita Bharati*. **Offers grants/fellowships:** No. **Awards prizes:** No. **Description:** In addition to annual conferences, ISHM aims at organizing seminars & symposia on the works of ancient, medieval and modern mathematics. Its journal, *Ganita Bharati*, publishes original research papers, articles, book reviews, news items and notices etc. within the ambit of history of mathematics. Papers, books and other items must be submitted in duplicate to the Editor (Dr. R.C. Gupta, R-20, Ras Bahar Colony, P.O. Lahar Gird, Jhansi-284003, India) or in electronic form to the Managing Editor as an E-mail attachment. Papers and articles intended for publication must comprise sufficiently novel and previously unpublished material. Author(s) will get 20 off-prints free.

International Academy of the History of Medicine (IAHM). Contact: Vivian Nutton, ucgavnu@ucl.ac.uk **Address:** Wellcome Institute for the History of Medicine, 183 Euston Road, London, NW1 2BP, United Kingdom. **Description:** This organization has been replaced by the European Association for the History of Medicine and Health.

International Academy of the History of Pharmacy. Membership is closed. A membership or subscription fee is charged. Meets annually with the primary meeting in September-November. **Contact:** Wolf D. Mueller-Jahncke, mueja@rz-online.de **Address:** Lindenstrasse 11, Kircher, 5754, Germany. **Tel:** 49.274.162.54 **Fax:** 49.274.16.735 **Founded:** 1952 **Purpose:** Encourage communication among professionals in the history of pharmacy. **Publications:** *Communications del' Academie Internationale de l'Histoire de la Pharmacie.* **Offers grants/fellowships:** No. **Awards prizes:** No.

International Academy of the History of Science. Founded: 1928 **Description:** See the entry for the International Union for the History and Philosophy of Science, Division of History of Science (IUHPS, DHS).

The International Commission for the History of Science in Islamic Civilization. Membership is open. Free. 250 members. Meets quadrennially. **Contact:** Gül A. Russell, garussell@tamu.edu **Address:** Dept. of Humanities in Medicine, 154 Reynolds Medical Building, College of Medicine, Texas A&M University System Health Science Center, College Station, TX, 77843-1114, United States. **Tel:** 1.979.845.6462 **Fax:** 1.979.845.8634 **Founded:** 1989 **Purpose:** To promote research through meetings, to exchange information through newsletters and a listserv, and to disseminate findings through publications. **Publications:** Newsletter (annually; hard copy and electronic). **Offers grants/fellowships:** No. **Awards prizes:** No. **Description:** The Commission's objectives are: (1) to provide a forum for up-to-date scholarly exchange and communication among historians of Graeco-Arabic and Islamic science as well as historians of science in related fields through publications, newsletters, listservs, and meetings; (2) to promote research of high standards; (3) encourage younger scholars through scholarships and participation at meetings; (4) to incorporate the subject into the general history of science education. (The new strategic plans include increased meetings and symposia, scholarships, awards, etc.) **Web site:** http://www.ou.edu/islamsci

International Commission on History of Meteorology (ICHM). Membership is open. Free. 150 members. Meets irregularly. **Contact:** Kris Harper, kharper@proaxis.com **E-mail:** jrflemin@colby.edu **Address:** c/o James Fleming, STS Program, Colby College, 5881 Mayflower Hill, Waterville, ME, 04901, United States. **Founded:** 2001 **Purpose:** To promote the history of meteorology, climatology, and related sciences. **Offers grants/fellowships:** No. **Awards prizes:** No. **Description:** Aims of ICHM: a) scholarly study of the history of meteorology and related sciences; b) international cooperation and communication; c) symposia at the ICHS and other venues; d) identification, collection, preservation, and access to historical materials; e) international historical bibliography; f) support the broader goals of DHS, IUHPS, and ICSU. **Web site:** http://www.meteohistory.org

International Commission on the History of Mathematics (ICHM). Membership is closed. Free. 65 members. Meets quarterly. **Contact:** Karen Virginia Hunger Parshall, khp3k@virginia.edu **E-mail:** hogend@math.uu.nl **Address:** Departments of History and Mathematics, University of Virginia, PO Box 400137, Charlottesville, VA, 22904-4137, United States. **Tel:** 1.434.924.1411 **Fax:** 1.434.982.3084 **Founded:** 1969 **Focus:** Mathematics. **Purpose:** The ICHM aims to promote the high-level study of the history of mathematics internationally. **Publications:** *Historia Mathematica.* **Offers grants/fellowships:** No. **Awards prizes:** Yes. **Description:** The International Commission for the History of Mathematics (ICHM) is an inter-union commission joining the International Mathematical Union (IMU) and the Division of the History of Science (DHS) of the International Union for the History and Philosophy of Science (IUHPS). The ICHM also sponsors or cosponsors scientific symposia at the International Congresses of the History of Science, at meetings of national history of science and mathematics societies, and at other conferences, in addition to maintaining a world-wide directory of historians of mathematics. **Web site:** http://www.math.uu.nl/ichm

International Commission on the History of the Geological Sciences. Membership is closed. Free. 180 members. Meets annually. **Contact:** David R. Oldroyd, d.oldroyd@unsw.edu.au **Address:** School of Science and Technology Studies, The University of New South Wales, Sydney, NSW 2052, Australia. **Tel:** 61.2.9449.5559 **Fax:** 61.2.9144.4529 **Founded:** 1967 **Focus:** Transcontinental, Earth Sciences. **Purpose:** The promotion and coordination of studies in the history of geosciences, worldwide. **Publications:** *INHIGEO Newsletter.* **Offers grants/fellowships:** Yes. **Awards prizes:** No. **Description:** The Commission is one of the Commissions of the International Union of Geological Sciences, and is affiliated with the International Union for the History and Philosophy of Science (Division of History of Science). It encourages and facilitates studies in the history of geosciences through its *Newsletter*, conferences, and publications. **Web site:** http://www.iugs.org

International Committee for the History of Technology (ICOHTEC). Membership is open. A membership or subscription fee is charged. 320 members. Meets annually with the primary meeting in June-August.

Contact: Hans-Joachim Braun, hjbraun@unibw-hamburg.de
Address: Universität der Bundeswehr Hamburg, Hamburg, 22039,
Germany. **Tel:** 49.40.6541.2794 **Fax:** 49.40.6541.2084 **Founded:** 1968
Focus: Technology. **Purpose:** Advancing worldwide international cooperation
in the history of technology through symposia, research projects, and
journals. **Publications:** *ICON: Journal of the International Committee for the
History of Technology.* **Offers grants/fellowships:** No. **Awards prizes:** No.
Description: ICOHTEC is a scientific section of IUHPS/DHS, but is based on
individual membership. Annual subscription, $25 for individuals and $100 for
institutions, includes the annual journal *ICON*. **Web site:** http://www.icohtec.org

International History, Philosophy, and Science Teaching Group (IHPST).
Membership is open. A membership or subscription fee is charged. 300
members. Meets biennially with the primary meeting in September-November.
Contact: Michael Robert Matthews, m.matthews@unsw.edu.au
Address: School of Education, University of New South Wales, Sydney, NSW,
2052, Australia. **Tel:** 61.2.9418.3665 **Fax:** 61.2.9385.1946 **Founded:** 1989
Focus: Education. **Purpose:** The application of history and philosophy of
science to science education pedagogy and theory. **Publications:** *Science
& Education.* **Offers grants/fellowships:** No. **Awards prizes:** No.
Description: IHPST aims to improve science education and science teacher
preparation by having theory, curriculum and pedagogy informed by the history
and philosophy of science. It has a particular interest in bringing these fields into
teacher-education programs. **Web site:** http://www.ihpst.org

**International Society for History of Arabic and Islamic Sciences and
Philosophy (SIHSPAI).** Membership is open. A membership or subscription
fee is charged. 175 members. Meets biennially with the primary meeting in
December-February. **Contact:** Ahmed Hasnaoui, ahmedhasnaoui@mageos.com
Address: Centre d'Histoire des Sciences et des Philosophies Arabes,
CNRS, 7 rue Guy Moquet, B.P. no. 8, Villejuif, Cedex, 94801, France.
Tel: 33.1.49.58.35.99 **Fax:** 33.1.49.58.35.47 **Founded:** 1989 **Focus:** 5th-13th
Century, Natural and Human History. **Purpose:** Promote the study of history
of Arabic and Islamic science and philosophy. **Publications:** *Arabic Sciences
and Philosophy*; *Newsletter of SIHSPAI*. **Offers grants/fellowships:** No.
Awards prizes: Yes. **Paid Staff:** *Ahmed Hasnaoui*. **Description:** The
geographical area for SIHSPAI is the Middle East and North Africa. Scholars
associated with the organization focus on the history of Arabic and Islamic
sciences and philosophy, with a special concentration on the medieval period.

International Society for the History of Medicine (ISHM). A membership or
subscription fee is charged. **E-mail:** aajet@noos.fr **Address:** 12 rue de l'Ecole
de Médecine, Paris, Cedex 06, Paris, 75270, France. **Tel:** 33.1.39.27.42.97
Founded: 1921 **Purpose:** To assist and support the historical study of
all questions relating to the medical and biomedical sciences and, more
generally, to all branches of the healing arts. **Publications:** *Vesalius.*

Description: Seeks to improve communication among individuals and professional groups throughout the world interested in the medical and biomedical sciences and healing arts. Aims to promote the teaching and spread of knowledge on these topics. The Society sponsors and oversees the organization of a biennial international congress in the history of medicine. **Web site:** http://www.bium.univ-paris5.fr/ishm/eng/

International Society for the History of Pharmacy (ISHP). Membership is open. A membership or subscription fee is charged. Meets biennially with the primary meeting in September-November. **Contact:** Axel Helmstaedter, helmstaedter@em.uni-frankfurt.de **Address:** c/o GOVI-Verlag, Carl-Mannich-Str. 26 , Eschborn, D-65760, Germany. **Tel:** 49.6196.928.262 **Fax:** 49.6196.928.203 **Founded:** 1926 **Purpose:** Umbrella organization for national societies devoted to the history of pharmacy. **Publications:** *ISHP Newsletter*. **Offers grants/fellowships:** Yes. **Awards prizes:** No. **Description:** ISHP is devoted to research and teaching the history of pharmacy worldwide. The Society forms an international center for handling all matters of a pharmaceutical historical nature without commercial interests. It endeavors to reach its goals by promoting research in writing and teaching the history of pharmacy, as well as disseminating knowledge and obtaining recognition for the discipline. Although individual membership is possible under some circumstances, ISHP is mainly an umbrella organization for the National Societies for Pharmaceutical History. At present, 16 nations are ISHP members. Every two years an international scientific conference is organized. Besides that there are several running projects aimed to increase and improve university teaching, to improve bibliographic resources and to support international research projects. **Web site:** http://www.histpharm.org

International Society for the History of the Neurosciences (ISHN). Membership is open. A membership or subscription fee is charged. Meets annually. **Contact:** Russell A. Johnson, rjohnson@library.ucla.edu **Address:** Neuroscience History Archives, Brain Research Institute, UCLA, Box 951761, Los Angeles, CA, 90095-1761, United States. **Tel:** 1.310.825.6940 **Fax:** 1.310.206.5855 **Founded:** 1995 **Purpose:** To improve communication among individuals and groups interested in the history of neuroscience, promote research in the history of neuroscience, and promote education in and stimulate interest for the history of neuroscience. **Publications:** *Journal of the History of the Neurosciences: Basic and Clinical Perspectives*. **Offers grants/fellowships:** No. **Awards prizes:** Yes. **Description:** The International Society for the History of the Neurosciences (ISHN) was founded in Montreal on May 14, 1995. The *Journal of the History of the Neurosciences: Basic and Clinical Perspectives* is the official organ of the society. Now in its eleventh year (2002) and increasing from three to four issues per volume, the journal provides a forum for investigations in the field of neuroscience history, broadly defined, including the history of basic, clinical, and behavioral neurosciences, ancient and non-western topics, and significant individuals, events, and technical advances. ISHN is a focal point for those interested in neuroscience history and offers, through its journal and annual meeting, opportunities to contribute to this rapidly growing field. **Web site:** http://www.ishn.org/

International Society for the History, Philosophy, and Social Studies of Biology (ISHPSSB). Membership is open. A membership or subscription fee is charged. 750 members. Meets biennially with the primary meeting in June-August. **Contact:** Chris Young, cyoung@aero.net **Address:** 1316 N. Astor Street, Milwaukee, WI, 53202, United States. **Founded:** 1990 **Focus:** Transcontinental, Biological Sciences. **Purpose:** International interdisciplinary scholarly exchange in studies of biology. **Publications:** *ISHPSSB Newsletter.* **Offers grants/fellowships:** Yes. **Awards prizes:** Yes. **Description:** ISHPSSB brings together scholars from diverse disciplines, including the life sciences, history, philosophy, and social studies of science. ISHPSSB summer meetings are known for innovative, transdisciplinary sessions, and for fostering informal, cooperative exchanges and ongoing collaborations. **Web site:** http://www.phil.vt.edu/ishpssb/

International Union of the History and Philosophy of Science (IUHPS). Membership is closed. A membership or subscription fee is charged. Meets biennially with the primary meeting in June-August. **Contact:** Dag Westerstahl, dag.westerstahl@phil.gu.se **Address:** Dept. of Philosophy, Goteborg University, Box 200, Goteborg, WI, SE, 405 30, Sweden. **Tel:** 46.31.773.45.73 **Fax:** 46.31.773.49.45 **Founded:** 1947-1956 **Focus:** Philosophy or Philosophy of Science. **Purpose:** Establish and promote international contacts among historians and philosophers of science and scientists interested in history and foundational problems of their discipline. **Offers grants/fellowships:** No. **Awards prizes:** No. **Description:** IUHPS has two Divisions: Division of History of Science (DHS) and Division of Logic, Methodology and Philosophy of Science (DLMPS). Each division has its own membership, including national members and scientific commissions. Each organizes international congresses at four-year intervals; hence there is a congress every two years. In intermediate years an International Joint DHS and DLMPS Conference is organized. The IUHPS currently has 79 member committees.

International Union of the History and Philosophy of Science (Division of History of Science). Membership is open. Free. Meets quadrennially with the primary meeting in June-August. **Contact:** Fabio Bevilacqua, bevilacqua@ fisicavolta.unipv.it **Founded:** 1947 **Purpose:** To facilitate international exchanges and understanding of history of science. **Offers grants/ fellowships:** No. **Awards prizes:** No. **Web site:** http://ppp.unipv.it/dhs

Japanese Association for the History of Geology. Membership is open. A membership or subscription fee is charged. 94 members. Meets tri-annually with the primary meeting in December-February. **Contact:** Michiko Yajima, pxi02070@nifty.ne.jp **Address:** #403, 2-24-1,

Minami-Ikebukuro, Toshima-ku, Tokyo, 171-0022, Japan. **Tel:** 81.3.3812.7039
Fax: 81.3.3812.7039 **Founded:** 1994 **Purpose:** Surveying the history of
the development of geological research in Japan. **Publications:** *JAHIGEO
Newsletter.* **Offers grants/fellowships:** No. **Awards prizes:** No.
Description: In March 1994, JAHIGEO was established to recognize
the centenary of the Geological Society of Japan. This is the only
organization promoting the history of the earth sciences in Japan.
Web site: http://www.geocities.co.jp/Technopolis/9866/jahigeo.html

The Linnean Society of London. Membership is open. A membership or
subscription fee is charged. 2500 members. Meets monthly. **Contact:** Gina
Lundy Douglas, gina@linnean.org **E-mail:** john@linnean.org **Address:**
Linnean Society of London, Burlington House, Piccadilly, London, W1J 0BF,
United Kingdom. **Tel:** 44.2.7434.4479 **Fax:** 44.2.7287.9364 **Founded:** 1788
Focus: Biological Sciences. **Purpose:** Promotion of natural history in all its
branches. **Publications:** *The Linnean Newsletter & Proceedings*; *Botanical
Journal of the Linnean Society*; *Biological Journal of the Linnean Society*;
Zoological Journal of the Linnean Society. **Offers grants/fellowships:** Yes.
Awards prizes: Yes. **Paid Staff:** *Gina Lundy Douglas*; *John Marsden.*
Description: The Linnean Society of London promotes all aspects of biology
(including the history of biology) but particularly those concerning the diversity
and interrelationships of organisms. The need for accurate identification of these,
through the tools of systematic biology, means that the Society covers all aspects
of the biological sciences as well as the historical foundations upon which
taxonomy is based. The role of the Society in providing the venue for the papers
by Darwin and Wallace on natural selection have led to a continued involvement
with evolutionary biology. **Web site:** http://www.linnean.org

Lone Star Historians of Science. Membership is open. Free. 25 members. Meets
annually with the primary meeting in March-May. **Contact:** Bruce J. Hunt,
bjhunt@mail.utexas.edu **Address:** Department of History, University of Texas,
Austin, TX, 78756, United States. **Tel:** 1.512.232.6109 **Fax:** 1.512.475.7222
Founded: 1988 **Purpose:** To bring together historians of science from around
Texas for an informal meeting each spring. **Offers grants/fellowships:** No.
Awards prizes: No. **Description:** The Lone Star group was founded in 1988.
Our constitution provides that there shall be "no officer, no by-laws, and no
dues," and we remain resolutely informal. We meet in a different city each
spring and, after listening to a talk by an invited speaker, enjoy a convivial
dinner together.

Maria Mitchell Association. Membership is open. A membership or subscription
fee is charged. Meets annually with the primary meeting in June-August.
Contact: Kathryn K. Pochman, kpochman@mmo.org **E-mail:** ahunt@mmo.org
Address: 4 Vestal Street, Nantucket, MA, 02554, United States.
Tel: 1.508.228.9198 **Fax:** 1.508.228.1031 **Founded:** 1902 **Focus:** North
America, Physical Sciences, Gender and Science. **Purpose:** Science education

and research; women in science. **Publications:** Annual Report; *The Comet*, quarterly newsletter. **Offers grants/fellowships:** No. **Awards prizes:** Yes. **Description:** Historic birthplace of America's first woman astronomer; natural science museum and aquarium; science library; public programs include environmental science education, field trips, star viewing, lectures. Annual Maria Mitchell Women in Science Award. Astronomical and biological research. **Web site:** http://www.mmo.org

Medical History Society of New Jersey. Membership is open. A membership or subscription fee is charged. 119 members. Meets semiannually. **Contact:** Frederick Skvara, MD, fcskvara@bellatlantic.net **Address:** Medical History Society of New Jersey, 2 Princess Road, Lawrenceville, NJ, 08648, United States. **Tel:** 1.609.896.1901 **Founded:** 1980 **Focus:** Medical Sciences. **Purpose:** Promotes interest, research and writing in medical history and allied fields. **Publications:** *Medical History Society of New Jersey Newsletter*. **Offers grants/fellowships:** No. **Awards prizes:** Yes. **Web site:** http://www.mhsnj.org/

National Association of Mining History Organizations (NAMHO). 80 members. **Contact:** Sallie Bassham, namho@silicondale.co.uk **Address:** Peak District Mining Museum, The Pavilion, Matlock Bath, Matlock, DE4 3NR, United Kingdom. **Tel:** 44.1629.583.834 **Founded:** 1979

New York Academy of Medicine (Historical Medicine Section). **Contact:** Jeremiah A. Barondess **E-mail:** info@nyam.org **Address:** 1216 Fifth Avenue, New York, NY, 10029, United States. **Tel:** 1.212.822.7202 **Founded:** 1847 **Purpose:** Foster education and communication about the history of medicine as part of a larger effort to enhance the health of the public. **Publications:** *Journal of Urban Health: The Bulletin of the New York Academy of Medicine*. **Offers grants/fellowships:** Yes. **Awards prizes:** Yes. **Description:** The New York Academy of Medicine, founded in 1847, is an independent non-profit institution committed to enhancing the health of the public with a particular emphasis on disadvantaged urban populations. The Historical Medicine section of the Academy promotes the study of the history of medicine. Its resources include the Academy's Rare Book Room, which contains approximately 49,000 volumes on the history of medicine and science, with materials dating from 1700 B.C. **Web site:** http://www.nyam.org

New York Academy of Sciences (History and Philosophy of Science Section). Membership is open. A membership or subscription fee is charged. 40,000 members. Meets semiannually. **Contact:** Bruce Chandler, bchandler@nyas.org **Address:** 2 E. 63rd St., New York, NY, 10021, United States. **Tel:** 1.212.838.0230 **Fax:** 1.212.753.3479 **Founded:** 1817 **Focus:** 14th-16th Century, Natural and Human History, Humanistic Relations of Science. **Purpose:** Advance the study of the history and philosophy of science worldwide. **Publications:** *Annals of the New York Academy of Sciences*; *Intersections*; *The Sciences*; *Update*;. **Offers grants/fellowships:** Yes.

Awards prizes: Yes. **Description:** The New York Academy of Sciences is an independent, nonprofit organization with nearly 40,000 members in more than 16 countries united by a commitment to promoting science and technology and their essential roles in fostering social and economic development. The History and Philosophy of Science section of the Academy supports research, education, and communication in this field. **Web site:** http://www.nyas.org

Organization of American Historians (OAH). Membership is open. A membership or subscription fee is charged. 10,000 members. Meets annually with the primary meeting in March-May. **Contact:** Lee W. Formwalt, oah@oah.org **Address:** 112 N. Bryan Avenue, Bloomington, IN, 47408, United States. **Tel:** 1.812.855.7311 **Fax:** 1.812.855.0696 **Founded:** 1907 **Focus:** North America. **Purpose:** Promote the study and research of American history. **Publications:** *OAH Magazine of History*; *OAH Newsletter*; *Journal of American History*; Annual Meeting Program. **Offers grants/fellowships:** Yes. **Awards prizes:** Yes. **Description:** The Organization of American Historians is the largest learned society devoted to the study of American history. Since its founding in 1907 as the Mississippi Valley Historical Association, the OAH has promoted the study and teaching of the American past through its many activities. The work of the organization is supported primarily through the contributions of its membership. **Web site:** http://www.oah.org

Pacific Circle. Contact: Peter H. Hoffenberg, peterh@hawaii.edu **Address:** Department of History, University of Hawaii, Honolulu, HI, 96822, United States. **Tel:** 1.808.956.8497 **Fax:** 1.808.956.9600 **Publications:** Newsletter. **Description:** This group brings together scholars interested in the history of science, technology and medicine around Oceania and the Pacific basin. In addition to publishing a newsletter, it regularly sponsors sessions at major scientific and history of science meetings.

Philosophy of Science Association (PSA). Membership is closed. A membership or subscription fee is charged. 1,000 members. Meets biennially with the primary meeting in September-November. **Contact:** George Gale, galeg@umkc.edu **Address:** Dept. of Philosophy, University of Missouri, Kansas City, Kansas City, MO, 64110-2499, United States. **Tel:** 1.816.235.2816 **Fax:** 1.816.235.2819 **Founded:** 1934 **Focus:** Philosophy or Philosophy of Science. **Purpose:** Further studies and free discussion from diverse standpoints in the field of philosophy of science. **Publications:** *Philosophy of Science*; *Philosophy of Science Association Newsletter*; *Proceedings of the Biennial Meeting*. **Offers grants/fellowships:** Yes. **Awards prizes:** Yes. **Description:** The Philosophy of Science Association aims to further studies and free discussion from diverse standpoints in the field of philosophy of science. To this end, the PSA engages in activities such as: publishing periodicals, essays and monographs in this field; sponsoring conventions and meetings; and awarding prizes for distinguished work. **Web site:** http://scistud.umkc.edu/psa/

Schweizerische Gesellschaft für Geschichte der Medizin und der Naturwissenschaften (Swiss Society for the History of Medicine and Science). Membership is open. A membership or subscription fee is charged. 270 members. Meets annually with the primary meeting in September-November. **Contact:** Urs Leo Gantenbein, ulganten@mhiz.unizh.ch **Address:** SGGMN at Medizinhistorisches Institut, Universität Zurich, Ramistrasse 71, Zurich, CH-8006, Switzerland. **Tel:** 0041.1.634.20.71 **Fax:** 0041.1.634.23.49 **Founded:** 1921 **Focus:** Natural and Human History. **Purpose:** To stress the importance and further the study of the history of medicine and science. **Publications:** *Gesnerus* (Swiss Journal of the History of Medicine and Science); *Gesnerus Supplementum.* **Offers grants/ fellowships:** No. **Awards prizes:** Yes. **Description:** The society unites friends and students of the history of medicine and science, connecting them to a wider public. It also stresses the importance of historical thinking in science, furthers the teaching and research regarding these subjects, and enhances international relationships. **Web site:** http://www.sggmn.ch

Schweizerische Paracelsus-Gesellschaft (Swiss Paracelsus Society). A membership or subscription fee is charged. 210 members. Meets annually with the primary meeting in September-November. **Contact:** Urs Leo Gantenbein, ulganten@mhiz.unizh.ch **Address:** SPG at Medizinhistorisches Institut, Universität Zurich, Raemistrasse 71, Zurich, CH-8006, Switzerland. **Tel:** 41.52.222.27.86 **Fax:** 41.52.223.03.32 **Founded:** 1942 **Focus:** 14th-16th Century, Natural and Human History, Humanistic Relations of Science. **Purpose:** To promote the study of the life, times, and teachings of Paracelsus. **Publications:** *Nova Acta Paracelsica.* **Offers grants/fellowships:** No. **Awards prizes:** No. **Description:** The society unites friends of Theophrastus of Hohenheim (1493/94-1541), called Paracelsus, studies his life and teachings in relation to his time, keeps contact with the leading scholars in the field, organizes annual scientific meetings, and publishes the results in its own scientific journal, *Nova Acta Paracelsica.* **Web site:** http://www.paracelsus-gesellschaft.ch

Skeptics Society. Membership is open. A membership or subscription fee is charged. 20,000 members. Meets monthly. **Contact:** Michael Shermer, skepticmag@aol.com **Address:** 2761 N. Marengo Avenue, Altadena, CA, 91001, United States. **Tel:** 1.626.794.3119 **Fax:** 1.626.794.1301 **Founded:** 1992 **Focus:** 20th and 21st Century, North America, Social Sciences, Science and Religion. **Purpose:** The investigation of extraordinary claims, revolutionary ideas, science and pseudo-science, and the promotion of science. **Publications:** *Skeptic* magazine. **Offers grants/fellowships:** No. **Awards prizes:** Yes. **Description:** *Skeptic* magazine is the quarterly publication of the Skeptics Society, affiliated with the California Institute of Technology where the monthly public science lecture series and annual scientific conferences are held. *Skeptic* features cutting-edge debates about scientific controversies in the form of special theme sections every issue, plus an interview with a world-class scientist, long and short book reviews of the latest publications in the world of science, and extensive research reports on the latest findings in science. Also included within each issue of *Skeptic* is *Jr. Skeptic* magazine, which presents science for kids. **Web site:** http://www.skeptic.com

Social Science History Association (SSHA). Membership is open. A membership or subscription fee is charged. 1000 members. Meets annually with the primary meeting in September-November. **Contact:** Erik W. Austin, erik@icpsr.umich.edu **Address:** Institute for Social Research, PO Box 1248, Ann Arbor, MI, 48106, United States. **Tel:** 1.734.998.9820 **Fax:** 1.734.998.9889 **Founded:** 1975 **Focus:** Transcontinental, Social Sciences. **Purpose:** Advance the study of the past by employing methods of the social sciences. **Publications:** *Social Science History* (quarterly journal); *SSHA News* (semiannual newsletter). **Offers grants/fellowships:** Yes. **Awards prizes:** Yes. **Description:** SSHA is the world's leading interdisciplinary professional association in the social sciences/history. **Web site:** http://www.ssha.org

Sociedad Española de Historia de las Ciencias y de las Técnicas (Spanish Society for the History of Science and Technology). Membership is open. A membership or subscription fee is charged. 400 members. Meets triennially with the primary meeting in September-November. **Contact:** Elena Ausejo, ichs@posta.unizar.es **E-mail:** presidente@sehcyt.unirioja.es **Address:** Departamento de Matemáticas y Computación, Universidad de La Rioja - C/ Luis de Ulloa s/n (Edificio Vives), Logroño, 26004, Spain. **Tel:** 34.9.7676.1119 **Fax:** 34.9.7676.1125 **Founded:** 1977 **Purpose:** Promote activities in the field of history of science and technology. **Publications:** *Llull: Revista de la Sociedad Española de Historia de las Ciencias y de las Técnicas.* **Web site:** http://www.unirioja.es/sehcyt

Sociedad Mexicana de Historia de la Ciencia y la Tecnología, A.C. (Mexican Society for the History of Science and Technology). Membership is open. A membership or subscription fee is charged. 190 members. Meets biennially with the primary meeting in September-November. **Contact:** Juan Jose Saldana, saldana@servidor.unam.mx **E-mail:** info@smhct.org **Address:** Sociedad Mexicana de Historia de la Ciencia y la Tecnología, Apartado Postal 21-873, Mexico DF, 04000, Mexico. **Tel:** 5255.5622.1864 **Fax:** 5255.5622.1864 **Founded:** 1964 **Purpose:** Promotion of history of science and technology, especially in Mexico. **Publications:** *Boletín de la Sociedad Mexicana de Historia de la Ciencia y la Tecnología.* **Offers grants/fellowships:** Yes. **Awards prizes:** Yes. **Description:** SMHCT is the oldest history of science organization in Latin America. Its main concern is the history of science and technology in Mexico and Latin America, but comparative studies in science and cultural diversity have been also developed in our conferences and publications. In 2001, SMHCT organized the XXIst International Congress of History of Science. Every two years SMHCT offers the "Enrique Beltran's Prize," consisting of a stipend and the publication of the prize-winning book. The SMHCT Web site is in Spanish and English. **Web site:** http://www.smhct.org

Sociedade Brasileira de História da Ciência (Brazilian Society of History of Science). Membership is open. A membership or subscription fee is charged. 538 members. Meets biennially with the primary meeting in September-November. **Contact:** Ana Maria Ribeiro de Andrade, anaribeiro@mast.br **E-mail:** sbhc@mast.br **Address:** Sociedade Brasileira de História da Ciência - presidência, Rua General Bruce, 586 (MAST/CHC), Rio de Janeiro, 20 921- 030, Brazil. **Tel:** 55.21.2580.7010 **Fax:** 55.21.2580.4531 **Founded:** 1983 **Purpose:** Brings together researchers from history of science, technological devices and processes, philosophy, science education, and social studies of science. **Publications:** *Revista da Sociedade Brasileira de História da Ciência*; *Boletim Eletrônico da SBHC*. **Offers grants/fellowships:** No. **Awards prizes:** Yes. **Description:** SBHC, an interdisciplinary organization, brings together researchers, scholars and students with common interest in all aspects of the history of science and technology, including philosophy, science education, and social studies of science. Members of the SBHC receive a biannual academic journal, the *Revista da Sociedade Brasileira de História da Ciência*. The *SBHC Newsletter* is free and is published in an electronic format, with information about the field, events, conferences, books and much more. The society also maintains a homepage with general information and links to Brazilian and international institutions and groups devoted to the history of science. **Web site:** http://www.mast.br/sbhc/inicio.htm

Society for History in the Federal Government (SHFG). Membership is open. A membership or subscription fee is charged. 450 members. Meets tri-annually. **Contact:** Raymond W. Smock, raysmock@aol.com **Address:** Box 14139, Benjamin Franklin Station, Washington, DC, 20044, United States. **Founded:** 1979 **Purpose:** The SHFG serves as the voice of the federal historical community. **Publications:** *SHFG Monthly Bulletin*; Occasional Papers Series; *The Federalist*. **Offers grants/fellowships:** No. **Awards prizes:** Yes. **Description:** The Society represents a broad cross section of historians, archivists, librarians, and curators representing all three branches of the federal government. Other members study the federal government from the historical perspective in an academic setting. Members include both civilian and military historians. Fields represented include science, medicine, politics, law, technology, agriculture, foreign relations, national security, and intelligence. **Web site:** http://www.shfg.org

Society for Medical History in Denmark. Membership is open. A membership or subscription fee is charged. 300 members. Meets bi-monthly. **Contact:** Henrik R. Wulff, h.r.wullff@image.dk **E-mail:** h.r.wulff@image.dk **Address:** Dansk Medicinsk-historisk Selskab, c/o Medicinsk Historisk Museum, Bredgade 62, Copenhagen, 1260 Copenhagen, Denmark. **Tel:** 45.35.32.38.00 **Fax:** 45.35.32.38.16 **Founded:** 1920 **Focus:** Europe, Medical Sciences. **Purpose:** Promotes the study of, and interest in, history of medicine, especially in Denmark. **Offers grants/fellowships:** No. **Awards prizes:** No. **Description:** The society organizes lecture evenings and excursions, especially for members of the medical profession interested in the history of medicine. It cooperates closely with the Department for the History of Medicine of the University of Copenhagen.

Society for Social Studies of Science (4S). Membership is open. A membership or subscription fee is charged. 1100 members. Meets annually with the primary meeting in September-November. **Contact:** Wesley Shrum, shrum@lsu.edu **Address:** c/o Wesley Shrum, Dept. of Sociology, Louisiana State University, Baton Rouge, LA, 70803, United States. **Founded:** 1975 **Purpose:** Promote communication in science and technology studies. **Publications:** *Science, Technology & Human Values*; *Technoscience*. **Offers grants/fellowships:** Yes. **Awards prizes:** Yes. **Web site:** http://www.lsu.edu/ssss

Society for the History of Natural History (SHNH). Membership is open. A membership or subscription fee is charged. 500 members. Meets semiannually with the primary meeting in March-May. **Contact:** John Charles Edwards, jedwards@talk21.com **Address:** c/o The Natural History Museum, Cromwell Road, SW7 5BD, United Kingdom. **Tel:** 44.020.8981.5812 **Founded:** 1936 **Purpose:** Historical and Bibliographic study of the growth the of natural history in all periods and cultures. **Publications:** *The Archives of Natural History*; Newsletter. **Offers grants/fellowships:** Yes. **Awards prizes:** Yes. **Description:** Through its meetings, the Society fosters the informal exchange of ideas. It focuses on all aspects of natural history, including art, literature, biography, bibliography, and investigative historical studies. **Web site:** http://www.shnh.org

Society for the History of Technology (SHOT). Membership is open. A membership or subscription fee is charged. 2000 members. Meets annually with the primary meeting in September-November. **Contact:** Stuart William Leslie, swleslie@jhu.edu **E-mail:** shot@jhu.edu **Address:** 216 B Ames Hall, Johns Hopkins University, Baltimore, MD, 21218, United States. **Tel:** 1.410.516.8349 **Fax:** 1.410.516.7502 **Founded:** 1958 **Focus:** Technology. **Purpose:** To encourage the study of the development of technology and its relations with society and culture. **Publications:** *Technology and Culture*; Newsletter (quarterly); SHOT-AHA Booklet Series (various titles). **Offers grants/fellowships:** Yes. **Awards prizes:** Yes. **Description:** An interdisciplinary organization, SHOT is concerned not only with the history of technological devices and processes, but also with the relations of technology to science, politics, social change, the arts and humanities, and economics. In addition to professional historians and museum curators, SHOT members include practicing scientists and engineers, anthropologists, librarians, political scientists, and economists. **Web site:** http://www.shot.jhu.edu

Society for the Social History of Medicine. Membership is open. Meets annually. **Contact:** Oonagh Walsh, o.walsh@abdn.ac.uk **Address:** Dept. of History, Univ. of Aberdeen, Meston Walk, King's College, Aberdeen, AB24 3FX, United

Kingdom. **Tel:** 44.01224.273884 **Fax:** 44.01224.272203 **Founded:** 1970
Purpose: Promote interdisciplinary study of the history of medicine.
Publications: *Social History of Medicine*; *Studies in the Social History of Medicine*; *The Gazette*. **Awards prizes:** Yes. **Description:** Since its inaugural meeting in 1970, the Society for the Social History of Medicine (SSHM) has pioneered inter-disciplinary approaches to the history of health, welfare, medical science and practice. Consequently, its membership consists of those interested in a variety of disciplines, including history, public health, demography, anthropology, sociology, social administration and health economics. **Web site:** http://www.sshm.org

Turkish Society for the History of Science (TBTK). Membership is closed. A membership or subscription fee is charged. 62 members. Meets biennially with the primary meeting in September-November. **Contact:** Ekmeleddin Mehmet Ihsanoglu, ircica@superonline.com **Address:** Turk Bilim Tarihi Kurumu, Barbaros Bulvari, Yildiz Sarayi, Musahip Agalar Dairesi, PO Box 234, 80700 Besiktas, Istanbul, Turkey. **Tel:** 90.212.260.0717 **Fax:** 90.212.258.4365 **Founded:** 1989 **Purpose:** To support research in the field of history of science in Turkey and to ensure cooperation among historians of science. **Publications:** *Newsletter of the Turkish Society for History of Science*. **Offers grants/fellowships:** No. **Awards prizes:** No. **Description:** The Society was established in 1989 in order to support research undertaken in the field of history of science in Turkey and to ensure cooperation among historians of science. In its capacity as a member in the International Union of the History and Philosophy of Science, Division of History of Science (IUHPS/DHS), TBTK represents Turkey in the international arena. For the purpose of exchanging ideas, input, and experiences among scholars from various disciplines, the Society organizes biennial scholarly meetings/symposia on themes of history of science and technology in Turkey. **Web site:** http://www.ircica.org

Journals and Newsletters

Preliminary research identified 212 potential serial publications in the history of science. 23 of these were no longer active. Of the remainder, 140 submitted data. 116 respondents described their publications as journals, while 14 returned information on newsletters. 10 respondents declined to classify themselves as a journal or newsletter. HSS staff members gathered data from web pages to augment entries for approximately a dozen journals that failed to respond. Publications are sorted alphabetically by their primary title. A list of inactive publications completes the section. Not all publications responded with complete data. All entries contain some portion of the following information:

Title
English translation of title, if applicable
Frequency
Type of publication (refereed or non-refereed journal, or newsletter)
International Standard Serial Number (ISSN)
Editor's name and e-mail address
Editorial office mailing address
Telephone number
Facsimile number
Name of publishing press
Primary subject of the publication, entered using standardized keywords
Year founded
Academic society or organization which produces the publication
Distribution format (print, print and electronic, web-based, or e-mail-based)
Circulation
Additional contents (professional news, book reviews, and/or bibliographies)
Degree of indexing
Availability of reprints
Is the publication electronically archived?
Brief description of the publication and its special features
URL of publication's Web site

Journals and Newsletters

Acta Musei Moraviae. Supplementum: Folia Mendeliana. Annual refereed journal. **ISSN:** 0085-0748. **Ed.** Anna Matalova, amatalova@mzm.cz **Primary E-mail:** genetika@mzm.cz **Address:** The Moravian Museum, Udolni 39, Brno, 659 37, Czech Republic. **Tel:** 42.5.4221.6216 **Fax:** 42.5.4221.2792 **Publishing Press:** 1965. **Founded:** 1965. **Produced By:** The Moravian Museum. **Format:** Print. **Circulation:** Less than 250. Includes professional news, book reviews, bibliographies. Partially indexed. Reprints available. Not archived electronically. **Description:** Analysis of archival sources pertaining to evolution of hereditary concepts and the early development of genetics focusing on Gregor Johann Mendel (1822-1884), the author of the first model of the transfer of genetic information from generation to generation. The journal is published in Brno where Mendel lived and worked and where his archival documents are kept in the Moravian Museum, the Augustinian Monastery, and the State Archives.

AIP History of Physics Newsletter. Semiannual newsletter. **Ed.** Spencer Weart, sweart@aip.org **Primary E-mail:** chp@aip.org **Address:** One Physics Ellipse, College Park, MD, 20740-3843, United States. **Tel:** 1.301.209.3165 **Fax:** 1.301.209.0882 **Publishing Press:** American Institute of Physics. **Founded:** 1964. **Produced By:** Center for History of Physics. **Format:** Print and Electronic. **Circulation:** 2500 to 5000. Includes professional news, bibliographies. Not indexed. Reprints available. Archived electronically. **Description:** Information on activities of the Center for History of Physics and other organizations, archives and individuals working to preserve and make known the history of modern physics, astronomy, geophysics and allied fields. Available free. **Web site:** http://www.aip.org/history/web-news.htm

Air Power History: The Journal of Air and Space History. Quarterly refereed journal. **ISSN:** 1044-016X. **Ed.** Jacob Neufeld, neufeldj@starpower.net **Address:** Air Power History, P. O. Box 10328, P.O. Box 10328, Rockville, MD, 20849-0328. **Tel:** 301.279.2718 **Fax:** 301.279.8861 **Publishing Press:** Sheridan Press. **Founded:** 1953. **Produced By:** Air Force Historical Foundation. **Format:** Print. **Circulation:** Greater than 5000. Includes professional news, book reviews. Partially indexed. Reprints available. Not archived electronically. **Description:** We publish quality articles in the field of military air and space history. Articles are based on sound scholarship, perceptive analysis and/or firsthand experience. The primary criterion is that the manuscript contribute to knowledge. Each issue includes between six and ten scholarly book reviews. Other departments feature letters, announcements, news, upcoming events, and a "History Mystery." **Web site:** http://www.afhistoricalfoundation.com

Ambix. Tri-annual refereed journal. **ISSN:** 0002-6980. **Ed.** Peter Morris, p.morris@nmsi.ac.uk **Address:** The Science Museum, Exhibition Road, London, SW7 2DD, United Kingdom. **Tel:** 44.0.20.7942.4167 **Fax:** 44.0.20.7942.4302

Publishing Press: Black Bear Press Limited. **Primary Subject:** Physical Sciences. **Founded:** 1937. **Produced By:** Society for the History of Alchemy and Chemistry. **Format:** Print. **Circulation:** 250 to 500. Includes professional news, book reviews. Partially indexed. Reprints available. Not archived electronically. **Description:** *Ambix*, which has an international editorial board, appears three times a year in March, July and November, and contains scholarly articles and reviews of books on all aspects of the history of the chemical sciences. **Web site:** http://www.open.ac.uk/Arts/HST/SHAC/ambix.htm

American Heritage of Invention & Technology. Non-refereed Journal. **ISSN:** 8756-7296. **Ed.** Frederick Allen **Publishing Press:** Forbes, Inc. **Founded:** 1985. **Format:** Print and Electronic. **Web site:** http://www.americanheritage.com/it/index.shtml

Analecta: Studies and Materials for the History of Science. Semiannual refereed journal. **ISSN:** 1230-1159. **Ed.** H. Lichocka **Primary E-mail:** ihn@ihnpan.waw.pl **Address:** IHN PAN, Nowy Swiat 72, Warszawa, 00-330, Poland. **Tel:** 65.72.864 ext. 245 **Fax:** 48.22.826.61.37 **Publishing Press:** Wydawnictwa IHN PAN. **Founded:** 1992. **Produced By:** Polska Akademia Nauk, Instytut Historii Nauki. **Format:** Print. Includes professional news, book reviews, bibliographies. Partially indexed. Reprints available. **Web site:** http://www.ihnpan.waw.pl/

Annals of Science: The History of Science and Technology. Quarterly refereed journal. **ISSN:** 0003-3790. **Ed.** Trevor H. Levere, annals.science@utoronto.ca **Address:** Institute for the History & Philosophy of Science & Technology, Room 316, Victoria College, University of Toronto, Toronto, Ontario, M5S 1K7, Canada. **Fax:** 1.416.978.3003 **Publishing Press:** Taylor & Francis Group. **Founded:** 1936. **Format:** Print and Electronic. **Circulation:** 500 to 1000. Includes book reviews. Partially indexed. Reprints unavailable. Not archived electronically. **Description:** *Annals of Science* is an international historical journal covering science and technology since classical antiquity, and welcomes articles in English, French, and German. Each issue includes an essay review and book reviews. It has an active international board. A unique feature is the reproduction of selected illustrations in color. **Web site:** http://www.tandf.co.uk/journals/tf/00033790.html

Arabic Sciences and Philosophy: A Historical Journal. Semiannual refereed journal. **ISSN:** 0957-4239. **Eds.** Ahmad Hasnaoui, hasnaoui@vjf.cnrs.fr; B. Musallam; Roshdi Rashed, rashed@paris7.jussieu.fr **Primary E-mail:** rashed@ paris7.jussieu.fr **Address:** CNRS-UMR 7062, 7 rue Guy Moquet - Bp n° 8, Villejuif Cedex, F-94801, France. **Tel:** 33.1.46.65.42.37 **Fax:** 33.1.46.65.42.37 **Publishing Press:** Cambridge University Press. **Founded:** 1991. **Format:** Print. **Circulation:** 250 to 500. Includes book reviews. Not indexed. Reprints available. Archived electronically. **Description:** *Arabic Sciences and Philosophy* is devoted to the history of the Arabic Sciences, mathematics and philosophy.

It publishes articles of the highest academic and editorial standards, focusing
on these disciplines in the world of Islam between the eighth and the eighteenth
centuries. It also covers the interrelations between the Arabic sciences and
philosophy and the Greek, Indian, Chinese, Latin, Byzantine, Syriac and
Hebrew sciences and philosophy. *Arabic Sciences and Philosophy* explores the
development of these disciplines within the social and ideological context in
which this growth took place. **Web site:** http://uk.cambridge.org/journals/asp/

Archaeoastronomy: The Journal of Astronomy in Culture. Annual refereed
journal. **ISSN:** 0190-9940. **Eds.** John B. Carlson, jcarlson@deans.umd.edu;
David S. P. Dearborn, dearborn2@llnl.gov; Constance H. McCluskey,
chmccluskey@msn.com; Stephen C. McCluskey, scmcc@wvu.edu; Clive L.
N. Ruggles, rug@le.ac.uk **Address:** The Center for Archaeoastronomy, P.O.
Box X, College Park, MD, 20741-3022, United States. **Tel:** 1.301.864.6637
Fax: 1.301.699.5337 **Publishing Press:** University of Texas Press, Austin.
Primary Subject: Humanistic Relations of Science. **Founded:** 1977.
Produced By: Center for Archaeoastronomy. **Format:** Print. **Circulation:**
500 to 1000. Includes book reviews, bibliographies. Partially indexed. Reprints
unavailable. Not archived electronically. **Description:** Beginning in 1977,
Archaeoastronomy has served as an outlet for refereed articles studying
the practice, use, and meaning of astronomy in cultural contexts. A broad
range of topics, including indigenous cosmologies, measurement systems,
calendrics, mathematics, geometry, surveying, navigation, and even settlement
planning, have found expression under the headings of archaeoastronomy,
ethnoastronomy, and history of astronomy. Archaeoastronomy (including
ethnoastronomy) is an interdisciplinary field incorporating the values,
methodologies and goals consistent with a number of established disciplines
such as astronomy, anthropology (including archaeology and ethnology), history
of science (history of astronomy), art history, geography and religious studies.
Web site: http://www.myriadennium.org

**Archives Internationales d'Histoire des Sciences (International Archives
of History of Science).** Semiannual refereed journal. **ISSN:** 0003-9810.
Ed. R. Halleux **Primary E-mail:** chst@ulg.ac.be **Address:** Archives
Internationales d'Histoire des Sciences, Administrative Secretariat: Centre
d'Histoire des Sciences, Université de Liège, 5 quai Banning, Liège, B-4000,
Belgium. **Tel:** 32.4.366.94.79 **Fax:** 32.4.366.94.47 **Founded:** 1919; New
Series 1972. **Produced By:** Académie internationale d'Histoire des Sciences.
Format: Print. **Circulation:** 250 to 500. Includes professional news, book
reviews, bibliographies. Not indexed. Reprints available. Not archived
electronically. **Description:** *Archives Internationales d'Histoire des Sciences*
is concerned with studies devoted to topics ranging from the earliest times to
the present, and covering extremely varied fields: astronomy, biology, physics,
and philosophy with a markedly interdisciplinary approach. The contributions
of scholars from all over the world are published in English, Italian, French,
and German, thus promoting exchange between those from a variety of cultural

backgrounds. Its international nature makes the *Archives* different from other publications in the field; it neither depends upon, nor is influenced by any particular local school of thought or historiographic methodology.

Archiwum Historii i Filozofii Medycyny. Quarterly refereed journal. **ISSN:** 0860-1844. **Eds.** Zdzislaw Gajda; Andrzej Sródka; Stanislaw Zwolski, surgo@interia.pl **Address:** Kopernika 7, Krakow, 31-034, Poland. **Tel:** 48.12.422.21.16 **Fax:** 48.12.422.21.16 **Founded:** 1924. **Produced By:** Polskie Towarzystwo Historii Medycyny i Farmacji. **Format:** Print. **Circulation:** 250 to 500. Includes professional news, book reviews. Partially indexed. Reprints unavailable. Not archived electronically.

Australian Society for the History of Medicine Newletter. Newsletter. **Address:** P.O. Box 8034, Mt. Pleasant, MacKay, Queensland, 4740, Australia.

Berichte zur Wissenschaftsgeschichte: Organ der Gesellschaft für Wissenschaftsgeschichte. Quarterly refereed journal. **ISSN:** 0170-6233. **Ed.** Fritz Adolf Krafft, krafft@mailer.uni-marburg.de **Address:** Philipps-Universität, Institut für Geschichte der Pharmazie, Roter Graben 10, Marburg, D-35032, Germany. **Tel:** 49.0.6421.282.28.29 **Fax:** 49.0.6421.282.28.78 **Publishing Press:** Wiley-VCH. **Founded:** 1978. **Produced By:** Gesellschaft für Wissenschaftsgeschichte. **Format:** Print. **Circulation:** 500 to 1000. Includes professional news, book reviews, bibliographies. Indexed. Reprints available. Archived electronically. **Description:** *Berichte zur Wissenschaftsgeschichte*, the official periodical of the German Society for the History of Science, founded and edited by Fritz Krafft, is a unique source for interdisciplinary historical subjects covering topics in the history of science, medicine, humanities, arts and engineering. The contributions are partly based on the papers of the annual symposia of the Society. Bridging the gap between the "two (or three) cultures," the journal also acquaints scientists with the history of their field. It illustrates the consequences of ideas and perceptions and shows the influence of theories and thoughts on the evaluation of historical facts. **Web site:** http://staff-www.uni-marburg.de/~krafft/Berichte.htm

Biologist. Bi-monthly refereed journal. **ISSN:** 0006-3347. **Ed.** Alison M. Bailey, a.bailey@iob.org **Primary E-mail:** biologist@iob.org **Address:** 20 Queensberry Place, London, SW7 2DZ, United Kingdom. **Tel:** 44.20.7581.8333 **Fax:** 44.20.7823.9409 **Publishing Press:** Institute of Biology. **Founded:** 1953. **Produced By:** Institute of Biology. **Format:** Print and Electronic. **Circulation:** Greater than 5000. Includes book reviews. Partially indexed. Reprints available. Archived electronically. **Description:** *Biologist* is a fully peer-reviewed and citation-listed journal that carries the full richness and diversity of biological research today. Science is brought to life with stimulating and authoritative review articles. Topical pieces discuss science policy, new developments or controversial issues. Aimed at professional biologists everywhere, its straightforward style makes it ideal for educators and students at all levels, as well as the interested amateur. If you feel able to convey your enthusiasm and expertise to a wide audience please contact the editor; feature submissions, historical articles, short opinion pieces and letters are welcome. **Web site:** http://www.iob.org/biologist

Biology and Philosophy. Quarterly refereed journal. **ISSN:** 0169-3867. **Ed.** Kim Sterelny, Kim.Sterelny@vuw.ac.nz **Primary E-mail:** editdept@wkap.nl **Address:** Biology and Philosophy, Kluwer, Journals Editorial Office, PO Box 990, Dordrecht, 3300 AZ, Netherlands. **Fax:** 31.78.6392555 **Publishing Press:** Kluwer Academic Publishers Group. **Primary Subject:** Biological Sciences. **Founded:** 1986. **Format:** Print and Electronic. **Circulation:** 500 to 1000. Includes book reviews. Partially indexed. Reprints unavailable. Not archived electronically. **Description:** *Biology and Philosophy* appears five times a year; one issue per year is a special issue with a guest editor. The journal does publish papers and review books with a significant historical component, but it neither publishes papers nor reviews works that are purely historical. As well as publishing book reviews, it also publishes area reviews. **Web site:** http://www.wkap.nl/journalhome.htm/0169-3867

Bollettino di Storia delle Scienze Matematiche. Semiannual refereed journal. **ISSN:** 0392-4432. **Ed.** Enrico Giusti, giusti@math.unifi.it **Address:** Dipartimento di Matematica, Viale Morgagni 67/a, Firenze, I-50134, Italy. **Tel:** 39.055.422.3275 **Fax:** 39.055.422.2695 **Publishing Press:** Istituti Editoriali e Poligrafici Internazionali. **Primary Subject:** Mathematics. **Founded:** 1981. **Produced By:** Il Giardino di Archimede. **Format:** Print. **Circulation:** 250 to 500. Includes book reviews. Reprints available. **Description:** The journal publishes correspondence and unpublished manuscripts of mathematicians of the past, bibliographical essays, and original papers concerning the history of mathematical sciences. **Web site:** http://www.math.unifi.it/archimede/archimede/bollettino/bollettino.html

British Journal for the History of Science. Quarterly refereed journal. **ISSN:** 0007-0874. **Ed.** Crosbie W. Smith, c.smith@ukc.ac.uk **Address:** Centre for History & Cultural Studies of Science, Rutherford College, University of Kent, Canterbury, Kent, CT2 7NX, United Kingdom. **Tel:** 44.1227.823791 **Fax:** 44.1227.827258 **Publishing Press:** Cambridge University Press. **Founded:** 1962. **Produced By:** British Society for the History of Science. **Format:** Print and Electronic. **Circulation:** 1000 to 2500. Includes book reviews. Indexed. Reprints available. Archived electronically. **Description:** *BJHS* is a research journal of international standing and provides an established and accessible resource for everyone concerned with the history of the sciences. While recognizing that the British Society for the History of Science is a "Broad Church" in matters historical, the editor seeks to encourage articles that treat the history of the sciences (including technology and medicine) in socio-cultural contexts. **Web site:** http://uk.cambridge.org/journals/bjh/

The British Journal for the Philosophy of Science. Quarterly refereed journal. **ISSN:** 0007-0882. **Ed.** Peter Clark **Primary E-mail:** bjps@st-andrews.ac.uk **Address:** The Editor, The British Journal for the Philosophy of Science, Department of Logic and Metaphysics, University of St. Andrews, Fife, KY16 9AL, United Kingdom. **Tel:** 44.0.1334.462459 **Fax:** 44.0.1334.462485 **Publishing Press:** Oxford University Press, Academic Division. **Founded:** 1950. **Produced By:** British Society for the Philosophy of Science. **Format:** Print and Electronic. **Circulation:** 1000 to 2500. Includes professional news, book reviews. Indexed. Reprints available. Archived electronically. **Description:** *The British Journal for the Philosophy of Science* is published for the British Society for the Philosophy of Science. The Society holds regular meetings and an annual conference. Society members can subscribe to the Journal at a reduced rate. **Web site:** http://www3.oup.co.uk/phisci/

Bulletin for the History of Chemistry. Semiannual refereed journal. **ISSN:** 1053-4385. **Ed.** Paul R. Jones, prjones@umich.edu **Address:** Department of Chemistry, University of Michigan, Ann Arbor, MI, 48109-1055, United States. **Fax:** 1.313.747.4865 **Founded:** 1988. **Produced By:** History of Chemistry Division, American Chemical Society. **Format:** Print. **Circulation:** 500 to 1000. Includes professional news, book reviews. Indexed. Reprints available. Not archived electronically. **Web site:** http://www.scs.uiuc.edu/~mainzv/HIST/bulletin/

Bulletin of Science, Technology, & Society. Bi-monthly refereed journal. **ISSN:** 0270-4676. **Ed.** Bill Vanderburg, vanderb@mie.utoronto.ca **Address:** University of Toronto, Dept of Mechanical and Industrial Engineering, 5 King's College Road, Toronto, ON, M5S 3G8, Canada. **Tel:** 1.416.978.2924 **Fax:** 1.416.978.3453 **Publishing Press:** Sage Science. **Founded:** 1981. **Format:** Print and Electronic. **Circulation:** 500 to 1000. Includes professional news, book reviews, bibliographies. Indexed. Reprints available. Archived electronically. **Description:** The journal is dedicated to understanding the role of science and technology in our world. Its aim is to apply this understanding to science and technology in order to ensure that the biosphere can support a full range of human values and aspirations. **Web site:** http://www.sagepub.co.uk/journals/details/j0246.html

Bulletin of the History of Medicine. Quarterly refereed journal. **ISSN:** 0007-5140. **Eds.** Gert H. Brieger; Jerome Bylebyl, jjbyleb@jhmi.edu **Primary E-mail:** sab@jhmi.edu **Address:** The Editors, Bulletin of the History of Medicine, Room 306, 1900 East Monument Street, Baltimore, MD, 21205, United States. **Tel:** 1.410.955.3179 **Fax:** 1.410.502.6819. **Publishing Press:** Johns Hopkins University Press. **Primary Subject:** Medical Sciences. **Founded:** 1933. **Produced By:** American Association for the History of Medicine. **Format:** Print and Electronic. **Circulation:** 1000 to 2500. Includes book reviews. Indexed. Reprints available. Archived electronically. **Description:** A leading journal in its field for more than half a century, the

Bulletin spans the social and scientific aspects of the history of medicine worldwide and includes reviews of recent, diverse books on medical history. AAHM members receive the *Bulletin* upon payment of the annual dues. **Web site:** http://muse.jhu.edu/journals/bulletin_of_the_history_of_medicine/

Canadian Bulletin of Medical History/Bulletin Canadien d'Histoire de la Medecine. Semiannual refereed journal. **ISSN:** 0823-2105. **Ed.** Cheryl Krasnick Warsh, warshc@mala.bc.ca **Address:** Dept of History, Malaspina University College, 900 Fifth Street, Nanaimo, British Columbia, V9R 5S5, Canada. **Tel:** 1.250.753.3245.ext.2113 **Fax:** 1.250.740.6459 **Publishing Press:** Wilfrid Laurier University Press. **Primary Subject:** Medical Sciences. **Founded:** 1984. **Produced By:** Canadian Society for the History of Medicine. **Format:** Print. **Circulation:** 250 to 500. Includes professional news, book reviews, bibliographies. Partially indexed. Reprints available. Archived electronically. **Description:** *Canadian Bulletin of Medical History/Bulletin Canadien d'Histoire de la Medecine* publishes articles, archival and research notes in either French or English on all aspects of the history of medicine, health care and related disciplines. While its focus is Canadian history, it publishes research from any geographical area and any temporal period. **Web site:** http://meds.queensu.ca/medicine/histm/cshmweb/publications.html

Centaurus (Copenhagen): International Magazine of the History of Mathematics, Science and Technology. Quarterly refereed journal. **ISSN:** 0008-8994. **Ed.** Ole Knudsen **Primary E-mail:** centaurus@ifa.au.dk **Address:** History of Science Department, Aarhus University, Ny Munkegade, Aarhus, DK-8000, Denmark. **Tel:** 45.8942.3512 **Fax:** 45.8942.3510 **Publishing Press:** Blackwell Munksgaard International Publishers Ltd. **Founded:** 1957. **Format:** Print and Electronic. **Circulation:** 250 to 500. Includes book reviews. Partially indexed. Reprints unavailable. Archived electronically. **Web site:** http://www.blackwellmunksgaard.com/centaurus

College of Physicians of Philadelphia. Transactions & Studies. Annual non-refereed journal. **ISSN:** 0010-1087. **Ed.** Edward T. Morman, emorman@collphyphil.org **Primary E-mail: Address:** College of Physicians of Philadelphia, 19 South 22nd Street, Philadelphia, PA, 19103-3097, United States. **Tel:** 1.215.563.3737.ext.265 **Fax:** 1.215.569.0356 **Publishing Press:** Watson Publishing International. **Primary Subject:** Medical Sciences. **Founded:** 1793. **Produced By:** College of Physicians of Philadelphia. **Format:** Print. **Circulation:** Less than 250. Not indexed. Reprints available. Not archived electronically. **Description:** First published in 1793, the journal *Transactions and Studies of the College of Physicians of Philadelphia* has been published annually since 1841. It contains annual reports of the president, sections, and program divisions, selected lectures presented at the College, articles which highlight the rich historical resources of the Library and Mütter Museum, and memoirs of deceased Fellows. **Web site:** http://www.collphyphil.org

Conecta: Boletin de Noticias sobre Historia de la Ciencia y la Tecnologia (Conecta: Bulletin of News in the History of Science, Medicine and Technology). Weekly newsletter. **ISSN:** 1576-4826. **Eds.** Francisco J. Martínez, fjmartinez@servet.uab.es; Rosa María Medina, rosam@ugr.es; Enrique Perdiguero, quique@umh.es **Address:** History of Science Division, Miguel Hernández University, San Juan, Alicante, 03550, Spain. **Tel:** 34.65919514 **Fax:** 34.65919551 **Founded:** 1995. **Produced By:** Miguel Hernández University. **Format:** Electronic (E-Mail Distribution). **Circulation:** Less than 250. Includes professional news, book reviews, bibliographies. Not indexed. Reprints unavailable. Archived electronically. **Description:** The publication is a newsletter about the history of science, medicine and technology and is devoted to spreading information in Spain and countries in Latin America, with special attention to the European context. **Web site:** http://dsp.umh.es/conecta

Configurations: A Journal of Literature, Science and Technology. Tri-annual refereed journal. **ISSN:** 1063-1801. **Eds.** James J. Bono, hischaos@acsu. buffalo.edu; T. Hugh Crawford, hugh.crawford@lcc.gatech.edu; Paula Findlen, pfindlen@leland.stanford.edu **Primary E-mail:** hischaos@acsu.buffalo.edu **Address:** Configurations, Georgia Institute of Technology, School of Literature, Communication, and Culture, Atlanta, GA, 30332-0165, United States. **Publishing Press:** Johns Hopkins University Press. **Primary Subject:** Humanistic Relations of Science. **Founded:** 1993. **Format:** Print and Electronic. **Circulation:** 500 to 1000. Includes book reviews, bibliographies. Indexed. Archived electronically. **Description:** *Configurations* is an interdisciplinary journal devoted to the interrelations among science, technology and culture. Special thematic issues are published periodically. **Web site:** http://muse.jhu.edu/journals/configurations/index.html

Cronos: Cuadernos Valencianos de Historia de la Medicina y de la Ciencia. Semiannual refereed journal. **ISSN:** 1139-711X. **Ed.** Jose L. Fresquet, jose.fresquet@uv.es **Primary E-mail:** iu.historia.ciencia.doc@uv.es **Address:** Universidad De Valencia, Instituto De Historia de la Ciencia y Documentacion, Valencia, 46010, Spain. **Tel:** 963864164 **Fax:** 963613975 **Founded:** 1962. **Produced By:** Universidad de Valencia, Instituto de Historia de la Ciencia y Documentacion. **Format:** Print. **Circulation:** 500 to 1000. Includes book reviews, bibliographies. Indexed. Reprints available. **Web site:** http://www.uv.es/IHCD/cronos/revista.html

Current Work in the History of Medicine: An International Bibliography. **ISSN:** 0011-3999. **Primary E-mail:** niscom@sirnetd.irnet.in **Address:** National Institute of Science Communication, Pusa Gate, K.S. Krishnan Marg, New Delhi, 110012, India. **Founded:** 1954.

Daedalus. **ISSN:** 0070-2528. **Ed.** Mats Hojeberg **Founded:** 1931. **Produced By:** Tekniska Museet. **Web site:** http://www.tekmu.se

Daedalus: Journal of the American Academy of Arts and Sciences. Quarterly refereed journal. **ISSN:** 0011-5266. **Primary E-mail:** daedalus@amacad.org **Address:** 136 Irving Street, Cambridge, MA, 02138, United States. **Tel:** 1.617.491.2600 **Fax:** 1.617.576.5088 **Publishing Press:** Cadmus Professional Communications. **Founded:** 1950. **Produced By:** American Academy of Arts and Sciences. **Format:** Print and Electronic. **Circulation:** Greater Than 5000. **Web site:** http://www.amacad.org/publications/daedalus.htm

Dialogues: Newsletter of the Canadian Science and Technology Historical Association. Irregular newsletter. **Primary E-mail:** di827@freenet.carleton.ca **Address:** 708 rue Morin, Ottawa, Ontario, KIK 3G9, Canada. **Tel:** 1.613.746.1914 **Founded:** 1995. **Produced By:** Canadian Science and Technology Historical Association. **Format:** Print and Electronic. **Circulation:** Less than 250. Includes professional news. Not indexed. Reprints available. Archived electronically. **Description:** *Dialogues* accepts articles in English and French which may be of interest to members of the Canadian Science and Technology Historical Association. The newsletter, however, publishes neither scholarly articles nor book reviews. **Web site:** http://www.physics.uoguelph.ca/hist/CSTHA.html

Dynamis: Acta Hispanica ad Medicinae Scientiarumque Historiam Illustrandam. Annual refereed journal. **ISSN:** 0211-9536. **Eds.** Jon Arrizabalaga, jonarri@bicat.csic.es; Teresa Ortiz-Gómez, tortiz@ugr.es **Primary E-mail:** tortiz@ugr.es **Address:** Dpto. Historia de la Medicina, Facultad de Medicina, Avda. de Madrid, Granada, 18008, Spain. **Tel:** 34.958.243512 **Fax:** 34.958.246116 **Publishing Press:** Editorial Universidad de Granada. **Founded:** 1981. **Format:** Print. **Circulation:** 250 to 500. Includes professional news, book reviews. Partially indexed. Reprints available. Archived electronically. **Description:** *Dynamis* is devoted to the history of medicine, health and science. Articles written in European languages are accepted. Most articles are published in Spanish or English; however, articles written in French, Italian and German also appear. **Web site:** http://www.ugr.es/~dynamis/

Early Science and Medicine: A Journal for the Study of Science, Technology and Medicine in the Pre-Modern Period. Quarterly refereed journal. **ISSN:** 1383-7427. **Ed.** Christoph Lüthy, luethy@phil.kun.nl **Address:** Center for Medieval and Renaissance Natural Philosophy, Faculty of Philosophy. University of Nijmegen. P.O. Box 9103, Nijmegen, 6500 HD, Netherlands. **Tel:** 31.24.361.5750 **Fax:** 31.24.361.5564 **Publishing Press:** Brill Academic Publishers. **Founded:** 1996. **Format:** Print. **Circulation:** 250 to 500. Includes book reviews. Partially indexed. Reprints unavailable. Archived electronically. **Description:** *Early Science and Medicine* is an international quarterly dedicated to the history of science, medicine and technology from the earliest times through the end of the eighteenth century. Particular interest is given to the text-based traditions that link antiquity with the Latin, Arabic, and Byzantine Middle Ages, the Renaissance and the early modern period up to the Enlightenment. It favors diachronic studies in a variety of forms, including commented text editions and monographic studies of historical figures and scientific questions or practices. The main language of the journal is English, although contributions in French and German are also accepted. **Web site:** http://www.phil.kun.nl/center/esm.html

Earth Sciences History. Semiannual refereed journal. **ISSN:** 0736-623X. **Ed.** Gregory A. Good, ggood@wvu.edu **Address:** History Department, West Virginia University, Morgantown, WV, 26506-6303, United States. **Tel:** 1.304.293.2421.extension.5247 **Fax:** 1.304.293.3616 **Publishing Press:** Allen Press. **Primary Subject:** Earth Sciences. **Founded:** 1982. **Produced By:** History of Earth Sciences Society. **Format:** Print. **Circulation:** 250 to 500. Includes book reviews, bibliographies. Partially indexed. Reprints unavailable. Archived electronically. **Description:** *Earth Sciences History* publishes articles, book reviews, and essay reviews in the history of geology, geography, geophysics, oceanography, meteorology, and other geosciences. Articles are refereed by at least two scholars. Maximum length of articles is approximately fifty typed, double-spaced pages. Articles may be submitted to the editor or to associate editors listed on the web page. **Web site:** http://www.historyearthscience.org

East Asian Science, Technology, and Medicine. Semiannual refereed journal. **ISSN:** 1562-918X. **Ed.** Hans Ulrich Vogel **Primary E-mail:** eastm@ uni-tuebingen.de **Address:** Seminar für Sinologie und Koreanistik, Eberhard Karls Universität Tübingen, Wilhelmstr. 133, Wilhelmstr. 133, Tübingen, 72074, Germany. **Tel:** 49.0.7071.297.2711 **Fax:** 49.0.7071.295733 **Founded:** 1975. **Produced By:** International Society for the History of East Asian Science, Technology, and Medicine. **Format:** Print. **Circulation:** 250 to 500. Includes professional news, book reviews, bibliographies. Not indexed. Reprints available. Archived electronically. **Description:** This journal includes studies on the science, technology, and medicine of traditional and contemporary East Asia. See studies based on original research using Chinese, Japanese, or Korean primary sources or artifacts, that elucidate the relationships and interactions of science, technology, and medicine with politics, society, economics, philosophy, culture, religion, historiography, as well as their disciplinary traditions, or throws light on the work of scientists, technologists, and physicians in East Asia. **Web site:** http://www.uni-tuebingen.de/sinologie/eastm/daten/about.html

Endeavour. Quarterly non-refereed journal. **ISSN:** 0160-9327. **Ed.** E. Henry Nicholls, endeavour@current-trends.com **Primary E-mail:** endeavour@ current-trends.com **Address:** 84 Theobald's Road, Holborn, London, WC1X 8RR, United Kingdom. **Tel:** 44.0.207.611.4400 **Fax:** 44.0.207.611.4401 **Publishing Press:** Elsevier Science London. **Founded:** 1977. **Format:** Print and Electronic. **Circulation:** 500 to 1000. Includes professional news, book reviews, bibliographies. Indexed. Reprints available. Archived electronically. **Description:** *Endeavour* publishes review articles in the history of science, technology and medicine. Unlimited in its coverage, the journal provides a critical forum for the interdisciplinary exploration and evaluation of specific subjects or periods of scientific endeavor. Each issue also features meeting reports and an extensive collection of book reviews. **Web site:** http://www.elsevier.nl/locate/issn/0160-9327/

Environmental History. Quarterly refereed journal. **ISSN:** 1084-5453. **Ed.** Adam Rome, axr26@psu.edu **Primary E-mail:** marochak@duke.edu **Address:** 701 Wm. Vickers Ave., Durham, NC, 27701, United States. **Tel:** 1.919.682.9319 **Fax:** 1.919.682.2349 **Publishing Press:** Sheridan Press. **Founded:** 1996. **Produced By:** American Society for Environmental History and Forest History Society. **Format:** Print. **Circulation:** 1000 to 2500. Includes professional news, book reviews, bibliographies. Partially indexed. Reprints available. Not archived electronically. **Description:** *Environmental History* is the leading international journal devoted to the history of human interaction with the non-human world. The field of environmental history is defined broadly, and work is published from scholars in all disciplines that provide insight into important issues in environmental history. The journal's articles, book reviews, and bibliographic information cover all time periods and all places. **Web site:** http://www.lib.duke.edu/forest/ehmain.html

Epistemologia: An Italian Journal for the Philosophy of Science. Semiannual refereed journal. **Ed.** Evando Agazzi, agazzi@nous.unige.it **Address:** Dipartimento di Filosofia, via Balbi 4, Genova, 16126, Italy. **Tel:** 39.0.10.2099795 **Fax:** 39.0.10.2099792 **Founded:** 1978. **Format:** Print. **Circulation:** 500 to 1000. Includes book reviews. Indexed. Reprints available. Archived electronically. **Web site:** http://www.tilgher.it/epistemologiae.html

Foundations of Science. Quarterly refereed journal. **ISSN:** 1233-1821. **Ed.** Diederik Aerts, diraerts@vub.ac.be **Address:** Free University Belgium, Krijgskundestraat 33, P.O.Box 17, Brussels, 1160, Belgium. **Tel:** 32.2.644.26.77 **Fax:** 32.2.644.07.44 **Publishing Press:** Kluwer Academic Publishers. **Founded:** 1996. **Format:** Print and Electronic. **Circulation:** 500 to 1000. Includes professional news, book reviews. Not indexed. Reprints available. Archived electronically. **Description:** *Foundations of Science* serves as a forum for the development and exchange of ideas of both a practical and a theoretical nature. Topics of interest include methodological, foundational, and philosophically important issues relating to science. **Web site:** http://www.kluweronline.com/issn/1233-1821

The Gazette. Tri-annual newsletter. **ISSN:** 0962-7839. **Ed.** Carsten Timmermann **Primary E-mail:** gazette@sshm.org **Address:** Centre for the History of Science, Technology and Medicine, University of Manchester, Mathematics Tower, Oxford Road, Manchester, M13 9PL, United Kingdom. **Tel:** 44.161.275.7950 **Fax:** 44.161.275.5699 **Founded:** 1990. **Produced By:** Society for the Social History of Medicine. **Format:** Print and Electronic.

Circulation: 250 to 500. Includes professional news. Not indexed. Reprints unavailable. Archived electronically. **Description:** The *Gazette* is the newsletter of the Society for the Social History of Medicine. Besides conference announcements and calls for papers, conference reports, obituaries and Society news, it also contains information on the annual SSHM essay competitions. **Web site:** http://www.sshm.org/gazette/gazette.html

Gesnerus: Swiss Journal of the History of Medicine and Sciences.
Quarterly refereed journal. **ISSN:** 0016-9161. **Ed.** Marcel H. Bickel, marcel.bickel@mhi.unibe.ch **Primary E-mail:** marcel.bickel@mhi.unibe.ch **Address:** Institut Universitaire Histoire de la Médecine, 1, ch. des Falaises, Lausanne, CH-1005, Switzerland. **Tel:** 41.021.314.7050 **Fax:** 41.021.314.7055 **Publishing Press:** Schwabe und Co. AG. **Primary Subject:** Technology. **Founded:** 1943. **Format:** Print. **Circulation:** 500 to 1000. Includes professional news, book reviews, bibliographies. Partially indexed. Reprints available. Archived electronically. **Description:** *Gesnerus* is the official journal of the Swiss Society for the History of Medicine and Sciences. It publishes original articles, short communications, news, activities, and book reviews. Languages used are German, English, French. **Web site:** http://www.schwabe.ch/docs/mags/9161-0.htm

Guide to History of Science Courses in Britain. **Ed.** Janet Browne, j.browne@ucl.ac.uk **Primary E-mail:** c.treacey@ic.ac.uk **Address:** Centre for the History of Science, Technology and Medicine, Sherfield Building, Imperial College, London, SW7 2AZ. **Founded:** 1992. **Produced By:** British Society for the History of Science. **Format:** Electronic (Web-Based). **Web site:** http://www.chstm.man.ac.uk/bshs/bshscour.htm

The Guide to the History of Science. Triennial non-refereed journal. **Ed.** Roger Turner, Roger@hssonline.org **Primary E-mail:** guide@hssonline.org **Address:** HSS Executive Office, Box 351330, University of Washington, Seattle, WA, 98195, United States. **Tel:** 1.206.543.9366 **Fax:** 1.206.685.9544 206.685.9544 **Publishing Press:** University of Chicago Press. **Founded:** 1983. **Produced By:** History of Science Society. **Format:** Print and Electronic. **Circulation:** 2500 to 5000. Indexed. Reprints unavailable. Not archived electronically. **Description:** The *Guide* indexes graduate programs, journals and newsletters, organizations, and museums involved in the history of science. A directory of members of the History of Science Society is also included. It is one of the most comprehensive international reference sources for current activities in HSTM. A searchable edition is available over the web. **Web site:** http://www.hssonline.org/guide

Historia Mathematica: International Journal of the History of Mathematics.
Quarterly refereed journal. **ISSN:** 0315-0860. **Eds.** Umberto Bottazzini, bottazzi@math.unipa.it; Craig Fraser, cfraser@chass.utoronto.ca **Primary E-mail:** hm@elsevier.com **Address:** 525 B Street, Suite 1900, San Diego,

CA, 92101-4495, United States. **Tel:** 1.619.699.6234 **Fax:** 1.619.699.6801 **Publishing Press:** Elsevier. **Primary Subject:** Mathematics. **Founded:** 1974. **Format:** Print. **Circulation:** 500 to 1000. Includes book reviews. Not indexed. Reprints available. Archived electronically. **Description:** *Historia Mathematica* publishes original research on the history of the mathematical sciences in all periods and cultures. **Web site:** http://www.chass.utoronto.ca/hm/

Historical Records of Australian Science. Semiannual refereed journal. **ISSN:** 0727-3061. **Ed.** Roderick W. Home, home@unimelb.edu.au **Address:** Department of History and Philosophy of Science, University of Melbourne, Parkville, Victoria, 3010, Australia. **Tel:** 61.3.8344.7037 **Fax:** 61.3.8344.7959 **Publishing Press:** CSIRO Publishing. **Primary Subject:** Australia and Oceania. **Founded:** 1966. **Produced By:** Australian Academy of Science. **Format:** Print and Electronic. **Circulation:** 250 to 500. Includes book reviews, bibliographies. Not indexed. Reprints available. Archived electronically. **Description:** *Historical Records of Australian Science* publishes peer-reviewed articles and book reviews on the history of science in Australia and the southwest Pacific, biographical memoirs of deceased Fellows of the Australian Academy of Science, and an annual bibliography of the history of Australian science. It is published in June and December. **Web site:** http://www.science.org.au/hras/index.htm

Historical Studies in Irish Science and Technology. Ed. R. Carol Power, carol.power@ds.ie **Address:** Royal Dublin Society, Ballsbridge, Dublin, 4, Ireland. **Founded:** 1980. **Produced By:** Royal Dublin Society.

Historical Studies in the Physical and Biological Sciences. Semiannual refereed journal. **ISSN:** 0890-9997. **Ed.** John L. Heilbron, john.heilbron@ dial.appleinter.net **Primary E-mail:** diana@socrates.berkeley.edu **Address:** Office for History of Science and Technology, 470 Stephens Hall, University of California, Berkeley, CA, 94720, United States. **Tel:** 1.510.642.4581 **Publishing Press:** University of California Press. **Primary Subject:** Physical Sciences. **Founded:** 1970. **Format:** Print. **Circulation:** 500 to 1000. Includes book reviews. Partially indexed. Reprints available. Archived electronically. **Description:** *HSPBS* specializes in archival studies of the history of physics and the biological sciences from around 1700 to the present. It covers both intellectual and institutional history. Contributors are invited to write to the editor concerning plans for their articles before submission. **Web site:** http://www.ucpress.edu/journals/hsps/

History and Philosophy of the Life Sciences. ISSN: 0391-9714. **Ed.** Bernardino Fantini, Bernardino.Fantini@medecine.unige.ch **Primary E-mail:** madams@sas.upenn.edu **Founded:** 1916. **Produced By:** Taylor & Francis Group. **Description:** An international journal devoted to the historical development of the life sciences and of their social and epistemological implications. The journal also covers the broader philosophical concerns of biology and medicine. The main interest of the journal is modern western scientific thought, although it also includes any period in history of the life sciences, (e.g. classical antiquity, the Middle Ages) and any cultural area, (e.g. Chinese and Indian medicine). **Web site:** http://www.tandf.co.uk/journals/tf/03919714.html

History and Technology. Quarterly refereed journal. **ISSN:** 0734-1512. **Ed.** John Krige, john.krige@hts.gatech.edu **Address:** School of History, Technology & Society, Georgia Institute of Technology, 685 Cherry Street, Atlanta, GA, 30332-0345, United States. **Tel:** 1.404.894.7765 **Fax:** 1.404.894.0535. **Publishing Press:** Routledge. **Founded:** 1984. **Format:** Print and Electronic. Includes book reviews. Indexed. Reprints available. Archived electronically. **Description:** *History and Technology* explores the links between technology and the scientific, cultural, economic, political and institutional contexts in which it is embedded. While not favoring any particular school or methodological approach, it particularly welcomes papers which explore the relationships between technology and society in new ways (e.g. social construction of technologies, technology and business history, large technical systems). **Web site:** http://www.tandf.co.uk/journals/titles/07341512.html

History of Psychiatry. Quarterly refereed journal. **ISSN:** 0957-154X. **Ed.** German E. Berrios, geb11@cam.ac.uk **Address:** University of Cambridge, Department of Psychiatry, Box 189, Addenbrooke's Hospital, Cambridge, CB2 2QQ, United Kingdom. **Tel:** 44.1223.336965 **Fax:** 44.1223.336968 **Publishing Press:** Alpha Academic. **Founded:** 1990. **Format:** Print. **Circulation:** 500 to 1000. Includes professional news, book reviews, bibliographies. Indexed. Reprints available. Archived electronically. **Description:** The journal maintains an eclectic historiographical position and publishes papers on all the narratives related to mental health and disorder, e.g. clinical, social, philosophical, political, theological and psychoanalytical. It encourages submissions from non-Anglophone countries and translations into English of seminal papers from all psychiatric cultures. **Web site:** http://www.alphaacademic.co.uk/hop.htm

History of Psychology. Quarterly refereed journal. **ISSN:** 1093-4510. **Ed.** Michael M. Sokal, msokal@wpi.edu **Address:** Department of Humanities and Arts, Worcester Polytechnic Institute, Worcester, MA, 01609, United States. **Tel:** 1.508.831.5712 **Fax:** 1.508.831.5932 **Publishing Press:** Educational Publishing Foundation. **Founded:** 1998. **Produced By:** American Psychological Association, Division of the History of Psychology. **Format:** Print and Electronic. **Circulation:** 500 to 1000. Includes professional news, book reviews, bibliographies. Indexed. Reprints available. Archived electronically. **Description:** *History of Psychology* serves as a forum for both psychologists and other interested scholars exploring the full range of current ideas and approaches relevant to the relationship between history and psychology. It features refereed scholarly articles addressing all aspects of psychology's past and of its interrelationship with the many contexts within which it has emerged and been practiced. It also publishes scholarly work in closely related areas, such as historical psychology (the history of consciousness and behavior), theory in psychology as it pertains to history, historiography, biography and autobiography, psychohistory, and the teaching of the history of psychology. *History of Psychology* also features essay reviews of thematically-related sets of books and other media addressing issues important for an understanding of this relationship. **Web site:** http://www.wpi.edu/~histpsy

History of Science: A Review of Literature and Research. Quarterly refereed journal. **ISSN:** 0073-2753. **Ed.** Rob Iliffe, r.iliffe@ic.ac.uk **Address:** Centre for History of Science, Technology and Medicine, Imperial College, London, SW7 2AZ, United Kingdom. **Tel:** 44.20.7594.9356 **Fax:** 44.20.7594.9353 **Publishing Press:** Science History Publications Ltd. **Founded:** 1962. **Format:** Print and Electronic. **Circulation:** 500 to 1000. Includes book reviews, bibliographies. Indexed. Reprints available. Archived electronically. **Description:** *History of Science* publishes articles that are concerned with the historiography and methodology of science, technology and medicine. Unless essays are relevant to these themes, it does not normally publish straightforward research articles. **Web site:** http://www.shpltd.co.uk/hs.html

History of Science Society Newsletter. Quarterly newsletter. **ISSN:** 0739-4934. **Ed.** Robert J. Malone **Primary E-mail:** info@hssonline.org **Address:** Box 351330, University of Washington, Seattle, WA, 98195, United States. **Tel:** 1.206.543.9366 **Fax:** 1.206.685.9544 **Founded:** 1972. **Produced By:** History of Science Society. **Format:** Print and Electronic. **Circulation:** 3,000. Includes professional news. Not indexed. Reprints unavailable. Not archived electronically. **Description:** The *HSS Newsletter* reaches 3,000 historians of science and institutions around the world. In addition to news, it regularly features articles on innovations in education. An annual survey of job openings and hiring patterns can be found in the October issue. **Web site:** http://www.hssonline.org

History of Technology. Annual refereed journal. **ISSN:** 0307-5451. **Primary E-mail:** vhiggs@continuumbooks.com **Address:** Institute of Historical Research, Senate House, University of London, London, WC1E 7HU, United Kingdom. **Publishing Press:** Continuum. **Founded:** 1976. **Format:** Print and Electronic. **Circulation:** 500 to 1000. Includes professional news, bibliographies. Not indexed. Reprints available. Archived electronically. **Description:** The journal takes a broad view of the history of technology, not merely incorporating both science and medicine, but all approaches to history, with special attention to social, cultural and institutional contexts. The range is from pre-classical to the present and is intentionally global and comparative. Planned special issues will embrace patents, the steam engine, and technology and arts and crafts, amongst others. The editor relies on an active international editorial board. **Web site:** http://www.cassell.co.uk

HOPOS (History of Philosophy of Science) Working Group Newsletter.
Semiannual newsletter. **ISSN:** 1527-9332. **Ed.** Saul Fisher, sf@mellon.org
Address: HOPOS Newsletter, The Oxford House, 3656 Johnson Avenue,
Bronx, NY, 10463, United States. **Tel:** 1.718.549.8417 **Primary Subject:**
Philosophy or Philosophy of Science. **Founded:** 1993. **Produced By:** History
of Philosophy of Science Working Group. **Format:** Electronic (Web-Based).
Circulation: 500 to 1000. Includes professional news, book reviews,
bibliographies. Partially indexed. Reprints available. Archived electronically.
Description: The *HOPOS Newsletter* is published by the History of Philosophy
of Science (HOPOS) Working Group, which is dedicated to the study of
historical topics in philosophy of science, from Aristotle to the very recent past.
The *Newsletter* features book reviews, announcements, conference reports, and
international surveys of resources. Current and past issues are located on the
HOPOS Working Group website. **Web site:** http://scistud.umkc.edu/hopos

Huntia: A Journal of Botanical History. Irregular refereed journal.
ISSN: 0073-4071. **Ed.** Scarlett Townsend, st19@andrew.cmu.edu
Address: Hunt Institute for Botanical Documentation, Carnegie Mellon
University, Pittsburgh, PA, 15213-3890, United States. **Tel:** 1.412.268.2434
Fax: 1.412.268.5677 **Publishing Press:** Allen Press. **Primary Subject:**
Biological Sciences. **Founded:** 1963. **Produced By:** Hunt Institute for
Botanical Documentation. **Format:** Print. **Circulation:** 250 to 500. Includes
book reviews, bibliographies. Not indexed. Reprints unavailable. Not archived
electronically. **Description:** *Huntia* publishes articles on all aspects of the
history of botany and is published irregularly in one or more numbers per
volume of approximately 200 pages. External contributions are welcome. Please
request the "Guidelines for Contributors" before submitting manuscripts for
consideration. **Web site:** http://huntbot.andrew.cmu.edu

HYLE: International Journal for Philosophy of Chemistry. Semiannual
refereed journal. **ISSN:** 1433-5158. **Ed.** Joachim Schummer **Primary E-mail:**
editor@hyle.org **Address:** Institute of Philosophy, University of Karlsruhe,
Karlsruhe, D-76128, Germany. **Tel:** 49.721.608.4774 **Fax:** 49.697.912.35861
Publishing Press: Institute of Philosophy, University of Karlsruhe. **Founded:**
1995. **Format:** Print and Electronic. Includes professional news, book
reviews, bibliographies. Indexed. Reprints available. Archived electronically.
Description: *HYLE* is a peer reviewed international journal dedicated to
all philosophical aspects of chemistry. Articles deal with epistemological,
methodological, foundational, and ontological problems of chemistry and
its subfields; the peculiarities of chemistry and relations to technology, other
scientific and non-scientific fields; aesthetical, ethical, and environmental
matters in chemistry; as well as philosophically relevant facets of the history,
sociology, linguistics, and teaching of chemistry. **Web site:** http://www.hyle.org

IEEE Annals of the History of Computing. Quarterly refereed journal.
ISSN: 1058-6180. **Ed.** Thomas Bergin, tbergin@american.edu

Primary E-mail: annals-ma@computer.org **Address:** IEEE Computer Society, 10662 Los Vaqueros Circle, 4400 Massachusetts Ave. N.W., Los Alamitos, CA, 90720, United States. **Tel:** 1.714.821.8380 **Fax:** 1.714.821.4010 **Primary Subject:** Technology. **Founded:** 1979. **Produced By:** Institute of Electrical and Electronics Engineers, Inc. **Format:** Print and Electronic. **Circulation:** 2500 to 5000. Includes professional news, book reviews, bibliographies. Indexed. Reprints unavailable. Archived electronically. **Description:** The *IEEE Annals of the History of Computing* serves as the archival journal for the history of computing. Its contents include historical papers and records concerned with the international computing and information processing field, and information on the heritage of computing and information processing for scholarly and educational purposes. The *Annals* is managed by an editor-in-chief, chosen by the Computer Society, who works with an appointed editorial board of scholars and practitioners. In addition to scholarly papers and practitioners' memoirs, *Annals* has departments which contain biographies, reviews, anecdotes, and current events of importance to computer historians. Prospective authors are strongly encouraged to visit the *Annals'* web page for more information. The web page also contains valuable information about the latest issue, including material which could not be accommodated due to space limitations. **Web site:** http://www.computer.org/annals/

Imago Mundi: The International Journal for the History of Cartography. Annual refereed journal. **ISSN:** 0308-5694. **Ed.** Catherine Delano Smith, c.delano-smith@qmw.ac.uk **Primary E-mail:** t.campbell@ockendon.clara.co.uk **Address:** Imago Mundi, 285 Nether Street, London, N3 IPD, United Kingdom. **Tel:** 44.20.8346.5112 **Founded:** 1935. **Format:** Print. **Circulation:** 500 to 1000. Includes professional news, book reviews, bibliographies. Not indexed. Reprints unavailable. Not archived electronically. **Description:** The only international scholarly journal solely concerned with all aspects of the study of early maps. **Web site:** http://ihr.sas.ac.uk/maps/imago.html

Institute for the History of Arabic Science Newsletter. **ISSN:** 0379-2927. **Ed.** Khaled Maghout **Founded:** 1977. **Produced By:** University of Aleppo, Institute for the History of Arabic Science. **Web site:** http://www.ou.edu/islamsci

International Studies in the Philosophy of Science. Tri-annual refereed journal. **ISSN:** 0269-8595. **Ed.** James W. McAllister, j.w.mcallister@let.leidenuniv.nl **Address:** Faculty of Philosophy, University of Leiden, Matthias de Vrieshof 4, P.O Box 9515, Leiden, 2300 RA, Netherlands. **Tel:** 31.71.527.2031 **Fax:** 31.71.527.2028 **Publishing Press:** Routledge. **Founded:** 1986. **Format:** Print and Electronic. **Circulation:** 250 to 500. Includes book reviews. Not indexed. Reprints available. Archived electronically. **Description:** *International Studies in the Philosophy of Science* publishes original research and reviews in philosophy of science and philosophically informed history and sociology of science. Its scope includes the foundations and methodology of the natural, social, and human sciences, philosophical implications of

particular scientific theories, and broader philosophical reflection on science.
Web site: http://www.tandf.co.uk/journals/routledge/02698595.html

Isis. Quarterly refereed journal. **ISSN:** 0021-1753. **Ed.** Margaret W. Rossiter,
isis@cornell.edu **Address:** 726 University Avenue, #201, Cornell University,
Ithaca, NY, 14850, United States. **Tel:** 1.607.254.4747 **Fax:** 1.607.255.0616
Publishing Press: University of Chicago Press, Journals Division.
Founded: 1912. **Produced By:** History of Science Society. **Format:** Print
and Electronic. **Circulation:** 2500 to 5000. Includes professional news,
book reviews. Indexed. Reprints unavailable. Archived electronically.
Description: *Isis* is the most widely circulating English language journal in the
history of science. It features scholarly articles on diverse topics, news of the
profession, and an extensive book review section. Every issue of the journal,
dating back to 1912, is now available online through J-STOR.
Web site: http://www.journals.uchicago.edu/Isis/

Issues in Science and Technology. Quarterly non-refereed journal. **Eds.** Kevin
Finneran, kfinnera@nas.edu; Bill Hendrickson, whendrick@aol.com **Primary
E-mail:** kfinnera@nas.edu **Address:** National Academy of Sciences, 2101
Constitution Av., Washington, DC, 20418, United States. **Tel:** 1.202.965.5648
Fax: 1.202.965.5649 **Founded:** 1984. **Produced By:** National Academies of
Science and Engineering and the University of Texas at Dallas. **Format:** Print
and Electronic. **Circulation:** Greater than 5000. Includes book reviews.
Indexed. Reprints unavailable. Archived electronically. **Description:** *Issues in
Science and Technology* provides a forum where academic experts, government
officials, business leaders, and others with a stake in public policy related to
science, technology, and medicine can address national problems with specific
proposals for action. **Web site:** http://www.nap.edu/issues

**JAHIGEO: Newsletter of the Japanese Association for the History
of Geology.** Annual newsletter. **ISSN:** 1442-3650. **Eds.** Yasumoto
Suzuki, suzuki@gerd.coi.jp; Michiko Yajima, pxi02070@nifty.ne.jp
Address: #403, 2-24-1, Minami-Ikebukuro, Toshima-ku, Tokyo, 171-0022,
Japan. **Tel:** 81.3.3812.7039 **Fax:** 81.3.3812.7039 **Founded:** 1999.
Produced By: The Japanese Association for the History of Geology.
Format: Print. **Circulation:** Less than 250. Includes professional news,
book reviews, bibliographies. Partially indexed. Reprints available. Not
archived electronically. **Description:** In March 1994, JAHIGEO was
established to recognize the centenary of the Geological Society of Japan.
JAHIGEO publishes an annual *Newsletter* in English, in order to inform
people at home and abroad about the study of the history of geology in Japan.
Web site: http://www.geocities.co.jp/Technopolis/9866/jahigeo1.html

Journal for General Philosophy of Science. Semiannual refereed journal.
ISSN: 0925-4560. **Eds.** Lutz Geldsetzer, geldsetzer@phil-fak.uni-
duesseldorf.de; Gert König **Primary E-mail:** cohnitz@phil-fak.uni-

duesseldorf.de **Address:** Philosophisches Institut der Heinrich-Heine-Universität Düsseldorf, Universitätsstrasse 1, Düsseldorf, D-40225, Germany. **Tel:** 49.211.81.11473 **Fax:** 49.211.81.11750 **Publishing Press:** Kluwer Academic Publishers. **Format:** Print and Electronic. Includes professional news, book reviews, bibliographies. Indexed. Reprints available. Archived electronically. **Description:** The *Journal for General Philosophy of Science* is a forum for the discussion of a variety of attitudes concerning the philosophy of science. It focuses on the methodological, ontological, epistemological, anthropological, and ethical foundations of the individual sciences. Particular emphasis is laid on bringing the natural, cultural, and technical sciences into a philosophical context. The historical presuppositions and conditions of current problems of the philosophy of science are also discussed. **Web site:** http://www.kluweronline.com/issn/0925-4560

Journal for the History of Astronomy. Quarterly refereed journal. **ISSN:** 0021-8286. **Ed.** Michael Hoskin, michael.hoskin@ntlworld.com **Address:** Churchill College, Cambridge, CB3 0DS, United Kingdom. **Tel:** 44.1223.840284 **Fax:** 44.1223.565532 **Publishing Press:** Science History Publications Ltd. **Primary Subject:** Astronomical Sciences. **Founded:** 1970. **Format:** Print. **Circulation:** 500 to 1000. Includes book reviews. Indexed. Not archived electronically. **Description:** *JHA* is devoted to the history of astronomy, astrophysics and cosmology, from the earliest civilizations to the present day, and to archaeoastronomy. Its subject matter extends to such allied fields as the history of navigation and timekeeping, and the use of historical records in the service of astronomy. **Web site:** http://www.shpltd.co.uk

Journal of Astronomical History and Heritage. Semiannual refereed journal. **ISSN:** 1440 2807. **Ed.** John L. Perdrix, astral@iinet.net.au **Address:** Astral Press, P.O. Box 107, Wembley, WA, 6913, Australia. **Tel:** 61.8.9387.4250 **Fax:** 61.8.9387.3981 **Publishing Press:** Astral Press. **Primary Subject:** Astronomical Sciences. **Founded:** 1998. **Format:** Print. **Circulation:** Less than 250. Includes book reviews, bibliographies. Partially indexed. Reprints available. Not archived electronically. **Description:** *JAHH* considers papers on all aspects of astronomical history, including studies which place the evolution of astronomy in political, economic, and cultural contexts. Papers on astronomical heritage may deal with historic telescopes and observatories and conservation projects. There are no page charges and authors receive 25 offprints free of costs. **Web site:** http://www.astralpress.com.au

Journal of Industrial History. Semiannual refereed journal. **ISSN:** 1463-6174. **Ed.** John F. Wilson, jjfwilson@aol.com **Address:** Nottingham University Business School, Jubilee Campus, Nottingham, Notts, NG8 1BB, United Kingdom. **Tel:** 44.0.115.846.7405 **Fax:** 440.115.846.6667 **Publishing Press:** Carnegie Publishing. **Founded:** 1998. **Produced By:** University of Nottingham International Business History Institute. **Format:** Print. **Circulation:** Less than 250. Includes professional news, book reviews. Not indexed. Reprints available.

Not archived electronically. **Description:** *JIH* covers industrial, business and management history, with no chronological or geographical boundaries. **Web site:** http://www.carnegiepub.co.uk/indhis.html

Journal of Medical Biography. Quarterly refereed journal. **ISSN:** 0967-7720. **Ed.** J. M. H. Moll **Primary E-mail:** delia.siedle@rsm.ac.uk **Address:** Journal of Medical Biography, Royal Society of Medicine Press, 1 Wimpole Street, London, W1G 0AE, United Kingdom. **Tel:** 44.20.7290.2923 **Fax:** 44.20.7290.2929 **Publishing Press:** Royal Society of Medicine Press. **Founded:** 1993. **Format:** Print. **Circulation:** 250 to 500. Includes book reviews. Indexed. Reprints available. Not archived electronically. **Description:** The *Journal of Medical Biography* is an international, peer-reviewed quarterly publication, indexed in MEDLINE, which focuses on the lives of people in or associated with the medical sciences, including those considered to be legendary as well as the less well known. Generously illustrated, it publishes papers in a variety of categories, such as Physicians, Surgeons, Investigators, Disorders, Patients, Truants and Entrepreneurs. **Web site:** http://www.rsm.ac.uk/pub/jmb.htm

Journal of the History of Biology. Tri-annual refereed journal. **ISSN:** 0022-5010. **Eds.** Garland E. Allen, allen@biology.wustl.edu; Jane Maienschein, maienschein@asu.edu **Primary E-mail:** jhb@asu.edu **Address:** Biology Department, PO Box 871501, Arizona State University, Tempe, AZ, 85287-1501, United States. **Tel:** 1.480.965.8644 **Fax:** 1.480.965.6684 **Publishing Press:** Kluwer Academic Publishers. **Primary Subject:** Biological Sciences. **Founded:** 1968. **Format:** Print and Electronic. **Circulation:** 500 to 1000. Includes book reviews, bibliographies. Indexed. Reprints available. Not archived electronically. **Description:** Generally devoted to the history of the life sciences in their intellectual, social and cultural context, including the interface of biology, medicine and technology. **Web site:** http://www.kluweronline.com/issn/0022-5010

Journal of the History of Dentistry. **ISSN:** 1089-6287. **Ed.** H. T. Loevy **Address:** American Academy of the History of Dentistry, 100 S. Vail Ave., Arlington Heights, IL, 60005, United States. **Founded:** 1952. **Produced By:** American Academy of the History of Dentistry.

Journal of the History of Ideas: An International Quarterly Devoted to Intellectual History. Refereed Journal. **ISSN:** 0022-5037. **Ed.** Donald R. Kelley, dkelley@rci.rutgers.edu **Address:** Journal of the History of Ideas, 88 College Avenue, Rutgers University, New Brunswick, NJ, 08901, United States. **Tel:** 1.732.932.1227 **Fax:** 1.732.932.8708 **Publishing Press:** Johns Hopkins. **Founded:** 1940. **Format:** Print and Electronic. Includes professional news, bibliographies. Indexed. Reprints available. Archived electronically. **Description:** Founded in 1940 by the philosopher Arthur O. Lovejoy, the *Journal of the History of Ideas* publishes articles in the history of philosophy, literature, the arts, the natural and human sciences, and religious, social, and political thought; it has become the world's leading journal of intellectual history. **Web site:** http://www.press.jhu.edu/press/journals/jhi/jhi.html

Journal of the History of Medicine and Allied Sciences. Quarterly refereed journal. **ISSN:** 0022-5045. **Ed.** Margaret E. Humphreys, meh@acpub.duke.edu **Address:** Box 90719, Department of History, Duke University, Durham, NC, 27708, United States. **Tel:** 1.919.684.2285 **Fax:** 1.919.681.7670 **Publishing Press:** Oxford University Press. **Primary Subject:** Medical Sciences. **Founded:** 1946. **Format:** Print and Electronic. **Circulation:** 1000 to 2500. Includes professional news, book reviews. Indexed. Reprints available. Archived electronically. **Web site:** http://www.jhmas.oupjournals.org

Journal of the History of the Behavioral Sciences. Quarterly refereed journal. **ISSN:** 0022-5061 (print); 1520-6696 (online). **Ed.** Raymond E. Fancher, fancher@yorku.ca **Address:** Department of Psychology, York University, 4700 Keele Street, Toronto, ON, M3J 1P3, Canada. **Tel:** 1.416.736.5115.ex t.66275 **Fax:** 1.416.736.5814 **Publishing Press:** John Wiley and Sons, Inc. **Founded:** 1965. **Produced By:** Published in affiliation with Cheiron (The International Society for the History of the Behavioral and Social Sciences), The European Society for the History of the Human Sciences; and the Forum for the History of the Human Sciences. **Format:** Print and Electronic. **Circulation:** 500 to 1000. Includes professional news, book reviews. Indexed. Reprints available. Archived electronically. **Description:** Devoted to the scientific, technical, institutional, and cultural history of the social and behavioral sciences. Coverage includes the core disciplines of psychology, anthropology, sociology, psychiatry and psychoanalysis, economics, linguistics, communications, political science, and the neurosciences, as well as related fields such as the history of science and medicine, historical theory, and historiography. **Web site:** http://www.interscience.wiley.com/jpages/0022-5061/

Journal of the History of the Neurosciences: Basic and Clincal Perspectives. Quarterly refereed journal. **ISSN:** 0964-704X. **Ed.** Stanley Finger, sfinger@ artsci.wustl.edu **Address:** Department of Psychology, Washington University, St. Louis, MO, 63130-4899, Usa. **Tel:** 1.314.935.6513 **Fax:** 1.314.935.7588 **Publishing Press:** Swets and Zeitlinger. **Primary Subject:** Medical Sciences. **Founded:** 1992. **Produced By:** International Society for the History of the Neurosciences. **Format:** Print and Electronic. **Circulation:** 250 to 500. Includes professional news, book reviews, bibliographies. Indexed. Reprints available. Archived electronically. **Description:** *JHN* is devoted to the history of the neurosciences, broadly defined. The subject matter includes the history of neurology, neurosurgery, neurophysiology, neurochemistry, etc. In addition to full-length papers, all refereed by at least two reviewers, *JHN* prints short papers and book reviews. Special columns (some to start in 2003) include: Anniversaries, Neurowords, Neurognostics (a quiz), The Arts, Stamps, and Instrumentation/Methods. **Web site:** http://www.szp.swets.nl/szp/journals/jh.htm

Journal of Transport History. Semiannual refereed journal. **Address:** 14 Bareli Street, St Mary's Road, Ealing, Tel Aviv, 69364, Israel. **Tel:** 972.3.699.5049 **Publishing Press:** Manchester University Press. **Founded:** 1953. **Format:** Print and Electronic. **Circulation:** 250 to 500. Includes book reviews, bibliographies. Partially indexed. Reprints unavailable. Archived electronically. **Description:** *JTH* is a fully refereed journal specializing in the economic, social and cultural aspects of transport and travel. Scholarly articles on any type of transport, in any country or period are considered for publication. **Web site:** http://www.manchesteruniversitypress.co.uk/jth.htm

Kagakushi Kenkyu. Quarterly refereed journal. **ISSN:** 0022-7692. **Address:** History of Science Society of Japan, West Pine Bldg., 2-15-19, Hirakawa-cho, Chiyoda-ku, Tokyo, 102-0093, Japan. **Tel:** 81.3.3239.0545 **Publishing Press:** Iwanami Shoten, Publishers. **Founded:** 1941. **Web site:** http://wwwsoc.nii.ac.jp/jshs/publ1.html

Kagakushi: The Journal of the Japanese Society for the History of Chemistry. Quarterly refereed journal. **ISSN:** 0386-9512. **Ed.** Yasu Furukawa, yfuru@cck.dendai.ac.jp **Primary E-mail:** kagakushi@jcom.home.ne.jp **Address:** c/o Makoto Ohno, Aichi Prefectural University, Kumabari, Nagakute, Aichi, 480-1198, Japan. **Tel:** 81.561.64.1111 **Fax:** 81.561.64.1107 **Primary Subject:** Physical Sciences. **Founded:** 1974. **Produced By:** The Japanese Society for the History of Chemistry. **Format:** Print. **Circulation:** 250 to 500. Includes professional news, book reviews, bibliographies. Not indexed. Reprints available. Not archived electronically. **Description:** Published by the Japanese Society for the History of Chemistry, *Kagakushi* deals exclusively with the history of chemistry and chemical technology. 100 issues have been published from 1974 to the fall of 2002. **Web site:** http://members.jcom.home.ne.jp/kagakushi/

Knowledge, Technology and Policy. Quarterly refereed journal. **ISSN:** 0897-1986. **Ed.** David Clarke, dsc@moted.org **Address:** 1309 West Walnut Street, Carbondale, IL, 62901, United States. **Publishing Press:** Transaction Publishers. **Founded:** 1986. **Format:** Print and Electronic. **Circulation:** 500 to 1000. Includes book reviews. Indexed. Reprints available. Archived electronically. **Description:** The editorial board of *Knowledge, Technology & Policy* is receptive to articles resting on the titular tripod. "Knowledge" means how technologies change the ways we think. Knowledge also refers to how we organize, access and use information–indeed, how we transform information into knowledge. "Policy" refers to what we should do about these things (if anything) as individuals, communities and governments. **Web site:** http://www.moted.org/kt&p/index.htm

Kwartalnik Historii Nauki i Techniki (Quarterly Journal of the History of Science and Technology). Quarterly refereed journal. **ISSN:** 0023-589X. **Ed.** Stefan Zamecki, ihn@ihnpan.waw.pl **Address:** IHN

PAN, Nowy Swiat 72, Warszawa, 00-330, Poland. **Tel:** 48.22.826.87.54
Fax: 48.22.826.61.37 **Publishing Press:** Wydawnictwo IHN PAN.
Founded: 1956. **Produced By:** Polska Akademia Nauk, Instytut Historii
Nauki. **Format:** Print. **Circulation:** 250 to 500. Includes professional
news, book reviews, bibliographies. Partially indexed. Reprints available.
Web site: http://www.ihnpan.waw.pl/

**Llull - Revista de la Sociedad Española de Historia de las Ciencias y de
las Técnicas (Llull - Journal of the Spanish Society for the History of
Science and Technology).** Tri-annual refereed journal. **ISSN:** 0210-8615.
Ed. Mariano Hormigón, hormigon@posta.unizar.es **Address:** S.E.H.C.Y.T.,
Facultad de Ciencias (Matemáticas), Ciudad Universitaria, Zaragoza,
E-50009, Spain. **Tel:** 34.9.7676.1119 **Fax:** 34.9.7676.1125 **Founded:** 1977.
Produced By: Sociedad Española de Historia de las Ciencias y de las Técnicas.
Format: Print. **Circulation:** 500 to 1000. Includes professional news, book
reviews, bibliographies. Indexed. Reprints available. Archived electronically.
Web site: http://www.unirioja.es/sehcyt

**Lychnos: Laerdomshistoriska Samfundets Aarsbok (Lychnos: Annual of
the Swedish History of Science Society).** Annual non-refereed journal.
ISSN: 0076-1648. **Ed.** Sven Widmalm, sven.widmalm@idehist.uu.se
Primary E-mail: lychnos@idehist.uu.se **Address:** Lychnos, Office for History
of Science, Box 256, Uppsala, SE-751 05, Sweden. **Tel:** 46.0.18.4711480
Fax: 46.0.18.108046 **Publishing Press:** Almqvist & Wiksell. **Founded:** 1936.
Produced By: The Swedish History of Science Society. **Format:** Print.
Circulation: 500 to 1000. Includes professional news, book reviews.
Partially indexed. Reprints unavailable. Not archived electronically.
Description: *Lychnos* is the only regular Swedish publication covering the
field of history of science. It contains scholarly articles and book reviews in the
history of science, intellectual history and cultural history. The content is mainly
in Swedish. **Web site:** http://www.vethist.idehist.uu.se/lychnos.html

**Manguinhos: Historia, Ciencias, Saude (Manguinhos: History, Sciences,
Health).** Tri-annual refereed journal. **ISSN:** 0104-5970. **Eds.** Jaime L. Benchimol,
jben@coc.fiocruz.br; Ruth Barbosa Martins, rmartins@coc.fiocruz,br
Primary E-mail: hscience@coc.fiocruz.br **Address:** Avenida Brasil 4365,
Prédio do Relógio, Rio de Janeiro, Rio de Janeiro, 21045-900, Brazil.
Tel: 55.21.2280.9241 **Fax:** 55.21.2598.4437 **Publishing Press:** Casa de
Oswaldo Cruz. **Founded:** 1994. **Produced By:** Fundacao Oswaldo Cruz.
Format: Print and Electronic. **Circulation:** 500 to 1000. Includes professional
news, book reviews, bibliographies. Indexed. Reprints available. Archived
electronically. **Description:** This journal links the social and life sciences.
Although the life sciences are its focal theme, it is also open to the historical
perspective in related fields. The essence of our journal is captured by three
words: history, sciences, health. By history we are referring to a way of
looking at things and interpreting and addressing them that is common to a

gamut of academic fields. Health, which might be considered our hallmark, embraces the issues that defines our place as scholars and as social actors. Sciences is plural because the interdisciplinary approach is sine qua non to progress in the biological sciences as much as in the social sciences. **Web site:** http://www.scielo.br/hcsm

Medical History. Quarterly refereed journal. **ISSN:** 0025-7273. **Ed.** William F. Bynum, w.bynum@ucl.ac.uk **Primary E-mail:** c.tonson-rye@ucl.ac.uk **Address:** Wellcome Trust Centre for the History of Medicine at UCL, 24 Eversholt St, London, NW1 1AD, United Kingdom. **Tel:** 44.0.20.7679.8107 **Fax:** 44.0.20.7679.8194 **Publishing Press:** The Wellcome Trust. **Founded:** 1957. **Format:** Print. **Circulation:** 500 to 1000. Includes professional news, book reviews. Indexed. Reprints unavailable. Not archived electronically. **Description:** *Medical History* publishes articles and notes on a whole range of medico-historical topics, from the development of medical ideas and therapies to public health, medical anthropology, and medical sociology. Its concern is to broaden and deepen the understanding of medicine through historical studies. **Web site:** http://www.ucl.ac.uk/histmed/publ.htm

Medizin, Gesellschaft, und Geschichte (Medicine, Society, and History). Annual refereed journal. **ISSN:** 0939-351X. **Ed.** Robert Jütte, robert.juette@ igm-bosch.de **Address:** Institut für Geschichte der Medizin der Robert Bosch Stiftung, Straussweg 17, Stuttgart, 70184, Germany. **Tel:** 49.711.4608.4171 **Fax:** 49.711.4608.4181 **Publishing Press:** Franz Steiner Verlag Stuttgart. **Primary Subject:** Medical Sciences, Social Relations of Science. **Founded:** 1984. **Produced By:** Institut für Geschichte der Medizin der Robert Bosch Stiftung. **Format:** Print. **Circulation:** 250 to 500. Includes professional news. Not indexed. Reprints unavailable. Not archived electronically. **Description:** *Medizin, Gesellschaft und Geschichte* is a forum for interdisciplinary approaches to medical history. It aims to broaden the scope of the social history of medicine. The journal is not only concerned with the history of regular medicine but also with alternative methods of healing. **Web site:** http://www.igm-bosch.de

Medizinhistorisches Journal (Medicohistorical Journal). Quarterly refereed journal. **ISSN:** 0025-8431. **Eds.** Johanna Bleker, johanna bleker@medizin. fu-berlin.de; Karl-Heinz Leven, leven@uni-freiburg.de; Georg Lilienthal, Georg.Lilienthal@LVB-Hessen.de; Andreas-Holger Maehle, a.h.maehle@ durham.ac.uk; Heinz-Peter Schmiedebach, geschmed@uni-greifswald.de; Paul Schoelmerich; Ursula Weisser, weisser@uke.uni-hamburg.de **Address:** Johanna Bleker, ZHGB, Klingsorstr. 119, 12203 Berlin, D-12203, Germany. **Tel:** 49.30.830092.20 **Fax:** 49.30.830092.37 **Publishing Press:** Urban & Fischer Verlag. **Primary Subject:** Medical Sciences. **Founded:** 1966. **Format:** Print. **Circulation:** 250 to 500. Includes bibliographies. Partially indexed. Reprints available. Not archived electronically. **Description:** *Medizinhistorisches* publishes scholarly, previously unpublished papers on the history of medicine and the life sciences, history of dentistry, history of veterinary medicine, and history of pharmacy. Papers can be submitted in German, English or French. Following an anonymous refereeing procedure, the editors decide jointly about acceptance or refusal of a manuscript for publication. **Web site:** http://www.urbanfischer.de/journals/medhistj/

Metascience: An International Review Journal for the History, Philosophy and Social Studies of Science. Tri-annual non-refereed journal. **ISSN:** 0815-0796. **Ed.** Nicolas Rasmussen, nicolas.rasmussen@unsw.edu.au **Address:** History & Philosophy Of Science, University of New South Wales, Sydney, NSW, 2052, Australia. **Fax:** 61.2.9385.8003 **Publishing Press:** Blackwells. **Founded:** 1984 (New Series, 1992). **Produced By:** Australasian Association for the History, Philosophy and Social Studies of Science (AAHPSSS). **Format:** Print and Electronic. **Circulation:** 250 to 500. Includes book reviews. Partially indexed. Reprints unavailable. Archived electronically. **Description:** *Metascience* publishes timely, in-depth book reviews, essay reviews, and author-critic symposia, on new and noteworthy books in the history, philosophy, and social studies of science, technology and medicine. Reviews are normally commissioned by the editorial board. **Web site:** http://www.arts.unsw.edu.au/sts/metascience/metascience.htm

Micrologus: Natura, Scienza e Societa Medievali (Micrologus: Nature, Sciences and Medieval Societies). Annual refereed journal. **Ed.** Agostino Paravicini Bagliani, agostino.paravicini@hist.unil.ch **Primary E-mail:** micrologus@hist.unil.ch **Address:** Section d'Histoire, Faculté des Lettres, Université de Lausanne, Lausanne, Vaud, 1015, Switzerland. **Tel:** 41.2.1692.2934 **Fax:** 41.2.1692.2935 **Publishing Press:** SISMEL, Edizioni del Galluzzo. **Founded:** 1993. **Format:** Print. **Circulation:** 500 to 1000. Indexed. Reprints unavailable. Not archived electronically. **Description:** *Micrologus* is devoted entirely to questions related to the history of nature, from antiquity to modern times. Topical issues (published annually) are conceived in an interdisciplinary manner, to foster collaboration among specialists from different research areas and traditions. Historians of medieval science, literature and culture join with specialists in art history and visual culture to discuss and reflect upon texts and themes of interest to researchers from an increasingly wide range of disciplines. **Web site:** http://sismelfirenze.it/micrologus

Minerva: A Review of Science, Learning and Policy. Quarterly refereed journal. **ISSN:** 0026-4695. **Ed.** Roy M. MacLeod, roy.macleod@history.usyd.edu.au **Address:** Department of History, University of Sydney, Sydney, NSW, 2006, Australia. **Tel:** 61.2.9351.2855 **Fax:** 61.2.9351.3918 **Publishing Press:** Kluwer Academic Publishers Group. **Founded:** 1962. **Format:** Print and Electronic. **Circulation:** 1000 to 2500. Includes book reviews. Indexed. Reprints available. Not archived electronically. **Description:** *Minerva* is devoted to the study of ideas, traditions, cultures and institutions in science, higher education and research. It is concerned no less with history than with present practice, and with the local as well as the global. It speaks to the scholar, the teacher, the policy-maker and the administrator. It features articles, essay reviews and "special" issues on themes of topical importance. It represents no single school of thought, but welcomes diversity, within the rules of rational discourse. Its contributions are peer-reviewed. Its audience is worldwide. **Web site:** http://www.kluweronline.com/issn/0026-4695

Neusis. Semiannual refereed journal. **ISSN:** 1106-6601. **Ed.** Jean Christianidis, ichrist@cc.uoa.gr **Address:** Department of History and Philosophy of Science, Athens University, University Campus, Ilissia, Athens, 15771, Greece. **Tel:** 30.10.727.5541 **Fax:** 30.10.727.5530 **Publishing Press:** NEFELI Publications. **Founded:** 1994. **Format:** Print. **Circulation:** 500 to 1000. Includes book reviews. Not indexed. Reprints unavailable. Not archived electronically. **Description:** *Neusis* is a Greek journal devoted to the history and philosophy of science and technology. All articles, book reviews etc. are in Greek. Articles written in other languages may also be submitted. If they meet the academic standards of the journal they will be translated into Greek before publication. **Web site:** http://sat1.space.noa.gr/hellinomnimon/neusis.htm

Newsletter for the History of Science in Southeastern Europe. Semiannual newsletter. **ISSN:** 1108-5630. **Ed.** Yannis Karas, ykaras@eie.gr **Primary E-mail:** gvlahakis@eie.gr **Address:** Institute for Neohellenic Research, 48, Vas. Constantinou Av., Athens, 11635, Greece. **Tel:** 30.10.727.3557 **Fax:** 30.10.724.6212 **Primary Subject:** Europe. **Founded:** 1999. **Produced By:** History of Science Programme of the Institute for Neohellenic Research, National Hellenic Research Foundation, Athens, Greece. **Format:** Print and Electronic. **Circulation:** 500 to 1000. Includes professional news, book reviews, bibliographies. **Description:** The *Newsletter for the History of Science in Southeastern Europe* is designed to foster closer contact among history of science groups in the countries of southeastern Europe, as well as to make known our research activities to the rest of the world. Alternate editor: Efthymios Nicolaidis. Editorial secretary: G.N. Vlahakis. Members of the editorial board are distinguished scholars: Miladin Apostolov (Bulgaria), Radu Iftimovici (Romania), Ekmeleddin Ihsanoglu (Turkey), Miloje Saric and Aleksandar Petrovic (Yugoslavia). **Web site:** http://www.eie.gr/institutes/kne/ife/

Newsletter of the Commission on History of Science and Technology in Islamic Civilization (IUHPS/DHS). Annual newsletter. **Ed.** Mercè Comes, comes@lingua.fil.ub.es **Address:** Departamento de Filologia Semítica (Àrab), Gran Via 585, Universitat de Barcelona, Barcelona, 08007, Spain. **Tel:** 34.93.403.57.04. **Fax:** 34.93.403.55.96. **Founded:** 1993. **Produced By:** Commission on History of Science and Technology in Islamic Civilization (IUHPS/DHS). **Format:** Print and Electronic. **Circulation:** 250 to 500. Includes professional news, bibliographies. Not indexed. Reprints unavailable. Archived electronically. **Description:** This free newsletter is about 16 pages in length. With its companion listserv (ISLAMSCI), it links historians of Islamic science in some 20 countries. **Web site:** http://www.ou.edu/islamsci/

Notes and Records of The Royal Society. Tri-annual refereed journal.
ISSN: 0035-9149. **Ed.** Alan Cook **Primary E-mail:** patrice.john@royalsoc.
ac.uk **Address:** 6-9 Carlton House Terrace, London, SW1Y 5AG, United
Kingdom. **Tel:** 44.0.20.7839.5561.ext.2628 **Fax:** 44.0.20.7976.1837
Publishing Press: The Royal Society. **Founded:** 1938. **Produced By:** Royal
Society of London. **Format:** Print and Electronic. **Circulation:** 1000 to 2500.
Includes book reviews. Not indexed. Archived electronically. **Description:**
Notes and Records is the Royal Society's journal on the history of science. Each
issue highlights fascinating examples of science shaping our lives, revealing
reminiscences and discoveries, and authoritative book reviews. *Notes and
Records* is a visible part of the Royal Society's commitment to the history of
science. **Web site:** http://www.pubs.royalsoc.ac.uk

**NTM: International Journal of History and Ethics of Natural Sciences,
Technology and Medicine.** Quarterly refereed journal. **ISSN:** 0036-6978.
Ed. Renate Tobies, tobies@mathematik.uni-kl.de **Address:** Universität
Kaiserslautern, Fachbereich Mathematik, Pf 3049, Kaiserslautern, PF 3049,
D-67653, Germany. **Publishing Press:** Birkhaeuser, Basel. **Founded:** 1960,
New Series 1993. **Format:** Print. **Circulation:** 500 to 1000. Includes
professional news, book reviews, bibliographies. Indexed. Reprints available.
Not archived electronically. **Description:** *NTM* is an international journal
for the history and ethics of natural sciences, technology and medicine. It
publishes original research papers, book reviews and news. Every issue
features a survey article of topical interest on current questions in the history
of science. By publishing a wide variety of research results on all phases of
the history of mankind. *NTM* seeks to intensify the reader's interest in the
lively development of the historiography of the sciences. Managing Editor:
Renate Tobies, Kaiserslautern; Editor of History of Mathematics and Sciences:
Menso Folkerts, Munich; Editor of History of Technology: Peter Zimmermann,
Munich; Editor of History of Medicine: Dietrich von Engelhardt, Lübeck.
Web site: http://www.birkhauser.ch/journals/4800/4800_tit.htm

Nuncius: Annali di Storia della Scienza. Semiannual refereed journal.
ISSN: 0394-7394. **Eds.** Marco Ciardi, ciardi@imss.fi.it; Paolo Galluzzi
Address: Istituto e Museo di Storia della Scienza, Piazza dei Giudici 1, Firenze,
50122, Italy. **Tel:** 39.055.265.311 **Fax:** 39.055.265.3130 **Publishing Press:**
Casa Editrice Leo S. Olschki. **Founded:** 1976. **Produced By:** Istituto e Museo
di Storia della Scienza. **Format:** Print. **Circulation:** 1000 to 2500. Includes
professional news, book reviews, bibliographies. Indexed. Reprints unavailable.
Not archived electronically. **Description:** *Nuncius* is the new series of the
Annali dell'Istituto e Museo di Storia della Scienza di Firenze, founded in 1976
by Maria Luisa Righini Bonelli. Galileo's *Sidereus Nuncius* proclaimed, in
the seventeenth century, new and extraordinary discoveries to the intellectual
community of the time. Today, *Nuncius*, paying homage to the great Galilean
tradition, provides historians of science with an efficient means of international
communication. It offers detailed bibliographical and archival documentation

and articles on all aspects of the discipline and the various methodological approaches which characterize it. *Nuncius* encourages critical, in-depth studies. **Web site:** http://www.imss.firenze.it/pubblic/enuncius.html

Osiris. Annual refereed journal. **ISSN:** 0369-7827. **Ed.** Kathryn Olesko, osiris@ georgetown.edu **Address:** BMW Center for German & European Studies, Georgetown University, 726 University Avenue, Washington, DC, 20057-1022, United States. **Tel:** 1.202.687.8300 **Fax:** 1.202.687.8359 **Publishing Press:** University of Chicago Press, Journals Division. **Founded:** 1936. **Produced By:** History of Science Society. **Format:** Print and Electronic. **Circulation:** 2500 to 5000. Indexed. Reprints available. Archived electronically. **Description:** *Osiris* is an annual thematic journal of the History of Science Society dedicated to mediating history and history of science. At the HSS annual meeting, the *Osiris* editorial board chooses volumes on the basis of proposals submitted by prospective guest editors four years in advance of publication. Guest editors assemble contributors to the volume and usually hold a conference on the theme of the volume. **Web site:** http://www.journals.uchicago.edu/Osiris/home.html

Perspectives in Biology and Medicine. Quarterly refereed journal. **ISSN:** 0031-5982. **Ed.** Robert Perlman, r-perlman@uchicago.edu **Address:** Editor, Perspectives in Biology and Medicine, Department of Pediatrics (MC 5058), The University of Chicago, 5841 S. Maryland Ave., Chicago, IL, 60637, United States. **Tel:** 1.773.702.6425 **Fax:** 1.773.702.9234 **Publishing Press:** Johns Hopkins University Press. **Primary Subject:** Biological Sciences. **Founded:** 1957. **Format:** Print and Electronic. **Circulation:** 1000 to 2500. Includes book reviews. Indexed. Reprints available. Archived electronically. **Description:** *Perspectives in Biology and Medicine* is an interdisciplinary scholarly journal. Readers include physicians, biologists, students, and scholars in other disciplines who are interested in the biological sciences. *PBM* publishes essays that place biological or medical issues in broader conceptual, historical, or cultural perspectives. **Web site:** http://muse.jhu.edu/journals/pbm/

Perspectives on Science: Historical, Philosophical, Social. Quarterly refereed journal. **ISSN:** 1063-6145. **Ed.** Joseph C. Pitt, jcpitt@vt.edu **Primary E-mail:** pos@vt.edu **Address:** Department of Philosophy, Virginia Tech, Blacksburg, VA, 24061, United States. **Tel:** 1.540.231.7879 **Fax:** 1.540.231.7879 **Publishing Press:** MIT Press. **Founded:** 1993. **Format:** Print and Electronic. **Circulation:** 500 to 1000. Includes book reviews. Indexed. Reprints available. Archived electronically. **Description:** *Perspectives on Science* is an interdisciplinary journal publishing articles that focus on the philosophical, historical and social aspects of science and technology. To emphasize the interdisciplinary nature of the journal, we require all articles to exhibit at least two of the three disciplinary foci. Each issue also contains an essay review of a number of books in a related area. **Web site:** http://mitpress.mit.edu/POSC

Pharmaceutical Historian. ISSN: 0079-1393. **Ed.** J. G. L. Burnby
Primary E-mail: bshpeditor@associationhq.org.uk **Address:** British Society
for the History of Pharmacy, 840 Melton Road, Thurmaston, Leicester, LE4
8BN, United Kingdom. **Founded:** 1967. **Produced By:** British Society for the
History of Pharmacy. **Web site:** http://www.bshp.org/

Pharmacy History Australia. Tri-annual non-refereed journal. **Ed.** G. C. Miller,
gcmiller@opera.iinet.net.au **Address:** 8 Leopold St., Nedlands, 6009, Australia.
Tel: 61.8.9386.6078 **Publishing Press:** Pharmceutical Society of Australia.
Primary Subject: Medical Sciences. **Founded:** 1997. **Format:** Print.
Circulation: 2500 to 5000. Includes professional news, book reviews,
bibliographies. Indexed. Reprints available. Not archived electronically.
Description: This journal showcases all aspects of the history of pharmacy. The
journal examines the individuals who have contributed to this history, through
social commentary, artifacts, biographies, conference papers and submitted
articles. **Web site:** http://www.psa.org.au/history/history.cfm

Pharmacy in History. Quarterly refereed journal. **ISSN:** 0031-7047.
Ed. Gregory Higby, gjh@pharmacy.wisc.edu **Primary E-mail:** ph@aihp.org
Address: School of Pharmacy, Univ. of Wisconsin, 425 N. Charter St.,
Madison, WI, 53705-2222. **Tel:** 1.608.262.5378 **Primary Subject:** Medical
Sciences. **Founded:** 1959. **Produced By:** American Institute of the History
of Pharmacy. **Format:** Print. **Circulation:** 500 to 1000. Includes professional
news, book reviews, bibliographies. Indexed. Reprints unavailable. Not archived
electronically. **Description:** *PH* is a quarterly devoted to exploring the place of
pharmacy in the history of world civilization. The journal commonly contains
two documented articles of 5,000 to 10,000 words, plus shorter notes, book
reviews, and bibliographic pieces. Continuing departments include "Historical
Images of the Drug Market," edited by William Helfand, and "In the Literature"
containing abstracts of articles related to the broad area of pharmaceutical
history. **Web site:** http://www.aihp.org

Philosophy and the History of Science. Biennial refereed journal.
ISSN: 1022-4874. **Eds.** Daiwie Fu, dwfu@hist.nthu.edu.tw; Wann-Sheng
Horng, horng@math.ntnu.edu.tw **Primary E-mail:** dwfu@mx.nthu.edu.tw
Address: Institute of History, National Tsing-Hua University, Hsinchu,
300, Taiwan. **Tel:** 886.35.715131.ext.4468 **Fax:** 886.35.716780
Publishing Press: Yuan-Liou Publishing Co., LTD. **Founded:** 1992.
Format: Print. **Circulation:** 250 to 500. Includes professional news, book
reviews. Not indexed. Reprints available. Not archived electronically.
Description: This journal has published 10 issues in total. It has now
reorganized itself into a new journal called: *Taiwanese Journal for Studies
of Science, Technology, and Medicine.* Please contact the editor of this
reorganized journal, Professor Chu Pingyi, at: kaihsin@pluto.ihp.sinica.edu.tw.
Web site: http://rpgs1.isa.nthu.edu.tw/faculty/dwfu/taiwan/

Philosophy of Science. Quarterly refereed journal. **ISSN:** 0031-8248. **Ed.** Noretta Koertge, koertge@indiana.edu **Primary E-mail:** philsci@indiana.edu **Address:** 123 Goodbody Hall, Department of History and Philosophy of Science, Indiana University, Bloomington, IN, 47405-7005, United States. **Tel:** 1.812.855.3229 **Fax:** 1.812.855.3631 **Publishing Press:** University of Chicago Press. **Founded:** 1934. **Produced By:** Philosophy of Science Association. **Format:** Print and Electronic. **Circulation:** 2500 to 5000. Includes book reviews. Indexed. Reprints available. Archived electronically. **Description:** *Philosophy of Science* publishes scholarly essays on the methodological, ontological, epistemological, and ethical premises and implications of all areas of scientific inquiry. We welcome ground breaking work on the conceptual issues that arise in various scientific disciplines, as well as studies of the structure and development of scientific theories. **Web site:** http://www.journals.uchicago.edu/PHILSCI/home.html

Physics In Perspective. Quarterly refereed journal. **ISSN:** 1422-6944. **Eds.** John S. Rigden, jsr@aip.org; Roger H. Stuewer, rstruewer@physics.spa.umn.edu **Primary E-mail:** rstruewer@physics.spa.umn.edu **Address:** Tate Laboratory of Physics, University of Minnesota, 116 Church Street SE, Minneapolis, MN, 55455, United States. **Tel:** 1.612.624.8073 **Fax:** 1.612.624.4578 **Publishing Press:** Birkhaeuser Verlag. **Founded:** 1999. **Format:** Print and Electronic. **Circulation:** 250 to 500. Includes book reviews. Indexed. Reprints available. Archived electronically. **Description:** *Physics in Perspective* is directed at a broad audience of historians and philosophers of physics, physicists, physics teachers, and the general public. It strives to bridge the gap between physicists and non-physicists by publishing historical and philosophical studies, firsthand accounts by physicists of their discoveries and achievements, biographical accounts of physicists and interviews with them, review articles, guides to sites of interest to physicists and other scientists in cities worldwide, scientific vignettes, and book reviews. All reveal the profound influence that physics has had on modern society and culture. **Web site:** http://link.springer.de/link/service/journals/00016/

Physis: Rivista Internazionale di Storia della Scienza. Semiannual refereed journal. **ISSN:** 0031-9414. **Ed.** Leo S. Olschki, celso@olschki.it **Address:** Casa editrice Leo S. Olschki, Cas. Post. 66, Firenze, 50100, Italy. **Tel:** 39.055.653.0684 **Fax:** 39.055.653.0214 **Founded:** 1959. **Format:** Print. **Circulation:** 500 to 1000. Includes book reviews, bibliographies. Partially indexed. **Description:** The journal brings together original contributions by both Italian and foreign authors on the history of science from ancient times to the present day. It covers a variety of disciplines and themes including the historiography of science, scientific museology, relationships between science and philosophy, science and technology and other interdisciplinary topics. **Web site:** http://www.olschki.it/cgi-bin/olscq2.pl?num=3444&lang=eng

Polhem: Tidskrift foer Teknikhistoria (Polhem: Journal for the History of Technology). Quarterly refereed journal. **ISSN:** 0281-2142. **Ed.** Hans Weinberger, hans@kth.se **Primary E-mail:** polhem@kth.se **Address:** Department of History of Science and Technology, Royal Institute of Technology, Stockholm, 100 44, Sweden. **Tel:** 46.8.790.87.99 **Fax:** 46.8.24.62.63 **Primary Subject:** Technology. **Founded:** 1983. **Produced By:** Svenska Nationalkommitten foer Teknikhistoria. **Format:** Print and Electronic. **Circulation:** 250 to 500. Includes professional news, book reviews. Partially indexed. Reprints unavailable. Not archived electronically. **Description:** *Polhem* publishes articles, reviews, debates, news and announcements in the history of technology. *Polhem* accepts contributions in Swedish, Danish, Norwegian and English. **Web site:** http://www.polhem.org

Psychoanalysis and History. **ISSN:** 1460-8235. **Ed.** Andrea Sabbadini, Sabbadini@dial.pipex.com **Address:** 38 Berkeley Road, London, N8 8RU, United Kingdom. **Publishing Press:** Artesian Books Ltd. **Founded:** 1998. **Web site:** http://www.artesianbooks.co.uk/pages/pah.htm

Public Understanding of Science. Quarterly refereed journal. **ISSN:** 0963-6625. **Ed.** Bruce V. Lewenstein, b.lewenstein@cornell.edu **Primary E-mail:** pubscience@cornell.edu **Address:** Department of Communication, Kennedy Hall, Cornell University, Ithaca, NY, 14853, United States. **Tel:** 1.607.255.7512 **Fax:** 1.607.254.1322 **Publishing Press:** Institute of Physics Publishing (UK). **Founded:** 1992. **Format:** Print and Electronic. **Circulation:** 250 to 500. Includes book reviews, bibliographies. Indexed. Reprints available. Archived electronically. **Description:** *Public Understanding of Science* is an international journal covering all aspects of the interrelationships between science (including technology and medicine) and the public. Topics include: surveys of public understanding of and attitudes towards science and technology, historical and contemporary case studies of public interaction with science, popular representations of science, science and the media, science in popular films and popular culture, and related topics. **Web site:** http://www.iop.org/journals/pu

Qorot: A Bulletin Devoted to the History of Medicine and Science. **ISSN:** 0023-4109. **Ed.** Shmuel Kottek, histmed@mdib.huji.ac.il **Address:** Magnes Press, Edmond Safra Campus, Giuat Ram, Jerusalem, 91904, Israel. **Publishing Press:** Magnes Press. **Founded:** 1952.

Quipu: Revista Latinoamericana de Historia de las Ciencias y la Tecnologia. **ISSN:** 0185-5093. **Ed.** Juan Jose Saldana **Primary E-mail:** quipu@servidor. unam.mx **Address:** Revista Quipu, Apartado Postal 21-023, Mexico D.F., 04000, Mexico. **Publishing Press:** Begona Garcia Gonzalez. **Founded:** 1984. **Produced By:** Sociedad Latinoamericana de Historia de las Ciencias y la Tecnologia.

Revista da Sociedade Brasileira de História da Ciência (Journal of the Brazilian Society for the History of Science). Semiannual refereed journal. **ISSN:** 0103-7188. **Eds.** Maria Rachel Fróes da Fonseca, froes@coc.fiocruz.br; Roberto de Andrade Martins, Rmartins@ifi.unicamp.br **Address:** Grupo de História e Teoria da Ciência, Caixa Postal 6059, Campinas, SP, 13084-971, Brazil. **Tel:** 55.19.3788.5516 **Fax:** 55.19.3788.5512 **Publishing Press:** Sociedade Brasileira de História da Ciência. **Founded:** 1985. **Produced By:** Sociedade Brasileira de História da Ciência. **Format:** Print and Electronic. **Circulation:** 250 to 500. Includes book reviews. Partially indexed. Reprints available. Archived electronically. **Description:** The *Revista da Sociedade Brasileira de História da Ciência* is a scholarly journal that publishes contributions on history of science and related fields. Most contributions are published in Portuguese, but all main modern Western languages are accepted. Abstracts in Portuguese and English. **Web site:** http://www.ifi.unicamp.br/~ghtc/sbhc.htm

Rittenhouse: The Journal of the American Scientific Instrument Enterprise. Semiannual refereed journal. **Ed.** Randall C. Brooks, rbrooks@nmstc.ca **Primary E-mail: Address:** Curator, Physical Sciences and Space, Canada Science and Technology Museum, PO Box 9724, Stn. T, Ottawa, ON, K1G 5A3, Canada. **Tel:** 1.613.990.2804 **Fax:** 1.613.990.3636 **Founded:** 1986. **Format:** Print. Includes professional news, book reviews. Indexed. Reprints available. **Description:** *Rittenhouse* aims to increase and diffuse knowledge about scientific instruments made and/or sold in the United States and the Americas. The areas covered include mathematical, optical and philosophical instruments, chemical, physical and electrical apparatus, sundials and globes. The time period covered is from the 17th to the mid-20th century. **Web site:** http://www.rittenhousejournal.org/

Saber y Tiempo: Revista de Historia de la Ciencia. Semiannual refereed journal. **ISSN:** 0328-6584. **Ed.** Nicolas Babini, babini@netex.com.ar **Address:** Saber y Tiempo, Asociacion Biblioteca Jose Babini, Buenos Aries, Capital Federal, 1059, Argentina. **Tel:** 54.11.4962.6174 **Fax:** 54.11.4962.6174 **Founded:** 1996. **Produced By:** Asociacion Biblioteca Jose Babini. **Format:** Print. **Circulation:** Less than 250. Includes book reviews, bibliographies. Not indexed. Reprints unavailable. Not archived electronically.

Science as Culture. Quarterly refereed journal. **ISSN:** 0950-5431. **Eds.** Les Levidow, l.levidow@open.ac.uk **Address:** 26 Freegrove Road, London, UK, N7 9RQ, United Kingdom. **Tel:** 44.020.7609.0507 **Fax:** 44.020.7609.4837 **Publishing Press:** Taylor & Francis. **Primary Subject:** Social Relations of Science. **Founded:** 1974. **Format:** Print and Electronic. **Circulation:** 250 to 500. Includes book reviews. Not indexed. Reprints available. Archived electronically. **Description:** *Science as Culture* explores how science mediates our cultural experience. Our culture is a scientific one, defining what is natural and what is rational. Its values can be seen in what are sought out as facts and made as artifacts, what are designed as processes and products, and what are forged as weapons and filmed as wonders. **Web site:** http://www.tandf.co.uk/journals/carfax/09505431.html

Science in Context. Quarterly refereed journal. **ISSN:** 0269-8897. **Eds.** Leo Corry, sic@post.tau.ac.il; Alexandre Metraux, Ametraux@aol.com; Juergen Renn, renn@mpiwg-berlin.mpg.de **Address:** The Cohn Institute for the History and Philosophy of Sciences and Ideas, Tel Aviv University, Ramat Aviv, Tel Aviv, 69 978, Israel. **Publishing Press:** Cambridge University Press. **Founded:** 1987. **Format:** Print and Electronic. **Circulation:** 500 to 1000. Not indexed. Reprints unavailable. Not archived electronically. **Description:** The journal is devoted to the study of the sciences from the points of view of comparative epistemology and historical sociology of knowledge. It is also committed to an interdisciplinary approach to the study of science and its cultural development; accepting contributions drawn from history, philosophy and sociology. **Web site:** http://uk.cambridge.org/journals/sic/

Science Studies: An Interdisciplinary Journal for Science and Technology Studies. Semiannual refereed journal. **ISSN:** 0786-3012. **Ed.** Marja Hayrinen-Alestalo, marja.alestalo@helsinki.fi **Primary E-mail:** aaro.tupasela@helsinki.fi **Address:** Department of Sociology, P. O. Box 18, University of Helsinki, 00014, Finland. **Tel:** 358.9.191.23964 **Fax:** 358.9.191.23967 **Publishing Press:** Valopaino Oy. **Founded:** 1988. **Produced By:** Finnish Society for Science and Technology Studies. **Format:** Print. **Circulation:** 250 to 500. Includes book reviews. Not indexed. Reprints available. Archived electronically. **Description:** *Science Studies* is a semi-annual peer-reviewed journal dedicated to science and technology studies. As an interdisciplinary journal, it covers a multitude of disciplines including philosophy, history, social policy and sociology. *Science Studies* seeks to develop an improved understanding of the nature of science itself, as an epistemological, utilitarian, ideological and political resource. **Web site:** http://pro.tsv.fi/stts/mag/index.html

Scientia Canadensis. Annual refereed journal. **ISSN:** 0829-2507. **Eds.** Michael Eamon, meamon@archives.ca; Robert Gagnon, gagnon.robert@uqam.ca **Primary E-mail:** meamon@archives.ca **Address:** Canadian Archives Branch, National Archives of Canada, Ottawa, Ontario, K1A 0N3, Canada. **Publishing Press:** Becker Associates. **Primary Subject:** North America. **Founded:** 1977. **Produced By:** Canadian Science and Technology Historical Association. **Format:** Print. **Circulation:** Less than 250. Includes book reviews, bibliographies. Reprints available. **Description:** *Scientia Canadensis* publishes original research contributions on the history of Canadian science, technology and medicine. **Web site:** http://www.er.uqam.ca/nobel/r20430/scientia_canadensis/

Scientometrics: An International Journal for all Quantitative Aspects of the Science of Science, Communication in Science and Science Policy. Monthly refereed journal. **ISSN:** 0138-9130. **Ed.** Tibor Braun

Primary E-mail: h1533bra@ella.hu **Address:** Institute of Inorganic and Analytical Chemistry, Eotvos Loránd University, Budapest, 1443 POB 123, Hungary. **Tel:** 36.1.311.5433 **Primary Subject:** 20th and 21st Century, Europe, Social Sciences, Science Policy. **Founded:** 1978. **Produced By:** Akademiai Kiado Rt. **Format:** Print and Electronic. Includes professional news, book reviews, bibliographies. Indexed. Reprints available. Archived electronically. **Description:** *Scientometrics* is the only journal in the world which deals with quantitative science studies and information science. **Web site:** http://www.kluweronline.com/issn/0138-9130

SIHSPAI: Newsletter of the International Society for History of Arabic Sciences and Philosophy. Tri-annual newsletter. **ISSN:** 1026-3977. **Ed.** Gül Russell, garussell@tamu.edu **Address:** Department of Humanities in Medicine, Texas A&M System Health Science Center, College of Medicine, 153 Reynolds Building, College Station, TX, 77843-1114. **Tel:** 1.979.845.6462 **Fax:** 1.979.845.8634 **Founded:** 1992. **Produced By:** Society for the History of Arabic/Islamic Science and Philosophy. **Format:** Print and Electronic. **Circulation:** 280. Includes professional news, bibliographies. Reprints unavailable. Archived electronically. **Description:** An up-to date scholarly reference tool to promote research and to keep SIHSPAI members abreast of meetings (full programs), publications (books, articles, collected volumes with full table of contents), manuscripts with descriptive illustrations (in color on the web), professional news, journals, research institutions, and electronic news. It covers all areas relevant to the history of science and philosophy in Islamic civilization: Graeco-Roman, Arabic/Persian/Ottoman, Hebrew, Latin; extending to Central Asian, Indian and Chinese. Cumulative bibliography in progress. **Web site:** http://phil-www.tamu.edu/sihspai/

Social Epistemology. Quarterly refereed journal. **ISSN:** 0269-1728. **Ed.** Joan Leach, jleach@pitt.edu **Primary E-mail:** jleach@pitt.edu **Address:** Rhetoric of Science Program, 1117 Cathedral of Learning, University of Pittsburgh, Pittsburgh, PA, 15260, United States. **Tel:** 1.412.648.7664 **Fax:** 1.412.624.1878 **Publishing Press:** Taylor and Francis. **Founded:** 1988. **Format:** Print and Electronic. **Circulation:** 250 to 500. Includes book reviews, bibliographies. Indexed. Reprints available. Archived electronically. **Description:** *Social Epistemology* continues its mission to publish innovative research and critical analyses of the social basis of knowledge. This includes historical accounts of knowledge production, dissemination and management. **Web site:** http://www.tandf.co.uk/journals/routledge/02691728.html

Social History of Medicine. Tri-annual refereed journal. **ISSN:** 0951-631X. **Eds.** Roger Davidson, roger.davidson@ed.ac.uk; Helen King, h.king@reading. ac.uk **Address:** Department of History, University of Reading, Whiteknights, Reading, RG6 6AA, United Kingdom. **Publishing Press:** Oxford University Press. **Founded:** 1970. **Produced By:** Society for the Social History of Medicine. **Format:** Print. **Circulation:** 250 to 500. Includes book reviews, bibliographies. Not indexed. Reprints available. Archived electronically. **Description:** *Social History of Medicine* is an international, interdisciplinary journal which publishes the latest research on every aspect of health, medicine and society across the ages. It also offers review articles, surveys of recent developments, critical assessments of archives and sources, and book reviews. **Web site:** http://www.shm.oupjournals.org

Social Studies of Science: An International Review of Research in the Social Dimensions of Science and Technology. Bi-monthly refereed journal. **ISSN:** 0306-3127. **Ed.** Michael E. Lynch, mel27@cornell.edu **Address:** Department of Science & Technology Studies, Cornell University, 632 Clark Hall, Ithaca, NY, 14853-2501, United States. **Tel:** 1.607.244.7294 **Fax:** 1.607.255.6044 **Publishing Press:** Sage Publications Ltd. **Founded:** 1971. **Format:** Print and Electronic. **Circulation:** 500 to 1000. Includes book reviews. Indexed. Reprints available. **Description:** *Social Studies of Science* is devoted mainly to the results of original research, whether empirical or theoretical, which bring fresh light to bear on the concepts, processes, mediations and consequences of modern natural science, technology, and medicine. It encourages contributions from political science, sociology, economics, history, philosophy, psychology, social anthropology, and the legal and educational disciplines. It welcomes studies of a full range of scientific and technological innovations from all countries. **Web site:** http://www.sagepub.co.uk/journals/details/j0005.html

Society for the History of Technology Newsletter. Newsletter. **Ed.** Stuart W. Leslie, swleslie@jhu.edu **Primary E-mail:** shot@jhu.edu **Address:** Department of the History of Science, 216B Ames Hall, Johns Hopkins University, MD, 21218, United States. **Web site:** http://shot.jhu.edu/news/index.htm

STS Network Japan Yearbook: A Journal for Science, Technology and Society/ Science and Technology Studies. Annual refereed journal. **Ed.** Sho Kasuga, office@stsnj.org **Address:** c/o Yuko Fujigaki's Research Office, The University of Tokyo, College of Arts and Sciences, 3-8-1, Komaba, Meguro-ku, Tokyo, 153-8902, Japan. **Fax:** 81.3.5454.6990 **Founded:** 1991. **Produced By:** STS Network Japan. **Format:** Print. **Circulation:** Less than 250. Includes professional news, book reviews, bibliographies. Not indexed. Reprints available. Not archived electronically. **Description:** STS Network Japan is an informal academic organization promoting the network of people interested in STS-related issues. *Year Book* includes referenced papers, detailed reports of symposiums, and summer school discussions. The publication is written almost entirely in Japanese. **Web site:** http://stsnj.org/

Studia Copernicana. Irregular refereed journal. **ISSN:** 0081-6701. **Ed.** Grazyna Rosinska, g.rosinska@astercity.net **Primary E-mail:** ihn@ihnpan.waw.pl **Address:** Institute for the History of Science, Nowy Swiat 72, Warszawa, 00-330, Poland. **Tel:** 48.22.65.72.836. **Fax:** 48.22.826.61.37

Publishing Press: Wydawnictwa IHN PAN. **Founded:** 1970. **Produced By:** Center for Copernican Studies, Institute for the History of Science, Polish Academy of Sciences. **Format:** Print. **Circulation:** 250 to 500. Not archived electronically. **Description:** Additional editors include: Marian Biskup, Jerzy Burchardt, Stanislaw Mossakowski, John D. North, Grazyna Rosinska, Alain Segonds, Pavel Spunar, Sabetai Unguru, Witold Wroblewski. **Web site:** http://www.ihnpan.waw.pl/

Studies in History and Philosophy of Science Part A. Quarterly refereed journal. **ISSN:** 0039-3681. **Eds.** Marina Frasca-Spada, mfs10@cam.ac.uk; Nicholas Jardine, nj103@cam.ac.uk **Address:** Department of History and Philosophy of Science, University of Cambridge, Free School Lane, Cambridge, CB2 3RH, United Kingdom. **Tel:** 44.1223.330466 **Fax:** 44.1223.334554 **Publishing Press:** Elsevier Science Ltd. **Founded:** 1970. **Format:** Print and Electronic. **Circulation:** 500 to 1000. Includes book reviews. Partially indexed. Reprints available. Archived electronically. **Description:** Devoted to the study of the history, philosophy and sociology of the sciences. The editors encourage contributions both in the long established areas of the history and the philosophy of the sciences, and in the topical areas of historiography, the material culture of science, the sciences in relation to gender, culture and society, and in relation to the arts. The journal is international in scope and content, and publishes papers from a wide range of countries and intellectual traditions. **Web site:** http://www.elsevier.nl/locate/shpsa

Studies in History and Philosophy of Science Part B: Studies in History and Philosophy of Modern Physics. Quarterly refereed journal. **ISSN:** 1355-2198. **Ed.** R. Clifton, rclifton@pitt.edu **Primary E-mail:** shpmp@pitt.edu **Address:** Department of Philosophy, University of Pittsburgh, 1001 Cathedral of Learning, Pittsburgh, PA, 15260, United States. **Tel:** 1.412.624.5793 **Fax:** 1.412.624.5377 **Publishing Press:** Elsevier Science Publishers. **Primary Subject:** 20th and 21st Century, Physical Sciences, Philosophy or Philosophy of Science. **Founded:** 1995. **Format:** Print and Electronic. **Circulation:** 1000 to 2500. Includes book reviews. Indexed. Reprints available. Archived electronically. **Description:** Devoted to all aspects of the history and philosophy of modern physics broadly understood, including physical aspects of astronomy, chemistry and other non-biological sciences. The primary focus is on physics from the late-nineteenth century to the present, the period of emergence of the kind of theoretical physics that has come to dominate the exact sciences in the twentieth century. The journal is internationally oriented with contributions from a wide range of perspectives. In addition to purely historical or philosophical papers, the editors particularly encourage papers that combine these two disciplines. The editors are also keen to publish papers of interest to physicists, as well as specialists in history and philosophy of physics. **Web site:** http://www.elsevier.nl/locate/issn/1355-2198/

Studies in History and Philosophy of Science Part C: Studies in History and Philosophy of Biological and Biomedical Sciences. Quarterly refereed journal. **ISSN:** 1369-8486. **Ed.** Nicholas Jardine, nj103@cam.ac.uk **Primary E-mail:** mfs10@cam.ac.uk **Address:** The Editors, Studies in History and Philosophy of Biological and Biomedical Sciences, Department of History and Philosophy of Science, University of Cambridge, Free School Lane, Cambridge, CB2 3RH, United Kingdom. **Tel:** 44.1223.334546 **Fax:** 44.1223.334554 **Publishing Press:** Elsevier Science Ltd. **Founded:** 1998. **Format:** Print and Electronic. **Circulation:** 500 to 1000. Includes book reviews. Partially indexed. Reprints available. Archived electronically. **Description:** Devoted to historical, sociological, philosophical and ethical aspects of the life and environmental sciences, of the sciences of mind and behavior, and of medical and biomedical sciences and their technologies. Contributions are from a wide range of countries, cultural traditions, and disciplines. Essays of interest to scientists and medics as well as to specialists in the history, philosophy and sociology of the sciences are also favored. **Web site:** http://www.elsevier.nl/locate/issn/1369-8486/

Studies in History of Medicine and Science. Annual refereed journal. **ISSN:** 0970-5562. **Eds.** S. M. Razaullah Ansari, Raza.Ansari@gmx.net; Altaf Ahmad Azmi; Hakeem Abdul Hameed **Address:** Hamdard Nagar, New Delhi, 110062, India. **Publishing Press:** Hamdard University Press, New Delhi, India. **Primary Subject:** Asia. **Founded:** 1977. **Produced By:** Jamia Hamdard. **Format:** Print. **Circulation:** 250 to 500. Includes book reviews. Indexed. Reprints available. Not archived electronically. **Description:** Volume IX(1985) to Volume XVI(2000) were based on primary sources in Arabic, Persian and Sanskrit. A number of texts were also published as supplements. The journal specialized in history of medicine and science of Afro-Asian cultural areas. Starting with Volume XVII, the editor has been Altaf A. Azmi. The journal is no longer based solely on primary sources.

Synthese. Monthly refereed journal. **ISSN:** 0039-7857. **Ed.** Jaakko J. Hintikka, hintikka@bu.edu **Primary E-mail:** synthese@bu.edu **Address:** Journals Editorial Office, Van Godewijckstraat 30, 3311 GX Dordrecht, P.O. Box 990, Dordrecht, 3300 AZ, Netherlands. **Tel:** 1.617.353.6806 **Fax:** 1.617.353.6805 **Publishing Press:** Kluwer. **Founded:** 1936. **Format:** Print and Electronic. Archived electronically. **Web site:** http://www.kluweronline.com/issn/0039-7857

Taiwanese Journal for Studies of Science, Technology, and Medicine. Ed. Chu Pingyi, kaihsin@pluto.ihp.sinica.edu.tw **Address:** Institute of History and Philology, Academia Sinica, Nankang, Taipei, Taiwan. **Fax:** 886.2.2786.8834 **Description:** A bilingual (Chinese and English) journal dedicated to interdisciplinary studies of Taiwanese and Chinese science, technology and medicine. It carries on the historical studies of *The Taiwanese Journal for Philosophy and History of Science* which will split into two journals after the tenth issue. We will not limit ourselves to historical aspects of Taiwanese and Chinese science, technology and medicine. Related sociological and anthropological studies are equally welcome. By publishing a bilingual journal, we wish to connect Western and Taiwanese scholars interested in the field. **Web site:** http://www.princeton.edu/~tomtso/mypage.htm

Technikgeschichte. Quarterly refereed journal. **ISSN:** 0040-117X.
Eds. Astrid Schürmann; Katharina Zeitz, katharina.zeitz@tu-berlin.de
Primary E-mail: technikgeschichte@tu-berlin.de **Address:** TU-Berlin TEL 12-1, Ernst-Reuter-Platz 7, Berlin, D-10587, Germany. **Tel:** 49.30.314.24069
Fax: 49.30.314.25962 **Publishing Press:** Verlag Kiepert KG.
Primary Subject: Technology. **Founded:** 1909. **Produced By:** Verein Deutscher Ingenieure. **Format:** Print. **Circulation:** 500 to 1000. Includes professional news, book reviews, bibliographies. Indexed. Reprints available. Archived electronically. **Web site:** http://www-philosophie.kgw.tu-berlin.de/ W3/philosophie/PhzsTechnikg/PhTechnikgeschichte.htm

Technology and Change in History. ISSN: 1385-920X. **Primary E-mail:** cs@brillusa.com **Address:** Brill Academic Publishers Inc., 112 Water Street, Suite 400, Boston, MA, 02109, United States. **Publishing Press:** Brill Academic Publishers. **Founded:** 1997.

Technology and Culture. Quarterly refereed journal. **ISSN:** 0040-165X.
Ed. John Michael Staudenmaier, S.J., staudejm@udmercy.edu
Primary E-mail: tac@chaos.press.jhu.edu **Address:** Technology and Culture Editorial Office, Henry Ford Museum and Greenfield Village, 20900 Oakwood Boulevard, Dearborn, MI, 48124, United States. **Tel:** 1.313.982.6083
Publishing Press: Johns Hopkins University Press. **Primary Subject:** Technology. **Founded:** 1960. **Produced By:** Johns Hopkins University Press, Journals Publishing Division. **Format:** Print and Electronic. **Circulation:** 2500 to 5000. Includes book reviews, bibliographies. Indexed. Reprints available. Archived electronically. **Description:** Features refereed articles studying technologies in historical contexts, such as: influences of contextual factors on the emergence and/or evolution of technologies; influences of technologies (broadly construed to include artifacts, systems, institutions etc.) on contexts either at time of emergence or during ordinary use; factors include those that are strategically defined by some actor(s) as well as symbolic, affective factors. Articles covering any geographical region or time period are welcome. **Web site:** http://www.shot.jhu.edu/tc.html

Technology in Society: An International Journal. Refereed Journal. **ISSN:** 0160-791X. **Eds.** George Bugliarello; A. George Schillinger, gschilli@poly.edu **Address:** Editors, Technology In Society, Box 693, Polytechnic University, 333 Jay Street, Brooklyn, NY, 11201. **Fax:** 1.718.260.3136 **Founded:** 1979. **Format:** Print and Electronic. Includes book reviews. Indexed. Reprints available. Archived electronically. **Web site:** http://www.elsevier.nl/inca/publications/store/3/8/4/

Technoscience: Newsletter of the Society for the Social Study of Science. Tri-annual newsletter. **Ed.** Linda Layne, laynel@rpi.edu **Primary E-mail:** technoscience-l@lists.rpi.edu**Tel:** 1.404.894.9050 **Fax:** 1.404.894.9372 **Founded:** 1990. **Produced By:** Society for Social Studies of Science. **Format:** Electronic (Web-Based). **Circulation:** 1000 to 2500. Includes professional news, book reviews. Not indexed. Reprints unavailable. Archived electronically. **Web site:** http://www.rpi.edu/dept/sts/technoscience

Vesalius. Acta Internationalia Historiae Medicinae (Journal of the International Society for the History of Medicine). Semiannual refereed journal. **ISSN:** 1373-4857. **Ed.** Diana Gasparon, dgasparo@ulb.ac.be **Address:** Erasmus University Hospital, Medical Museum, 808 route de Lennik, Brussels, Belgium, B-1070, Belgium. **Tel:** 32.2.555.34.31 **Fax:** 32.2.555.34.71 **Primary Subject:** Medical Sciences, Humanistic Relations of Science. **Founded:** 1995. **Produced By:** International Society for the History of Medicine. **Format:** Print. **Circulation:** 500 to 1000. Includes professional news, book reviews, bibliographies. Partially indexed. Reprints available. Not archived electronically. **Web site:** http://www.bium.univ-paris5.fr/ishm/eng/

Wellcome History. Tri-annual non-refereed journal. **Ed.** Sanjoy Bhattacharya, sanjoy.bhattacharya@ucl.ac.uk **Address:** Wellcome Trust Centre for the History of Medicine at UCL, University College London, Euston House, 24 Eversholt Street, London, NW1 1AD, United Kingdom. **Tel:** 44.207.679.8155 **Fax:** 44.207.679.8192 **Publishing Press:** Publications Department, The Wellcome Trust. **Founded:** 1996. **Produced By:** The Wellcome Trust, UK. **Format:** Print and Electronic. **Circulation:** 2500 to 5000. Includes professional news, book reviews, bibliographies. Partially indexed. Reprints unavailable. Archived electronically. **Description:** Aims to increase interaction amongst historians of medicine and science worldwide, on any theme, any geographical area or any period. Book reviews are welcome, as are submissions of books for review. The journal also welcomes feature articles of not more than 3000 words in length. (Authors should contact the the editor before preparing/submitting pieces). *Wellcome History* also advertises funding information for students and researchers, information on job openings and conference announcements. **Web site:** http://www.wellcome.ac.uk/en/1/awtpubhomwhi.html

Würzburger Medizinhistorische Mitteilungen. Annual refereed journal. **ISSN:** 0177-5227. **Ed.** Gundolf Keil, gesch.med@mail.uni-wuerzburg.de **Address:** Institut für Geschichte der Medizin der Universität Würzburg, Oberer Neubergweg 10a, Würzburg, Bavaria, D-97074, Germany. **Tel:** 49.931.79678.0 **Fax:** 49.931.79678.78 **Publishing Press:** Königshauen & Neumann, Würzburg. **Founded:** 1983. **Produced By:** Würzburger Medizinhistorische Gesellschaft. **Format:** Print. Includes professional news, book reviews, bibliographies. Partially indexed. Reprints unavailable. Not archived electronically. **Web site:** http://www.uni-wuerzburg.de/medizingeschichte/wmm.html

Inactive Publications

The following journals and newsletters are no longer published. Where known, the year of the final issue is given in parentheses.

Academie Internationale d'Histoire des Sciences. Collection des Travaux
Acta Historiae Rerum Naturalium nec non Technicarum (1992)
American Lectures in the History of Medicine and Science (1986)
British Museum (Natural History) Bulletin
Caduceus: A Humanities Journal for Medicine and the Health Sciences (1998)
Currents in Science, Technology and Society
Dejiny Ved a Techniky (1993)
Hippokrates: Informationen aus der Medizinischen Wissenschaft und Praxis (1979)
History of Medicine (1976)
International Congress of History of Medicine. Proceedings (1986)
Islamic Thought and Scientific Creativity (1996)
Janus: Revue Internationale de l'Histoire des Sciences, de la Medecine, de la Pharmacie et de la Technique (1990)
Ocherki po Istorii Estestvoznaniya i Tekhniki: Respublikanskii Mezhvedomstvennyi Sbornik Nauchnykh Trudov
Opusculum: Kirja- ja Oppihistoriallinen Aikakauskirja. Bok- Och Laerdomshistorisk Tidskrift (1998)
Psychologie und Geschichte
Revista Finlay
Rivista di Storia della Scienza
Science in History
Technologia: Historical and Social Studies in Science, Technology and Industry (1991)
Tractrix: Yearbook for the History of Science, Medicine, Technology, and Mathematics (1993)
Wissenschaft in den Medien (1993)
Yale Studies in the History of Science and Medicine

Index

This index covers instutions, organizations, and publications (pages 188-381).
Graduate programs are generally indexed by the name of the university, rather than
by the name of the department or division.

Country Code List

Algeria 213	Greece 30	Panama...................507
Argentina 54	Guatemala 502	Paraguay 595
Australia 61	Honduras 504	Peru 51
Austria 43	Hong Kong 852	Philippines 63
Bangladesh 880	Hungary 36	Poland 48
Belgium 321	Iceland 354	Portugal 351
Bolivia 591	India 91	Romania 40
Bosnia &	Indonesia 62	Russia 7
Herzegovina 387	Iran 98	Saudi Arabia 966
Botswana 267	Iraq 964	Singapore 65
Brazil 55	Ireland 353	Slovakia 42
Bulgaria 359	Israel 972	Slovenia 386
Cameroon 237	Italy 39	South Africa 27
Canada 1	Japan 81	South Korea 52
Chile 56	Jordan 962	Spain 349
China 86	Kenya 254	Sweden 46
Colombia 57	Kuwait 965	Switzerland 41
Congo.....................243	Latvia 371	Syria 963
Costa Rica 506	Libya 218	Taiwan 886
Croatia 385	Lithuania 370	Thailand 66
Cuba 53	Luxembourg 352	Tunisia 216
Cyprus 357	Malaysia 60	Turkey 90
Czech Republic 42	Mexico 52	Ukraine 380
Denmark 45	Monaco 339	United Arab
Ecuador 592	Morocco 212	Emirates 971
Egypt 20	Nepal 977	United Kingdom 44
El Salvador 503	Netherlands 31	Uruguay 598
Estonia 372	New Zealand 64	United States
Ethiopia 251	Nicaragua 505	of America...................1
Finland 358	Nigeria....................234	Venezuela 58
France 331	Norway 47	Yugoslavia 381
Germany 49	Pakistan 92	Zimbabwe 263

Studies in History and Philosophy of Science Part A

Editors:
N. Jardine and
M. Frasca-Spada,
*University of
Cambridge, Free School
Lane, Cambridge, UK*

Year 2002 ● Volume 33, ● 4 issues
ISSN: 0039-3681

**Ranked 2nd out of 28 in the 2000 ISI
subject category for Studies in History
and Philosophy of Science**

Studies in History and Philosophy of Science
is devoted to the integrated study of the
history, philosophy and sociology of the
sciences. The editors encourage
contributions both in the long-established
areas of the history of the sciences and the
philosophy of the sciences and in the topical
areas of historiography of the sciences, the
sciences in relation to gender, culture and
society and the sciences in relation to arts.
The Journal is international in scope and
content and publishes papers from a wide
range of countries and cultural traditions.

Audience
For historians and philosophers of science,
mathematicians, physicists and natural
scientists.

Abstracting / Indexing
ASSIA, Arts and Humanities Citation Index,
BIOSIS, Current Contents/Arts & Humanities,
Historical Abstracts, Mathematical Reviews, Research
Alert, Science Citation Index, Social Sciences
Citation Index, Sociological Abstracts.

Studies in History and Philosophy of Science

Part C: Studies in History and Philosophy of Biological and Biomedical Sciences

Editors:
N. Jardine and
M. Frasca-Spada,
*University of Cambridge,
Free School Lane,
Cambridge, UK*

Year 2002 ● Volume 33C, ● 4 issues
ISSN: 1369-8486

*Studies in History and Philosophy of
Biological and Biomedical Sciences* is
devoted to historical, sociological,
philosophical and ethical aspects of the life
and environmental sciences, of the sciences
of mind and behaviour, and of the medical
and biomedical sciences and technologies.
The period covered is from the middle of the
nineteenth century (the time of the so-called
"laboratory revolution" in medicine and the
life sciences) to the present.

Audience
For historians, philosophers and sociologists
of science, as well as medics, biologists and
psychologists with an interest in the
development of biological and biomedical
sciences.

Abstracting / Indexing
ASSIA, BIOSIS, Current Contents/Arts &
Humanities, Current Contents/Social and
Behavioral Sciences, Historical Abstracts,
Mathematical Reviews, Research Alert,
Sociological Abstracts.

Available online via ScienceDirect

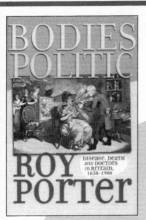

CAMBRIDGE
UNIVERSITY PRESS

is Pleased to Announce

THE CAMBRIDGE HISTORY OF SCIENCE

The Cambridge History of Science comprises eight volumes, the first four arranged chronologically from antiquity through the eighteenth century, the latter four organized thematically and covering the nineteenth and twentieth centuries. Eminent scholars from Europe and North America, who together form the editorial board for the series, edit the respective volumes.

General Editors:

David C. Lindberg, *Hilldale Professor of the History of Science, University of Wisconsin, Madison*

and

Ronald L. Numbers, *Hilldale and William Coleman Professor of the History of Science and Medicine, University of Wisconsin, Madison*

Volume 1: **ANCIENT SCIENCE**
Editor: **Alexander Jones,** *University of Toronto*
Publication: Forthcoming

Volume 2: **MEDIEVAL SCIENCE**
Editors: **David C. Lindberg** and
Michael H. Shank, *University of Wisconsin, Madison*
Publication: 2004

Volume 3: **EARLY MODERN SCIENCE**
Editors: **Lorraine J. Daston,** *Max Planck Institute, Berlin,* and
Katharine Park, *Wellesley College*
Publication: Forthcoming

Volume 4: **THE EIGHTEENTH CENTURY**
Editor: **Roy Porter,** *Wellcome Institute for the History of Medicine*
Publication: 2001

Volume 5: **THE MODERN PHYSICAL AND MATHEMATICAL SCIENCES**
Editor: **Mary Jo Nye,** *Oregon State University*
Publication: 2002

Volume 6: **THE MODERN BIOLOGICAL AND EARTH SCIENCES**
Editors: **Peter Bowler,** *Queen's University, Belfast,* and
John Pickstone, *University of Manchester*
Publication: 2004

Volume 7: **THE MODERN SOCIAL AND BEHAVIORAL SCIENCES**
Editors: **Dorothy Ross,** *Johns Hopkins University* and
Theodore Porter, *University of California, Los Angeles*
Publication: 2002

Volume 8: **MODERN SCIENCE IN NATIONAL AND INTERNATIONAL CONTEXT**
Editors: **David N. Livingstone,** *Queen's University, Belfast* and
Ronald L. Numbers, *University of Wisconsin, Madison*
Publication: 2003

Call toll-free 800-872-7423
www.cambridge.org

Guide Survey

Your feedback will make the *Guide* more useful. After you become familiar with the *Guide*, please take a moment to complete this survey and mail it back to us. Thank you!

1. How do you use the *Guide*? (Check any)
- ☐ Undergraduate advising
- ☐ Contacting colleagues
- ☐ Learning about HSS
- ☐ For employment purposes
- ☐ Graduate student advising
- ☐ Evaluating programs & institutions
- ☐ Learning about other scholarly societies
- ☐ Finding potential teachers/schools
- ☐ Discovering research resources
- ☐ Finding publication opportunities
- ☐ Browsing to learn about the profession
- ☐ Other _____

2. How comprehensive is the *Guide*?

1	2	3	4	5
Incomplete		Adequate		Comprehensive

3. Overall, how would you rate your reaction to the *Guide*?

1	2	3	4	5
Displeased		Pleased		Very pleased

4. How many times a month do you use this printed edition of the *Guide*? _____

5. How many times a month do you use the Internet edition of the *Guide*, at www.hssonline.org? _____

6. If the *Guide* were not an HSS membership benefit, how much would you pay for a printed copy?
- ☐ Wouldn't buy it
- ☐ Would pay this amount: _____

7. Which phrase best describes you?

Undergraduate	Graduate Student
Junior Faculty Member	Public Historian
Senior Faculty Member	Independent Scholar
Interested Non-scholar	Administrator

8. Comments: _____
_____
_____

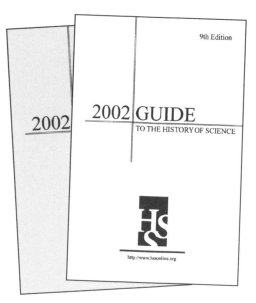

9th Edition

2002

2002 GUIDE

TO THE HISTORY OF SCIENCE

http://www.hssonline.org

Is one copy really enough?

Do you need your *own* copy?

More than just a membership directory, the **Guide** includes valuable information about study and research programs, scholarly periodicals, and organizations related to the history of science. To make sure a copy of the **Guide** is available to your students and colleagues, please forward this form to your librarian, with a note urging the librarian to order it.

HSS Executive Office
Box 351330
University of Washington
Seattle, WA 98195
USA

Guide to the History of Science

Yes! Please send me the *Guide to the History of Science*, as indicated below:

☐ Paper edition (ISBN: 0-226-81709-1) $39.00*
☐ Cloth edition (ISBN: 0-226-81708-3) $50.00*

*Please add $4.50 for shipping and handling; outside USA, please add $5.50. Canadian residents, please also add 7% GST to your order.

Members who have already ordered second copies of the *Guide* need not reorder at this time. Please allow three weeks for delivery of your second copy.

Payment Options   VISA MasterCard

☐ **Charge** my   ☐ MasterCard   ☐ Visa   Exp. date _____
   Acct. no. _____
   Signature _____
☐ **Check** enclosed, payable to the *University of Chicago Press*
☐ **Purchase order** enclosed

Name _____

Address _____

City/State/Zip _____

Please return this form with your payment to the **University of Chicago Press, Journals Division, P.O. Box 37005, Chicago, IL 60637.**